Progress in Botany

Volume 79

Series Editors
Ulrich Lüttge, TU Darmstadt,
FB Biologie (10), Schnittspahnstraße 3–5,
64287 Darmstadt, Germany

Francisco M. Cánovas
Universidad de Málaga, Depto. Biología
Molecular y Campus de Teatinos, 29071
Málaga, Spain

Rainer Matyssek
Technische Universität München
Hans-Carl-von-Carlowitz-Platz
Wissenschaftszentrum Weihenstephan
85354 Freising, Germany

More information about this series at http://www.springer.com/series/124

Francisco M. Cánovas • Ulrich Lüttge •
Rainer Matyssek
Editors

Progress in Botany Vol. 79

Springer

Editors
Francisco M. Cánovas
Depto. Biología Molecular y Bioquímica
Universidad de Málaga
Málaga, Spain

Ulrich Lüttge
Department of Biology
TU Darmstadt
Darmstadt, Germany

Rainer Matyssek
Ecophysiology of Plants, School of Life
Sciences Weihenstephan
Technische Universität München LS
Ökophysiologie der Pflanzen
Freising, Bayern
Germany

ISSN 0340-4773 ISSN 2197-8492 (electronic)
Progress in Botany
ISBN 978-3-319-71412-7 ISBN 978-3-319-71413-4 (eBook)
https://doi.org/10.1007/978-3-319-71413-4

Library of Congress Control Number: 2017959590

Printed on acid-free paper

This Springer imprint is published by Springer Nature
The registered company is Springer International Publishing AG
The registered company address is: Gewerbestrasse 11, 6330 Cham, Switzerland

Contents

Contributors

Ebenezer Ajewole Department of Biology, University of Western Ontario, London, ON, Canada

London Research and Development Centre, Agriculture and Agri-Food Canada, London, ON, Canada

Robert B. Allen Private Researcher, Lincoln, New Zealand

Ana Arnaiz Centro de Biotecnologia y Genomica de Plantas, Universidad Politecnica de Madrid (UPM), Instituto Nacional de Investigacion y Tecnologia Agraria y Alimentaria (INIA), Madrid, Spain

John S. Boyer Division of Plant Sciences, College of Agriculture, Food and Natural Resources, 52 Agriculture Laboratory, University of Missouri, Columbia, MO, USA

Kornel Burg Austrian Institute of Technology, Vienna, Austria

Inmaculada Couso Instituto de Bioquímica Vegetal y Fotosíntesis, Consejo Superior de Investigaciones Científicas (CSIC)-Universidad de Sevilla, Sevilla, Spain

José L. Crespo Instituto de Bioquímica Vegetal y Fotosíntesis, Consejo Superior de Investigaciones Científicas (CSIC)-Universidad de Sevilla, Sevilla, Spain

Isabel Diaz Centro de Biotecnologia y Genomica de Plantas, Universidad Politecnica de Madrid (UPM), Instituto Nacional de Investigacion y Tecnologia Agraria y Alimentaria (INIA), Madrid, Spain

Mercedes Diaz-Mendoza Centro de Biotecnologia y Genomica de Plantas, Universidad Politecnica de Madrid (UPM), Instituto Nacional de Investigacion y Tecnologia Agraria y Alimentaria (INIA), Madrid, Spain

Mónica Domínguez-González Instituto de Bioquímica Vegetal y Fotosíntesis, Consejo Superior de Investigaciones Científicas (CSIC)-Universidad de Sevilla, Sevilla, Spain

Andrea Gomez-Sanchez Centro de Biotecnologia y Genomica de Plantas, Universidad Politecnica de Madrid (UPM), Instituto Nacional de Investigacion y Tecnologia Agraria y Alimentaria (INIA), Madrid, Spain

Pablo Gonzalez-Melendi Centro de Biotecnologia y Genomica de Plantas, Universidad Politecnica de Madrid (UPM), Instituto Nacional de Investigacion y Tecnologia Agraria y Alimentaria (INIA), Madrid, Spain

Matthew A. Jones School of Biological Sciences, University of Essex, Colchester, UK

Shrikaar Kambhampati Department of Biology, University of Western Ontario, London, ON, Canada

London Research and Development Centre, Agriculture and Agri-Food Canada, London, ON, Canada

Andrzej Kornaś Institute of Biology, Pedagogical University, Kraków, Poland

Mikhail V. Kozlov Section of Ecology, Department of Biology, University of Turku, Turku, Finland

Ute Krämer Molecular Genetics and Physiology of Plants, Ruhr University Bochum, Bochum, Germany

Elżbieta Kuźniak Department of Plant Physiology and Biochemistry, Faculty of Biology and Environmental Protection, University of Lodz, Łódź, Poland

Michael Lakatos Applied Pharmacy, University of Applied Sciences Kaiserslautern, Pirmasens, Germany

Stéphane D. Lemaire Laboratoire de Biologie Moléculaire et Cellulaire des Eucaryotes, Institut de Biologie Physico-Chimique, Sorbonne Universités, UPMC Univ Paris 06, CNRS, UMR8226, Paris, France

Maria A. Lo Gullo Dipartimento di Scienze Chimiche, Biologiche, Farmaceutiche e Ambientali, Università di Messina, Messina, Italy

Frédéric Marsolais Department of Biology, University of Western Ontario, London, ON, Canada

London Research and Development Centre, Agriculture and Agri-Food Canada, London, ON, Canada

Manuel Martinez Centro de Biotecnologia y Genomica de Plantas, Universidad Politecnica de Madrid (UPM), Instituto Nacional de Investigacion y Tecnologia Agraria y Alimentaria (INIA), Madrid, Spain

Peter Millard Landcare Research, Lincoln, New Zealand

Zbigniew Miszalski Institute of Plant Physiology, Polish Academy of Sciences, Kraków, Poland

Małopolska Centre of Biotechnology, Jagiellonian University, Kraków, Poland

Andrea Nardini Dipartimento di Scienze della Vita, Università di Trieste, Trieste, Italy

María Esther Pérez-Pérez Instituto de Bioquímica Vegetal y Fotosíntesis, Consejo Superior de Investigaciones Científicas (CSIC)-Universidad de Sevilla, Sevilla, Spain

Sarah J. Richardson Landcare Research, Lincoln, New Zealand

Maria Estrella Santamaria Centro de Biotecnologia y Genomica de Plantas, Universidad Politecnica de Madrid (UPM), Instituto Nacional de Investigacion y Tecnologia Agraria y Alimentaria (INIA), Madrid, Spain

Tadeja Savi Dipartimento di Scienze della Vita, Università di Trieste, Trieste, Italy

Anna Schulten Molecular Genetics and Physiology of Plants, Ruhr University Bochum, Bochum, Germany

Dorina Strieth University of Kaiserslautern, Kaiserslautern, Germany

Patrizia Trifilò Dipartimento di Scienze Chimiche, Biologiche, Farmaceutiche e Ambientali, Università di Messina, Messina, Italy

Blanca Velasco-Arroyo Centro de Biotecnologia y Genomica de Plantas, Universidad Politecnica de Madrid (UPM), Instituto Nacional de Investigacion y Tecnologia Agraria y Alimentaria (INIA), Madrid, Spain

Elena L. Zvereva Section of Ecology, Department of Biology, University of Turku, Turku, Finland

Curriculum Vitae

John Strickland Boyer College of Agriculture, Food and Natural Resources
302-645-1618
University of Missouri (boyerjs@missouri.edu)

Professional Preparation

1959	A.B. Biology, Swarthmore College
1961	M.S. Botany, Univ. Wisconsin-Madison
1964	Ph.D. Botany, Duke Univ.

Appointments

1964–1965	Visiting Asst. Professor, Duke Univ.
1965–1966	Assistant Scientist, Connecticut Agricultural Experiment Station
1966–1978	Assistant, Associate, and Full Professor, Dept. Botany and Dept. Agronomy, Univ. Illinois at Urbana-Champaign

1978–1984	Plant Physiologist, USDA/ARS, Univ. Illinois at Urbana-Champaign
1984–1987	Prof. Dept Soil and Plant Sci. and Dept. Biochemistry and Biophysics, Texas A&M Univ.
1987–2005	E. I. du Pont Prof. Marine Plant Biochemistry/Biophysics, Coll. Marine Studies and Coll. Agriculture, Univ. Delaware
1998–2002	Program Director, Marine Biology-Biochemistry, Univ. Delaware
2005–present	E. I. du Pont Prof. Biochemistry/Biophysics Emeritus, Univ. Delaware
2014–present	Distinguished Research Professor, Univ. Missouri

Awards

1961–1964	James B. Duke I Fellow, Duke Univ.
1972	Climate Laboratory Fellow, DSIR, Palmerston North, New Zealand
1977	Shull Award, Am. Soc. Plant Physiologists
1983–1984	Senior Scientist, von Humboldt Foundation
1985	Citation Classic, ISI (Boyer. 1970. Plant Physiol 46:233)
1985	Fellow, Crop Science Society of America
1986	Fellow, American Society of Agronomy
1987	Distinguished Lecturer, National Science Council, Taiwan
1987	Walker Lecturer, University of Washington, Seattle
1988	Fellow, Japan Society for Promotion of Science
1988	International Lecturer, Japan Crop Science Society
1989	Tansley Lecturer, Univ. Edinburgh, Univ. Bangor, Univ. Lancaster, UK
1990	Member National Academy of Sciences, USA
1993	Kramer Lecturer, Duke University
1995	Fellow, Research School of Biological Sciences, Austral. Nat. Univ.
1997	Bullard Fellow, Organismal and Evolutionary Biology, Harvard Univ.
1999	Francis Alison Award, University of Delaware
2002	Listed, 250 most highly cited scientists in the plant and animal sciences
2005	Fellow Australian Academy of Science
2007	Fellow American Society of Plant Biologists
2007	Barnes Lifetime Membership Award, Am. Soc. Plant Biologists
2007	Sir Frederick McMaster Research Fellowship, CSIRO, Canberra

Plant Water Relations: A Whirlwind of Change

John S. Boyer

Abstract Water is increasingly recognized as a limiting resource consumed by plants in copious amounts. Its large role in agriculture has awakened interest in how plants use it and how they conserve it. Fortunately the study of plant water relations underwent major changes starting about 50 years ago with thermodynamically based methods for measuring plant water status. The methods allowed conditions to be repeated, biochemical events to be repeatedly detected, and responses to water deficits understood more fully. This was followed by the realization that enzymes were not responding directly to water deficits nor was photosynthesis responding only by closing stomata (which conserved water and limited CO_2 entry) but also by diminishing CO_2 demand. The decrease in demand suggested that photosynthesis might acclimate to water deficits, and tests showed not only that this occurred but also that cellular Mg^{2+} concentrations were a central controller of the acclimation. Osmotic adjustment was discovered in a form that used photosynthetic products to maintain turgor and allowed growth where otherwise none would occur. It was found that water potentials were induced by the growth process itself and were important controllers of growth rates in land plants. In an alga surrounded by water and unaffected by these potentials, the chemistry of pectin determined cell enlargement and is now being explored in land plants most of which contain pectin in their cell walls. Ultimately, it became possible to reverse reproductive failure during a drought by feeding photosynthetic products to the plant, thus identifying biochemical origins of failure and gene targets to enhance tolerance to those environments. In fact, commercial agriculture increasingly sees drought tolerance as an important plant character and it is gratifying that reproductive reversal may be contributing to

---

Communicated by Rainer Matyssek

J.S. Boyer (✉)
Division of Plant Sciences, College of Agriculture, Food and Natural Resources,
52 Agriculture Laboratory, University of Missouri, Columbia, MO 65211, USA
e-mail: boyerjs@missouri.edu

Progress in Botany (2018) 79: 1–32, DOI 10.1007/124_2017_3,
Published online: 14 April 2017

this effort. Although plant water relations are inherently multigenic and complex, these findings demonstrate benefits of understanding plant water relations that were scarcely imagined 50 years ago.

Contents

1 Introduction

My brother and I grew up on a beef cattle farm that we worked for our parents as we aimed for university training. At the university, we became amazed at how little we knew about the basic biology of farming. When a chance came to go to graduate school, I took it because the application of biology to farming was so fascinating. Raising enough food for the animals had been difficult, so I focused on plant sciences. Aside from the importance of soils, it seemed that mineral nutrition and water availability were two key factors, and I resolved to learn more about both. A couple of years with Gerry C. Gerloff at University of Wisconsin acquainted me with soils and the genetics of mineral nutrition and then an opportunity to work with water availability led me to the lab of Paul J. Kramer at Duke.

No sooner had I arrived in 1961 than it became apparent that water availability was a tough subject. The availability of soil water was continually changing and the literature was scarce. Those papers that were published were based mostly on the time after water was withheld. This meant the soil type, atmospheric conditions, and plant size controlled the results and could scarcely be repeated. In addition, only a few methods were available to characterize conditions in the plant and soil. Terms like "diffusion pressure deficit" and "osmotic pressure" were opaque. Many hours were spent delving into these concepts with the other students.

Then it occurred to me that using NaCl to vary salinity would combine the aquatic culture methods of mineral nutrition with a study of water availability that would allow steady, repeatable conditions to be achieved. It would then be possible to test the new ideas of Bernstein (1961) that certain plant species grow in saline conditions because they absorb the external salt and use it internally to improve the osmotic potential for water uptake. Bernstein termed this process "osmotic adjustment." I chose cotton for my study because it was moderately salinity tolerant. Would osmotic adjustment allow water uptake but also permit stomata to function normally and photosynthesis to occur when plants adjusted to the salinized medium?

The answer was yes, stomata could open and close normally and photosynthesis took place even though water availability was limited by the saline conditions (Fig. 1). The leaves had adjusted to the salinity and the stomata followed suit. I had grown the plants in saline conditions from the time of seedling establishment but an oversight was that I neglected to mention this fact in my first paper (Boyer 1965). Papers from other labs sometimes dumped salt on established plants and generally found incomplete adjustment (e.g., Gale et al. 1967). I had also observed this behavior but should have addressed it in my paper to avoid confusion.

Later, I had the chance to revisit osmotic adjustment. Bob Meyer was an early graduate student at University of Illinois and he discovered osmotic adjustment in water-deficient plants (Meyer and Boyer 1972). Because NaCl was absent, the plants could not use salt to adjust osmotically and instead used photosynthetic products. The phenomenon was particularly noticeable in enlarging tissues that accumulated the products of photosynthesis extensively (Meyer and Boyer 1972, 1981; Michelena and Boyer 1982). The accumulation allowed water uptake with the result that turgor was

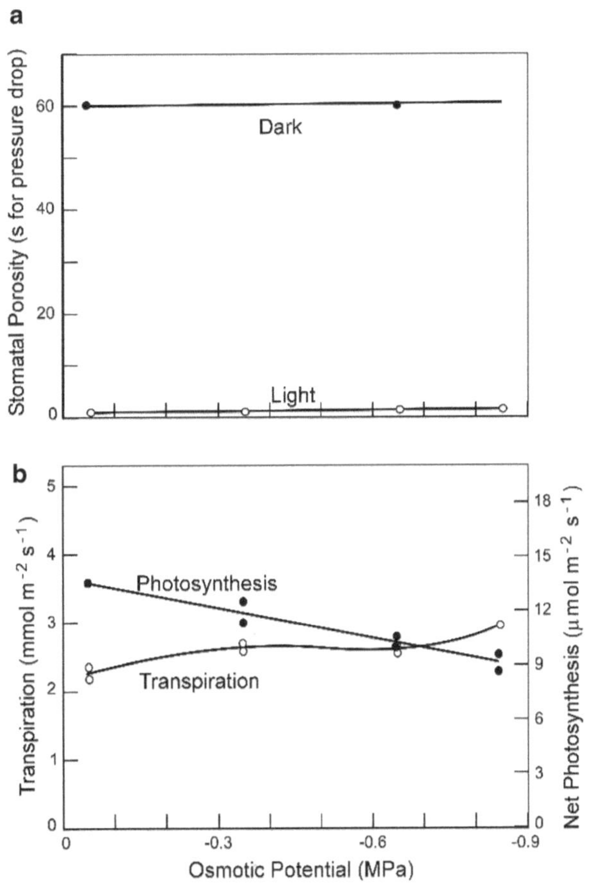

Fig. 1 Stomatal porosity (a), transpiration and net photosynthesis (b) in cotton salinized from the time of seedling establishment. Highest salinity was about 1/3 that of seawater (osmotic potential of −0.85 MPa). Porosity was measured with a porometer that forced air through the leaf (stomata on both sides). Less time indicated more open stomata, which opened fully in light regardless of salinity, and closed in the dark. Transpiration increased as salinity increased, but photosynthesis tended to diminish. Redrawn from Boyer (1965). Copyright American Society of Plant Biologists

nearly fully maintained (Fig. 2). Because turgor had to be above a minimum in order for growth to occur, this helped the cells continue growing. If the adjustment was prevented, turgor fell and growth was more inhibited than when the adjustment occurred. Therefore, osmotic adjustment was more general than Bernstein (1961) ever imagined.

But returning to my graduate work at Duke, Ralph Slatyer visited in 1963 to collaborate with Paul Kramer on a water relations textbook. Ralph was well-known for his pioneering work in micrometeorology and he knew physics and thermodynamics. I

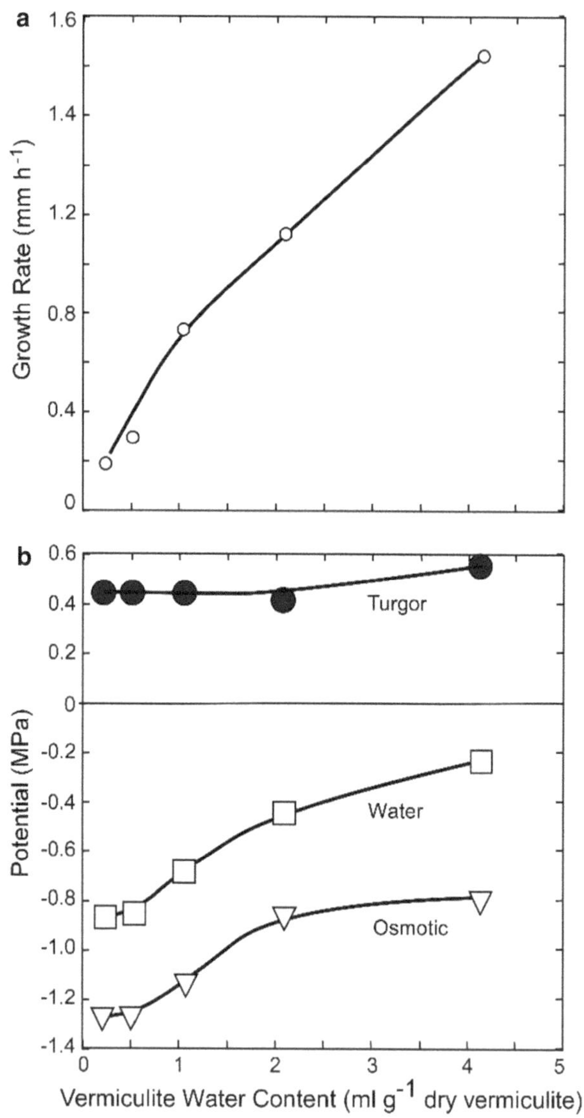

Fig. 2 Growth (**a**) and osmotic adjustment (**b**) in the elongating region of soybean hypocotyls in vermiculite having various water contents. Osmotic potential essentially kept pace with the water potential in the elongating region so that turgor decreased only slightly. No NaCl was present and instead the intact seedlings used photosynthetic products from the cotyledons to adjust the osmotic potential. Note that growth was strongly inhibited by the water deficit (**a**) and indicated that something in addition to turgor (**b**) was inhibiting growth. Redrawn from Meyer and Boyer (1972)

was enthralled. He suggested that we should abandon "diffusion pressure deficits" and relate plant work to thermodynamics using potentials as described by J. Willard Gibbs (1875–1876). In fact, Ralph Slatyer and Sterling Taylor (1960) had recently published a paper suggesting that soil and plant water status could be unified by this approach. I was persuaded and resolved to write in those terms (Boyer 1965). Ralph's influence on the water relations and plant transport fields cannot be overestimated, as is apparent also in the essay of Ulrich Lüttge (2016).

While my cotton paper was in press, Steve Rawlins (1964) published that the vapor pressure method I used to characterize salinity conditions was in error by up to 90%!! I was distraught until it occurred to me that his argument rested on the calibration necessary for the method. Thus, if I used the solution indicated by the calibration, its vapor pressure should be the same as the tissue. It wasn't! Steve was right!

From that experiment, we developed the "isopiestic" method of measuring the water potential meaning equal vapor pressures (Boyer and Knipling 1965; Boyer 1969, 1995; Kramer and Boyer 1995). Vapor methods measure plant water status by placing a tissue sample in a small sealed chamber and letting the liquid evaporate to the air, raising the humidity until no net vapor exchange occurs, that is, until the vapor activity matches the liquid activity in the sample. It is then only necessary to measure the vapor activity to determine the liquid activity. The simplest way is to expose a standard solution to the vapor in the chamber. If it neither evaporates nor condenses, its vapor activity is the same as the vapor in the chamber and thus the liquid in the sample.

A thermocouple detects whether the standard evaporates (cooler) or condenses (warmer). The approach is to expose a standard to the vapor and then follow it with a second standard closer to the vapor activity in the chamber. This allows extrapolation to the standard neither evaporating nor condensing, i.e., having no net vapor exchange with the sample. Because the extrapolated standard is in equilibrium with liquid in the sample, the liquid activity is identified and no calibration is required. The extrapolated standard is unaffected by the arrangement of the sample in the chamber, the tortuosity of the vapor path through the sample, or any waxes coating the tissue surface. All other measurements are affected by these factors. In addition, the thermocouple can be used to test isothermal conditions around the sample important for isopiestic measurements. This is the power of thermodynamic equilibrium.

The method is similar to the isopiestic method used in physical chemistry to determine the vapor pressure of solutions. The availability of this method for soils and plant tissues and even intact plants has been central to all of our subsequent experiments for my entire career. Not only could it measure tissue water potential but also the osmotic potential (Ehlig 1962) from which turgor pressure could be determined. So much of plant performance depends on these three parameters that the method opened many avenues of investigation for us.

While this was going on, Paul Kramer received a manuscript from Science for review. Written by Scholander et al. (1965), it described a pressure chamber to measure tension on water in the conducting xylem of trees. Paul asked me to review it and I recommended publication. I also suggested to Scholander in the review that, because he had already published that the solute concentrations in the xylem were low, the tensions measured with the pressure chamber ought to be nearly the same as the water potential. Scholander did not accept my suggestion and published without making the connection between his pressure chamber results and the water potential.

Shortly thereafter, I had an offer from Connecticut Agricultural Experiment Station in New Haven and found myself in Steve Rawlins' old position. For 11 months, I devoted myself to a comparison of the pressure chamber and the isopiestic psychrometer. The comparison was very favorable so that the pressure chamber essentially measured the water potential (Boyer 1966, 1967), which was the first time this had been demonstrated (Fig. 3). It confirmed that pressure (tension during the daytime) was the main xylem component in plants and also indicated that water in soil and plants experiences the same thermodynamics despite being in vastly different systems.

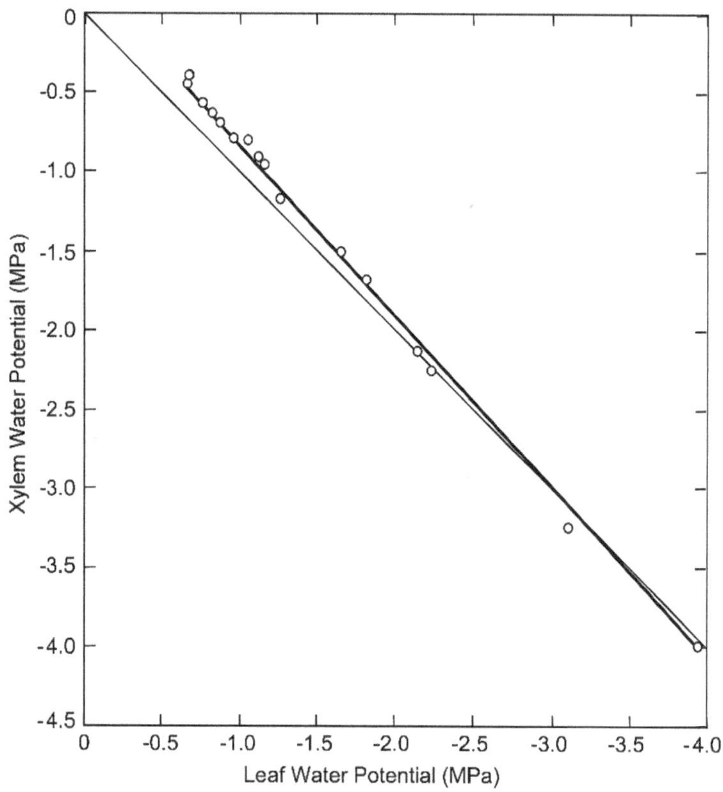

Fig. 3 Xylem water potential measured on branches with a pressure chamber in comparison with water potential measured on leaves of the same branches with an isopiestic thermocouple psychrometer in yew (*Taxus cuspidata*). Xylem water potential was the sum of the pressure chamber result and the osmotic potential of xylem solution that was always dilute. This indicates that water potential in xylem was mostly tension. The *solid line* indicates a 1:1 correspondence between the two methods. Similar results were obtained in several other spp. Redrawn from Boyer (1967). Copyright American Society of Plant Biologists

Gibbs (1875–1876) had originally conceived of chemical potentials as the Gibbs free energy per mole of substance but Slatyer and Taylor (1960) suggested that the concept could be converted to pressure units by dividing Gibbs' chemical potential by the partial volume of a mole of water (partial indicates that other substances can be present and for water the partial volume is 18.05 cm^3 mol^{-1} at 20°C, considered constant in most biological systems). Expressed this way, the chemical potential was called the "water potential" and had units of pressure, i.e., megaPascals where 1 MPa = 10^6 N m^{-2} = 10 bar = 9.87 atm. Compared to pure water that acted as a reference (water potential of zero), forces that contributed to the water potential were pressure (positive or negative), solute (negative because it spread water molecules apart), matrices (negative because solid media like soil or cell walls also spread water apart), and gravitational pull (important over large distances). The isopiestic method and pressure chamber became major ways to measure these forces (Boyer 1995).

A few years later, Hüsken et al. (1978) developed the pressure probe that could measure turgor directly in individual plant cells. In retrospect, it seems remarkable that these three methods and their associated terminology developed so quickly. I think having a thermodynamic basis helped in their development. They made soil and plant water relations accessible to scientists in other disciplines but even more importantly, they allowed experiments to be repeated. In effect, knowing the thermodynamic state of water in plants and soils allowed a scientist to return to that state over and over again. This laid the groundwork for physiological and biochemical investigations and, more recently, molecular genetic work.

Soon (1966) opportunity arose in the form of an offer from the University of Illinois and I took it. The attraction was interaction with exceptional colleagues like Jack Hanson, Dick Hageman, Jim Gerdemann, Larry Vanderhoef, Govindjee, Gregorio Weber, and many others with their students. Through the years, we published together and their students added rich diversity to our efforts. The trade-off was heavy teaching and administrative responsibilities that restricted my research to evenings and weekends. My wife Jean, herself a plant scientist, was fully supportive. I was very grateful and to conserve time, my efforts focused only on nitrogen metabolism, photosynthesis, growth, and reproduction because they seemed particularly relevant to agriculture. The remainder of this essay is devoted to how water relations affect these areas.

2 Nitrogen Metabolism

With the recent application of thermodynamics to water relations, a central question was how plant metabolism responded to water availability. Did enzymes react to changes in turgor pressure? Or water potential? Nitrogen metabolizing enzymes were candidates because nitrogenase activity had recently been detected in vivo in nitrogen fixing species (Hardy et al. 1968; Fishbeck et al. 1973). Also, the synthesis of nitrate reductase had just been shown to respond to the water status of the plant (Morilla et al. 1973).

Beginning with nitrogen fixing soybean, Chi-Ying Huang found that nitrogenase activity could be detected in the nodules of completely intact plants growing in soil (Huang et al. 1975a). When water was withheld, plant water potential declined and photosynthesis was inhibited in the shoot. In the root, Fig. 4 shows nitrogenase immediately lost activity but could be partly recovered by increasing photosynthesis using high CO_2 (Huang et al. 1975b). At the same time, Janet Sprent in Scotland found that the activity could be partly recovered by increasing the oxygen concentration around the nodules. Although nodules normally shield nitrogenase from oxygen (Avenhaus et al. 2016), the gas is required for the respiratory activity of the cells around the nitrogenase (Pankhurst and Sprent 1976). Water deficits had increased the oxygen barrier so that the concentration in the nodule became too low for the necessary respiration. The ability to reverse the effects of limited water indicated that the flux of carbon skeletons from shoot photosynthesis together

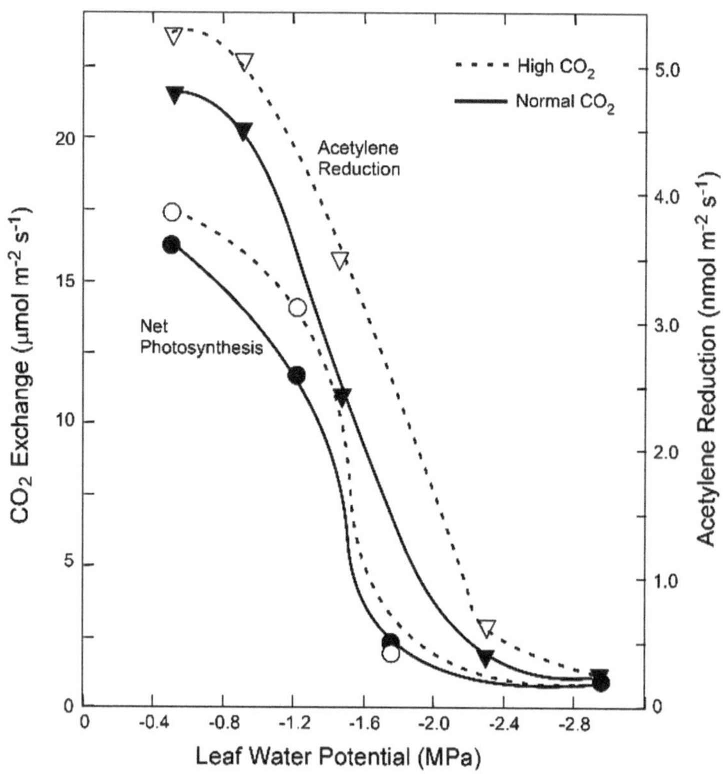

Fig. 4 Net photosynthesis and acetylene reduction at various leaf water potentials in intact soybean plants growing in soil. Normal CO_2 concentration was 300 μmol mol^{-1} and high CO_2 concentration was 600 μmol mol^{-1}. Note that measurements were in situ without disturbing the plants. High CO_2 concentration caused increased photosynthesis in the shoot and thus increased acetylene reduction in the root nodules. Redrawn from Huang et al. (1975b). Copyright American Society of Plant Biologists

with the diffusion of oxygen from the soil controlled the rate of nitrogen fixation in drying soil. In effect, water deficits were affecting these processes through physiological means rather than by altering the enzyme nitrogenase directly. This was the first time that enzyme activity had been monitored in a completely intact plant exposed to limited water and was the closest we ever came to monitoring the biochemistry of whole plants.

Moving on to nitrate reductase, Dale Shaner repeated the work of Morilla et al. (1973) that shoots of maize seedlings lost reductase activity when dehydrated (Shaner and Boyer 1976b). But the controls lost activity too, even if they were kept in water at high humidity in the light (Shaner and Boyer 1976a). Dale was intrigued and tried adding nitrate to the water of the controls. The nitrate prevented the lost activity. Could the flux of nitrate be important for the control shoots?

We imagined that the molecular budget for the reaction would be similar to a mass or energy budget in which INPUT + OUTGO + STORAGE = 0. The INPUT would be the flux of the substrate nitrate (positive). The OUTGO would be the flux of the product nitrite (negative). The STORAGE would be the difference between the two represented by the enzyme reaction. As a sequence of enzymatic events, the budget would then be shown as:

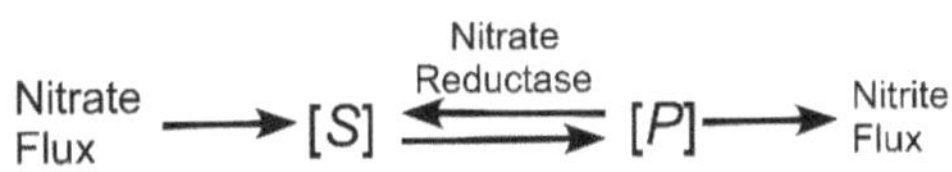

where [*S*] is the substrate concentration in the cytoplasm (nitrate) and [*P*] is the product concentration in the cytoplasm (nitrite). The pools of *S* and *P* would be small because of the small volume of the cytoplasmic compartment. *S* would be quickly controlled by the Nitrate Flux coming from the xylem. Likewise, *P* would be controlled by the Nitrite Flux.

We knew substantial amounts of nitrate were in the vacuoles but this large pool would be sequestered and have little effect on the enzyme in the cytoplasm. Basically because of the small cytoplasmic pool the enzyme activity would appear to be determined by the flux of nitrate when in fact the enzyme responded to the local concentration of nitrate (Shaner and Boyer 1976a). Consequently, *S* (nitrate) served two roles for the enzyme: its flux altered the rate of enzyme synthesis (Morilla et al. 1973) and controlled the enzyme activity (Shaner and Boyer 1976a).

The flux to the cytoplasm diminished as water potentials became low in large part because the soil delivered less nitrate to the root. The flux could be partially recovered by feeding higher concentrations of nitrate to the soil before water was withheld (Shaner and Boyer 1976b). This doubled the flux of nitrate to the leaves during dehydration. Figure 5 shows that the higher flux improved the enzyme activity. Again, the enzyme response was controlled by a physiological process (nitrate flux) rather than the enzyme being directly affected by the water deficit.

There is no doubt that water is required for enzyme activity. Substrate must diffuse to the active site through water, and the reaction at the active site depends on the motion of enzyme subunits or domains in the aqueous medium. Dehydration restricts this motion. However, the amount of water necessary to restrict enzyme motion is much less than in a typical water limited cell. For example, urea when free in soil decreases its activity below water potentials equivalent to −15 MPa where the water would approach only monomolecular layers (Skujins and McLaren 1967). But nitrogenase and nitrate reductase in cells lost all activity at water potentials equivalent to −1 to −2 MPa in intact plants (Figs. 4 and 5). With this result, we abandoned the hypothesis that low water potentials acted directly on cellular enzymes. We began to emphasize instead the physiological impacts of low water potentials.

After this work was done, others found that light/dark transitions and phosphorylation also alter nitrate reductase activity (e.g., Lillo et al. 2003). Although the water potential had no direct effect, the advent of molecular genetics allowed many additional control and signaling systems to be discovered. It will be exciting to see how these affect this important enzyme for nitrogen entrance into plants.

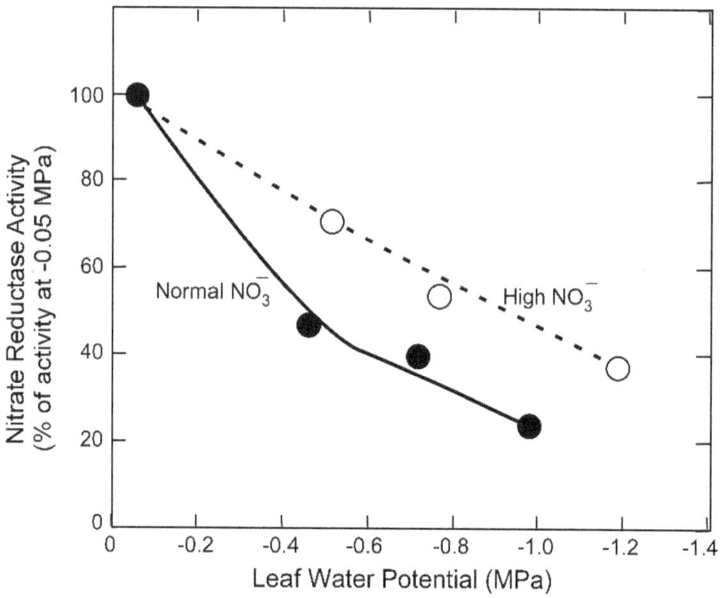

Fig. 5 Effect of nitrate flux on activity of nitrate reductase at various leaf water potentials. Normal NO_3^- was 15 mM, High NO_3^- was 45 mM. Flux of nitrate to the site of nitrate reductase synthesis was increased by feeding the higher concentration of nitrate to soil immediately prior to exposure to low leaf water potentials. Activity is expressed as a percent of the control at leaf water potential of −0.05 MPa. Redrawn from Shaner and Boyer (1976b). Copyright American Society of Plant Biologists

3 Photosynthesis

Plants encountering water deficits lose photosynthetic activity much more than they lose respiratory activity (Fig. 6). In an ecological context, this makes sense because the plant remains alive and potentially able to propagate itself for the next season despite minimal photosynthesis. In view of the previous section on enzyme function, we wondered if photosynthetic enzymes or membranes might be inhibited by water deficits. Boyer and Bowen (1970) found large losses in Photosystem 2 activity during water deficits, and Boyer and Potter (1973) and Potter and Boyer (1973) reported that turgor changes (and thus water potential) did not cause the photosynthesis alterations. Instead the osmotic potential tracked the changes. In other words, water contents were important (because they control the osmotic potential) or the composition of solutes had changed. Once again physiological conditions were more important than direct effects of water deficits on the enzyme in the chloroplast.

If water was withheld from sunflower leaves and chloroplasts were isolated from them, further tests showed less activity for Photosystems 1 and 2 and photophosphorylation compared to control tissue supplied with water (Keck and Boyer 1974). Apparently, withholding water exposed the chloroplasts to conditions that limited their activity even when they were isolated and assayed in aqueous media without a water deficit. This was surprising, so we tested if there were chloroplast changes in

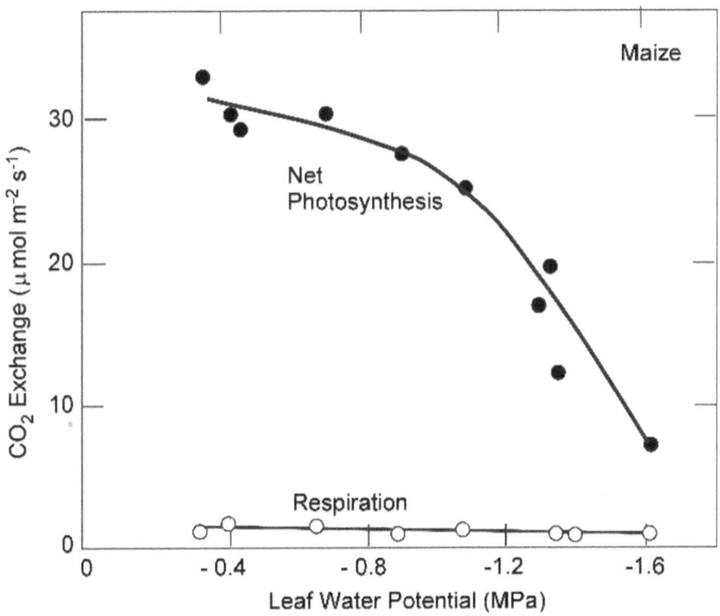

Fig. 6 Net photosynthesis and respiration in maize at various leaf water potentials. Activities were measured in the entire shoot of intact plants grown in soil for 4–5 weeks. Redrawn from Boyer (1970). Copyright American Society of Plant Biologists

intact plants and found lower quantum yields (Mohanty and Boyer 1976; Matthews and Boyer 1984) and altered membrane structure when chloroplasts were viewed with an electron microscope (Fellows and Boyer 1976). Chloroplasts were obviously losing activity in the intact leaves.

Further work in sunflower eliminated photoinhibition as a reason for the loss at least in sunflower (Sharp and Boyer 1987) and indicated that photosynthesis could acclimate to prolonged exposure to low water potentials (Matthews and Boyer 1984). Rao et al. (1987) implicated changes in Mg^{2+} which is largely free in the cytoplasm and chloroplasts despite its structural presence in chlorophyll (Portis and Heldt 1976; Portis 1981). Inorganic ions concentrate as water is lost from cells but photophosphorylation (Younis et al. 1979, 1983) and protein synthesis (Rubin et al. 1979) require Mg^{2+} to be in a narrow concentration range for maximum activity. Our current hypothesis is that water deficits cause Mg^{2+} to concentrate outside of this range (Rao et al. 1987). It would be useful to test further if Mg^{2+} homeostasis confers drought tolerance in plants.

Photosynthesis requires CO_2 as a substrate and land plants inevitably trade water for CO_2 because the CO_2 must diffuse into leaves and dissolve in water wetting the surfaces of the leaf cells. In fact, leaves usually lose water at least 100 times faster than CO_2 enters, and the gateways for this exchange are the stomatal pores. Stomata open by accumulating osmotica in the guard cells, particularly potassium (Fischer 1968; Fischer and Hsiao 1968; Mansfield and Jones 1971). During a water deficit, stomata close and conserve water. Ehret and Boyer (1979) found that the closure was caused by potassium loss from the guard cells. Figure 7 shows the loss and indicates that it is reversed when water is resupplied to the leaf. Therefore, two events seemed important for photosynthesis during a water deficit: stomata closed and restricted water loss and CO_2 entry but also photosynthetic metabolism appeared to diminish and possibly demand less CO_2.

In order to determine which process was more important, we measured the concentration of CO_2 inside leaves (c_i). If stomata closed during a water deficit, photosynthesis might be starved for CO_2 and c_i would decrease. On the other hand, if demand diminished, c_i would increase or at least remain constant. The c_i was thus a key for determining the limitation of photosynthesis during a water deficit. In sunflower, Lauer and Boyer (1992) used a method pioneered by Sharkey et al. (1982) to measure c_i directly and avoid possible errors in calculations of c_i (see below). The c_i increased and indicated that photosynthesis was demanding less CO_2 than the stomata were allowing in.

We also developed a gas exchange system to detect photosynthesis at CO_2 concentrations as high as 50,000 μmol mol^{-1} (5% CO_2). This allowed us to force CO_2 into the leaf despite stomatal closure (Graan and Boyer 1990). We developed gentle ways to peel the epidermis to determine photosynthesis in the complete absence of closed stomata (Tang et al. 2002). These methods confirmed the c_i findings that photosynthetic metabolism was more directly limiting than the restricted entry of CO_2 into sunflower leaves.

It would be helpful to extend this type of work to various species because the c_i balance is likely to be a key response to dehydrating environments. A major difficulty

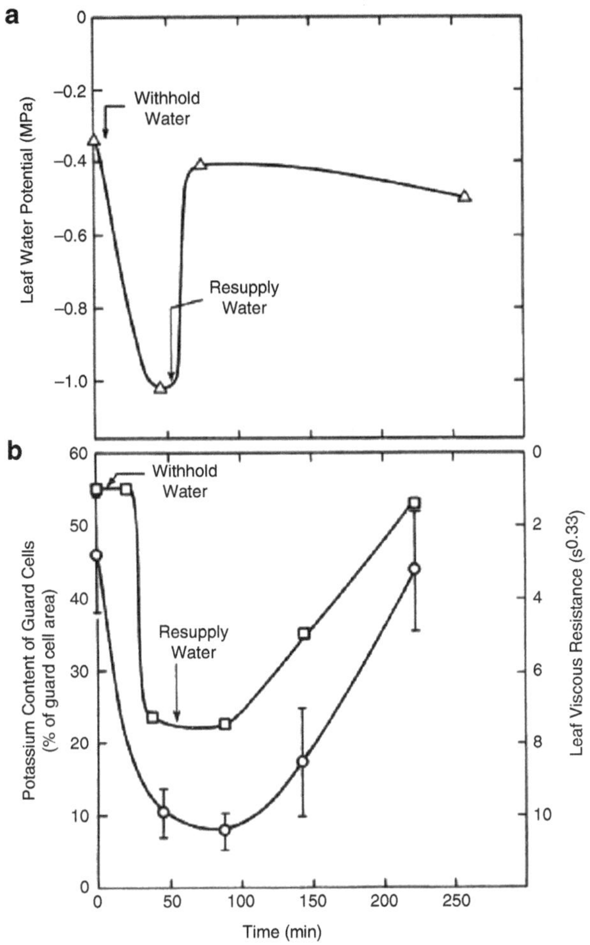

Fig. 7 Leaf water potentials (**a**), potassium content of stomatal guard cells (**b**, *squares*), and leaf viscous resistance to air flow through the leaf (**b**, *circles*) at various times after withholding water, then resupplying it in sunflower. The leaves have stomata on both surfaces, so viscous flow of air through the leaf indicates the degree of stomatal opening. When the viscous flow becomes higher, stomata are opening. From Ehret and Boyer (1979)

is knowing c_i accurately because it is usually calculated instead of being directly measured. In order to determine c_i directly, we incorporated into our gas exchange apparatus a cup that could be sealed to the abaxial (lower) epidermis of a leaf (Boyer and Kawamitsu 2011). The c_i equilibrated with the air in the cup and gave a measure of c_i similar to that pioneered by Sharkey et al. (1982). We compared the measured c_i with the calculated one described by Moss and Rawlins (1963) and their later variants (von Caemmerer and Farquhar 1981; Boyer and Kawamitsu 2011):

$$c_i = c_a - 1.6\frac{A_s}{E_s}(w_i - w_a) \tag{1}$$

where c_a is the CO_2 concentration in the bulk air outside of the leaf (mol mol^{-1}), A_s and E_s are the rate of CO_2 assimilation and transpiration through stomata, respectively (mol m^{-2} s^{-1}), w_i and w_a are the water vapor concentrations inside and outside of the leaf (mol mol^{-1}), and 1.6 is the ratio of diffusivities for water vapor and CO_2 in air.

Notice that calculated c_i is determined mostly from the ratio A_s/E_s because c_a and w_a are atmospheric properties, w_i is determined from leaf temperature, and 1.6 is a constant. In effect, the equation uses water loss as a tracer for CO_2 entry into the leaf. This has the great advantage that the calculation can be made from standard gas exchange measurements without otherwise disturbing the leaf. But as stomata close, E_s and A_s are increasingly determined by cuticle properties, not stomata. Cuticle properties are not considered in the Moss/Rawlins relation.

We began to investigate how much the cuticle altered c_i. The investigation showed cuticle to be 20–40 times more conductive for water vapor than for CO_2 (Boyer et al. 1997; Boyer 2015a). Consequently, by using water as a tracer for CO_2 (Eq. 1), the cuticle overestimated the amount of CO_2 entering the leaf (Boyer 2015b). As a result, calculated c_i were too high especially when stomata closed and cuticle transport dominated the gas exchange of the leaf. At night of course, this reversed and calculated c_i became too low (Hanson et al. 2016). If calculated c_i was corrected for these cuticle effects, the calculations came closer to c_i measured directly (Boyer 2015b). This indicated that calculations of c_i need to include cuticle properties.

Ideally, it would be best to have a simple method of incorporating cuticle properties into the calculation. This would preserve the advantage that the leaf is undisturbed. But for now this remains in the future.

4 Plant Enlargement

I arrived at University of Illinois with only my isopiestic psychrometer and soon developed a new system to measure the water potential of whole leaves on intact plants (Boyer 1968). The intact leaves gave the same water potentials as samples (disks) taken from the same leaves. But no matter how much water was supplied, the potentials never were higher than −0.2 MPa. I began to think that the psychrometer was in error and decided to wait for up to 30 h to allow time for the water potential to change. During this extended measurement, the potentials remained stable. However, the leaves absorbed water slowly throughout this time and when the leaves were removed from the psychrometer, they were larger. The leaves had grown. Even recently "mature" leaves grew.

Considering how little water enters the leaf as it grows, the potential of −0.2 MPa was surprising. Water enters leaves much more rapidly for transpiration than for growth (often by at least 100-fold) but leaf water potentials are only slightly more negative (say −0.4 MPa). Why were water potentials associated with growth so close to those for transpiration when water was moving so much faster for transpiration? In other words, if water potential differences drive water uptake, why were they so similar for vastly different uptakes?

After monitoring the rapid water movement in individual cells isolated from the leaves, we began to think that perhaps water flowed along two leaf paths. Forces for the movement would differ for the two paths, at least in sunflower leaves. Water for growth had to reach all the cells including the epidermis. Water for transpiration might bypass many of those cells and evaporate early, deep in the leaf (Boyer 1977). If so, water uptake for transpiration might encounter a low resistance and the water potential differences driving the flow could be small. On the other hand, water uptake for growth might encounter a high resistance because water would have to traverse many tissues to reach the epidermises. Water potential differences for growth would then be large.

These concepts forced us to envision water potentials associated with the growth process itself (when transpiration was not occurring), and we termed them "growth-induced water potentials" because: (1) they were not present in mature tissues and (2) they were induced by the growth process (Molz and Boyer 1978; Cavalieri and Boyer 1982; Boyer 1985). Growth induced these potentials because the walls were softened (loosened) and yielded to the turgor pressure (Boyer 2001). The turgor was held lower than it otherwise would be. As a result, the cell compartment was expanded to a larger size.

The water potential of the cell was transmitted to the apoplast as a tension and could be measured with a pressure chamber illustrated in Fig. 8. The tension moved water into the growing cells and would have a size that depended on the anatomy of the growing organ. Basically, the water would move from the vascular system through intervening small cells to the ultimate cells on the surface. Consequently, Mark Westgate found that the growth-induced water potentials differed depending on which organ was measured in the maize plant because the organs differed in size and vascular supply (Westgate and Boyer 1985b), as seen in Fig. 9. John Passioura similarly found that the anatomy of the tissue contributed to the development of growth-induced water potentials in soybean hypocotyls (Passioura and Boyer 2003), and Wendy Silk and her colleagues elegantly showed the effect in maize roots of various sizes (Wiegers et al. 2009).

In fact, Hiroshi Nonami in our lab conceptualized these potentials in three dimensions as growth-induced "fields" (Fig. 10). Using a pressure probe both to measure the turgor and sample osmotic potentials in individual cells, he reported a growth-induced water potential field in three dimensions for the growing region of soybean hypocotyls (Nonami and Boyer 1993). The field was negligible in the mature basal tissues of the same hypocotyls. Moreover, it could be reversed in only a few cells next to the xylem and cause rapid responses of growth rates (Nonami and Boyer 1990; Tang and Boyer 2003). Figure 11 shows this kind of reversal (red) and indicates that the field is in the wrong direction for extracting water from the xylem. Consequently, rapid changes in xylem water status have immediate effects on growth rates for the whole organ (Nonami and Boyer 1990; Maruyama and Boyer 1994; Passioura and Boyer 2003; Tang and Boyer 2003).

Theoretically, growth-induced water potentials should allow water to be extracted from nearby mature tissue because the potential in the growing region would be lower than in the mature region. Rainer Matyssek et al. (1991b) illustrated this effect by

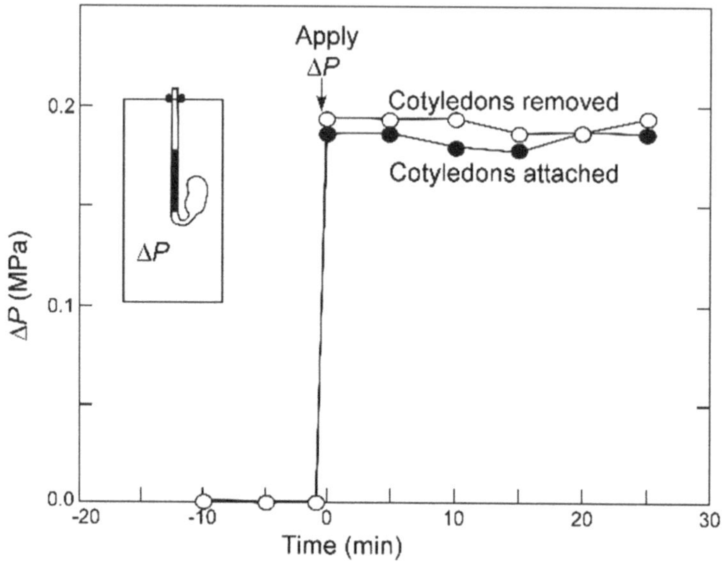

Fig. 8 Measuring growth-induced water potential with a pressure chamber in soybean hypocotyls. In this example, the hypocotyl had been growing rapidly before the roots were excised for the pressure chamber measurement. The growth-induced water potential was determined as the balancing pressure necessary to keep xylem solution at the cut surface (0.2 MPa). This result indicated that the potential existed mostly as a tension (−0.2 MPa) in the apoplast of the elongating region (*black area*). Results are shown for plants with cotyledons but were the same if the cotyledons were removed and thus could not be attributed to cotyledons. The tension measured this way disappeared when growth was fully prevented by pressurizing the hypocotyl of the intact seedling before roots were removed. This illustrated that the growth process itself had caused the growth-induced water potential. Redrawn from Boyer (2001)

excising various plant tissues, as shown in Fig. 12. Only when no mature tissue was attached to the elongating region did growth cease. This showed unequivocally that water in the growing region had a lower potential than in the mature region. The water moved from mature to growing tissue because of the potential difference between the two tissues (Matyssek et al. 1991a, b).

If one considers that growth occurs while transpiration occurs, the competition between the two processes becomes important. For example, a typical grass leaf has the growing region at the base and the more mature blade exposed to the atmosphere. Water moves into the growing region but also through the growing region to be lost in the exposed blade by transpiration. The two processes were investigated first by Mark Westgate who showed that the growth-induced water potential had to be lower than the water potential in the xylem in order for water to enter the base of the leaf for growth (Westgate and Boyer 1984). As water moved through the growing region to the exposed blade, it followed a shallow gradient in water potentials along the xylem. An-ching Tang explored how this gradient related to the anatomy of the maize leaf and found that the protoxylem passed right through the growing region and

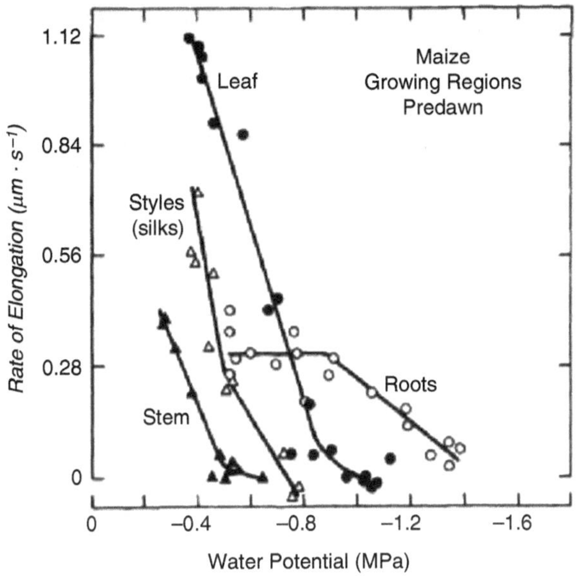

Fig. 9 Growth-induced water potentials in growing regions of various organs of the maize plant. As shown, the growth-induced potentials were −0.3 MPa (Stem), −0.4 MPa (Styles and Leaf), and −0.55 MPa (Roots) during rapid growth. By contrast in adjacent mature tissues the potentials were close to those in the soil (−0.07 to −1.7 MPa, not shown here for clarity but details are in Westgate and Boyer (1985b)). When water was withheld from the soil, growth-induced water potentials became lower and growth ceased at potentials of −0.6 MPa (Stem), −0.75 MPa (Styles), and −1.0 MPa (Leaf). Roots continued growing slowly at −1.4 MPa. All potentials were measured predawn when transpiration was essentially zero. From Westgate and Boyer (1985b)

transported water to the mature blade (Tang and Boyer 2002). The growing region had to extract water from this protoxylem and did so by forming a growth-induced water potential field around each xylem vessel. The field extracted small amounts of water for the growth process while the bulk made its way to the exposed leaf for transpiration.

Growth-induced water potentials are all around us even though we may not notice them. Matyssek et al. (1991a, b) had found that a growing region could extract water from nearby basal mature tissues, so this explains why potatoes can sprout in storage even though no external water is present. The growth-induced water potential of the sprout moves water from the surrounding mature potato tissues into the growing cells. It is only necessary for the cell walls of the bud to become more extensible and prevent the turgor from rising as much as in a mature cell, nicely shown by Sachio Maruyama (Maruyama and Boyer 1994). Despite the fact that water transport for growth is slow compared to transpiration, growth-induced water potentials explain everyday phenomena such as how greens become crisp in the cold or how leaves grow on a tree previously cut down and having no external water supply.

Cosgrove and Cleland (1983) and Cosgrove et al. (1984) viewed these processes differently. At first, Cosgrove and Cleland (1983) obtained solute from growing tissue

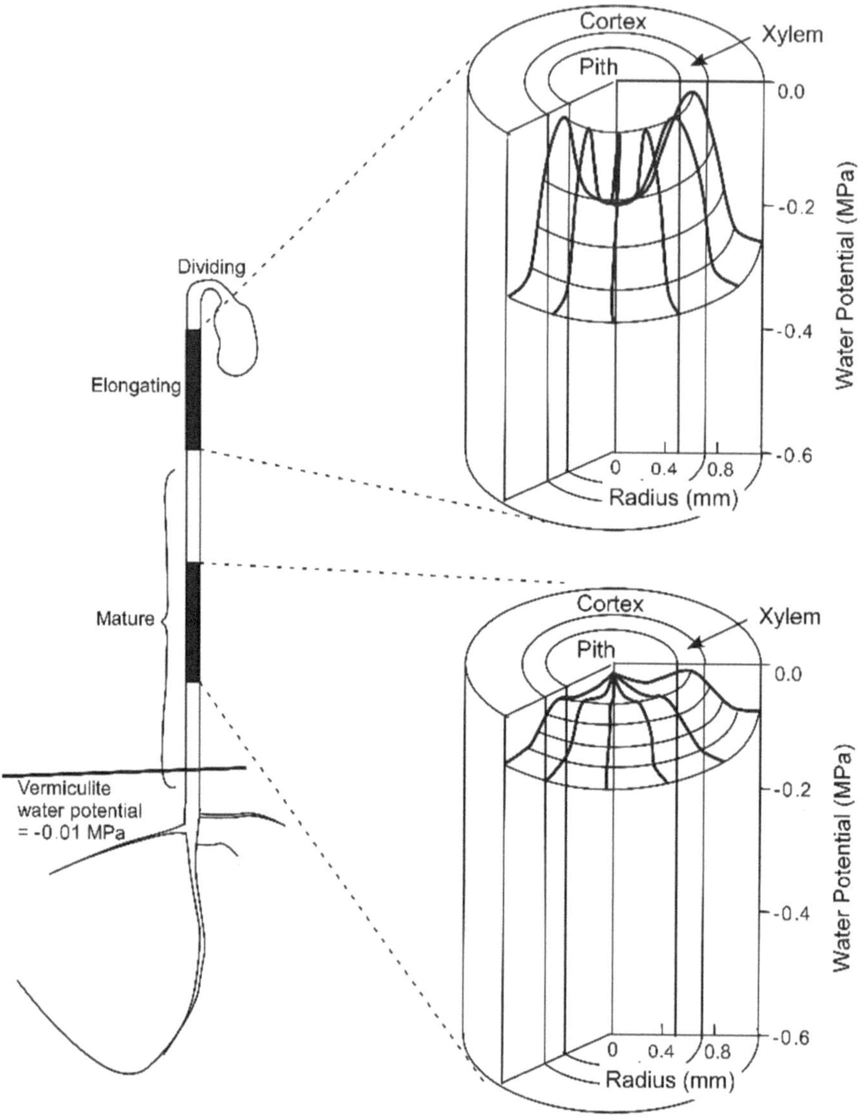

Fig. 10 Growth-induced water potential field in elongating region compared with water potential field in mature region of the same plants. The hypocotyls of these plants were elongating at about 1.5 mm h^{-1}. Fields show the highest potential in xylem and lower potentials in pith and cortex. Note that growth-induced one is sizable in comparison with that in the mature zone. Fields were directly measured in intact plants with microcapillary of a pressure probe. Redrawn from Nonami and Boyer (1993). Copyright American Society of Plant Biologists

and thought it came from the apoplast and thus kept turgor low in the growing cells. However, we succeeded in obtaining apoplast solution from completely intact seedlings and it contained little solute (Nonami and Boyer 1987). Subsequently, Cosgrove et al. (1984) considered our measurements to be artifacts of tissue excision that

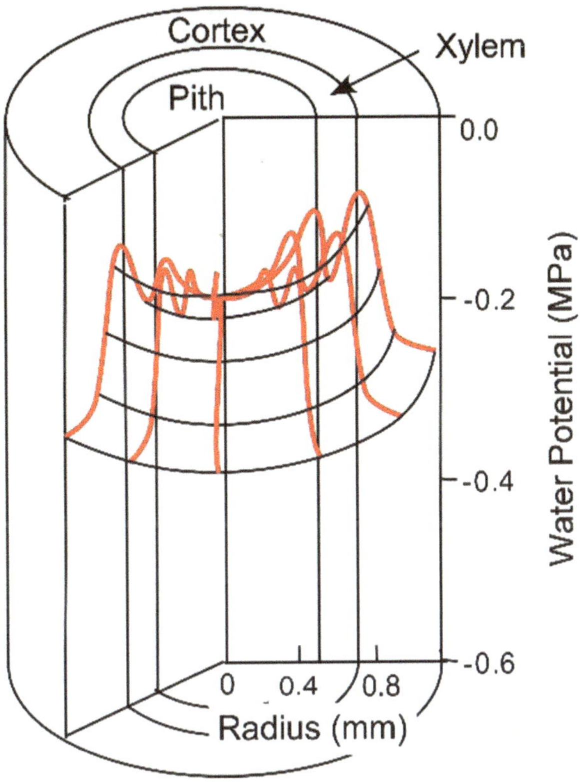

Fig. 11 Growth-induced water potential after xylem tension is increased (*red*) compared to Fig. 10. Field is reversed next to the xylem and prevents water from being extracted. The reversal thus stops growth immediately until the potentials in the field move down to reestablish a new one similar to the one in Fig. 10 but lower. Because of the small volume, xylem tension can change rapidly and affect growth just as rapidly

actually measured cell wall relaxation, but they included attached mature tissue that served as a water reservoir for the growing tissues (Matyssek et al. 1988). When the mature tissue was removed, relaxation was minor (Boyer et al. 1985). Moreover, Boyer (1968), Boyer et al. (1985), and Matyssek et al. (1991a) detected growth-induced water potentials in completely intact plants (no excision).

Eventually others (Fricke and Flowers 1998; Martre et al. 1999; Fricke 2002) detected growth-induced water potentials too and Wendy Silk and I wrote a review about them (Boyer and Silk 2004). There is increasing recognition that growth hydraulics may be important factors in breeding programs and I look forward to seeing these concepts develop further.

5 Chemistry of Cell Enlargement

The mechanism of cell enlargement has fascinated scientists for more than a century. Pfeffer discovered how turgor developed in plants and knew it was somehow required for the growth process (Pfeffer 1900). Our move to the Biology/Biochemistry

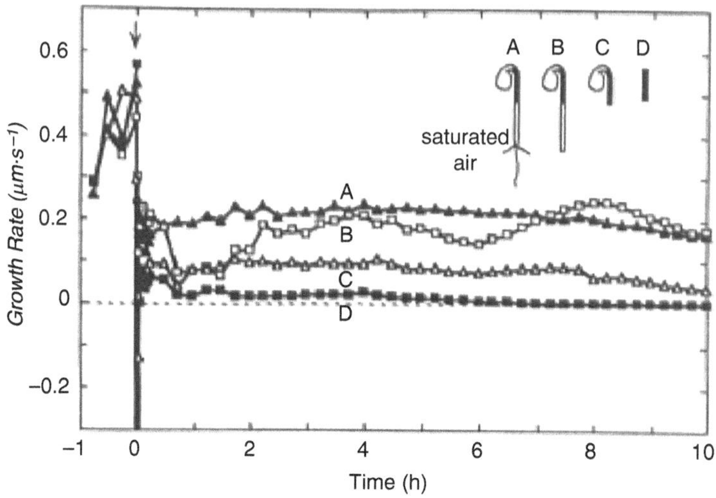

Fig. 12 If water potentials are induced by growth, they should be capable of withdrawing water from nearby mature tissue because the potential is lower than in the mature tissue. Initially, the hypocotyls grew rapidly in intact plants supplied with hydrated vermiculite in saturated air (before time zero). At the *arrow*, the vermiculite water source was removed (A) or roots (B), mature region of the hypocotyl (C), and all mature tissues (D) were also removed. Note that growth occurred whenever the elongating region was attached to mature tissue as a water reservoir. Growth ceased in (D) when all mature tissue had been removed. Elongating region is shown in *black* and is where all the measurements were made. From Matyssek et al. (1991b)

Program in the Marine College at the University of Delaware (1987) allowed me to work further on this with marine plants having large enough cells for physiological work. We were intrigued by the findings of Paul Green and his colleagues in *Nitella*, a large-celled alga that could be used to measure growth while turgor was continuously monitored (Green et al. 1971). Paul did not want to continue that work despite being urged to do so. We thought we could continue Paul's work with newer technologies developed after Paul did his original experiments.

Using *Chara corallina*, a close relative of *Nitella*, we were able to monitor turgor pressures with a pressure probe and growth with a position transducer simultaneously. It was also possible to isolate the cell walls without removing them from the medium in which the plants were grown. Gluing the isolated wall onto a capillary made it possible to use the pressure probe to inject oil into the lumen vacated by the cytoplasm and pressurize it to the same pressures as in the live cells. Anything could be injected into the cell or lumen at any pressure without ever removing the cell or wall from the growth medium.

We found that the isolated walls could grow as fast as the live cell on the intact plant, but only for 1 or 2 h (Proseus and Boyer 2006a). After that, the isolated walls stopped growing while the live cells continued to grow. But if we supplied pectin to the growth medium, the walls resumed growth as fast as the live cell for several hours (Fig. 13). Pectin is a wall constituent in the alga and in land plants, so its activity came

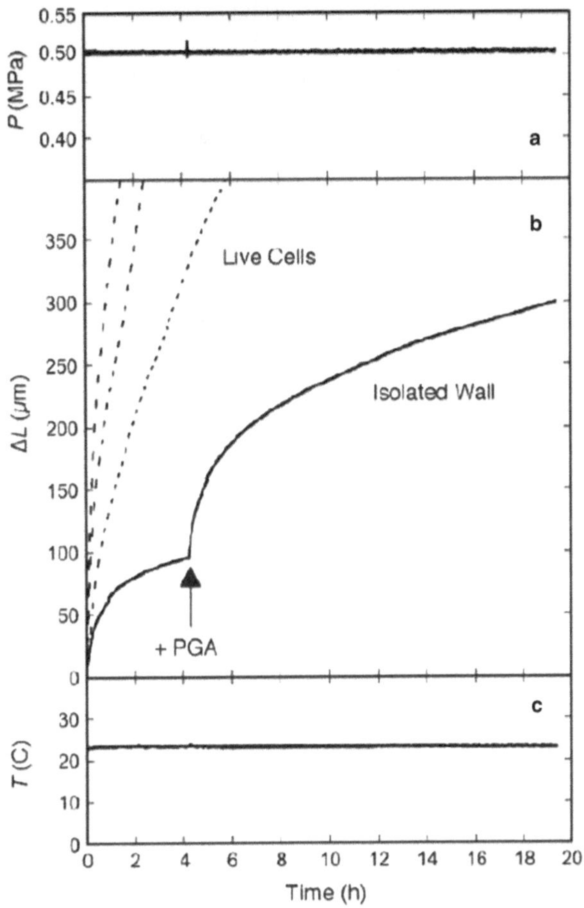

Fig. 13 Growth (a, ΔL), pressure (b, P), and temperature (c, T) for live cells and isolated cell walls of *Chara corallina* internodes in the medium in which the plants had been grown. At the *arrow*, pectin as polygalacturonic acid (PGA) was added to the medium. Note that the addition caused growth of the isolated wall to resume at the rate of the live cells. Reprinted from Proseus and Boyer (2006a)

as a great surprise. It was shown that the pectin acted as a chelator to remove calcium from the cell wall (Proseus and Boyer 2006b). Since calcium acts as a cross-bridge to bind pectin polymers together, it seemed that calcium removal may have allowed the walls to expand. Pectin also had a similar action in live cells. This suggested that calcium cross-bridging was central to the growth process. Not only central but likely to be enzymeless because the chemistry occurred whether isolated walls were boiled or not (Proseus and Boyer 2006a). Moreover, the pectin lost its activity if the turgor dropped below a minimum (Proseus and Boyer 2007). The cells still appeared turgid but not sufficiently to cause growth of the wall or live cell.

How could a cross-bridging mechanism for growth be sensitive to turgor? Hints came if more than 95% of the calcium was removed from isolated walls. They were weakened so much that they no longer could hold normal turgor (about 0.5 MPa) (Proseus and Boyer 2006b). Also, in live cells deprived of calcium, the wall became too weak to hold turgor and the wall burst (Proseus and Boyer 2012). Another fact became apparent when pectin was placed in tension (Morikawa et al. 1974; Proseus and Boyer 2007). The cross-bridging became distorted and weakened. The turgor had distorted the pectin in the existing wall and weakened its calcium cross-bridging so that new undistorted pectin could target the wall calcium. This suggested that the normal release of pectin produced in the cytoplasm could remove calcium from the existing wall and thereby cause the wall to expand.

Putting these findings together, a chemical mechanism was proposed (Fig. 14) for wall expansion that involved a calcium pectate cycle in the wall (Boyer 2009). As long

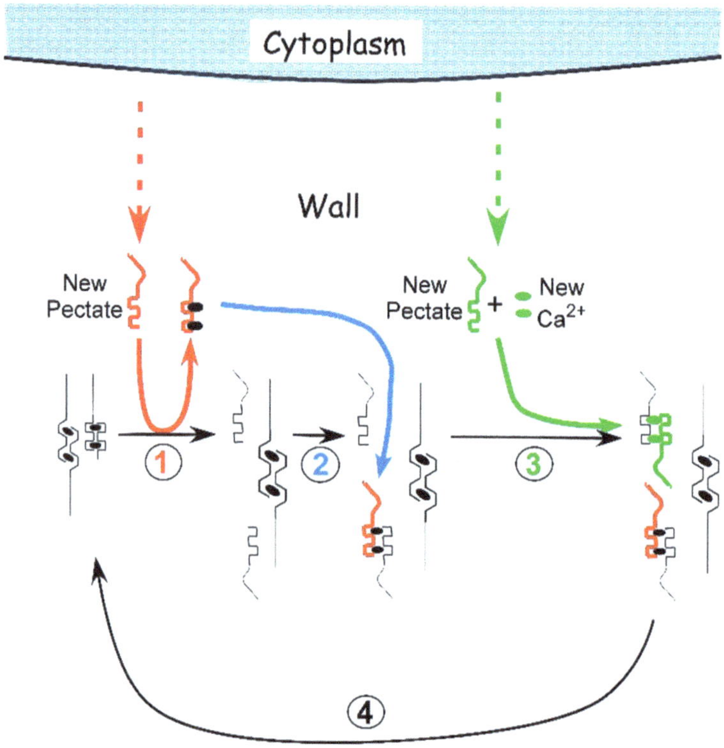

Fig. 14 Proposed mechanism of cell enlargement in *Chara*. The diagram shows the calcium pectate cycle occurring in the cell wall for two calcium pectate cross-bridges (*black ovals* in antiparallel pectate molecules, *left side* of figure). Turgor pressure is high enough to distort the eggbox in one of the pair, weakening its bonds with calcium (*left pectate* in pair). New pectate from the cytoplasm (*dashed red arrow*) is undistorted and preferentially removes calcium from the distorted pectate (step 1, *red*). This load-bearing pectate relaxes after its cross-bridging calcium is removed. The wall elongates incrementally, shifting the load to the other member of the pectate pair, which distorts. The remaining steps 2–4 follow by depositing calcium pectate (step 2, *blue*) and new calcium from the medium plus new pectate from the cytoplasm (step 3, *green*), resulting in a cycle (step 4, *black*). The net result is elongation plus wall deposition. Although shown for only two cross-bridged pectate molecules, the same principles apply to larger numbers of cross-bridges. Note that in *Chara* the cycle occurs in the medium in which the cells are grown (0.6 mM Ca^{2+}). Also note that the rate of growth depends on the rate of pectate release from cytoplasm to wall by exocytosis (*red and green dashed arrows*). Each step in the cycle was demonstrated by experiments. From Boyer (2016)

as pectin, calcium, and sufficient turgor were available, the cycle could take place in pectin-containing walls. Because *Chara* and *Nitella* are some of the closest relatives to the progenitors of land plants, it seems possible that this algal chemistry might occur in the more advanced species. Boyer (2016) reviewed evidence for such a mechanism and concluded that despite considerable evidence in favor of a similar mechanism, definitive experiments are still needed for the advanced species. It will be exciting to learn whether the algal chemistry has greater relevance than now known.

6 Reproduction

About 75% of the harvested acreage in US agriculture comes from plant reproductive structures. Unfortunately, the process is complex and easily disrupted. For example, losses can be irreversible if unfavorable environments such as water deficits occur around the time of pollination. After pollination, they typically cause smaller or lighter fruiting structures that partially recover if conditions improve and photosynthesis resumes. Salter and Goode (1967) summarized this behavior in a range of crops and concluded that early stages of reproduction were especially vulnerable. At the time, few explanations were available but indicated that there were major time-dependent activities controlling the process.

Maize is an example and is mostly harvested for its reproductive structures. When we grew the plants in the field, pollination could be carried out under favorable conditions but water deficiency developing afterward caused smaller, lighter kernels. During this time (6-week drought until maturity), the plants had less photosynthetic activity and transported stored sugars from stems and leaves to the ears (Jurgens et al. 1978). But with less stored sugars and starch than photosynthesis would normally supply, kernels were small (McPherson and Boyer 1977; Boyer and Westgate 2004).

By contrast, a water deficit around the time of pollination irreversibly disrupted kernel development. Sometimes the irreversibility was attributed to pollen being shed uselessly because silks (stigmas and styles) developed less rapidly than pollen (Tatum and Kehr 1951; Herrero and Johnson 1981; Oury et al. 2016a, b). Later, it was discovered that the delay in silk development was controlled by only a few genes (Edmeades et al. 1993). When these genes were incorporated into commercial hybrids, the silks grew early and were able to be pollinated.

The silks were always wetter than the pollen, and the pollen used the water in the silks to grow toward the embryo sac (Westgate and Boyer 1986a). Even if silks were exposed to pollen, however, the ears were entirely barren at the end of the season during a water deficit (Westgate and Boyer 1985a). There was an undeveloped embryo in every floret in the middle of the ear indicating that pollination had succeeded and fertilization had occurred (Westgate and Boyer 1986b). In effect, the embryos had aborted. This indicated that maize yields could not be drought-proofed simply with precocious silks.

Interestingly, shortening the deficit to only a few days around pollination gave the same abortion as a long-term drought (Westgate and Boyer 1986b). We wondered if we could increase the carbon supply to the ear during that short time, and Mark Boyle succeeded in feeding photosynthetic products to stems of water-deficient maize (Fig. 15). The feeding prevented 70–80% of the barrenness (Boyle et al. 1991b). Abortion thus had a biochemical origin!

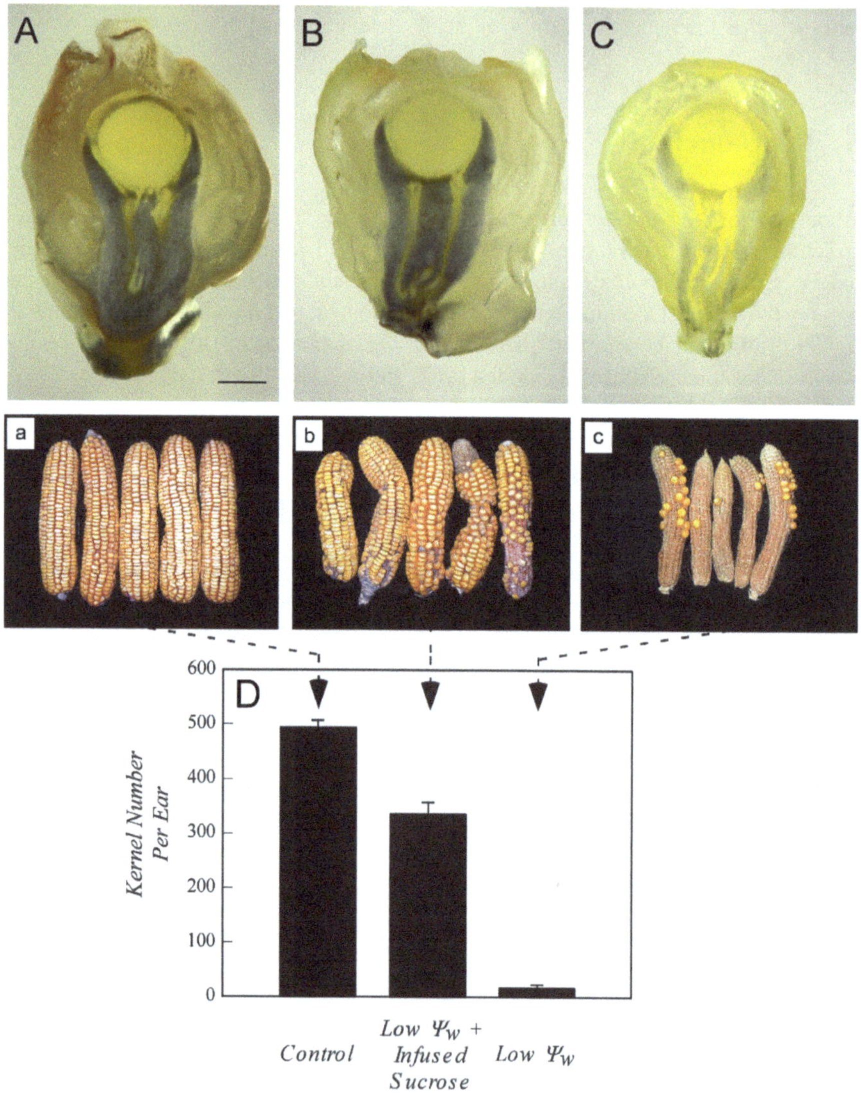

Fig. 15 Ovary starch during pollination and ears at maturity in maize. (**A, a**) hydrated controls, (**B, b**) water-deficient plants whose stems were fed sucrose in photosynthetic quantities, and (**C, c**) the same water deficiency but unfed. Kernel numbers at maturity are shown in (**D**). Sucrose feeding largely prevented irreversible abortion. Starch in ovaries at pollination determined fate of ear development. Ears were slightly deformed at maturity because of ovary sampling. Redrawn from McLaughlin and Boyer (2004a, b)

We are indebted to Mark for his perseverance in developing the feeding method since he had to get the equivalent of 5 g of photosynthetic product into the plant every day to make up for the diminished photosynthesis in the water-deficient plants. Many attempts failed until he fed the product in a solution to a newly cut well in the stem and attached a reservoir (Boyle et al. 1991a). The open xylem in the well then took up the solution.

Later, Chris Zinselmeier and Mike Lauer found that sucrose was the active ingredient among Mark's photosynthetic products (Zinselmeier et al. 1995). Invertase hydrolyzed the sucrose to glucose and fructose (Zinselmeier et al. 1999). John McLaughlin developed a way to image the glucose (Fig. 16) in the developing florets (McLaughlin and Boyer 2004a). When photosynthesis diminished because of a water deficit, glucose disappeared from the images in the florets (McLaughlin and Boyer 2004b). If sucrose was fed to the plant stems, invertase hydrolyzed the sugar to glucose and it reappeared in the ovaries despite the water deficit. Abortion was largely prevented.

At first, we thought that the developing florets were simply starving for glucose but John McLaughlin assayed several genes for their expression in the water-deficient florets (McLaughlin and Boyer 2004b). Genes for invertase were generally downregulated but for senescence were upregulated. Moreover, sucrose fed to the stems prevented the downregulation and upregulation. Consequently, we had to change our mind about the starvation hypothesis. From these results, florets were aborting because genes in their ovaries were monitoring ovary sugar status and were actively inducing senescence when the status became low.

In recent years, there has been a sea change in how plant breeders view water use by plants. It was once thought that plants selected for superior performance in favorable environments automatically selected for superior performance during drought. Unfortunately, this approach missed genes that were activated only during

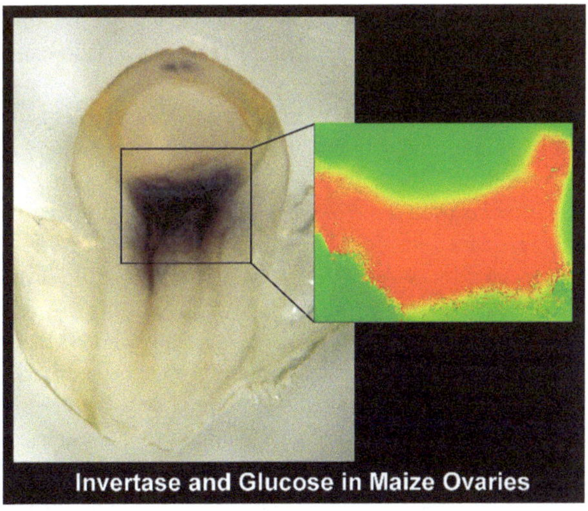

Fig. 16 Images of cell wall invertase activity (*black* in ovary) and glucose (*red* in inset) of ovary of maize on the day of pollination. Glucose is a product of invertase activity and used to make starch (Fig. 15). During water deficits, the glucose is depleted in the ovary. From McLaughlin and Boyer, unpublished

a water deficit. Genes for osmotic adjustment or ovary senescence are examples. The upregulation of senescence genes would be entirely missed if superior performance was judged only in favorable conditions. We are gratified that certain seed companies are now exposing maize to drought during their selection programs and are using our sucrose feeding results to develop more drought tolerant maize.

7 Retrospectives

There are many students and other associates who could not be mentioned in this essay but whose effort added greatly to the education and science from our lab. I am indebted to all of them not only for their hard work but also for what they accomplished. My wonderful colleagues and their students were also invigorating and I owe much to them as well as to the universities and USDA good enough to hire me. It is amazing to have the privilege of being a scientist in such settings!

When I started work on plant water relations, it was a scientific backwater. There was confusion about important concepts and even basic principles like the ability to repeat experiments. These drawbacks undoubtedly drove scientists away. With the advent of thermodynamic concepts and measurement methods, repeatability became possible. Furthermore, the public began to realize that plants use copious amounts of an increasingly scarce resource. Irrigation is an answer but is often unavailable and expensive. These views were expressed in a review (Boyer 1982) and more emphasis is now placed on how plants use water and whether it can be conserved (Kramer and Boyer 1995). Seed companies are now developing drought tolerance programs because they can make a profit. My scientific life has thus witnessed a major whirlwind-like change.

Still, it surprises me that more scientists have not used isopiestic psychrometry not only for water relations but also for basic physiology. The water potential, osmotic potential, and turgor are important for so many fundamental processes that the ability to measure them seems essential. Turgor is a good example since it is necessary for cell enlargement.

Two barriers seem to exist for these measurements. First, physics, physical chemistry, and thermodynamics are rarely part of the biology curriculum. Consequently, the power of these methods is often unappreciated. Our psychrometry and gas exchange experiments used some of the equilibrium thermodynamic principles and allowed us to observe plant behavior often missed by others. But a downside is that findings need to be repeated by others and risk being ignored if the instruments are unavailable. Secondly, building the apparatus and practicing its use requires a substantial scientific commitment. Many plant scientists simply cannot make such a large commitment. In an effort to alleviate some of the problem, we marketed our isopiestic system for 15 years and it is now available more widely than before. But the problem persists.

Molecular genetics offers revolutionary opportunities to extend our understanding of plants. Embraced by plant scientists, it is increasingly used to study plant water relations. But as with most methodologies, there are limitations to its use especially when confronted with the multigenic aspects of water relations. By combining molecular genetic methods and fundamental physiological measurements, our understanding will accelerate. I look forward to seeing more combinations of these disciplines.

References

Avenhaus U, Cabeza RA, Liese R, Lingner A, Dittert K, Salinas-Riester G, Pommerenke C, Schulze J (2016) Short-term molecular acclimation processes of legume nodules to increased external oxygen concentration. Front Plant Sci. doi:10.3389/fpls.2015.01133

Bernstein L (1961) Osmotic adjustment of plants to saline media. I. Steady state. Am J Bot 48:909–918

Boyer JS (1965) Effects of osmotic water stress on metabolic rates of cotton plants with open stomata. Plant Physiol 40:229–234

Boyer JS (1966) Isopiestic technique: measurement of accurate leaf water potentials. Science 154: 1459–1460

Boyer JS (1967) Leaf water potentials measured with a pressure chamber. Plant Physiol 42:133–137

Boyer JS (1968) Relationship of water potential to growth of leaves. Plant Physiol 43:1056–1062

Boyer JS (1969) Measurement of the water status of plants. Annu Rev Plant Physiol 20:351–364

Boyer JS (1970) Leaf enlargement and metabolic rates in corn, soybean, and sunflower at various leaf water potentials. Plant Physiol 46:233–235

Boyer JS (1977) Regulation of water movement in whole plants. Symp Soc Exp Biol 31:455–470

Boyer JS (1982) Plant productivity and environment. Science 218:443–448

Boyer JS (1985) Water transport. Annu Rev Plant Physiol 36:473–516

Boyer JS (1995) Measuring the water status of plants and soil. Academic Press, San Diego, CA

Boyer JS (2001) Growth–induced water potentials originate from wall yielding during growth. J Exp Bot 52:1483–1488

Boyer JS (2009) Cell wall biosynthesis and the molecular mechanism of plant enlargement. Funct Plant Biol 36:383–394

Boyer JS (2015a) Turgor and the transport of CO_2 and water across the cuticle (epidermis) of leaves. J Exp Bot 66:2625–2633

Boyer JS (2015b) Impact of cuticle on calculations of the CO_2 concentration inside leaves. Planta 242:1405–1412

Boyer JS (2016) Enzyme-less growth in *Chara* and terrestrial plants. Front Plant Sci. doi:10.3389/fpls.2016.00866

Boyer JS, Bowen BL (1970) Inhibition of oxygen evolution in chloroplasts isolated from leaves with low water potentials. Plant Physiol 45:612–615

Boyer JS, Kawamitsu Y (2011) Photosynthesis gas exchange system with internal CO_2 directly measured. Environ Control Biol 49:193–207

Boyer JS, Knipling EB (1965) Isopiestic technique for measuring leaf water potentials with a thermocouple psychrometer. Proc Natl Acad Sci U S A 54:1044–1051

Boyer JS, Potter JR (1973) Chloroplast response to low leaf water potentials. I. Role of turgor. Plant Physiol 51:989–992

Boyer JS, Silk WK (2004) Hydraulics of plant growth. Funct Plant Biol 31:761–773

Boyer JS, Westgate ME (2004) Grain yields with limited water. J Exp Bot 55:2385–2394

Boyer JS, Cavalieri AJ, Schulze E-D (1985) Control of the rate of cell enlargement: excision, wall relaxation, and growth-induced water potentials. Planta 163:527–543

Boyer JS, Wong SC, Farquhar GD (1997) CO_2 and water vapor exchange across leaf cuticle (epidermis) at various water potentials. Plant Physiol 114:185–191

Boyle MG, Boyer JS, Morgan PW (1991a) Stem infusion of maize plants. Crop Sci 31:1241–1245

Boyle MG, Boyer JS, Morgan PW (1991b) Stem infusion of liquid culture medium prevents reproductive failure of maize at low water potentials. Crop Sci 31:1246–1252
Cavalieri AJ, Boyer JS (1982) Water potentials induced by growth in soybean hypocotyls. Plant Physiol 69:492–496
Cosgrove DJ, Cleland RE (1983) Solutes in the free space of growing stem tissues. Plant Physiol 72:326–331
Cosgrove DJ, Van Volkenburgh E, Cleland RE (1984) Stress relaxation of cell walls and the yield threshold for growth: demonstration and measurement by micro-pressure probe and psychrometer techniques. Planta 162:46–54
Edmeades GO, Bolaños J, Hernández M, Bello S (1993) Causes for silk delay in a lowland tropical maize population. Crop Sci 33:1029–1035
Ehlig CF (1962) Measurement of energy status of water in plants with a thermocouple psychrometer. Plant Physiol 37:288–290
Ehret DL, Boyer JS (1979) Potassium loss from stomatal guard cells at low water potentials. J Exp Bot 30:225–234
Fellows RJ, Boyer JS (1976) Structure and activity of chloroplasts of sunflower leaves having various water potentials. Planta 132:229–239
Fischer RA (1968) Stomatal opening: role of potassium uptake by guard cells. Science 168:784–785
Fischer RA, Hsiao TC (1968) Stomatal opening in isolated epidermal strips of *Vicia faba*. II. Responses to KCl concentration and the role of potassium absorption. Plant Physiol 43:1953–1958
Fishbeck K, Evans HJ, Boersma LL (1973) Measurement of nitrogenase activity of intact legume symbionts *in situ* using the acetylene reduction assay. Agron J 65:429–433
Fricke W (2002) Biophysical limitation of cell elongation in cereal leaves. Ann Bot 90:157–167
Fricke W, Flowers TJ (1998) Control of leaf cell elongation in barley. Generation rates of osmotic pressure and turgor, and growth-associated water potential gradients. Planta 206:53–65
Gale J, Kohl HC, Hagan RM (1967) Changes in the water balance and photosynthesis of onion, bean and cotton plants under saline conditions. Physiol Plant 20:408–420
Gibbs JW (1875–1876) On the equilibrium of heterogeneous substances. Trans Conn Acad Arts Sci III:108–248
Graan T, Boyer JS (1990) Very high CO_2 partially restores photosynthesis in sunflower at low water potentials. Planta 181:378–384
Green PB, Erickson RD, Buggy J (1971) Metabolic and physical control of cell elongation rate. *In vivo* studies in *Nitella*. Plant Physiol 47:423–430
Hanson DT, Stutz SS, Boyer JS (2016) Why small fluxes matter: the case and approaches for improving measurements of photosynthesis and (photo)respiration. J Exp Bot 67:3027–3039
Hardy RWF, Holsten RD, Jackson EK, Burns RC (1968) The acetylene-ethylene assay for N_2 fixation: laboratory and field evaluation. Plant Physiol 43:1185–1207
Herrero MP, Johnson RR (1981) Drought stress and its effects on maize reproductive systems. Crop Sci 21:105–110
Huang CY, Boyer JS, Vanderhoef LN (1975a) Acetylene reduction (nitrogen fixation) and metabolic activities of soybean having various leaf and nodule water potentials. Plant Physiol 56:222–227
Huang CY, Boyer JS, Vanderhoef LN (1975b) Limitation of acetylene reduction (nitrogen fixation) by photosynthesis in soybean having low water potentials. Plant Physiol 56:228–232
Hüsken D, Steudle E, Zimmermann U (1978) Pressure probe technique for measuring water relations of cells in higher plants. Plant Physiol 61:158–163
Jurgens SK, Johnson RR, Boyer JS (1978) Dry matter production and translocation in maize subjected to drought during grain fill. Agron J 70:678–682
Keck RW, Boyer JS (1974) Chloroplast response to low leaf water potentials. III. Differing inhibition of electron transport and photophosphorylation. Plant Physiol 53:474–479
Kramer PJ, Boyer JS (1995) Water relations of plants and soils. Academic Press, San Diego, CA
Lauer MJ, Boyer JS (1992) Internal CO_2 measured directly in leaves: abscisic acid and low leaf water potential cause opposing effects. Plant Physiol 98:1310–1316

Lillo C, Lea US, Leydecker M-T, Meyer C (2003) Mutation of the regulatory phosphorylation site of tobacco nitrate reductase results in constitutive activation of the enzyme *in vivo* and nitrite accumulation. Plant J 35:566–573

Lüttge U (2016) Transport processes: the key integrators in plant biology. Prog Bot 77:3–65

Mansfield TA, Jones RJ (1971) Effects of abscisic acid on potassium uptake and starch content of stomatal guard cells. Planta 101:147–158

Martre P, Bogeat-Triboulot MB, Durand JL (1999) Measurement of a growth-induced water potential gradient in tall fescue leaves. New Phytol 142:435–439

Maruyama S, Boyer JS (1994) Auxin action on growth in intact plants: threshold turgor is regulated. Planta 193:44–50

Matthews MA, Boyer JS (1984) Acclimation of photosynthesis to low water potentials. Plant Physiol 74:161–166

Matyssek R, Maruyama S, Boyer JS (1988) Rapid wall relaxation in elongating tissues. Plant Physiol 86:1163–1167

Matyssek R, Maruyama S, Boyer JS (1991a) Growth-induced water potentials may mobilize internal water for growth. Plant Cell Environ 14:917–923

Matyssek R, Tang A-C, Boyer JS (1991b) Plants can grow on internal water. Plant Cell Environ 14:925–930

McLaughlin JE, Boyer JS (2004a) Glucose localization in maize ovaries when kernel number decreases at low water potential and sucrose is fed to the stems. Ann Bot 94:75–86

McLaughlin JE, Boyer JS (2004b) Sugar-responsive gene expression, invertase activity, and senescence in aborting maize ovaries at low water potentials. Ann Bot 94:675–689

McPherson HG, Boyer JS (1977) Regulation of grain yield by photosynthesis in maize subjected to a water deficiency. Agron J 69:714–718

Meyer RF, Boyer JS (1972) Sensitivity of cell division and cell elongation to low water potentials in soybean hypocotyls. Planta 108:77–87

Meyer RF, Boyer JS (1981) Osmoregulation, solute distribution, and growth in soybean seedlings having low water potentials. Planta 151:482–489

Michelena VA, Boyer JS (1982) Complete turgor maintenance at low water potentials in the elongating region of maize leaves. Plant Physiol 69:1145–1149

Mohanty P, Boyer JS (1976) Chloroplast response to low leaf water potentials. IV. Quantum yield is reduced. Plant Physiol 57:704–709

Molz FJ, Boyer JS (1978) Growth-induced water potentials in plant cells and tissues. Plant Physiol 62:423–429

Morikawa H, Tanizawa K, Senda M (1974) Infrared spectra of *Nitella* cell walls and orientation of carboxylate ions in the walls. Agric Biol Chem 38:343–348

Morilla CA, Boyer JS, Hageman RH (1973) Nitrate reductase activity and polyribosomal content of corn (*Zea mays* L.) seedlings having low leaf water potentials. Plant Physiol 51:817–824

Moss DN, Rawlins SL (1963) Concentration of carbon dioxide inside leaves. Nature 197:1320–1321

Nonami H, Boyer JS (1987) Origin of growth-induced water potential: solute concentration is low in apoplast of enlarging tissues. Plant Physiol 83:596–601

Nonami H, Boyer JS (1990) Primary events regulating stem growth at low water potentials. Plant Physiol 93:1601–1609

Nonami H, Boyer JS (1993) Direct demonstration of a growth-induced water potential gradient. Plant Physiol 102:13–19

Oury V, Tardieu F, Turc O (2016a) Ovary apical abortion under water deficit is caused by changes in sequential development of ovaries and in silk growth in maize. Plant Physiol 171:986–996

Oury V, Caldeira CF, Prodhomme D, Pichon J-P, Gibon Y, Tardieu F, Turc O (2016b) Is change in ovary carbon status a cause or a consequence of maize ovary abortion in water deficit during flowering? Plant Physiol 171:997–1008

Pankhurst CE, Sprent JI (1976) Effects of temperature and oxygen tension on the nitrogenase and respiratory activities of turgid and water-stressed soybean and French bean root nodules. J Exp Bot 27:1–9

Passioura JB, Boyer JS (2003) Tissue stresses and resistance to water flow conspire to uncouple the water potential of the epidermis from that of the xylem of elongating plant stems. Funct Plant Biol 30:325–334
Pfeffer W (1900) The physiology of plants, 2nd edn. English translation by AJ Ewart. Clarendon Press, Oxford
Portis AR Jr (1981) Evidence of low stromal Mg^{2+} in intact chloroplasts in the dark. I. Studies with the ionophore A23187. Plant Physiol 67:985–989
Portis AR Jr, Heldt HW (1976) Light-dependent changes of the Mg^{2+} concentration in the stroma in relation to the Mg^{2+} dependence of CO_2 fixation in intact chloroplasts. Biochim Biophys Acta 449:439–446
Potter JR, Boyer JS (1973) Chloroplast response to low leaf water potentials. II. Role of osmotic potential. Plant Physiol 51:993–997
Proseus TE, Boyer JS (2006a) Identifying cytoplasmic input to the cell wall of growing *Chara corallina*. J Exp Bot 57:3231–3242
Proseus TE, Boyer JS (2006b) Calcium pectate chemistry controls growth rate of *Chara corallina*. J Exp Bot 57:3989–4002
Proseus TE, Boyer JS (2007) Tension required for calcium pectate chemistry to control growth in *Chara corallina*. J Exp Bot 58:4283–4292
Proseus TE, Boyer JS (2012) Calcium deprivation disrupts enlargement of *Chara corallina* cells. Further evidence for the calcium pectate cycle. J Exp Bot 63:3953–3958
Rao IM, Sharp RE, Boyer JS (1987) Leaf magnesium alters photosynthetic response to low water potentials in sunflower. Plant Physiol 84:1214–1219
Rawlins SL (1964) Systematic error in leaf water potential measurements with a thermocouple psychrometer. Science 146:644–646
Rubin AH, Terasaki M, Sanui H (1979) Major intracellular cations and growth control: correspondence among magnesium content, protein synthesis, and the onset of DNA synthesis in BALB/c3T3 cells. Proc Natl Acad Sci U S A 76:3917–3921
Salter PJ, Goode JE (1967) Crop responses to water at different stages of growth. Commonwealth Agricultural Bureau, Farnham Royal, Bucks
Scholander PF, Hammel HT, Bradstreet ED, Hemmingsen EA (1965) Sap pressure in vascular plants. Science 148:339–346
Shaner DL, Boyer JS (1976a) Nitrate reductase activity in maize (*Zea mays* L.) leaves. I. Regulation by nitrate flux. Plant Physiol 58:449–504
Shaner DL, Boyer JS (1976b) Nitrate reductase activity in maize (*Zea mays* L.) leaves. II. Regulation by nitrate flux at low leaf water potentials. Plant Physiol 58:505–509
Sharkey TD, Imai K, Farquhar GD, Cowan IR (1982) A direct confirmation of the standard method of estimating intercellular partial pressure of CO_2. Plant Physiol 69:657–659
Sharp RE, Boyer JS (1987) Photosynthesis at low water potentials in sunflower: lack of photoinhibitory effects. Plant Physiol 82:90–95
Skujins JJ, McLaren AD (1967) Enzyme reaction rates at limited water activities. Science 158:1569–1570
Slatyer RO, Taylor SA (1960) Terminology in plant- and soil-water relations. Nature 187:922–924
Tang A-C, Boyer JS (2002) Growth-induced water potentials and the growth of maize leaves. J Exp Bot 53:489–503
Tang A-C, Boyer JS (2003) Root pressurization affects growth-induced water potentials and growth in dehydrated maize leaves. J Exp Bot 54:2479–2488
Tang A-C, Kawamitsu Y, Kanechi M, Boyer JS (2002) Photosynthetic oxygen evolution at low water potential in leaf disks lacking an epidermis. Ann Bot 89:861–870
Tatum LA, Kehr WR (1951) Observations on factors affecting seed set with inbred strains of dent corn. Agron J 43:270–275
von Caemmerer S, Farquhar GD (1981) Some relationships between the biochemistry of photosynthesis and the gas exchange of leaves. Planta 153:376–387

Westgate ME, Boyer JS (1984) Transpiration- and growth-induced water potentials in maize. Plant Physiol 74:882–889

Westgate ME, Boyer JS (1985a) Carbohydrate reserves and reproductive development at low water potentials in maize. Crop Sci 25:762–769

Westgate ME, Boyer JS (1985b) Osmotic adjustment and the inhibition of leaf, root, stem and silk growth at low water potentials in maize. Planta 164:540–549

Westgate ME, Boyer JS (1986a) Silk and pollen water potentials in maize. Crop Sci 26:947–951

Westgate ME, Boyer JS (1986b) Reproduction at low silk and pollen water potentials in maize. Crop Sci 26:951–956

Wiegers BS, Cheer AY, Silk WK (2009) Modeling the hydraulics of root growth in three dimensions with phloem water sources. Plant Physiol 150:2092–2103

Younis HM, Boyer JS, Govindjee (1979) Conformation and activity of chloroplast coupling factor exposed to low chemical potential of water in cells. Biochim Biophys Acta 548:328–340

Younis HM, Weber G, Boyer JS (1983) Activity and conformational changes in chloroplast coupling factor induced by ion binding: formation of a magnesium-enzyme-phosphate complex. Biochemistry 22:2505–2512

Zinselmeier C, Lauer MJ, Boyer JS (1995) Reversing drought-induced losses in grain yield: sucrose maintains embryo growth in maize. Crop Sci 35:1390–1400

Zinselmeier C, Jeong B-R, Boyer JS (1999) Starch and the control of kernel development in maize at low water potentials. Plant Physiol 121:25–35

Molecular Markers for Genetic Diversity

Kornel Burg

Abstract Markers are essential for categorising the living creatures. Starting from using visible features, science made the hidden characters visible, eventually using the material of inheritance, the DNA as source of markers. This chapter allows an insight into the different types of DNA based marker systems, guiding the reader through marker systems either with or without preliminary sequence information, with low and the nowadays so powerful high throughput technologies.

Contents

Communicated by Francisco M Cánovas

K. Burg (✉)
Retired Scientist, Austrian Institute of Technology, Vienna, Austria
e-mail: kornel.burg@aon.at

Progress in Botany (2018) 79: 33–48, DOI 10.1007/124_2017_9,
Published online: 2 March 2017

1 Introduction

1.1 Evolution of Markers: A Brief History

The need for categorising the living organisms is as old as mankind. To be able to do that, features should be identified, making differentiation of the organisms possible. The first differentials were based on those features, visible characters, which could be simply observed: colour, morphology, taste, etc. Later, the observations that characters are transmitted from parents to offspring following particular rules marked the advent of genetics. With the development of biology and chemistry, more and more characters of the living organisms were discovered, which could serve as markers in describing an organism. Here, we consider a character as marker if it is well visible or assayable. This development led to the establishment of markers on different molecular levels such as metabolites (resins in conifers), proteins (isoenzymes, and structural and storage proteins) and eventually the material of inheritance, the DNA. Recognition of DNA as the molecule of inheritance initiated strong technological development for the analysis of these molecules. Identification of restriction enzymes, establishment of DNA sequencing and polymerase chain reaction (PCR) revolutionized the development and use of markers based on the genetic material itself. These technologies allowed the generation of virtually infinite number of markers.

1.2 The Plant Cell Genome

The eukaryotic plant genome harbours a large nuclear and two small organellar genomes, chloroplast and mitochondria. The nuclear genome follows the Mendelian rules of inheritance, while the organellar genomes show cytoplasmic/parental type inheritance. Higher plants mostly inherit the chloroplasts from only one parent; most angiosperms inherit chloroplasts maternally via the female gamete, whereas many gymnosperms inherit chloroplasts via the male pollen. The inheritance of mitochondria generally shows similar regimes; most angiosperms have maternal inheritance for mitochondria while this is paternal in gymnosperms.

1.2.1 Nuclear Genome

The nuclear genome is the largest component of the plant cell and its inheritance follows the Mendelian laws. The nuclear genome is organized in several DNA molecules, in chromosomes (Heslop-Harrison and Schmidt 2012). The set of unique DNA molecules is called as complement, which is the basic genetic information of the species. The cells of sexually reproducing plants generally contain two complements (diploid) but higher ploidy is also frequently observed. Sequence analysis of the nuclear DNA revealed the presence of unique and repetitive DNA sequences, where unique sequences represent mostly genes with their regulatory elements, while repetitive DNA is represented by different transposon and structural (mini- and microsatellite and centromere and telomere repeats) DNA sequence elements. All the analysed plant nuclear genomes contain about 30–40 thousand different genes. The size of the nuclear genome, however, varies greatly in higher plants from 61 Mbp (*Genlisea tuberosa*) to 150 Gbp (*Paris japonica*). This enormous difference is the result of the presence of different amounts of repetitive elements (mostly transposons) and duplication (polyploidization) of the DNA during evolution. All these different sequence components may serve as markers for different goals.

1.2.2 Chloroplast Genome

All higher plants contain one or more chloroplasts in their cells. The chloroplast, the organelle for photosynthesis, has its own genome which is about 120–160 kb long circular DNA present in several up to several thousand copies per organelle (Bendich 1987). The chloroplast genome contains genes related to photosynthesis but also has genes for its own protein synthesis. These are ribosomal RNA and ribosomal protein genes as well as tRNA genes. All these elements are of prokaryotic type reflecting the cyanobacterial origin of the chloroplast. The inheritance of the chloroplast genome does not follow Mendelian rules, since it is transmitted uniparentally. Generally chloroplasts are maternally inherited, but in Pines they are paternally transmitted via pollens. As a result of the lack of the possibility for recombination, chloroplast genome shows high level of conservation throughout the whole plant kingdom. Because of the high copy number of this chloroplast DNA in the plant cell (may sum up to 10–15% of the total DNA), these sequences will behave as 'repetitive' DNA in total DNA preparations.

1.2.3 Mitochondrial Genome

The plant cell harbours several mitochondria, which are the sites for oxidative phosphorylation. Similar to chloroplasts, they have their own genome, showing large size variation in the different plant species from about 0.2 Mbp in *Oenothera* to over 2 Mbp in *Cucurbitaceae*. The mitochondrial genome shows internal recombination resulting in sub-genomic forms (Rasmusson and Møller 2015) and its genes may have exon intron structure (Oda et al. 1992). The mitochondria have their own protein synthetizing machinery represented by ribosomal proteins, transfer RNAs and ribosomal RNAs reflecting their proteobacterial origin. Evolutionarily, the mitochondrial DNA is less conserved throughout the plant kingdom, compared to chloroplast DNA. However, mtDNA gene sequences evolve very slowly within species having point mutations rarely. Probably this is because plant mitochondria contain an active DNA recombination system that allows copy correction of mutations. Mitochondria mostly show maternal inheritance in both *Angiosperms* and *Gymnosperms* but there are examples for paternal inheritance as well (Fauré et al. 1994). This DNA behaves also as 'repetitive' DNA component in total DNA preparations.

2 DNA Based Markers

The phenotype of an organism is the composite of its observable characters, e.g. morphology, biochemical or physiological properties. However, the phenotype is the cumulative result of genes operating in the organisms' cells residing in the genome. The set of genes together with the noncoding elements building up an organism is called the genotype. Comparing individual genotypes of the same species reveals that there are several differences in their DNA sequence. This sequence variation of the genomes is the source for nearly an infinite number of DNA based marker sites, which are also called as genetic markers. Genetic markers are DNA sequences with known or unknown physical locations on the genomic DNA. They are sites of sequence variation that can be used to identify individuals or species, or may be used to associate an inherited character with a gene through genetic linkage with nearby but possibly unidentified or uncharacterized genes.

With the present-day technologies, it is not feasible to use all existing differences in genome comparisons. Therefore, techniques were developed by reducing the genomic region for marker establishment. This way, the costs and labour requirements could be kept at an affordable level; however, the number of synoptically investigated markers is increasing drastically with technology development.

Dependent on the technology used for detecting DNA sequence polymorphisms, the markers may show dominant/recessive or co-dominant inheritance as far as the nuclear genome is concerned.

Technically, the markers may be grouped based on whether a priori sequence information is necessary for the marker development. Whenever quick results are necessary for non-studied organisms, marker systems with no a priori sequence information is the choice.

2.1 Medium and Low Throughput Techniques

2.1.1 DNA Based Markers with No Preliminary Sequence Information

Multilocus Marker Systems

These techniques generate numerous marker sites in a single experiment, by generating either differentially sized PCR fragments or site specific DNA sequence information. Comparative analysis of either the banding patterns or the sequence information will identify polymorphic sites of the genome.

Random Amplification of the Genomic DNA

This is the choice if sequence information of an organism is not available and fast results are necessary. The technique may provide several marker sites even within a single PCR reaction. The technique is based on PCR amplification of arbitrary regions of genomic DNA. At 'fortunate' sites, the primer finds two close enough locations in inverted orientation, which allows the formation of a PCR product. Based on the nature of the applied primer, there are two versions of random amplification. Either one can use random PCR primers (RAPD) or may use a certain sequence component of the genomic DNA as source for primer sequence, e.g. tandem repeat, mostly simple sequence repeat (SSR) regions.

Random Amplified Polymorphic DNA

Random Amplified Polymorphic DNA (RAPD) is a single primer PCR reaction (Williams et al. 1990). Mostly ten-base long primers are used; however, different primer sizes were also described (Caetano-Anollés et al. 1991). Sets of ten-base long primers are commercially available. Since the applied primers are short, therefore the annealing temperature used in the PCR reaction is low (35–40°C). This also favours the incomplete attachment of the primers to the genomic DNA resulting in secondary bands appearing by chance in the reactions. Therefore, at primer selection it is advisable to run several parallel reactions to identify the stable bands. Preliminary testing for informative power of primers is necessary. RAPD is the simplest multilocus marker system; however, the repeatability is bad, especially between laboratories. The obtained banding pattern is very much dependent on the quality of the genomic DNA (purity and fragmentation), on the used DNA polymerase and even on the PCR machine used. Therefore, nowadays it is not accepted as a single marker system in publications.

RAPD is a dominant marker system. Using total DNA isolates, the organellar fragments may be over-represented (Weising et al. 2005) especially in leaf DNA preparations.

Inter Simple Sequence Repeat PCR

Inter Simple Sequence Repeat PCR (ISSR-PCR) is also a single primer PCR reaction (Zietkiewicz et al. 1994). In this case, the sequence of a repetitive genomic component is used as primer, e.g. the core sequence of SSR (with other name: Microsatellite or STR Short Tandem Repeat) regions is used as PCR primer (see for review Reddy et al. 2002). The primer length is about 16–25 base which results in a more specific PCR reaction compared to RAPD. The primers may be anchored by adding random bases to the 3′ end of primer, which may also serve to reduce the number of different PCR products for better analysis.

ISSR is a dominant marker system, not influenced by the presence of organellar DNA.

Restriction Enzyme Site Based Marker Systems of the Genomic DNA

Amplified Fragment Length Polymorphism

The method is based on the digestion of the genomic DNA by two different restriction endonucleases, using a frequent cutter (four-base cutter) and six-base cutter enzymes (Vos et al. 1995). The generated sticky ends are adaptor ligated to add PCR primer sites to the generated DNA fragments thus generating PCR amplicons. The PCR reaction is driven by two primers fitting the adaptors. The system preferentially amplifies fragments with two different ends since fragments with identical ends form stem loop structures because of the identical adaptor sequences, therefore having disadvantage during PCR amplification (Vos et al. 1995). For visualization of the fragments by electrophoresis, one primer is either fluorescent or radioactively labelled. For better visualization of the banding patterns by reducing the number of generated fragments, the PCR primers are generally elongated at the 3′ ends by one to three selected nucleotides corresponding to the adjacent genomic sequence. The fragments are separated by their length on denaturing polyacrylamide gels or preferably via automated capillary sequencing instruments. Despite the length driven separation, the banding patterns are scored on the presence/absence of bands at a given length. Amplified Fragment Length Polymorphism (AFLP) is also a dominant marker system.

Conversion of Variable Fragments to Defined PCR Amplicons

Sequence Characterized Amplified Region

The advantage of the previously discussed marker systems is that in a single PCR reaction they deliver several polymorphic marker sites. Disadvantage is, however, that these marker sites (fragment presence/absence) behave as dominant markers. Important/interesting polymorphic sites identified either by RAPD, ISSR or AFLP fragments may be converted to defined PCR amplicons by subsequent isolation and sequencing of the polymorph fragment. This way, the sequence specific PCR primers may be designed allowing a more detailed analysis of the polymorph genomic region. By reidentifying polymorphic site(s) within the Sequence Characterized Amplified Region (SCAR) region, the marker may be converted to a co-dominant one, e.g. cleaved amplified polymorphic sequence (CAPS) or single nucleotide polymorphism (SNP) marker (see later).

2.1.2 Preliminary Sequence Information Based Marker Systems

Restriction Fragment Length Polymorphism

It was the first DNA based marker system used widespread for genetic analysis. The method is based on the detection of restriction site variations by fragment length driven gel electrophoresis followed by filter hybridization of the transferred fragments either by radioactively or fluorescently labelled probes in homologous samples. The method also predicts that at least partial sequence information of the region for analysis, which will be used as probe, should be obtained by cloning and sequencing. This method is laborious and slow; however, PCR based version is still in use.

Cleaved Amplified Polymorphic Sequence

Alternative name is PCR-Restriction Fragment Length Polymorphism (RFLP). Based on the known sequence information, generally 1–2 kb long PCR amplicons are designed and the generated PCR fragments are digested with the proper restriction enzyme and subsequently size separated by gel electrophoresis to visualize the restriction site variation at a given position of the genome. The PCR based RFLP analysis is less laborious and allows the parallel analysis of high number of samples for the particular genomic site. Non-denaturing polyacrylamide gel separation of the fragments reveals also DNA sequence differences (Diekmann 1989). Based on the conserved sequences of the uniparentally inherited organellar genomes, PCR-RFLP based universal marker system was established by nesting PCR primers in the organellar genes and amplifying the intergene regions (Taberlet et al. 1991; Demesure et al. 1995). Universal (consensus) primers have the ability to amplify selected DNA regions across a broad range of species. Later, a database was developed for the putative primer sites of the chloroplast genome (Heinze 2007).

2.1.3 Repeated Sequence DNA as Source of Markers

In the eukaryotic genomes, there are different classes of repeated DNA sequence elements. Tandem repeated sequences of 2–6 bp are called microsatellites, longer elements as minisatellites and kilobase sized repeats mostly transposons.

Simple Sequence Repeats as Markers

Short simple sequence stretches are present as highly repetitive interspersed elements in all eukaryotic genomes (Tautz 1989). These elements show length variation and were suggested for using as markers in relationship studies of individuals within and between populations and as a general source for polymorphic DNA markers for genome mapping and linkage studies (Tautz 1989).

Using the microsatellite or simple sequence regions and their genomic flanking sequence for developing PCR amplicons results in a multiallelic co-dominant marker system, and this is a very popular marker type. Not all SSR regions are variable; however, variable SSR regions with more than ten length variants in populations can be frequently identified. SSR regions with genomic flanks may be obtained by direct cloning (Hayden and Sharp 2001) but nowadays partial or full genomic sequencing may also provide the necessary sequence information (Zalapa et al. 2012).

Monotonous Poly A tracks with variable length are found in the chloroplast genome (Ebert and Peakall 2009), which are used as markers, called chloroplast SSR (CpSSR).

SSR regions found at 3′ end of genes are called as EST-SSR and used as gene related SSR marker.

SSR is a co-dominant marker system.

Recently, there are efforts to use NGS for high throughput analysis of SSR length variation (Vartia et al. 2016).

Sequence Specific Amplified Polymorphism

This is a related technique to AFLP. The genomic DNA is digested by either one or two restriction endonucleases and the developed sticky ends are ligated to adaptors harbouring a PCR primer site (Syed and Flavell 2006). This primer site is paired with a primer planned in a known genomic sequence. If this latter site is within an adapted fragment, it will result in PCR product. To develop a multilocus marker system, known repeat elements of the genome are used as source, for example, the long terminal repeats (LTR) of Copia transposons. This way a multilocus genotyping system will be built reflecting transposon distribution and movement in the genome (Syed and Flavell 2006). Sequence Specific Amplified Polymorphism (SSAP) is also a dominant marker system.

2.1.4 Nucleotide Sequence as Marker Source

Genome comparisons on the nucleic acid sequence level resulted in considering minor sequence differences as putative marker sites (Rafalski 2002). Two different minor sequence alterations are generally considered: SNPs and insertions/deletions (Indels).

Single Nucleotide Polymorphism

SNPs are the most abundant polymorphisms of the plant genome. Their average frequency ranges between 1 in couple of 10 bp to couple of 100 bp. The actual measured frequency depends on the number of compared genomes/populations, geographic/genetic distance, coding or noncoding regions and ploidy. Generally speaking, one SNP can be expected within 100 base pairs, thus providing millions of potential marker sites per genome. The SNP marker system behaves as co-dominant, since the alleles are differentiated by their sequence. SNP markers are generally biallelic, since the occurrence of even triallelic sites is rare (Cao et al. 2015).

Insertion/Deletion

An Indel per definition spans from one to thousands of bases. However, for technical reasons only small, mostly few nucleotides long Indels are considered. These mutations are less frequent compared to SNPs (Mullaney et al. 2010). They have special significance in evolutionary studies because of their much lower reverse mutation frequency.

2.2 *High Throughput Technologies*

The rapid development of nucleic acid sequencing technologies (next-generation sequencing: NGS; Behjati and Tarpey 2013) made possible the mass identification of SNPs and Indels. When the present overview is written, the capacity of these systems reached 5 billion (5×10^9) sequence reads (100–300 bases long) resulting up to 1,500 gigabase (Gb) sequence information in a single run. For comparison, the size of the rice genome is about 420 megabase (Mb). These systems are the so-called Sequencing By Synthesis systems where the DNA fragment to be sequenced is linked/ligated to an adaptor; therefore, identification of the sequenced fragment is solved by including a 'barcode' sequence in the adaptor, which flags the sequenced fragment. By barcoding, several DNA samples may be sequenced parallelly allowing even population studies with this technology (Fumagalli et al. 2013).

2.2.1 High Throughput Single Nucleotide Polymorphism Discovery

This is based on the application of NGS techniques. In most cases, re-sequencing of the full genome is not necessary and also not possible in most of the cases, because whole genome sequence is not available. Therefore, several genome-wide sampling sequencing (GWSS) methods have been developed for reducing the complexity of the genomes. Such reduced genome complexity libraries could be obtained as follows.

Single Nucleotide Polymorphism Discovery in Targeted Regions

Transcriptome Sequencing

The preferential method is RNA-seq, also called whole transcriptome shotgun sequencing (WTSS). This is the direct sequencing of the RNA population of the cells by NGS technology via cDNA synthesis (Wang et al. 2009) and was also used for identifying SNP polymorphisms in the plant transcriptome (Pootakham et al. 2014). RNA-seq is limited to transcribed regions; consequently, it will only discover sequence variations in exon regions. Therefore, regulatory elements residing in introns are omitted; however, regulatory elements found in exons are included. RNA-seq also provides precise data on expression level of different transcripts even discriminating isoforms.

Hypomethylome Sequencing: Single Nucleotide Polymorphism Discovery in the Gene Space

By proper selection of the restriction enzyme, it is possible to target the genome reduction to selected regions. By selecting methylation sensitive restriction endonuclease, preferentially the gene space will be represented and the repeat, mostly transposon, elements excluded in the genomic reduction sequence library (Barbazuk et al. 2007; Gore et al. 2009; Elshire et al. 2011; Wischnitzki et al. 2015) since the gene space is hypomethylated and the repeated part is highly methylated in plants (Rabinowicz et al. 2005).

Genome-Wide Distributed Single Nucleotide Polymorphisms

Sequencing the Flanking Regions of Restriction Endonucleases

Most of the restriction sites are conserved within a species apart of those which provide the base for RFLP analysis (see before). Different methodological versions for obtaining the DNA sequence flanking particular restriction sites are the most popular ways for genome complexity reduction (Jiang et al. 2016). For example, Restriction site Associated DNA sequencing (RAD-seq) was developed by Baird et al. (2008), while Maughan et al. (2009, 2010) published Genomic Reduction based on Restriction-Site Conservation (GR-RSC) protocol. See also Hypomethylome Sequencing section above.

Genotyping by Sequencing (GBS)

The ultra-high capacity of the present-day NGS technologies make possible the comparative analysis of genomes by combining molecular marker discovery and genotyping in one shot (Elshire et al. 2011; Deschamps et al. 2012). Barcoding at NGS allows the parallel analysis of several genomes; therefore, the system may be applied in all areas of plant breeding, where high density marker systems are needed like QTL mapping (Verma et al. 2015; Balsalobre et al. 2017) or Marker Assisted Selection (MAS) (He et al. 2014).

2.3 Detection Methodologies

Basically, there are two ways to visualize the DNA based marker sites. Methods like RAPD, AFLP, SSAP, SSR, RFLP and CAPS are based on separation of the generated

DNA fragments according to their length by electrophoretic systems. Scoring either records the length variation of the fragments (SSR, RFLP and CAPS) or records the presence/absence of a fragment in a given size category (RAPD, ISSR and AFLP). The most common gel systems are agarose and polyacrylamide gels allowing the analysis of moderate sample numbers (several hundreds daily). Higher throughputs (thousands daily) are achieved by the use of capillary electrophoresis systems; however, fluorescent labelling of the fragments is prerequisite for this type of analysis.

The appearance of DNA sequence based marker systems (SNP and Indel) allowed the mass recovery of these marker sites as described earlier but also initiated the development of mass detection systems. However, SNP/Indels may also be detected at low scale by allele specific PCR by nesting the 3′ end of one of the PCR primers at the SNP/Indel site. One base difference is not always sufficient for a proper differentiation of the alleles; therefore, methods to achieve more precise priming were achieved by further modification of the 3′ end of the primer (Liu et al. 2012).

2.3.1 Microarray Based High Throughput Single Nucleotide Polymorphism Detection

Nucleic Acid Hybridization Based Detection of Single Nucleotide Polymorphisms

Synthetic oligonucleotide copies (15–20 bases) of the sequence versions of identified SNPs are immobilized on a surface. The sample genomic DNA is fragmented and fluorescent labelled and hybridized to the deposited oligos. The endpoint of the hybridization is monitored by fluorescent scanning. The determination of identity of the SNP is based on the difference in fluorescent intensity caused by 1 base pair mismatch of the hybridization signal. The capacity of the system is about 9×10^5 SNP per experiment (Affymetrix array).

Primer Extension Based Single Nucleotide Polymorphism Detection

This method (Chen and Sullivan 2003) may also be applied as medium scale technology by using several oligonucleotides of different length (Rogers et al. 2012). The sample DNA is used as template and after annealing the primers to the template a primer extension reaction is made with fluorescent labelled nucleotides. The results can be visualized by use of capillary electrophoresis. At large-scale detection, oligonucleotide preceding the SNP site is synthetized and deposited on bead surface (Shen et al. 2005). The sample DNA is fragmented and hybridized onto the deposited oligos. The hybridized sample DNA serves as template for a one base primer extension reaction with fluorescent labelled nucleotides. The incorporated label will be monitored by fluorescent scanning and the identity of the incorporated base established.

2.4 Bioinformatic Analysis

In the past decade(s), appropriate informatics systems were developed for the analysis of the enormous datasets generated by the high throughput sequencing and by the SNP mass detection systems. Generally, each provider (Illumina, Ion Torrent, Affymetrix, etc.) supplies also the necessary bioinformatic background for the at least preliminary analysis (sorting according to barcodes, grouping and alignment of sequences) of the data. Additional analysis packages either may be purchased or available free in internet (for example: https://omictools.com/, http://simgene.com/index.jsp).

3 Concluding Remarks

In this concise chapter, I intended to give a brief insight into the instruments of marker development and testing based on the DNA molecule itself. Principally, variation of each base pair of the DNA may serve as marker site as we have seen it in case of SNPs. However, there was always a tenor to balance the analysed portion of the genome and minimum requirement of dataset to answer the goals of biological problems. Frequently, budgeting may also have a significant bottleneck in selecting the methodology. Mentioning only few fields, DNA based markers are used at a wide scale to develop high density genetic linkage maps of different genomes (Verma et al. 2015), analyse population genetic diversity and their adaptation to the environment (Davey and Blaxter 2010) and track the genetic milestones in species evolution (Twyford and Ennos 2012). The higher the density of the markers in a genetic map, the better the chance to locate and identify genes (Genome-Wide Association Studies; GWAS) beyond quantitative traits (Quantitative Trait Loci; QTL). This is very important in plant breeding, since following/selecting certain gene variants/alleles and/or allele combinations may speed up breeding programmes by using MAS. Population analysis with high throughput technologies using high sample numbers and high number of marker sites with, e.g. Radseq (Davey and Blaxter 2010), along with transcriptomics (Evans 2015) will allow the identification of genes and alleles important in population adaptation to the changing environment.

On the technical point of view, the appearance of the high throughput technologies on the market brought along the appearance of companies selling services with these technologies making these expensive methods available for smaller labs as well. Thus, investments for quickly developing machinery became unnecessary. These services generally also include the preliminary data analysis.

Additionally, SNP arrays for analysis of selected species are available (Ganal et al. 2012) also commercially (see, e.g. Illumina and Affymetrix).

References

Baird NA, Etter PD, Atwood TS, Currey MC, Shiver AL, Lewis ZA, Selker EU, Cresko WA, Johnson EA (2008) Rapid SNP discovery and genetic mapping using sequenced RAD markers. PLoS One 3:e3376

Balsalobre TWA, da Silva Pereira G, Margarido GRA et al (2017) GBS-based single dosage markers for linkage and QTL mapping allow gene mining for yield-related traits in sugarcane. BMC Genomics 18:72. doi:10.1186/s12864-016-3383-x

Barbazuk WB, Emrich SJ, Chen HD, Li L, Schnable PS (2007) SNP discovery via 454 transcriptome sequencing. Plant J 51:910–918

Behjati S, Tarpey PS (2013) What is next generation sequencing? Arch Dis Child Educ Pract Ed 98:236–238. doi:10.1136/archdischild-2013-304340

Bendich AJ (1987) Why do chloroplasts and mitochondria contain so many copies of their genome? Bioessays 6:279–282

Caetano-Anollés G, Bassam BJ, Gresshoff PM (1991) DNA amplification fingerprinting using very short arbitrary oligonucleotide primers. Nat Biotechnol 9:553–557. doi:10.1038/nbt0691-553

Cao M, Shi J, Wang J, Hong J, Cui B, Ning G (2015) Analysis of human triallelic SNPs by next-generation sequencing. Ann Hum Genet 79:275–281. doi:10.1111/ahg.12114

Chen X, Sullivan PF (2003) Single nucleotide polymorphism genotyping: biochemistry, protocol, cost and throughput. Pharmacogenomics J 3:77–96. doi:10.1038/sj.tpj.6500167

Davey JW, Blaxter ML (2010) RADSeq: next-generation population genetics. Brief Funct Genomics 9:416–423. doi:10.1093/bfgp/elq031

Demesure B, Sodzi N, Petit R (1995) A set of universal primers for amplification of polymorphic non-coding regions of mitochondrial and chloroplast DNA in plants. Mol Ecol 4:129–134

Deschamps S, Llaca V, May GD (2012) Genotyping-by-sequencing in plants. Biology 1:460–483. doi:10.3390/biology1030460

Diekmann S (1989) The migration anomaly of DNA fragments in polyacrylamide gels allows the detection of small sequence-specific DNA structure variations. Electrophoresis 10:354–359

Ebert D, Peakall R (2009) Chloroplast simple sequence repeats (cpSSRs): technical resources and recommendations for expanding cpSSR discovery and applications to a wide array of plant species. Mol Ecol Resour 9:673–690. doi:10.1111/j.1755-0998.2008.02319.x

Elshire RJ, Glaubitz JC, Sun Q, Poland JA, Kawamoto K, Buckler ES, Mitchell SE (2011) A robust, simple genotyping-by-sequencing (GBS) approach for high diversity species. PLoS One 6:e19379. doi:10.1371/journal.pone.0019379

Evans TG (2015) Considerations for the use of transcriptomics in identifying the 'genes that matter' for environmental adaptation. J Exp Biol 218:1925–1935. doi:10.1242/jeb.114306

Fauré S, Noyer JL, Carreel F, Horry JP, Bakry F, Lanaud C (1994) Maternal inheritance of chloroplast genome and paternal inheritance of mitochondrial genome in bananas (*Musa acuminata*). Curr Genet 25:265–269

Fumagalli M, Vieira FG, Korneliussen TS, Linderoth T, Huerta-Sánchez E, Albrechtsen A, Nielsen R (2013) Quantifying population genetic differentiation from next-generation sequencing data. Genetics 195:979–992. doi:10.1534/genetics.113.154740

Ganal MW, Polley A, Graner EM, Plieske J, Wieseke R, Luerssen H, Durstewitz G (2012) Large SNP arrays for genotyping in crop plants. J Biosci 37(5):821–828

Gore MA, Wright MH, Ersoz ES, Bouffard P, Szekeres ES, Jarvie TP, Hurwitz BL, Narechania A, Harkins TT, Grills GS, Ware DH, Buckler ES (2009) Large-scale discovery of gene-enriched SNPs. Plant Gen 2:121–133

Hayden MJ, Sharp PJ (2001) Sequence-tagged microsatellite profiling (STMP): a rapid technique for developing SSR markers. Nucleic Acids Res 29:E43–E43

He J, Zhao X, Laroche A, Lu Z-X, Liu HK, Li Z (2014) Genotyping-by-sequencing (GBS), an ultimate marker-assisted selection (MAS) tool to accelerate plant breeding. Front Plant Sci. doi:10.3389/fpls.2014.00484

Heinze B (2007) A database of PCR primers for the chloroplast genomes of higher plants. Plant Methods 3:4. doi:10.1186/1746-4811-3-4

Heslop-Harrison JS, Schmidt T (2012) Plant nuclear genome composition. In: eLS. Wiley Online Library. doi:10.1002/9780470015902.a0002014.pub2

Jiang Z, Wang H, Michal JJ, Zhou X, Liu B, Woods LCS, Fuchs RA (2016) Genome wide sampling sequencing for SNP genotyping: methods, challenges and future development. Int J Biol Sci 12:100–108. doi:10.7150/ijbs.13498

Liu J, Huang S, Sun M, Liu S, Liu Y, Wang W, Xiurong Zhang X, Wang H, Hua W (2012) An improved allele-specific PCR primer design method for SNP marker analysis and its application. Plant Methods 8:34. doi:10.1186/1746-4811-8-34

Maughan PJ, Yourstone SM, Jellen EN, Udall JA (2009) SNP discovery via genomic reduction, barcoding and 454-pyrosequencing in amaranth. Plant Gen 2:260–270

Maughan PJ, Yourstone SM, Byers RL, Smith SM, Udall JA (2010) Single-nucleotide polymorphism genotyping in mapping populations via genomic reduction and next-generation sequencing: proof of concept. Plant Gen. doi:10.3835/plantgenome2010.07.0016

Mullaney JM, Mills RE, Pittard WS, Devine SE (2010) Small insertions and deletions (INDELs) in human genomes. Hum Mol Genet 19:R131–R136. doi:10.1093/hmg/ddq400

Oda K, Yamato K, Ohta E, Nakamura Y, Takemura M, Nozato N, Akashi K, Kanegae T, Ogura Y, Kohchi T, Ohyama K (1992) Gene organization deduced from the complete sequence of liverwort *Marchantia polymorpha* mitochondrial DNA. J Mol Biol 223:1–7. doi:10.1016/0022-2836(92)90708-R

Pootakham W, Shearman JR, Ruang-areerate P et al (2014) Large-scale SNP discovery through RNA sequencing and SNP genotyping by targeted enrichment sequencing in cassava (*Manihot esculenta* Crantz). PLoS One 9(12):e116028. doi:10.1371/journal.pone.0116028

Rabinowicz PD, Citek R, Budiman MA et al (2005) Differential methylation of genes and repeats in land plants. Genome Res 15:1431–1440

Rafalski A (2002) Applications of single nucleotide polymorphisms in crop genetics. Curr Opin Plant Biol 5:94–100. doi:10.1016/S1369-5266(02)00240-6

Rasmusson AG, Møller IM (2015) The genetic system in plant mitochondria has several special features. In: Taiz L, Zeiger E, Møller IM, Murphy A (eds) Plant physiology and development, 6th edn. Topic 12.6 published by Sinauer Associates

Reddy MP, Sarla N, Siddiq EA (2002) Inter simple sequence repeat (ISSR) polymorphism and its application in plant breeding. Euphytica 128:9–17

Rogers SM, Payton M, Allen RW, Melcher U, Carver J, Fletcher J (2012) Method: a single nucleotide polymorphism genotyping method for wheat streak mosaic virus. Investigative Genet 3:10. doi:10.1186/2041-2223-3-10

Shen R, Fan J-B, Campbell D, Jing W, Chang W, Chen J, Doucet D, Yeakley J, Bibikova M, Wickham Garcia ME, McBride C, Steemers F, Garcia F, Kermani BG, Gunderson K, Oliphant A (2005) High-throughput SNP genotyping on universal bead arrays. Mutat Res 573:70–82. doi:10.1016/j.mrfmmm.2004.07.022

Syed NH, Flavell AJ (2006) Sequence-specific amplification polymorphisms (SSAPs): a multi-locus approach for analyzing transposon insertions. Nat Protoc 1:2746–2752

Taberlet P, Gielly L, Pautou G, Bouvet J (1991) Universal primers for amplification of three non-coding regions of chloroplast DNA. Plant Mol Biol 17:1105–1109

Tautz D (1989) Hypervariability of simple sequences as a general source for polymorphic DNA markers. Nucleic Acids Res 17:6463–6471

Twyford AD, Ennos RA (2012) Next-generation sequencing as a tool for plant ecology and evolution. Plant Ecol Divers 5:411–413. doi:10.1080/17550874.2012.754513

Vartia S, Villanueva-Cañas JS, Finarelli J, Farrell ED et al (2016) A novel method of microsatellite genotyping-by-sequencing using individual combinatorial barcoding. R Soc Open Sci 3:150565. doi:10.1098/rsos.150565

Verma S, Gupta S, Bandhiwal N, Kumar T, Bharadwaj C, Bhatia S (2015) High-density linkage map construction and mapping of seed trait QTLs in chickpea (*Cicer arietinum* L.) using genotyping-by-sequencing (GBS). Sci Rep 5:17512. doi:10.1038/srep17512

Vos P, Hogers R, Bleeker M et al (1995) AFLP: a new technique for DNA fingerprinting. Nucleic Acids Res 23:4407–4414. doi:10.1093/nar/23.21.4407

Wang Z, Gerstein M, Snyder M (2009) RNA-seq: a revolutionary tool for transcriptomics. Nat Rev Genet 10:57–63. doi:10.1038/nrg2484

Weising K, Nybom H, Wolff K, Kahl G (2005) Properties of RAPD markers. In: DNA fingerprinting in plants. CRC Press, Taylor & Francis Group, p 37

Williams JGK, Kubelik AR, Livak KJ, Rafalski JA, Tingey SV (1990) DNA polymorphisms amplified by arbitrary primers are useful as genetic markers. Nucleic Acids Res 18:6531–6535

Wischnitzki E, Sehr EM, Hansel-Hohl K, Berenyi M, Burg K, Fluch S (2015) How to isolate a plant's hypomethylome in one shot. Biomed Res Int 2015:570568. doi:10.1155/2015/570568

Zalapa JE, Cuevas H, Zhu H, Steffan S, Senalik D, Zeldin E, McCown B, Harbut R, Simon P (2012) Using next-generation sequencing approaches to isolate simple sequence repeat (SSR) loci in the plant sciences. Am J Bot 99:193–208. doi:10.3732/ajb.1100394

Zietkiewicz E, Rafalski A, Labuda D (1994) Genome fingerprinting by simple sequence repeat (SSR)-anchored polymerase chain reaction amplification. Genomics 20:176–183

Advances in Asparagine Metabolism

Shrikaar Kambhampati, Ebenezer Ajewole, and Frédéric Marsolais

Abstract Asparagine is a key compound in the nitrogen metabolism of plants. It acts as a major form of nitrogen storage and transport to sink tissues. In the past 20 years, tremendous progress has been made in our understanding of the enzymology of asparagine metabolic pathways and of the biological role of asparagine metabolic enzymes in higher plants, by using a variety of approaches, ranging from protein crystallography to the use of mutants and overexpression lines in combination with molecular analysis and metabolite profiling. Most of the genes and enzymes involved in asparagine metabolism have been identified and characterized, including asparagine synthetase, β-cyanoalanine nitrilase/hydratase, asparaginase, serine:glyoxylate aminotransferase, and ω-amidase. In leaf, asparagine biosynthesis by asparagine synthetase is stimulated in the dark, a situation associated with carbon deficit. Asparagine appears to participate in photorespiration, providing an input of nitrogen to balance an output of serine or glycine to other pathways. Manipulating asparagine biosynthesis and degradation can alter seed development and composition, and has been used to mitigate the problem of acrylamide formation in food. There is an apparent positive association between the levels of asparagine present in reproductive organs and seed protein accumulation. This chapter presents a perspective on the field of asparagine metabolism, highlighting possible areas of future investigation.

Communicated by Francisco M. Cánovas

S. Kambhampati, E. Ajewole, and F. Marsolais (✉)
Department of Biology, University of Western Ontario, London, ON, Canada

London Research and Development Centre, Agriculture and Agri-Food Canada, London, ON, Canada
e-mail: Frederic.Marsolais@agr.gc.ca

Progress in Botany (2018) 79: 49–74, DOI 10.1007/124_2017_13,
Published online: 11 July 2017

Keywords β-Cyanoalanine nitrilase/hydratase • ω-Amidase • Acrylamide • Asparaginase • Asparagine • Asparagine aminotransferase • Asparagine synthetase • Plants • Seed protein concentration

Contents

1 Introduction

Asparagine (Asn) is a central intermediate in amino acid metabolism and an important form of stored or transported nitrogen (N) in higher plants. To date, most of the genes and enzymes involved in Asn biosynthesis and degradation have been identified, however, due to the highly branched and interlaced nature of amino acid metabolism and its dependence on nutrient status, photoperiod, and several other factors, the mechanisms involved in the precise regulation of these enzymatic steps are still unclear. While efforts are underway in characterizing these mechanisms, this chapter aims to highlight the research done so far and point towards some of the key areas or questions requiring further investigation. The chapter will begin by reviewing the history of the discovery of Asn and its metabolism, providing a context for the description of subsequent studies. This will be followed by examining recent findings on its biosynthesis and catabolism, including its relationship with cyanide metabolism. The role of Asn as a transport and storage form of N and its regulatory aspects will then be discussed along with its function in N metabolism and carbon:N relationships. These aspects are intimately linked with the relationship between N-assimilates as Asn transported to sink tissues and the accumulation of seed or grain protein. Finally, some applications in crops will be presented related to the mitigation of acrylamide in fried food. The field of Asn metabolism has been reviewed recently by Gaufichon et al. (2016b), with an emphasis on Arabidopsis, and by Duff (2015) who focused on asparagine synthetase (AS). To avoid redundancy, the reader will sometime be directed to these excellent reviews.

Asn metabolism is part of the broader context of N metabolism. N is one of the fundamental inputs essential for plant growth and development and plays a key role in determining crop productivity. The basic processes involved in N metabolism include: uptake by roots usually in an inorganic form as nitrate, conversion of nitrate to ammonium by nitrate reductase and nitrite reductase and ammonium assimilation to organic forms like glutamine and glutamate via the glutamine synthetase/glutamine:oxoglutarate aminotransferase (GS/GOGAT) cycle. These processes are very well elucidated at both the biochemical and molecular levels (Lea and Ireland 1999; Hodges 2002; Wang et al. 2004; Hirel and Lea 2001). Subsequently, glutamine and glutamate are metabolized into aspartate and Asn via the action of aspartate aminotransferase (AspAT) and AS, respectively (Gaufichon et al. 2016b; Lea et al. 2007). Of these first amino acids produced the amide containing glutamine and Asn act as efficient storage and partitioning compounds and are utilized for further downstream synthesis of amino acids.

2 The History of the Discovery of Asparagine and Its Metabolic Pathways

Although this chapter focuses on recent advances, it seems useful to start by placing the discovery of each reaction of Asn metabolism in its historical context. The discovery of L-Asn dates back over 200 years with its identification from natural sources by Delaville (1802) and first isolation by French chemists Vauquelin and Robiquet (1806) from spears of *Asparagus sativus*. Not only was Asn the first amino acid identified, it was one of the first examples of the preparation of a D-amino acid by Piutti (1886). Piutti was also credited with the determination of the chemical structure of Asn, and the first observation of enantioselectivity of a biological receptor, for his discovery of a difference in taste between D- and L-Asn. It was rapidly discovered that Asn is present in all higher plants, and Schulze and Winterstein (1910) were the first to show that, although present in small quantities in green plants, it accumulates under carbohydrate deficiency or starvation in general. Asn was also reported to be formed as a temporary N reserve during aberrations in normal protein metabolism, when excess ammonium is formed (Prianischnikov 1922). Furthermore, the role of Asn as a translocated nitrogenous substance in a form suitable for subsequent re-synthesis from one organ of a plant to another was demonstrated by Chibnall (1924). Murneek (1935) summarized the findings of several researchers at the time and reported that under carbohydrate depleted conditions excess protein unused by the plant is hydrolyzed by means of proteolytic enzymes and hence amino acids are formed including Asn.

Etiolated lupin (*Lupinus albus*) seedlings have a considerable accumulation of Asn and were hence used for many subsequent studies to identify the mechanism of Asn biosynthesis (Vickery et al. 1942; Meiss 1952). The first successful demonstration of Asn biosynthesis via amidation in the presence of ATP was done using [^{14}C]-labelled aspartate (Webster and Varner 1955). Glutamine-dependent Asn biosynthesis via AS was reported, thereafter, in lupin nodules by Scott et al. (1976). The authors suggested that the plant fraction of nodules catalyzed the glutamine-dependent while the bacterial fraction performed the ammonium-dependent Asn biosynthesis. A possible deamidation of excess Asn by asparaginase (ASPG) upon translocation to the sink tissues and the incorporation of the amide group into amino acids to be used as building blocks for protein synthesis were suggested from the results of incorporation studies using [^{14}C]- and [^{15}N]-labelled Asn in lupin cotyledons (Atkins et al. 1975). Streeter (1977) reported on Asn aminotransferase activity in soybean leaves and a further catalytic activity responsible for deamidation of the aminotransferase product. By determining the distribution of ASPG and Asn aminotransferase activities in pea (*Pisum sativum*) and through inhibitor studies, Ireland and Joy (1981) showed that ASPG predominates in developing seed and young expanding leaves, whereas transamination remains stable throughout leaf development. It was later confirmed by Ta et al. (1984a, b) using [^{15}N]-labelled Asn and a combination of aminotransferase and GS inhibitors, that both N groups of Asn are readily used by expanding leaves, and that the major routes of Asn catabolism are the deamidation of Asn to aspartate and ammonia and transamination to produce α-ketosuccinamate that is rapidly converted to α-hydroxysuccinamate. They further concluded that the ammonia released by deamidation is reassimilated by the GS/GOGAT cycle. Based on enzymatic, subcellular localization and [^{15}N] labelling studies in the presence of inhibitors of photorespiration (Ta et al. 1985; Ireland and Joy 1981, 1983b), Asn aminotransferase was hypothesized to be serine-glyoxylate aminotransferase. This was confirmed when mutants of serine-glyoxylate aminotransferase in barley and tobacco were shown to lack Asn aminotransferase activity (Havir and McHale 1988; Murray et al. 1987).

3 Asparagine Biosynthesis and Degradation

A major route for Asn biosynthesis is via the ATP-dependent transfer of the amide group of glutamine to the β-carboxyl group of aspartate by the action of AS (Fig. 1). So far, two types of ASs, AsnA and AsnB, have been identified. While prokaryotes utilize AsnA type ASs that require ammonia as an amide donor as well as AsnB type ASs that can catalyze the reaction using either ammonium or glutamine as an amide donor, most eukaryotes only use AsnB type ASs (Gaufichon et al. 2010; Duff 2015). AS is notably difficult to assay from plant tissues (Romagni and Dayan 2000). AsnB type ASs are members of the N-terminal nucleophile hydrolase (Ntn) group of glutamine amidotransferases (Larsen et al. 1999; Massière and Badet-Denisot 1998). They are characterized by an N-terminal cysteine nucleophilic residue producing a cysteinyl-glutamine tetrahedral intermediate from which ammonium is abstracted. Glutamate is released by hydrolysis of the resulting γ-glutamyl thioester intermediate. The ammonia is tunneled to a C-terminal transferase domain. This domain activates aspartate through ATP hydrolysis as a β-aspartyl AMP intermediate. Nucleophilic attack by the ammonia results in cleavage and release of Asn. There are two groups of AS enzymes in higher plants designated as class I and II. Results of kinetic analyses of recombinant maize AS enzymes indicated that class I enzymes may have specialized functions as they can

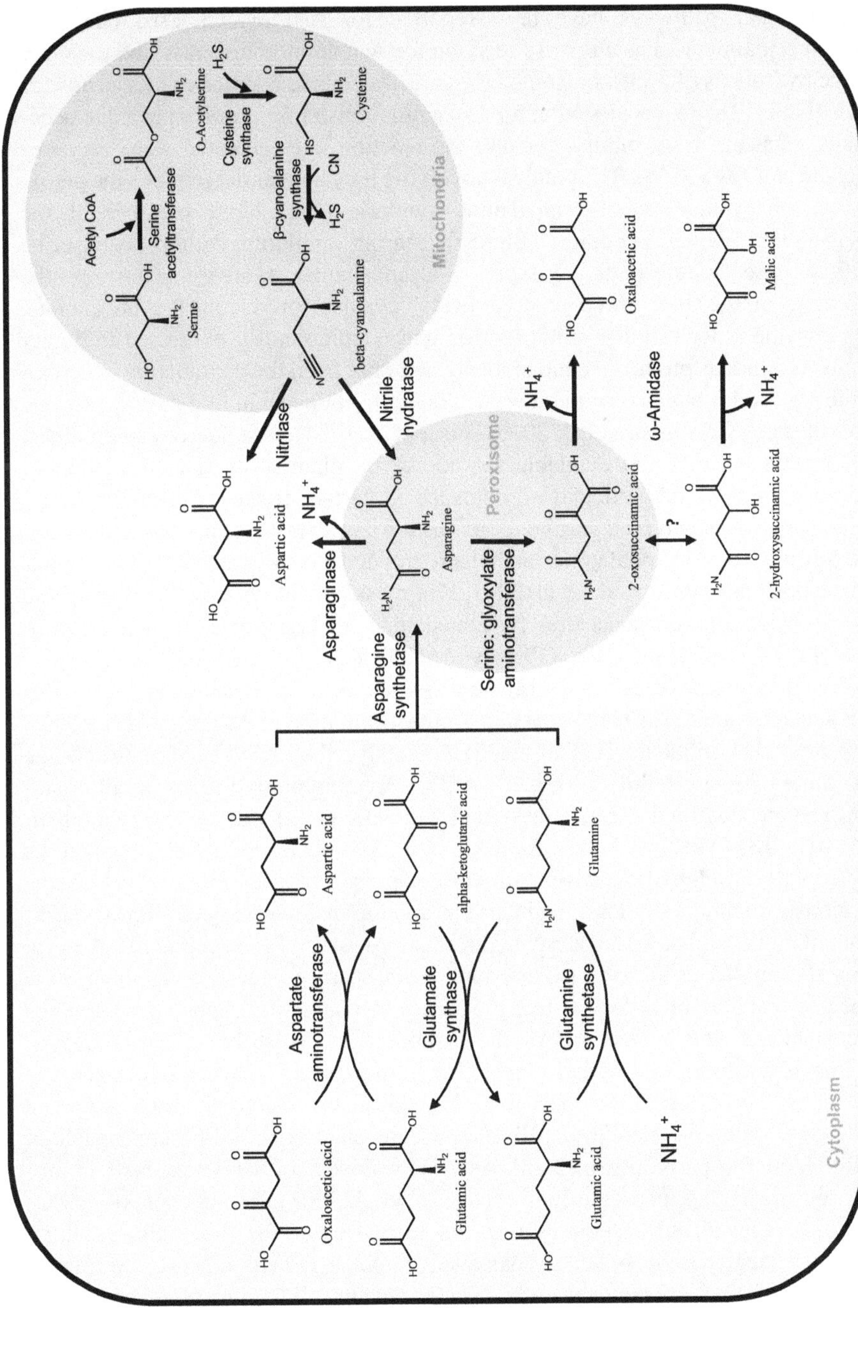

Fig. 1 Asn biosynthesis and degradation pathway representing all the known and unknown enzymes involved in its metabolism

have higher affinity for glutamine and their expression is restricted to specific tissues (Duff et al. 2011).

Two other pathways have been reported for Asn biosynthesis, one from 2-oxosuccinamic acid as a reverse reaction for Asn aminotransferase and the other by the hydrolysis of β-cyanoalanine using nitrilase/nitrile hydratase (Sieciechowicz et al. 1988c). The reverse reaction for Asn aminotransferase is not expected to occur at any great extent *in planta*, because the reaction rates detected were very low (Ireland and Joy 1983a). The pathway involving β-cyanoalanine requires the detoxication of cyanide by β-cyanoalanine synthase (Fig. 1), a member of the β-substituted alanine synthases (BSASs), sharing similarity with *O*-acetylserine sulfhydrylase synthesizing cysteine. β-Cyanoalanine synthase catalyzes the exchange of the thiol group of cysteine for cyanide, producing β-cyanoalanine. The enzyme is localized in mitochondria and is ubiquitously expressed in higher plants. Cyanide is produced endogenously, not only from the degradation of cyanogenic glycosides but also from other metabolic reactions including during ethylene biosynthesis (Machingura et al. 2016; Peiser et al. 1984). Cyanide is also produced by bacteria in soils and available for uptake by plant roots. The formation of β-cyanoalanine is of particular significance in certain tissues, especially during root development and seed germination. The nitrilase NIT4 acts both as a hydratase and nitrilase for β-cyanoalanine catabolism producing Asn or aspartate and ammonia, respectively (Piotrowski et al. 2001; Piotrowski and Volmer 2006). Transgenic Arabidopsis expressing a bacterial β-cyanoalanine nitrilase showed increased flux of carbon from cyanide into Asn (O'Leary et al. 2014).

Two major routes exist for Asn catabolism in plants. Deamidation and release of Asp and ammonium is a necessary step for transformation of Asn from its N storage form into other amino acids. This reaction is catalyzed by ASPG and represents the first route of Asn catabolism (Fig. 1). ASPGs are commonly present in all organisms and are classified as bacterial (type I – dimeric and II – tetrameric) and plant-type (type III) (Michalska and Jaskolski 2006). The plant-type ASPGs contain the nucleophile aminohydrolase domain. Proteins sharing this domain include the previously mentioned Class II glutamine amidotransferases, penicillin acylases, proteasome subunits, glycosylasparagines and γ-glutamyl transpeptidase-like proteins (Brannigan et al. 1995). Plant-type ASPGs are produced as a polypeptide precursor capable of self-cleavage, producing two subunits, exposing a threonine nucleophile at the N-terminus of the β-subunit. The distribution of plant-type ASPGs is not limited to higher plants. They are present in bacteria, for example, *Escherichia coli* EcAIII (Michalska et al. 2005) and in mammals, such as human hASNase3 (Nomme et al. 2012). Beside Asn, the plant-type ASPGs are capable of hydrolyzing β-aspartyl dipeptides. β-Aspartyl residues arise spontaneously in proteins during ageing, particularly in seed (Ogé et al. 2008). In this context, ASPGs constitute specialized peptidases required to hydrolyze the dipeptides resulting from the degradation of these damaged proteins. Plant ASPGs are further subdivided into K^+-dependent and K^+-independent ASPGs, in reference to their catalytic activation by K^+ (Sodek et al. 1980). Whereas K^+-dependent ASPGs have a substrate preference towards Asn, K^+-independent ASPGs have a comparable activity towards Asn and β-aspartyl dipeptides (Bruneau et al. 2006; Gabriel et al.

2012; Borek et al. 2004). In fact, a distinguishing feature of K^+-dependent ASPGs is their high level of catalytic activity and catalytic efficiency with Asn as substrate, as compared with the K^+-independent ASPGs. In legumes, which rely on Asn as a major transport form, expression of K^+-dependent ASPGs predominates over that of K^+-independent ASPGs, a notable exception being lupine. Recently, the crystal structure of a K^+-dependent ASPG from common bean (*Phaseolus vulgaris*), PvAspG1, was solved, shedding light on the mechanism of K^+ activation (Bejger et al. 2014). The main features of the active site are the N-terminal nucleophile, Thr^{196} and the Arg^{224} residue coordinating the carboxylate group of the substrate through its side chain. There are two potassium ions per heterodimer. One is present in place of a sodium ion first observed in the structure of lupine K^+-independent ASPG (Michalska et al. 2006), coordinated by a series of residues in the α-subunit, Leu^{58}-Arg^{68} termed the stabilization loop. The second loop, spanning residues Val^{111}-Ser^{118}, is called the activation loop. This loop is distant from the active site, which is consistent with the mode of activation K^+ as an allosteric effector, or type II activation (Page and Di Cera 2006). The availability of crystal structures in the presence of sodium or potassium highlighted a catalytic switch mechanism affecting the position of three residues, His^{117}, Arg^{224}, and Glu^{250}. In the K^+-bound form of the enzyme, the side chain of Glu^{250} positions the side chains of Arg^{224} and His^{117} in the activation loop through hydrogen bonding. In the Na^+ bound form, the side chain of His^{117} has moved towards the protein core, pushing away Arg^{224} into a position where it can no longer interact with the substrate. The Glu^{250} side chain also points away from the active site. Current efforts are focused on understanding the molecular basis of the difference in K^+ activation between the two subfamilies of plant ASPGs. The plant K^+-dependent and K^+-independent ASPGs share approximately 60% identity in amino acid sequence. Within the activation loop, most amino acid residues are conserved, except Ser^{118}, which is replaced by an Ile residue in K^+-independent ASPGs. Ser^{118} positions one of two water molecules which are part of the coordination sphere of the K^+ ion. In addition, in one active site configuration, Ser^{118} interacts with the side chain of Glu^{250}. In general, potassium activation is observed in intracellular enzymes and sodium activation in secreted enzymes. The plant ASPGs constitute an interesting example of the emergence of K^+ activation in a cytosolic enzyme, resulting in enhanced catalytic activity and substrate specificity.

A secondary focus of this research is to improve the K^+-dependent ASPGs for use as a chemotherapeutic agent. Bacterial ASPGs, including *Escherichia coli* ASPG II (EcAII) and *Erwinia carotovora* ASPG, are notable in being used as chemotherapeutic agents to treat acute lymphoblastic leukemia (Lopes et al. 2017). These drugs work by depleting Asn in the blood circulation, starving off cancer cells. However, there are significant potential side effects associated with them. Part of the toxicity of the bacterial ASPGs may arise from their secondary glutaminase activity (Chan et al. 2014). In this regard, the plant-type ASPGs are devoid of this activity. The bacterial ASPGs have a higher affinity for Asn than the plant-type ASPGs by approximately two orders of magnitude; therefore, this characteristic requires improvement.

The second route of Asn catabolism involves transamination of the α-amino group of Asn to a 2-oxo acid acceptor, producing 2-oxosuccinamate, followed by its hydrolysis to form oxaloacetate and ammonia via ω-amidase (Fig. 1). 2-Oxosuccinamate can also be rapidly converted to 2-hydroxysuccinamate in soybean and pea leaves (Lloyd and Joy 1978). As previously mentioned, Asn transaminase corresponds to the photorespiratory enzyme, serine:glyoxylate aminotransferase. Zhang et al. (2013) used a recombinant Arabidopsis serine: glyoxylate aminotransferase (AGT1) to show that the enzyme exhibited a higher catalytic activity with Asn than with its prototypical substrates, serine or alanine. No catalytic activity was detected for the reverse reactions leading to Asn formation. The following step in the pathway is catalyzed by ω-amidase. The significance of this step is also related to glutamine metabolism. Glutamine transamination produces α-ketoglutaramate. This reaction and the subsequent hydrolysis of α-ketoglutaramate by ω-amidase are involved in the methionine salvage pathway in plastids (Ellens et al. 2015). The transamination step is involved in converting α-ketomethylthiobutyric acid to methionine. Zhang and Marsolais (2014) used sequence homology to mammalian ω-amidase NIT2 to identify and characterize the corresponding gene and enzyme in Arabidopsis. The gene is annotated as a nitrilase/cyanide hydratase and apolipoprotein *N*-acyltransferase family protein in the Araport 11 assembly, and has been referred to as nitrilase-like protein 3 (Piotrowski et al. 2003). It shares 56% identity in amino acid sequence with mouse NIT2. The following hits to mouse NIT2 were nitrilase-like protein 2, an enzyme of unknown function, and nitrilase-like protein 1, encoding a carbamoylputrescine amidohydrolase (Piotrowski et al. 2003). The next hit was nitrilase 2, part of a small family of enzymes including nitrilases 1–3, acting on intermediates of glucosinolate catabolism, which are located in tandem on chromosome 3 (Vorwerk et al. 2001; Hillebrand et al. 1998). These are structurally related with the previously mentioned NIT4, a β-cyanoalanine nitrilase/nitrile hydratase. Interestingly, a mutant having a T-DNA insertion localized in the first exon of the Arabidopsis ω-amidase gene produced a transcript encoding a predicted protein encoded from the downstream Met^{63}, missing a 60 amino acid predicted chloroplast transit peptide (Zhang and Marsolais 2014). This indicates that the downstream initiation codon is likely used for translation initiation, potentially resulting in a dual localization in the cytosol or plastid. The plastidial localization of ω-amidase is coherent with a function in the methionine salvage pathway. The cytosolic localization implies that α-ketosuccinamate produced by serine:glyoxylate aminotransferase might be exported out of the peroxisome for subsequent hydrolysis by an as yet unknown transporter. Alternatively, it has been hypothesized that channels or pores might be involved in the movement of metabolites across the peroxisomal membrane, allowing for high rates of transport (for a review, see Hodges et al. 2016). Recombinant ω-amidase exhibited comparable catalytic activities with α-ketosuccinamate, α-ketoglutaramate, and α-hydroxysuccinamate, with K_m values in the low mM range, similar to K_m values of ASPGs and serine:glyoxylate aminotransferase for Asn (Zhang and Marsolais 2014). By contrast, mammalian ω-amidase has a K_m value in the low micromolar range for α-ketosuccinamate

(Jaisson et al. 2009), which is considered a reactive, potentially toxic metabolite, subject to spontaneous cyclization and dimerization (Cooper et al. 1985; Stephani and Meister 1971). The Arabidopsis ω-amidase T-DNA insertion mutant was found to overaccumulate α-ketosuccinamate and α-hydroxysuccinamate by two to three-fold, both under control conditions and after treatment with exogenous Asn (Zhang and Marsolais 2014). The enzyme responsible for the reduction of α-ketosuccinamate to α-hydroxysuccinamate is yet to be identified and characterized from higher plants (Fig. 1). Hariharan et al. (2017) indicated that mammalian lactate dehydrogenase catalyzes this reaction.

ω-Amidase belongs to the superfamily of nitrilases, which also includes different families of amidases and *N*-acyltransferases (Brenner 2002). These enzymes use a catalytic triad of glutamic acid, lysine, and cysteine residues. Within the superfamily, Arabidopsis nitrilases NIT1-4 belong to family 1, while ω-amidase belongs to family 10. It is remarkable that Arabidopsis nitrilase 4 and ω-amidase, which are structurally related, both participate to the Asn metabolic pathway. Recently, Mammalian NIT1, a close relative of ω-amidase (NIT2), has been characterized as a repair enzyme hydrolyzing deaminated and carbonylated glutathione, a metabolite formed by transamination, into α-ketoglutarate and cysteinylglycine (Peracchi et al. 2017). A crystal structure is available for the yeast homolog of mammalian NIT2 (Liu et al. 2013). The enzyme was shown to form a thioester bond between the catalytic Cys^{169} residue and α-ketoglutarate or oxaloacetate, the products of α-ketoglutaramate or α-ketosuccinamate hydrolysis. However, the yeast enzyme has extremely low activity with α-ketosuccinamate as compared with mammalian and plant ω-amidases (Zhang and Marsolais 2014; Jaisson et al. 2009; Krasnikov et al. 2009). Interestingly, an active site C169S mutant was found to bind glutathione in the active site, providing a clue to the function of NIT1 (Liu et al. 2013). The availability of a crystal structure of the plant ω-amidase would shed light on the specificity and catalytic mechanism of this important enzyme of the glutamine and Asn catabolic pathway. Finally it is worth mentioning that, from the results of labelling studies, Asn is implicated as a precursor of 3-isoxazolin-5-one derivatives in plants, which includes the neurotoxic non-protein amino acid β-*N*-oxalyl-L-α,β-diaminopropionic acid in grass pea (reviewed by Becker et al. (2017) and Xu et al. (2017)). The enzymatic reactions leading from Asn to the 3-isoxazolin-5-one ring are unknown.

4 Role of Asparagine in Nitrogen Metabolism and Carbon/Nitrogen Relationships

Photoperiod is an important regulator of Asn metabolism. Transcripts of pea *AS1*, encoding a class I AS, are induced in leaves in the dark (Tsai and Coruzzi 1990). This response results in transient accumulation of Asn at night, under conditions of low carbon:N ratio (Harmer et al. 2005). The opposite happens with starch reserves, which are mobilized during the night. A similar situation is observed in

Arabidopsis, where expression of the orthologous *ASN1* is induced in leaves at night and correlates with Asn concentration (reviewed in Lam et al. 1995; Coruzzi 2003). In soybean, transcripts of *GmAS2*, encoding a class I AS, oscillate with the day and night cycle (Wan et al. 2006). Similarly, Arabidopsis *ASN2*, a class II AS gene, was shown to exhibit higher transcript levels during the night period (Gaufichon et al. 2013). This model predicts that ASPG activity should peak at dawn or early during the light period. In pea, ASPG activity was found to be high in the light and low in the dark consistent with this model (Sieciechowicz et al. 1985). This was specific to young, expanding leaves, and ASPG activity peaked 4 h into the light period (Sieciechowicz et al. 1988a). Regulation did not appear to involve protein phosphorylation and is probably transcriptional (Sieciechowicz et al. 1988b). It would be interesting to confirm the mode and timing of ASPG regulation by examining the corresponding transcript, protein, and metabolite levels during leaf development.

Biosynthesis of Asn is primarily dependent on the N status and C to N ratio of the plant and is regulated through the expression of AS genes. In nodulated soybean roots, AS activity under N stress is decreased, resulting in reduced Asn concentration in the xylem sap (Lima and Sodek 2003). The change in Asn concentration is reversible and the recovery closely parallels AS activity in nodules (*Glycine max*) (Lima and Sodek 2003). All three soybean *AS* genes are repressed in root under N deficit (Antunes et al. 2008). Nitrate is involved as a signal in the up-regulation of *AS* genes in Arabidopsis (Wang et al. 2003, 2004; Scheible et al. 2004). Cytokinins act as a secondary signal of nitrate availability. The expression of cytokinin biosynthetic and signaling genes is up-regulated by nitrate (Krouk 2016; Sakakibara et al. 2006). In turn, cytokinins up-regulate the expression of *ASN1*, *GDH1*, and *GDH2* (Brenner et al. 2005). Transcript levels of *ASN1* in Arabidopsis are regulated by the carbon:N ratio. Sucrose suppresses the accumulation of *ASN1* transcript in dark-grown plants, however, Asn, glutamine or glutamate prevent this effect (Lam et al. 1994). Similar results were obtained in maize root tips (Chevalier et al. 1996). Arabidopsis *ASN2* is regulated differently. It is induced by sucrose, and repressed by Asn, glutamine or glutamate (Lam et al. 1998). However, by profiling transcripts and performing network analysis under treatments with inorganic N as nitrate and ammonium or organic N as glutamate or glutamine, in the presence or absence of the GS inhibitor methionine sulfoximine, Gutiérrez et al. (2008) showed that *ASN1* responds to glutamate or a downstream metabolite, in concert with *GDH1*, whereas the K^+-dependent *ASPGB1* had an opposite regulation. *ASN2* positively responds to inorganic N. The transcription factor BZIP1 regulates *ASN1* (Hanson et al. 2008; Para et al. 2014).

Cadmium stress leads to reduced ammonia assimilation in leaves of tomato, and was found to result in increased levels of class I AS transcripts and Asn accumulation in roots (Chaffei et al. 2004). Ariz et al. (2013) compared the effect of ammonium stress under low and high illumination. The latter condition relieves stress by balancing the carbon:N ratio. The combination of ammonium and low light increased Asn accumulation in leaf. This was paralleled by elevated *PsAS1* transcript, and accompanied by increased activity of anaplerotic enzymes and glutamate dehydrogenase in both leaf and root. The role of these enzymes is to generate carbon skeletons for the tricarboxylic acid cycle. Under high illumination, the excess N accumulated as polyamines rather than amino acids.

Another related condition leading to Asn accumulation is leaf senescence which can be mimicked by continuous dark exposure. Arabidopsis *ASN1* is part of a group of dark-inducible genes (*din6*) expressed during leaf senescence (Fujiki et al. 2001). In asparagus spears and detached leaves, a rapid decline in sucrose is associated with increased expression of class I AS (Davies et al. 1996). This is also observed in broccoli florets postharvest (Downs and Somerfield 1997). Its promoter contains a *cis* regulatory element, TTGCACG, involved in regulation by senescence, and negative regulatory elements for sucrose regulation (Winichayakul et al. 2004). Yet another situation related to leaf senescence occurs when bacterial pathogen infection triggers increased expression of class I AS and Asn accumulation in tomato leaf (Olea et al. 2004). This response appears to involve a remobilization of N, as it also involved increased expression of cytosolic GS1, and AS expression was localized in phloem cells of leaf vascular bundles. More recently it was observed that silencing of *CaAS1* in pepper led to increased susceptibility to bacterial pathogens (Hwang et al. 2011). Conversely, overexpression of *CaAS1* in transgenic Arabidopsis was associated with enhanced disease tolerance. It therefore appears as though AS-mediated N remobilization may be essential for disease tolerance under these conditions. The relationship between N metabolism and pathogen infection in plants has been reviewed by Fagard et al. (2014).

Arabidopsis *ASN2* is associated with ammonium assimilation under light conditions (Wong et al. 2004). An *ASN2* mutant was impaired in salt stress induced, ammonium-dependent Asn accumulation (Maaroufi-Dguimi et al. 2011). Careful analysis of *ASN2* loss-of-function mutants revealed that the gene is involved in N assimilation and Asn export to the phloem. The mutant plants were characterized by impaired growth, increased ammonium accumulation, reduced Asn, and increased γ-aminobutyric acid and alanine concentration in leaves (Gaufichon et al. 2013). This was interpreted as a consequence of increased flux through the γ-aminobutyric acid shunt, leading to alanine biosynthesis via the transamination of pyruvate. The authors indicated that decreased aspartate utilization for Asn biosynthesis was likely to reduce the flux from phosphoenolpyruvate to oxaloacetate, resulting in higher availability of pyruvate and fumarate. *ASN2* expression is localized in leaf phloem companion cells, and the *ASN2* mutants displayed reduced Asn concentration in phloem exudates, and delayed senescence. Loss of function of the closely related *ASN3* led to a relatively milder phenotype (Gaufichon et al. 2016a). Similarly in rice, the class II AS, OsAS1, is involved in ammonium assimilation in roots, and appears to work in conjunction with cytosolic glutamine synthetase1;2 and glutamate synthase1 (Ohashi et al. 2015). Class II AS *PvAS3* is broadly expressed in young tissues of common bean, and its expression is repressed by nitrate in nodules and root (Gálvez-Valdivieso et al. 2013).

The general picture emerging from these data is that class I AS is involved in the regulation of Asn levels in response to the carbon:N ratio, likely in concert with a glutamate dehydrogenase breaking down the glutamate generated by the AS reaction, generating α-ketoglutarate to balance carbon metabolism. Class II AS is generally involved in N assimilation.

Asn transamination is implicated in photorespiration through the involvement of serine:glyoxylate aminotransferase. Its significance is related to an input of N to balance an output of serine or glycine used for other pathways. However, to a large extent this remains an open question. For example, the lack of impact of *ASN2* mutations on photosynthetic rate was interpreted as evidence that Asn is not implicated in photorespiratory N flux (Gaufichon et al. 2013). Nevertheless, the decrease in Asn was correlated with lower glycine and serine concentration in the mutants, which could be interpreted as a consequence of a lack of N input from Asn. The authors argued that the incorporation of [^{15}N] from Asn into glycine measured by Ta et al. (1985) could have happened after degradation by ASPG. However, this does not take into account the fact that the amino N was specifically labelled in these experiments. More recently, Arabidopsis plants overexpressing serine: glyoxylate aminotransferase exhibited a decrease in serine and Asn concentration in leaf at the end of day, which suggested a higher flux of Asn to compensate for increased serine turnover (Modde et al. 2016). In the Arabidopsis *nit4* mutant lacking β-cyanoalanine nitrilase/hydratase activity, which has a reduced flux of [^{13}C]-labelled cyanide into Asn, treatment with potassium cyanide led to decreased levels of glycine and serine as compared with wild-type (O'Leary et al. 2014). Conversely, treatment of a line overexpressing β-cyanoalanine nitrilase/hydratase led to increased levels of glycine and serine. These results suggested that N flux from Asn into photorespiration occurs downstream from cyanide detoxication.

Whether Asn acts as a metabolite signal and if so how it is perceived is an open area of investigation. Beside glutamate and glutamine discussed above, histidine has been recently implicated in the regulation of abscisic acid biosynthesis and fatty acid oxidation (Ma and Wang 2016). Understanding of amino acid signaling in plants lags behind that of other organisms. Proteins involved in amino acid signaling in plants include the chloroplastic PII, involved in the regulation of Arg biosynthesis and lipid accumulation, and a glutamine sensor, except in the Brassicaceae (Chellamuthu et al. 2014; Feria Bourrellier et al. 2010); the protein kinase general control non-derepressible 2 (GCN2), involved in the perception of amino acid deficiency through binding of uncharged tRNAs, which phosphorylates eIF2α (α-subunit of eukaryotic translation initiation factor 2) (Li et al. 2013; Wang et al. 2017); the protein kinase target of rapamycin (TOR), a positive regulator of translation and growth processes in response to carbon and N nutrients (Dobrenel et al. 2016); an aspartyl-tRNA synthetase involved in the perception of the immunomodulatory β-aminobutyric acid (Luna et al. 2014); and glutamate receptor homologs, one of which was recently implicated in the control of stomatal opening by acting as a calcium channel in response to methionine binding (Kong et al. 2016). Activation of mammalian TOR Complex 1 (mTORC1) requires amino acid sensors such as CASTOR1 for arginine and Sestrin2 for leucine (Saxton et al. 2016; Wolfson et al. 2016). Asn was recently identified in a group of amino acids involved in priming mTORC1 for activation by other amino acids including leucine (Dyachok et al. 2016). In mammalian cancer cells, Asn recently was shown to act as a regulator of amino acid uptake, particularly serine, arginine, and histidine, to coordinate nucleotide and protein synthesis and to be an essential metabolite for cell proliferation (Krall et al. 2016). Clearly, more work is required to understand the role and perception of Asn as a potential metabolite signal of N status in higher plants.

5 Asparagine as a Transport and Storage Form of Nitrogen

The N requirements of actively growing roots and leaves are mostly met by inorganic sources, NO_3^- and NH_4^+ that are primarily transported via xylem, and a restricted group of organic sources that include Asn transported through both xylem and phloem (Sieciechowicz et al. 1988c). Asn is very well established as a primary N transport compound from its relative abundance in xylem and phloem sap of lupins and its high N:C ratio (1:2) (Atkins et al. 1983). In its soluble form, Asn is a substrate for only a few enzymatic reactions, and has a low net charge under physiological conditions, which makes it an ideal storage compound (Lea et al. 2007). Early work by Pate et al. (1981) in lupine showed that Asn was the major amino acid in all plant parts especially in nodulated roots, leaves, and pods where it could account for 60–80% of total amino acid content. A thorough investigation of N sources in xylem sap of different legume species revealed that although ureides form the majority of N transport compounds in nodulated *Glycine max*, *Vigna unguiculata*, and *Phaseolus vulgaris*, amino acids and nitrate predominate under non-N fixing conditions (do Amarante et al. 2006). Only amino acids and nitrate were found in xylem sap of other legume species viz., *Crotalaria juncea*, *Pisum sativum*, and *Lupinus albus* under both nodulated and non-nodulated conditions. Further analysis of amino acid fractions of these species also revealed that Asn was the major form of N transport under most conditions, whereas glutamine represents a marker of symbiotic activity. The same study highlighted the importance of Asn as an indicator of N status regardless of whether N fixation takes place, because N deficiency resulted in a decrease of the Asn:aspartic acid ratio. These observations were extended to several legumes and non-legume species, indicating that the Asn:aspartic acid ratio in xylem is a key indicator of N status. Furthermore, a comprehensive review of Asn in plants, by Lea et al. (2007), highlighted the literature that shows evidence of Asn as a storage and transport form of N in several perennials. It is also noteworthy that glutamine and arginine may carry out the same role as Asn depending on species and type of tissue (Lea et al. 2007).

The pattern of expression of Asn biosynthetic genes, including *ASN2* discussed above, and catabolic genes support a role of Asn as a major source of N transported to sink tissues. A fusion between a lupine ASPG promoter and the β-glucuronidase reporter gene (GUS) directed expression to the apical meristem and developing seeds in both transgenic tobacco and lupine (Grant and Bevan 1994). A similar experiment was performed for Arabidopsis ASPGs (Ivanov et al. 2012). Late seed maturation is associated with the accumulation of free amino acids which are stored as a source of readily available N for use during germination prior to the degradation of storage proteins (Fait et al. 2006). Interestingly, during germination,

ASPGA1 expression was observed in endosperm cells prior to its rupture, starting at the micropylar end. After the rupture of the endosperm, expression was present throughout the hypocotyl and root, including the transition zone and immature root hairs. At day 2 after germination, expression was present in the hypocotyl and elongation zone of the root, and was restricted to root tip at day 3. In mature plants, expression was observed in quiescent lateral root primordia, immature trichomes, leaf vasculature, apical meristem, and inflorescence stem. During reproductive development, expression was observed in developing pollen grain and in the tegument of developing seeds. *ASPGB1* expression is less intense and was restricted to developing pollen grains. In rice, expression of the ASPG gene *OsAsnase2* was localized to the mesophyll companion cells of flag leaves and dorsal vascular bundles and nucellar projections of developing grains (Yabuki et al. 2017).

Conifers constitute well understood model systems of ammonium assimilation (Cantón et al. 2005). Pine species prefer ammonium as a N source (Castro-Rodríguez et al. 2016). During seedling germination, a large quantity of ammonium liberated by the degradation of storage proteins is exported as Asn in the hypocotyl (Cañas et al. 2006). This correlates with the expression of a class I AS gene, *PsAS1*, in hypocotyl. A K^+-dependent ASPG gene, *PsASPG*, is expressed later in the same tissue during development (Cañas et al. 2007). Expression coincides with the development of the secondary vascular system, characterized by a high demand for phenylalanine for the biosynthesis of lignin precursors (Pascual et al. 2016). *PsAS1* and *PsASPG* transcripts were localized to the cambial cells in the vasculature, suggesting a role in the assimilation of transported Asn in the hypocotyl (Cañas et al. 2007).

6 Relationship Between Asparagine and Seed or Grain Protein Content

The relationship between Asn transported from source tissues and protein accumulated in the developing seed or grain has been studied especially in legumes. This topic has been recently reviewed by Herman (2014) who highlighted the genetic control of protein concentration. The developing seed is a differentiated tissue with different dynamics of Asn catabolism. In pea, ASPG activity is high in seed coat during early development, while activity in cotyledon increases in later stages (Sodek et al. 1980). In soybean cotyledons grown in culture, free Asn was found to be the highest in quantity (33–49%) and using [^{13}C] and [^{15}N] labelled Asn in NMR studies, it was determined that about half was incorporated directly into protein (Schaefer et al. 1981), providing evidence for both direct incorporation and incorporation after catabolism. In a time course study, the amino N of Asn was incorporated more efficiently into protein, suggesting a lag in catabolism of the amide group by ASPG and subsequent ammonia re-assimilation (Skokut et al. 1982). Likely for this reason, glutamine is a preferred source of N over Asn for soybean cotyledons grown in culture (Haga and Sodek 1987). ASPG activity responds to the N source, with glutamine exerting an inhibitory effect (Tonin and Sodek 1990). More recent work examining N flux in soybean cotyledons in culture has confirmed that Asn is relatively inert as compared with glutamine (Allen and Young 2013).

In soybean seed, a correlation has been observed between free Asn concentration in the developing cotyledons at mid-maturation and protein concentration at maturity (Hernández-Sebastià et al. 2005). This was observed in genetically related cultivars differing in protein concentration throughout the photoperiod, as well as in field grown, genetically unrelated cultivars. An examination of Asn metabolism in different tissues of soybean seed indicated a high expression of AS early during seed coat development which rapidly declined, suggesting pre-conditioning of N-assimilates that are transmitted to the developing embryo (Pandurangan et al. 2012). The high levels of AS in seed coat probably serve to establish the high Asn concentration in cotyledon. In parallel, K^+-dependent ASPGB levels progressively increased and were correlated with seed protein concentration, suggestive of higher N flux in a high protein genotype. Quantitative reverse transcription PCR results suggested that a specific isoform, *ASPGB1a*, was differentially expressed in seed coat. ASPGB levels were undetectable by Western in cotyledons, whereas AS was readily detectable, also likely to contribute to the high levels of free Asn in this organ. In a more recent study, the effect of overexpressing PvAspG1 in soybean cotyledons was examined (Pandurangan et al. 2015). The lines had a reduced concentration of free Asn as a percentage of total free amino acids, with a corresponding increase in free aspartic acid. Two lines had a decreased seed N concentration by ca. 10% with reduced extractible globulins by 10–30% and a corresponding increase in lipid concentration. A third line exhibited no change in seed composition. In this line, the decrease in free Asn appeared to be compensated by an increase in free glutamic acid levels. Overall, the results supported a possible role of Asn as a metabolite signal in cotyledon stimulating protein accumulation. The increased ASPG activity in cotyledon did not result in enhanced uptake of Asn, highlighting the critical regulatory control of amino acid transporters in developing cotyledons and in phloem loading (Sanders et al. 2009; Zhang et al. 2010).

Soybean AS1 and AS2 are both class I AS enzymes. Wan et al. (2006) have shown that the levels of *AS1* transcript in soybean leaves are correlated with seed protein concentration. Whereas *AS2* showed the expected pattern of gene expression during the night, *AS1* transcript levels did not differ markedly during the photoperiod. This regulation is similar in common bean (Osuna et al. 2001). Most crop species have two or more genes coding for class I AS, unlike Arabidopsis (Duff 2015). The findings described above indicate a possible specialization of class I ASs, with AS1 being involved in asparagine export to the phloem. The lack of photoperiodic regulation of *AS1* raises the question of whether the activity of the enzyme might be controlled post-translationally in the light. In Arabidopsis, constitutive overexpression of *ASN1* led to increased seed protein concentration (Lam et al. 2003). However, it was not clear whether this was due to increased Asn transported to the seed, or enhanced biosynthesis in seed tissue. A recent study by

Gaufichon et al. (2017) highlighted the activation of Asn metabolism and expression of related genes, including *ASN1*, *ASPGA1* and *ASPGB1*, in floral organs at the onset of fertilization. *ASN1* transcript was localized to the companion cells in minor leaf veins. In reproductive organs, *ASN1* protein was localized to the vasculature of the silique wall and septum, as well as inner and outer seed integuments. *ASN1* mutants were shown to be affected in embryo development. Plants constitutively expressing *ASN1* exhibited increased *ASN1* transcript and Asn concentration in floral organs, whereas the timing of expression under a seed storage protein promoter was too late to influence seed protein concentration. Overall these results suggest that local or in situ biosynthesis in flower tissue early during embryo development conditions protein accumulation later in seed. It raises the possibility that even earlier expression, in floral buds or apical meristem, could influence seed protein accumulation, if the transcript or protein is long lived. The correlation between high Asn in leaf and phloem and grain protein concentration was previously reported not only in legumes but also in cereals like barley and maize (Dembinski et al. 1991; Lohaus et al. 1998). Increased Asn to Gln ratio in cobs of maize was also suggested to have a role in conditioning kernel development (Seebauer et al. 2004).

The importance of Asn in remobilization of N, pod and seed development, and seed storage protein accumulation was demonstrated in model legume *Lotus japonicus* mutants lacking the ASPG NSE1 (Credali et al. 2013). Both seed weight and seed storage protein accumulation were affected in the mutants. By contrast, an Arabidopsis mutant lacking both ASPG genes had a relatively mild phenotype (Ivanov et al. 2012). These plants displayed elevated levels of free Asn in mature seed only under conditions of low illumination leading to a low carbon:N ratio.

7 Agronomic Applications

Free amino acid pools of the tubers of potatoes (*Solanum tuberosum*) are known to be constituted with a high amount of Asn (Koch et al. 2003). As acrylamide, a neurotoxin and a probable carcinogen present in heat-processed food like French-fries and potato chips are formed from Asn and sugars (Blank et al. 2005), efforts were made to reduce Asn concentration in tubers of potatoes (Chawla et al. 2012). As described above, distinct regulatory mechanisms of *AS* gene family members based on environmental stimuli, C/N metabolic status, developmental cues, and tissue specificity have already been characterized. Based on high sequence similarity with *Arabidopsis ASN1*, *StAst1* was predicted to have a primary role in Asn storage in tubers and was silenced in tubers to produce potatoes with low Asn (20% compared to control) concentration (Chawla et al. 2012). The same study highlighted the importance of specificity in roles played by different homologs of AS by demonstrating that *StAst1*/*StAst2*-silenced tubers of potato displayed a stunted growth and reduced tuber yield per plant and indicated that *StAst2* had a greater impact on field growth of potato plants (Chawla et al. 2012; Rommens et al.

2008; Zhu et al. 2016). The reduced Asn trait is part of the Innate® potatoes developed by the J. R. Simplot Company (Halterman et al. 2016). These potatoes are expected to substantially reduce the risk of dietary exposure to acrylamide (Tran et al. 2016). A similar approach was taken recently to address the problem of acrylamide formation in baked wheat and rye products (Curtis and Halford 2016). It turns out that adequate sulfur nutrition plays an important role in ensuring that free Asn levels are not elevated (Postles et al. 2016). As in barley, wheat TaASN1 transcripts were elevated in developing grain in response to sulfur deficiency (Gao et al. 2016).

8 Conclusions

There have been tremendous advances in our understanding of the function of Asn in plant metabolism. Asn is a central intermediate in amino acid metabolism, involved in multiple processes in plant development. Under conditions of low carbon:N ratio, such as in the dark, Asn biosynthesis mediated by class I AS is coupled with the breakdown of glutamic acid by glutamate dehydrogenase. Class II AS is generally involved in N assimilation and transport, in conjunction with cytosolic GS/GOGAT. Recent results highlight that beyond the role of AS in phloem loading in source leaves, in situ expression in reproductive tissues also influences seed protein accumulation. The contribution of Asn to photorespiration warrants further investigation, particularly to evaluate whether Asn biosynthesis is coupled with photorespiratory N flux. In the Asn catabolic pathway, the gene responsible for the reduction of α-ketosuccinamate remains to be identified. It is anticipated that the detailed characterization of mutant and transgenic plants having altered Asn metabolism, coupled with gene co-expression analysis, will enable further insights into these questions. In addition, the possible role of Asn as a metabolite signal is yet to be elucidated. Having a deeper understanding of Asn metabolism may facilitate applications to improve crop characteristics, including nitrogen use efficiency, despite the challenges associated with the development of these so-called second generation traits (Rothstein et al. 2014). The manipulation of Asn biosynthesis in potato tubers to mitigate the risk of toxic acrylamide formation in fried food provides an interesting example of such applications.

Acknowledgments Research on Asn in the authors' laboratory was supported by a Discovery grant from the Natural Sciences and Engineering Research Council.

References

Allen DK, Young JD (2013) Carbon and nitrogen provisions alter the metabolic flux in developing soybean embryos. Plant Physiol 161:1458–1475. doi:10.1104/pp.112.203299

Antunes F, Aguilar M, Pineda M, Sodek L (2008) Nitrogen stress and the expression of asparagine synthetase in roots and nodules of soybean (*Glycine max*). Physiol Plant 133:736–743. doi:10.1111/j.1399-3054.2008.01092.x

Ariz I, Asensio AC, Zamarreño AM, García-Mina JM, Aparicio-Tejo PM, Moran JF (2013) Changes in the C/N balance caused by increasing external ammonium concentrations are driven by carbon and energy availabilities during ammonium nutrition in pea plants: the key roles of asparagine synthetase and anaplerotic enzymes. Physiol Plant 148:522–537. doi:10.1111/j.1399-3054.2012.01712.x

Atkins CA, Pate JS, Sharkey PJ (1975) Asparagine metabolism – key to the nitrogen nutrition of developing legume seeds. Plant Physiol 56:807–812. doi:10.1104/pp.56.6.807

Atkins CA, Pate JS, Peoples MB, Joy KW (1983) Amino acid transport and metabolism in relation to the nitrogen economy of a legume leaf. Plant Physiol 71:841–848. doi:10.1104/pp.71.4.841

Becker T, Pasteels J, Weigel C, Dahse H-M, Voigt K, Boland W (2017) A tale of four kingdoms – isoxazolin-5-one- and 3-nitropropanoic acid-derived natural products. Nat Prod Rep 34: 343–360. doi:10.1039/C6NP00122J

Bejger M, Imiolczyk B, Clavel D, Gilski M, Pajak A, Marsolais F, Jaskolski M (2014) Na^+/K^+ exchange switches the catalytic apparatus of K-dependent plant L-asparaginase. Acta Crystallogr D Biol Crystallogr 70:1854–1872. doi:10.1107/S1399004714008700

Blank I, Robert F, Goldmann T, Pollien P, Varga N, Devaud S, Saucy F, Huynh-Ba T, Stadler RH (2005) Mechanisms of acrylamide formation: Maillard-induced transformation of asparagine. Adv Exp Med Biol 561:171–189. doi:10.1007/0-387-24980-X_14

Borek D, Michalska K, Brzezinski K, Kisiel A, Podkowinski J, Bonthron DT, Krowarsch D, Otlewski J, Jaskolski M (2004) Expression, purification and catalytic activity of *Lupinus luteus* asparagine β-amidohydrolase and its *Escherichia coli* homolog. Eur J Biochem 271: 3215–3226. doi:10.1111/j.1432-1033.2004.04254.x

Brannigan JA, Dodson G, Duggleby HJ, Moody PCE, Smith JL, Tomchick DR, Murzin AG (1995) A protein catalytic framework with an N-terminal nucleophile is capable of self-activation. Nature 378:416–419

Brenner C (2002) Catalysis in the nitrilase superfamily. Curr Opin Struct Biol 12:775–782. doi:10.1016/S0959-440X(02)00387-1

Brenner WG, Romanov GA, Kollmer I, Burkle L, Schmulling T (2005) Immediate-early and delayed cytokinin response genes of *Arabidopsis thaliana* identified by genome-wide expression profiling reveal novel cytokinin-sensitive processes and suggest cytokinin action through transcriptional cascades. Plant J 44:314–333. doi:10.1111/j.1365-313X.2005.02530.x

Bruneau L, Chapman R, Marsolais F (2006) Co-occurrence of both L-asparaginase subtypes in *Arabidopsis*: At3g16150 encodes a K^+-dependent L-asparaginase. Planta 224:668–679. doi:10.1007/s00425-006-0245-9

Cañas RA, De La Torre F, Cánovas FM, Cantón FR (2006) High levels of asparagine synthetase in hypocotyls of pine seedlings suggest a role of the enzyme in re-allocation of seed-stored nitrogen. Planta 224:83–95. doi:10.1007/s00425-005-0196-6

Cañas RA, De La Torre F, Cánovas FM, Cantón FR (2007) Coordination of *PsAS1* and *PsASPG* expression controls timing of re-allocated N utilization in hypocotyls of pine seedlings. Planta 225:1205–1219. doi:10.1007/s00425-006-0431-9

Cantón FR, Suárez MF, Cánovas FM (2005) Molecular aspects of nitrogen mobilization and recycling in trees. Photosynth Res 83:265–278

Castro-Rodríguez V, Assaf-Casals I, Pérez-Tienda J, Fan X, Avila C, Miller A, Cánovas FM (2016) Deciphering the molecular basis of ammonium uptake and transport in maritime pine. Plant Cell Environ 39:1669–1682. doi:10.1111/pce.12692

Chaffei C, Pageau K, Suzuki A, Gouia H, Ghorbel MH, Masclaux-Daubresse C (2004) Cadmium toxicity induced changes in nitrogen management in *Lycopersicon esculentum* leading to a metabolic safeguard through an amino acid storage strategy. Plant Cell Physiol 45:1681–1693

Chan WK, Lorenzi PL, Anishkin A, Purwaha P, Rogers DM, Sukharev S, Rempe SB, Weinstein JN (2014) The glutaminase activity of L-Asparaginase is not required for anticancer activity against ASNS-negative cells. Blood 123:3596–3606. doi:10.1182/blood-2013-10-535112

Chawla R, Shakya R, Rommens CM (2012) Tuber-specific silencing of asparagine synthetase-1 reduces the acrylamide-forming potential of potatoes grown in the field without affecting tuber shape and yield. Plant Biotechnol J 10:913–924. doi:10.1111/j.1467-7652.2012.00720.x

Chellamuthu VR, Ermilova E, Lapina T, Luddecke J, Minaeva E, Herrmann C, Hartmann MD, Forchhammer K (2014) A widespread glutamine-sensing mechanism in the plant kingdom. Cell 159:1188–1199. doi:10.1016/j.cell.2014.10.015

Chevalier C, Bourgeois E, Just D, Raymond P (1996) Metabolic regulation of asparagine synthetase gene expression in maize (*Zea mays* L.) root tips. Plant J 9:1–11

Chibnall AC (1924) Investigations on the nitrogenous metabolism of the higher plants. VI. The rôle of asparagine in the metabolism of the mature plant. Biochem J 18:395–404

Cooper AJL, Duffy TE, Meister A (1985) α-Keto acid ω-amidase from rat liver. Methods Enzymol 113:350–358

Coruzzi GM (2003) Primary N-assimilation into amino acids in *Arabidopsis*. Arabidopsis Book 2: e0010. doi:10.1199/tab.0010

Credali A, García-Calderón M, Dam S, Perry J, Díaz-Quintana A, Parniske M, Wang TL, Stougaard J, Vega JM, Márquez AJ (2013) The K^+-dependent asparaginase, NSE1, is crucial for plant growth and seed production in *Lotus japonicus*. Plant Cell Physiol 54:107–118. doi:10.1093/pcp/pcs156

Curtis TY, Halford NG (2016) Reducing the acrylamide-forming potential of wheat. Food Energy Secur 5:153–164. doi:10.1002/fes3.85

Davies KM, Seelye JF, Irving DE, Borst WM, Hurst PL, King GA (1996) Sugar regulation of harvest-related genes in Asparagus. Plant Physiol 111:877–883

Delaville M (1802) Sur les sèves d'asperges et de choux. Ann Chim 41:298

Dembinski E, Rafalski A, Wisniewska I (1991) Effect of long-term selection for high and low protein content on the metabolism of amino acids and carbohydrates in maize kernel. Plant Physiol Biochem 29:549–557

do Amarante L, Lima JD, Sodek L (2006) Growth and stress conditions cause similar changes in xylem amino acids for different legume species. Environ Exp Bot 58:123–129. doi:10.1016/j.envexpbot.2005.07.002

Dobrenel T, Caldana C, Hanson J, Robaglia C, Vincentz M, Veit B, Meyer C (2016) TOR signaling and nutrient sensing. Annu Rev Plant Biol 67:261–285. doi:10.1146/annurev-arplant-043014-114648

Downs CG, Somerfield SD (1997) Asparagine synthetase gene expression increases as sucrose declines in broccoli after harvest. N Z J Crop Hortic Sci 25:191–195

Duff SMG (2015) Asparagine synthetase. In: D'Mello JPF (ed) Amino acids in higher plants. CAB International, Wallingford, pp 100–128

Duff SMG, Qi Q, Reich T, Wu X, Brown T, Crowley JH, Fabbri B (2011) A kinetic comparison of asparagine synthetase isozymes from higher plants. Plant Physiol Biochem 49:251–256. doi:10.1016/j.plaphy.2010.12.006

Dyachok J, Earnest S, Iturraran EN, Cobb MH, Ross EM (2016) Amino acids regulate mTORC1 by an obligate two-step mechanism. J Biol Chem 291:22414–22426. doi:10.1074/jbc.M116.732511

Ellens KW, Richardson LGL, Frelin O, Collins J, Ribeiro CL, Hsieh Y-F, Mullen RT, Hanson AD (2015) Evidence that glutamine transaminase and omega-amidase potentially act in tandem to close the methionine salvage cycle in bacteria and plants. Phytochemistry 113:160–169. doi:10.1016/j.phytochem.2014.04.012

Fagard M, Launay A, Clément G, Courtial J, Dellagi A, Farjad M, Krapp A, Soulié M-C, Masclaux-Daubresse C (2014) Nitrogen metabolism meets phytopathology. J Exp Bot 65: 5643–5656. doi:10.1093/jxb/eru323

Fait A, Angelovici R, Less H, Ohad I, Urbanczyk-Wochniak E, Fernie AR, Galili G (2006) Arabidopsis seed development and germination is associated with temporally distinct metabolic switches. Plant Physiol 142:839–854

Feria Bourrellier AB, Valot B, Guillot A, Ambard-Bretteville F, Vidal J, Hodges M (2010) Chloroplast acetyl-CoA carboxylase activity is 2-oxoglutarate-regulated by interaction of PII with the biotin carboxyl carrier subunit. Proc Natl Acad Sci U S A 107:502–507. doi:10.1073/pnas.0910097107

Fujiki Y, Yoshikawa Y, Sato T, Inada N, Ito M, Nishida I, Watanabe A (2001) Dark-inducible genes from *Arabidopsis thaliana* are associated with leaf senescence and repressed by sugars. Physiol Plant 111:345–352. doi:10.1034/j.1399-3054.2001.1110312.x

Gabriel M, Telmer PG, Marsolais F (2012) Role of asparaginase variable loop at the carboxyl terminal of the alpha subunit in the determination of substrate preference in plants. Planta 235:1013–1022. doi:10.1007/s00425-011-1557-y

Gálvez-Valdivieso G, Alamillo JM, Fernández J, Pineda M (2013) Molecular characterization of *PVAS3*: an asparagine synthetase gene from common bean prevailing in developing organs. J Plant Physiol 170:1484–1490. doi:10.1016/j.jplph.2013.06.002

Gao R, Curtis TY, Powers SJ, Xu H, Huang J, Halford NG (2016) Food safety: structure and expression of the asparagine synthetase gene family of wheat. J Cereal Sci 68:122–131. doi:10.1016/j.jcs.2016.01.010

Gaufichon L, Reisdorf-Cren M, Rothstein SJ, Chardon F, Suzuki A (2010) Biological functions of asparagine synthetase in plants. Plant Sci 179:141–153. doi:10.1016/j.plantsci.2010.04.010

Gaufichon L, Masclaux-Daubresse C, Tcherkez G, Reisdorf-Cren M, Sakakibara Y, Hase T, Clément G, Avice JC, Grandjean O, Marmagne A, Boutet-Mercey S, Azzopardi M, Soulay F, Suzuki A (2013) *Arabidopsis thaliana ASN2* encoding asparagine synthetase is involved in the control of nitrogen assimilation and export during vegetative growth. Plant Cell Environ 36:328–342. doi:10.1111/j.1365-3040.2012.02576.x

Gaufichon L, Marmagne A, Yoneyama T, Hase T, Clément G, Trassaert M, Xu X, Shakibaei M, Najihi A, Suzuki A (2016a) Impact of the disruption of *ASN3*-encoding asparagine synthetase on Arabidopsis development. Agronomy 6:12. doi:10.3390/agronomy6010012

Gaufichon L, Rothstein SJ, Suzuki A (2016b) Asparagine metabolic pathways in Arabidopsis. Plant Cell Physiol 57:675–689. doi:10.1093/pcp/pcv184

Gaufichon L, Marmagne A, Belcram K, Yoneyama T, Sakakibara Y, Hase T, Grandjean O, Gilles C, Citerne S, Boutet-Mercey S, Masclaux-Daubresse C, Chardon F, Soulay F, Xu X, Trassaert M, Shakiebaei M, Najihi A, Suzuki A (2017) *ASN1*-encoded asparagine synthetase in floral organs contributes to nitrogen filling in Arabidopsis seeds. Plant J. doi:10.1111/tpj.13567

Grant M, Bevan MW (1994) Asparaginase gene expression is regulated in a complex spatial and temporal pattern in nitrogen-sink tissues. Plant J 5:695–704

Gutiérrez RA, Stokes TL, Thum K, Xu X, Obertello M, Katari MS, Tanurdzic M, Dean A, Nero DC, McClung CR, Coruzzi GM (2008) Systems approach identifies an organic nitrogen-responsive gene network that is regulated by the master clock control gene *CCA1*. Proc Natl Acad Sci U S A 105:4939–4944

Haga KI, Sodek L (1987) Utilization of nitrogen sources by immature soybean cotyledons in culture. Ann Bot 59:597–602

Halterman D, Guenthner J, Collinge S, Butler N, Douches D (2016) Biotech potatoes in the 21st century: 20 years since the first biotech potato. Am J Potato Res 93:1–20. doi:10.1007/s12230-015-9485-1

Hanson J, Hanssen M, Wiese A, Hendriks MM, Smeekens S (2008) The sucrose regulated transcription factor bZIP11 affects amino acid metabolism by regulating the expression of ASPARAGINE SYNTHETASE1 and PROLINE DEHYDROGENASE2. Plant J 53: 935–949. doi:10.1111/j.1365-313X.2007.03385.x

Hariharan VA, Denton TT, Paraszcszak S, McEvoy K, Jeitner TM, Krasnikov BF, Cooper AJ (2017) The enzymology of 2-hydroxyglutarate, 2-hydroxyglutaramate and 2-hydroxysuccinamate and their relationship to oncometabolites. Biology 6:E24. doi:10.3390/biology6020024

Harmer SL, Covington MF, Bläsing O, Stitt M (2005) Circadian regulation of global gene expression and metabolism. In: Hall AJW, McWatters HG (eds) Endogenous plant rhythms, Annual plant reviews, vol 21. Blackwell, Oxford, pp 133–165

Havir EA, McHale NA (1988) A mutant of *Nicotiana sylvestris* lacking serine:glyoxylate aminotransferase: substrate specificity of the enzyme and fate of [2-^{14}C]glycolate in plants with genetically altered enzyme levels. Plant Physiol 87:806–808

Herman EM (2014) Soybean seed proteome rebalancing. Front Plant Sci 5:437. doi:10.3389/fpls.2014.00437

Hernández-Sebastià C, Marsolais F, Saravitz C, Israel D, Dewey RE, Huber SC (2005) Free amino acid profiles suggest a possible role for asparagine in the control of storage-product accumulation in developing seeds of low- and high-protein soybean lines. J Exp Bot 56:1951–1963. doi:10.1093/jxb/eri191

Hillebrand H, Bartling D, Weiler EW (1998) Structural analysis of the *nit2/nit1/nit3* gene cluster encoding nitrilases, enzymes catalyzing the terminal activation step in indole-acetic acid biosynthesis in *Arabidopsis thaliana*. Plant Mol Biol 36:89–99

Hirel B, Lea PJ (2001) Ammonia assimilation. In: Lea PJ, Morot-Gaudry J-F (eds) Plant nitrogen. Springer-Verlag/INRA Editions, Paris, pp 79–99

Hodges M (2002) Enzyme redundancy and the importance of 2-oxoglutarate in plant ammonium assimilation. J Exp Bot 53:905–916. doi:10.1093/jexbot/53.370.905

Hodges M, Dellero Y, Keech O, Betti M, Raghavendra AS, Sage R, Zhu XG, Allen DK, Weber AP (2016) Perspectives for a better understanding of the metabolic integration of photorespiration within a complex plant primary metabolism network. J Exp Bot 67:3015–3026. doi:10.1093/jxb/erw145

Hwang IS, An SH, Hwang BK (2011) Pepper asparagine synthetase 1 (CaAS1) is required for plant nitrogen assimilation and defense responses to microbial pathogens. Plant J 67:749–762. doi:10.1111/j.1365-313X.2011.04622.x

Ireland RJ, Joy KW (1981) Two routes for asparagine metabolism in *Pisum sativum* L. Planta 151: 289–292. doi:10.1007/bf00395182

Ireland RJ, Joy KW (1983a) Purification and properties of an asparagine aminotransferase from *Pisum sativum* leaves. Arch Biochem Biophys 223:291–296

Ireland RJ, Joy KW (1983b) Subcellular localization of asparaginase and asparagine aminotransferase in *Pisum sativum* leaves. Plant Physiol 72:1127–1129

Ivanov A, Kameka A, Pajak A, Bruneau L, Beyaert R, Hernández-Sebastià C, Marsolais F (2012) *Arabidopsis* mutants lacking asparaginases develop normally but exhibit enhanced root inhibition by exogenous asparagine. Amino Acids 42:2307–2318. doi:10.1007/s00726-011-0973-4

Jaisson S, Veiga-da-Cunha M, Van Schaftingen E (2009) Molecular identification of ω-amidase, the enzyme that is functionally coupled with glutamine transaminases, as the putative tumor suppressor Nit2. Biochimie 91:1066–1071

Koch W, Kwart M, Laubner M, Heineke D, Stransky H, Frommer WB, Tegeder M (2003) Reduced amino acid content in transgenic potato tubers due to antisense inhibition of the leaf H^+/amino acid symporter StAAP1. Plant J 33:211–220

Kong D, Hu HC, Okuma E, Lee Y, Lee HS, Munemasa S, Cho D, Ju C, Pedoeim L, Rodriguez B, Wang J, Im W, Murata Y, Pei ZM, Kwak JM (2016) L-Met activates *Arabidopsis* GLR Ca^{2+} channels upstream of ROS production and regulates stomatal movement. Cell Rep 17:2553–2561. doi:10.1016/j.celrep.2016.11.015

Krall AS, Xu S, Graeber TG, Braas D, Christofk HR (2016) Asparagine promotes cancer cell proliferation through use as an amino acid exchange factor. Nat Commun 7:11457. doi:10.1038/ncomms11457

Krasnikov BF, Chien CH, Nostramo R, Pinto JT, Nieves E, Callaway M, Sun J, Huebner K, Cooper AJL (2009) Identification of the putative tumor suppressor Nit2 as ω-amidase, an enzyme metabolically linked to glutamine and asparagine transamination. Biochimie 91: 1072–1080

Krouk G (2016) Hormones and nitrate: a two-way connection. Plant Mol Biol 91:599–606. doi:10.1007/s11103-016-0463-x

Lam HM, Peng SSY, Coruzzi GM (1994) Metabolic regulation of the gene encoding glutamine-dependent asparagine synthetase in *Arabidopsis thaliana*. Plant Physiol 106:1347–1357

Lam HM, Coschigano K, Schultz C, Melo-Oliveira R, Tjaden G, Oliveira I, Ngai N, Hsieh MH, Coruzzi G (1995) Use of Arabidopsis mutants and genes to study amide amino acid biosynthesis. Plant Cell 7:887–898. doi:10.1105/tpc.7.7.887

Lam HM, Hsieh MH, Coruzzi G (1998) Reciprocal regulation of distinct asparagine synthetase genes by light and metabolites in *Arabidopsis thaliana*. Plant J 16:345–353

Lam HM, Wong P, Chan HK, Yam KM, Chen L, Chow CM, Coruzzi GM (2003) Overexpression of the *ASN1* gene enhances nitrogen status in seeds of Arabidopsis. Plant Physiol 132:926–935. doi:10.1104/pp.103.020123

Larsen TM, Boehlein SK, Schuster SM, Richards NGJ, Thoden JB, Holden HM, Rayment I (1999) Three-dimensional structure of *Escherichia coli* asparagine synthetase B: a short journey from substrate to product. Biochemistry 38:16146–16157. doi:10.1021/bi9915768

Lea PJ, Ireland RJ (1999) Nitrogen metabolism in higher plants. In: Singh BK (ed) Plant amino acids: biochemistry and biotechnology. Marcel Dekker, New York, pp 1–44

Lea PJ, Sodek L, Parry MAJ, Shewry PR, Halford NG (2007) Asparagine in plants. Ann Appl Biol 150:1–26. doi:10.1111/j.1744-7348.2006.00104.x

Li MW, Auyeung WK, Lam HM (2013) The GCN2 homologue in Arabidopsis thaliana interacts with uncharged tRNA and uses Arabidopsis eIF2α molecules as direct substrates. Plant Biol 15:13–18. doi:10.1111/j.1438-8677.2012.00606.x

Lima JD, Sodek L (2003) N-stress alters aspartate and asparagine levels of xylem sap in soybean. Plant Sci 165:649–656. doi:10.1016/S0168-9452(03)00251-6

Liu H, Gao Y, Zhang M, Qiu X, Cooper AJ, Niu L, Teng M (2013) Structures of enzyme-intermediate complexes of yeast Nit2: insights into its catalytic mechanism and different substrate specificity compared with mammalian Nit2. Acta Crystallogr D Biol Crystallogr 69: 1470–1481. doi:10.1107/s0907444913009347

Lloyd NDH, Joy KW (1978) 2-Hydroxysuccinamic acid: a product of asparagine metabolism in plants. Biochem Biophys Res Commun 81:186–192. doi:10.1016/0006-291X(78)91647-9

Lohaus G, Büker M, Hußmann M, Soave C, Heldt H-W (1998) Transport of amino acids with special emphasis on the synthesis and transport of asparagine in the Illinois low protein and Illinois high protein strains of maize. Planta 205:181–188. doi:10.1007/s004250050310

Lopes AM, Oliveira-Nascimento LD, Ribeiro A, Tairum CA Jr, Breyer CA, Oliveira MAD, Monteiro G, Souza-Motta CMD, Magalhães PDO, Avendaño JGF, Cavaco-Paulo AM, Mazzola PG, Rangel-Yagui CDO, Sette LD, Converti A, Pessoa A (2017) Therapeutic L-asparaginase: upstream, downstream and beyond. Crit Rev Biotechnol 37:82–99. doi:10.3109/07388551.2015.1120705

Luna E, Van Hulten M, Zhang Y, Berkowitz O, López A, Pétriacq P, Sellwood MA, Chen B, Burrell M, Van De Meene A, Pieterse CMJ, Flors V, Ton J (2014) Plant perception of β-aminobutyric acid is mediated by an aspartyl-tRNA synthetase. Nat Chem Biol 10: 450–456. doi:10.1038/nchembio.1520

Ma H, Wang S (2016) Histidine regulates seed oil deposition through abscisic acid biosynthesis and β-oxidation. Plant Physiol 172:848–857. doi:10.1104/pp.16.00950

Maaroufi-Dguimi H, Debouba M, Gaufichon L, Clément G, Gouia H, Hajjaji A, Suzuki A (2011) An Arabidopsis mutant disrupted in *ASN2* encoding asparagine synthetase 2 exhibits low salt stress tolerance. Plant Physiol Biochem 49:623–628

Machingura M, Salomon E, Jez JM, Ebbs SD (2016) The β-cyanoalanine synthase pathway: beyond cyanide detoxification. Plant Cell Environ 39:2329–2341. doi:10.1111/pce.12755

Massière F, Badet-Denisot MA (1998) The mechanism of glutamine-dependent amidotransferases. Cell Mol Life Sci 54:205–222

Meiss AN (1952) The formation of asparagine in etiolated seedlings of *Lupinus albus* L. Bulletin 553, The Connecticut Agricultural Experiment Station
Michalska K, Jaskolski M (2006) Structural aspects of L-asparaginases, their friends and relations. Acta Biochim Pol 53:627–640
Michalska K, Brzezinski K, Jaskolski M (2005) Crystal structure of isoaspartyl aminopeptidase in complex with L-aspartate. J Biol Chem 280:28484–28491
Michalska K, Bujacz G, Jaskolski M (2006) Crystal structure of plant asparaginase. J Mol Biol 360:105–116. doi:10.1016/j.jmb.2006.04.066
Modde K, Timm S, Florian A, Michl K, Fernie AR, Bauwe H (2016) High serine:glyoxylate aminotransferase activity lowers leaf daytime serine levels, inducing the phosphoserine pathway in Arabidopsis. J Exp Bot 68:643–656. doi:10.1093/jxb/erw467
Murneek AE (1935) Physiological rôle of asparagine and related substances in nitrogen metabolism of plants. Plant Physiol 10:447–464
Murray AJS, Blackwell RD, Joy KW, Lea PJ (1987) Photorespiratory N donors, aminotransferase specificity and photosynthesis in a mutant of barley deficient in serine:glyoxylate aminotransferase activity. Planta 172:106–113
Nomme J, Su Y, Konrad M, Lavie A (2012) Structures of apo and product-bound human L-asparaginase: insights into the mechanism of autoproteolysis and substrate hydrolysis. Biochemistry 51:6816–6826. doi:10.1021/bi300870g
O'Leary B, Preston GM, Sweetlove LJ (2014) Increased β-cyanoalanine nitrilase activity improves cyanide tolerance and assimilation in *Arabidopsis*. Mol Plant 7:231–243. doi:10.1093/mp/sst110
Ogé L, Bourdais G, Bove J, Collet B, Godin B, Granier F, Boutin JP, Job D, Jullien M, Grappin P (2008) Protein repair L-isoaspartyl methyltransferasel is involved in both seed longevity and germination vigor in arabidopsis. Plant Cell 20:3022–3037. doi:10.1105/tpc.108.058479
Ohashi M, Ishiyama K, Kojima S, Konishi N, Nakano K, Kanno K, Hayakawa T, Yamaya T (2015) Asparagine synthetase1, but not asparagine synthetase2, is responsible for the biosynthesis of asparagine following the supply of ammonium to rice roots. Plant Cell Physiol 56:769–778. doi:10.1093/pcp/pcv005
Olea F, Pérez-García A, Cantón FR, Rivera ME, Cañas R, Ávila C, Cazorla FM, Cánovas FM, de Vicente A (2004) Up-regulation and localization of asparagine synthetase in tomato leaves infected by the bacterial pathogen *Pseudomonas syringae*. Plant Cell Physiol 45:770–780. doi:10.1093/pcp/pch092
Osuna D, Gálvez-Valdivieso G, Piedras P, Pineda M, Aguilar M (2001) Cloning, characterization and mRNA expression analysis of *PVAS1*, a type I asparagine synthetase gene from *Phaseolus vulgaris*. Planta 213:402–410. doi:10.1007/s004250000513
Page MJ, Di Cera E (2006) Role of Na^+ and K^+ in enzyme function. Physiol Rev 86:1049–1092
Pandurangan S, Pajak A, Molnar SJ, Cober ER, Dhaubhadel S, Hernández-Sebastià C, Kaiser WM, Nelson RL, Huber SC, Marsolais F (2012) Relationship between asparagine metabolism and protein concentration in soybean seed. J Exp Bot 63:3173–3184. doi:10.1093/jxb/ers039
Pandurangan S, Pajak A, Rintoul T, Beyaert R, Hernandez-Sebastia C, Brown DC, Marsolais F (2015) Soybean seeds overexpressing asparaginase exhibit reduced nitrogen concentration. Physiol Plant 155:126–137. doi:10.1111/ppl.12341
Para A, Li Y, Marshall-Colorí A, Varala K, Francoeur NJ, Moran TM, Edwards MB, Hackley C, Bargmann BOR, Birnbaum KD, Richard McCombie W, Krouk G, Coruzzi GM (2014) Hit-and-run transcriptional control by bZIP1 mediates rapid nutrient signaling in Arabidopsis. Proc Natl Acad Sci U S A 111:10371–10376. doi:10.1073/pnas.1404657111
Pascual MB, El-Azaz J, de la Torre FN, Cañas RA, Avila C, Cánovas FM (2016) Biosynthesis and metabolic fate of phenylalanine in conifers. Front Plant Sci 7:1030. doi:10.3389/fpls.2016.01030
Pate JS, Atkins CA, Herridge DF, Layzell DB (1981) Synthesis, storage, and utilization of amino compounds in white lupin (*Lupinus albus* L.) Plant Physiol 67:37–42. doi:10.1104/pp.67.1.37

Peiser GD, Wang TT, Hoffman NE, Yang SF, Liu HW, Walsh CT (1984) Formation of cyanide from carbon 1 of 1-aminocyclopropane-1-carboxylic acid during its conversion to ethylene. Proc Natl Acad Sci U S A 81:3059–3063

Peracchi A, Veiga-da-Cunha M, Kuhara T, Ellens KW, Paczia N, Stroobant V, Seliga AK, Marlaire S, Jaisson S, Bommer GT, Sun J, Huebner K, Linster CL, Cooper AJL, Van Schaftingen E (2017) Nit1 is a metabolite repair enzyme that hydrolyzes deaminated glutathione. Proc Natl Acad Sci U S A 114:E3233–E3242. doi:10.1073/pnas.1613736114

Piotrowski M, Volmer JJ (2006) Cyanide metabolism in higher plants: cyanoalanine hydratase is a NIT4 homolog. Plant Mol Biol 61:111–122

Piotrowski M, Schönfelder S, Weiler EW (2001) The *Arabidopsis thaliana* isogene *NIT4* and its orthologs in tobacco encode β-cyano-L-alanine hydratase/nitrilase. J Biol Chem 276: 2616–2621

Piotrowski M, Janowitz T, Kneifel H (2003) Plant C-N hydrolases and the identification of a plant *N*-carbamoylputrescine amidohydrolase involved in polyamine biosynthesis. J Biol Chem 278: 1708–1712

Piutti A (1886) Ein neues Asparagin. Ber Dtsch Chem Ges 19:1691–1695

Postles J, Curtis TY, Powers SJ, Elmore JS, Mottram DS, Halford NG (2016) Changes in free amino acid concentration in rye grain in response to nitrogen and sulfur availability, and expression analysis of genes involved in asparagine metabolism. Front Plant Sci 7:917. doi:10.3389/fpls.2016.00917

Prianischnikov D (1922) Das ammoniak als anfangs- und endprodukt des stickstoffumsatzes in den pflanzen. Landwirtsch Vers-Stat 99:267–286

Romagni JG, Dayan FE (2000) Measuring asparagine synthetase activity in crude plant extracts. J Agric Food Chem 48:1692–1696

Rommens CM, Yan H, Swords K, Richael C, Ye J (2008) Low-acrylamide French fries and potato chips. Plant Biotechnol J 6:843–853. doi:10.1111/j.1467-7652.2008.00363.x

Rothstein SJ, Bi YM, Coneva V, Han M, Good A (2014) The challenges of commercializing second-generation transgenic crop traits necessitate the development of international public sector research infrastructure. J Exp Bot 65:5673–5682. doi:10.1093/jxb/eru236

Sakakibara H, Takei K, Hirose N (2006) Interactions between nitrogen and cytokinin in the regulation of metabolism and development. Trends Plant Sci 11:440–448. doi:10.1016/j.tplants.2006.07.004

Sanders A, Collier R, Trethewy A, Gould G, Sieker R, Tegeder M (2009) AAP1 regulates import of amino acids into developing Arabidopsis embryos. Plant J 59:540–552. doi:10.1111/j.1365-313X.2009.03890.x

Saxton RA, Chantranupong L, Knockenhauer KE, Schwartz TU, Sabatini DM (2016) Mechanism of arginine sensing by CASTOR1 upstream of mTORC1. Nature 536:229–233. doi:10.1038/nature19079

Schaefer J, Skokut TA, Stejskal EO, McKay RA, Varner JE (1981) Asparagine amide metabolism in developing cotyledons of soybean. Proc Natl Acad Sci U S A 78:5978–5982

Scheible WR, Morcuende R, Czechowski T, Fritz C, Osuna D, Palacios-Rojas N, Schindelasch D, Thimm O, Udvardi MK, Stitt M (2004) Genome-wide reprogramming of primary and secondary metabolism, protein synthesis, cellular growth processes, and the regulatory infrastructure of Arabidopsis in response to nitrogen. Plant Physiol 136:2483–2499

Schulze E, Winterstein E (1910) Handbuch der biochemischen Arbeitsmethoden, vol 2. Berlin Urban & Schwarzenberg, Berlin, p 510

Scott DB, Farnden KJF, Robertson JG (1976) Ammonia assimilation in lupin nodules. Nature 263: 703–705

Seebauer JR, Moose SP, Fabbri BJ, Crossland LD, Below FE (2004) Amino acid metabolism in maize earshoots. Implications for assimilate preconditioning and nitrogen signaling. Plant Physiol 136:4326–4334. doi:10.1104/pp.104.043778

Sieciechowicz KA, Ireland RJ, Joy KW (1985) Diurnal variation of asparaginase in developing pea leaves. Plant Physiol 77:506–508

Sieciechowicz K, Joy KW, Ireland RJ (1988a) Diurnal changes in asparaginase activity in pea leaves. I. The requirement for light for increased activity. J Exp Bot 39:695–706

Sieciechowicz KA, Ireland RJ, Joy KW (1988b) Diurnal changes in asparaginase activity in pea leaves. II. Regulation of activity. J Exp Bot 39:707–721

Sieciechowicz KA, Joy KW, Ireland RJ (1988c) The metabolism of asparagine in plants. Phytochemistry 27:663–671. doi:10.1016/0031-9422(88)84071-8

Skokut TA, Varner JE, Schaefer J, Stejskal EO, McKay RA (1982) [^{15}N]NMR Determination of asparagine and glutamine nitrogen utilization for synthesis of storage protein in developing cotyledons of soybean in culture. Plant Physiol 69:308–313

Sodek L, Lea PJ, Miflin BJ (1980) Distribution and properties of a potassium-dependent asparaginase isolated from developming seeds of *Pisum sativum* and other plants. Plant Physiol 65:22–26

Stephani RA, Meister A (1971) Structure of the dimeric α-keto acid analogue of asparagine. J Biol Chem 246:7115–7118

Streeter JG (1977) Asparaginase and asparagine transaminase in soybean leaves and root nodules. Plant Physiol 60:235–239. doi:10.1104/pp.60.2.235

Ta TC, Joy KW, Ireland RJ (1984a) Amino acid metabolism in pea leaves: utilization of nitrogen from amide and amino groups of [^{15}N]asparagine. Plant Physiol 74:822–826

Ta TC, Joy KW, Ireland RJ (1984b) Utilization of the amide groups of asparagine and 2-hydroxysuccinamic acid by young pea leaves. Plant Physiol 75:527–530

Ta TC, Joy KW, Ireland RJ (1985) Role of asparagine in the photorespiratory nitrogen metabolism of pea leaves. Plant Physiol 78:334–337

Tonin GS, Sodek L (1990) Asparaginase, allantoinase and glutamine synthetase activities in soybean cotyledons grown in vitro. Phytochemistry 29:2829–2831

Tran NL, Barraj LM, Collinge S (2016) Reduction in dietary acrylamide exposure-impact of potatoes with low acrylamide potential. Risk Anal. doi:10.1111/risa.12709

Tsai FY, Coruzzi GM (1990) Dark-induced and organ-specific expression of two asparagine synthetase genes in *Pisum sativum*. EMBO J 9:323–332

Vauquelin LN, Robiquet PJ (1806) La découverte d'un nouveau principe végétal dans le suc des asperges. Ann Chim 57:88–93

Vickery HB, Pucher GW, Deuber CG (1942) The preparation of asparagine. J Biol Chem 145: 45–53

Vorwerk S, Biernacki S, Hillebrand H, Janzik I, Muller A, Weiler EW, Piotrowski M (2001) Enzymatic characterization of the recombinant *Arabidopsis thaliana* nitrilase subfamily encoded by the *NIT*2/*NIT*1/*NIT*3-gene cluster. Planta 212:508–516

Wan TF, Shao GH, Shan XC, Zeng NY, Lam HM (2006) Correlation between *AS1* gene expression and seed protein contents in different soybean (*Glycine max* [L.] Merr.) cultivars. Plant Biol 8:271–276. doi:10.1055/s-2006-923876

Wang RC, Okamoto M, Xing XJ, Crawford NM (2003) Microarray analysis of the nitrate response in Arabidopsis roots and shoots reveals over 1,000 rapidly responding genes and new linkages to glucose, trehalose-6-phosphate, iron, and sulfate metabolism. Plant Physiol 132:556–567

Wang R, Tischner R, Gutiérrez RA, Hoffman M, Xing X, Chen M, Coruzzi G, Crawford NM (2004) Genomic analysis of the nitrate response using a nitrate reductase-null mutant of Arabidopsis. Plant Physiol 136:2512–2522. doi:10.1104/pp.104.044610

Wang L, Li H, Zhao C, Li S, Kong L, Wu W, Kong W, Liu Y, Wei Y, Zhu JK, Zhang H (2017) The inhibition of protein translation mediated by AtGCN1 is essential for cold tolerance in *Arabidopsis thaliana*. Plant Cell Environ 40:56–68. doi:10.1111/pce.12826

Webster GC, Varner JE (1955) Aspartate metabolism and asparagine synthesis in plant systems. J Biol Chem 215:91–99

Winichayakul S, Moyle RL, Ryan DJ, Farnden KJF, Davies KM, Coupe SA (2004) Distinct cis-elements in the Asparagus officinalis asparagine synthetase promoter respond to carbohydrate and senescence signals. Funct Plant Biol 31:573–582. doi:10.1071/FP03198

Wolfson RL, Chantranupong L, Saxton RA, Shen K, Scaria SM, Cantor JR, Sabatini DM (2016) Sestrin2 is a leucine sensor for the mTORC1 pathway. Science 351:43–48. doi:10.1126/science.aab2674

Wong HK, Chan HK, Coruzzi GM, Lam HM (2004) Correlation of *ASN2* gene expression with ammonium metabolism in Arabidopsis. Plant Physiol 134:332–338

Xu Q, Liu F, Chen P, Jez JM, Krishnan HB (2017) β-*N*-Oxalyl-L-α,β-diaminopropionic acid (β-ODAP) content in *Lathyrus sativus*: the integration of nitrogen and sulfur metabolism through β-cyanoalanine synthase. Int J Mol Sci 18:526. doi:10.3390/ijms18030526

Yabuki Y, Ohashi M, Imagawa F, Ishiyama K, Beier MP, Konishi N, Umetsu-Ohashi T, Hayakawa T, Yamaya T, Kojima S (2017) A temporal and spatial contribution of asparaginase to asparagine catabolism during development of rice grains. Rice 10:3. doi:10.1186/s12284-017-0143-8

Zhang Q, Marsolais F (2014) Identification and characterization of omega-amidase as an enzyme metabolically linked to asparagine transamination in Arabidopsis. Phytochemistry 99:36–43. doi:10.1016/j.phytochem.2013.12.020

Zhang L, Tan Q, Lee R, Trethewy A, Lee YH, Tegeder M (2010) Altered xylem-phloem transfer of amino acids affects metabolism and leads to increased seed yield and oil content in Arabidopsis. Plant Cell 22:3603–3620. doi:10.1105/tpc.110.073833

Zhang Q, Lee J, Pandurangan S, Clarke M, Pajak A, Marsolais F (2013) Characterization of Arabidopsis serine:glyoxylate aminotransferase, AGT1, as an asparagine aminotransferase. Phytochemistry 85:30–35. doi:10.1016/j.phytochem.2012.09.017

Zhu X, Gong H, He Q, Zeng Z, Busse JS, Jin W, Bethke PC, Jiang J (2016) Silencing of vacuolar invertase and asparagine synthetase genes and its impact on acrylamide formation of fried potato products. Plant Biotechnol J 14:709–718. doi:10.1111/pbi.12421

Redox Control of Autophagy in Photosynthetic Organisms

María Esther Pérez-Pérez, Inmaculada Couso, Mónica Domínguez-González, Stéphane D. Lemaire, and José L. Crespo

Abstract Autophagy is a major catabolic pathway by which eukaryotic cells deliver unnecessary or damaged cytoplasmic material to the vacuole for its degradation and recycling in order to maintain cellular homeostasis. Mounting evidence suggests the existence of a strong link between autophagy and the production of reactive oxygen species in several systems. Here, we review recent studies in algae and plants describing redox control of autophagy and discuss the conserved regulatory proteins that may transmit redox signals to the autophagic machinery.

Contents

Communicated by Francisco M. Cánovas

M.E. Pérez-Pérez, I. Couso, M. Domínguez-González, and J.L. Crespo (✉)
Instituto de Bioquímica Vegetal y Fotosíntesis, Consejo Superior de Investigaciones Científicas (CSIC)-Universidad de Sevilla, Avda. Américo Vespucio, 49, Sevilla 41092, Spain
e-mail: crespo@ibvf.csic.es

S.D. Lemaire
Laboratoire de Biologie Moléculaire et Cellulaire des Eucaryotes, Institut de Biologie Physico-Chimique, Sorbonne Universités, UPMC Univ Paris 06, CNRS, UMR8226, Paris 75005, France

Progress in Botany (2018) 79: 75–88, DOI 10.1007/124_2017_6,
Published online: 4 April 2017

1 Introduction

Autophagy (also called macroautophagy) is a widely conserved catabolic process by which eukaryotic cells degrade and recycle intracellular material. During autophagy, a membrane structure called the phagophore is formed which expands to sequester portions of the cytoplasm in a double membrane-bound compartment called the autophagosome. Activation of autophagy thus allows cells to recycle macromolecular constituents under adverse nutritional conditions and to eliminate damaged proteins/organelles or toxic components generated during stress, in order to maintain the cellular homeostasis or to cope with unfavorable conditions. Thus, autophagy functions as a pro-survival process in the cell rather than a pro-death pathway (He and Klionsky 2009; Mizushima et al. 2011; Li and Vierstra 2012; Liu and Bassham 2012). Autophagy was initially considered to be a nonselective degradation process but an increasing number of studies performed in yeasts, mammals, and plants demonstrated that certain organelles and protein aggregates are selectively cleared via autophagy. For instance, removal of damaged mitochondria via mitophagy, inactive proteasome via proteaphagy, or photo-damaged chloroplasts via chlorophagy has been reported among other selective forms of autophagy (Okamoto et al. 2009; Marshall et al. 2015; Izumi et al. 2017).

Autophagy is mediated by a set of proteins coded by *ATG* (autophagy-related) genes, which were originally identified through genetic screens for autophagy-defective mutants in yeasts (Tsukada and Ohsumi 1993). The high evolutionary conservation of many of these genes allowed the identification of *ATG* orthologs in the genome of other eukaryotes including plants and algae (Thompson and Vierstra 2005; Bassham et al. 2006; Diaz-Troya et al. 2008; Shemi et al. 2015). ATG proteins catalyze and regulate the formation of the autophagosome and its fusion to the vacuole. Based on their participation in the process of autophagosome formation, ATG proteins can be clustered in different categories that include: (1) the initiation complex, constituted by the ATG1 kinase and the regulatory proteins ATG11, ATG13, ATG17, ATG29, and ATG31, regulates the first steps in autophagosome formation; (2) the phosphatidylinositol (PtdIns) 3-kinase complex (VPS15, VPS30/Atg6, VPS34, and ATG14), which is involved in vesicle nucleation; (3) ATG9 and its recycling system (ATG2, ATG9, and ATG18), required for the expansion of the phagophore; and (4) the ATG12 (ATG5, ATG7, ATG10, ATG12, and ATG16) and ATG8 (ATG3, ATG4, ATG7, and ATG8) ubiquitin-like conjugation systems, which function in vesicle expansion, autophagosome formation, cargo recognition and autophagosome targeting to the vacuole (Mizushima et al. 2011).

Control of autophagy has been associated with the production of reactive oxygen species (ROS) in several systems. Upon starvation, human cells generate ROS, specifically H_2O_2, that seem to be essential for autophagosome biogenesis and autophagic degradation (Scherz-Shouval et al. 2007). In the model yeast *Saccharomyces cerevisiae*, autophagy has also been linked with the cellular redox state. Incubation of yeast cells with H_2O_2 triggers autophagy, whereas activation of autophagy by the TOR inhibitor rapamycin results in a pronounced increase of the total glutathione (reduced glutathione [GSH] + oxidized glutathione [GSSG]) pool, a marker of redox unbalance (Perez-Perez et al. 2014). Here, we review recent studies in plants and algae describing redox control of autophagy and discuss the conserved regulatory proteins that may transmit redox signals to the autophagic machinery.

2 Activation of Autophagy by Oxidative Stress in Plants and Algae

Excessive ROS may cause irreversible oxidative damage to proteins, lipids, and nucleic acids (Apel and Hirt 2004). At high levels, this over-oxidized material can be toxic and must be cleared away in order to maintain cell viability. The chloroplast is known to be one of the main sources of ROS in plants and algae. Superoxide anions ($O_2^{\bullet-}$) are generated as by-products of photosynthetic electron transport and readily converted into hydrogen peroxide (H_2O_2) inside the chloroplast through chemical and enzymatic reactions. Singlet oxygen ($^1O_2$) and hydroxyl radicals are also produced during photosynthesis and can cause oxidative damage. In addition to the chloroplast, ROS are produced in plant cells in mitochondria, peroxisomes, and at the plasma membrane by NADPH-oxidases (NOX) (Fig. 1). Recent studies in photosynthetic organisms described the activation of autophagy in response to several stimuli that increase ROS generation, regardless of the origin and location of ROS production in the cell (Xiong et al. 2007; Liu et al. 2009; Perez-Perez et al. 2010, 2012a, b).

Treatment of Arabidopsis plants with H_2O_2 results in severe oxidative stress and leads to the induction of autophagy (Xiong et al. 2007). Under these conditions, different types of irreversible oxidation of proteins occur, including carbonylation, sulfinic/sulfonic acid formation, or Tyr nitration. The finding that mutant plants defective in autophagy are hypersensitive to H_2O_2 and accumulate carbonylated proteins demonstrated that this degradative process is required for the cellular adaptation to oxidative stress. Similar to H_2O_2, treatment of plants with the ROS-producing agent methylviologen (MV) also triggers autophagy (Xiong et al. 2007) (Fig. 2). MV is able to intercept electrons from PSI in the chloroplast or from the respiratory electron transport chain in the plant mitochondria and then reacts with oxygen to produce $O_2^{\bullet-}$, which in turn is rapidly converted to H_2O_2. Therefore, MV generates ROS in the chloroplast and/or in the mitochondria in plants and may induce autophagy by oxidative damage caused in different organelles.

A similar effect of H_2O_2 and MV on autophagy activation has been shown in the model green alga *Chlamydomonas reinhardtii*. The ability of Chlamydomonas to grow in the dark using acetate as carbon source provided a unique advantage to dissect the origin of redox signals in response to oxidative stress for the control of autophagy. Treatment of Chlamydomonas cells with H_2O_2 triggers autophagy both

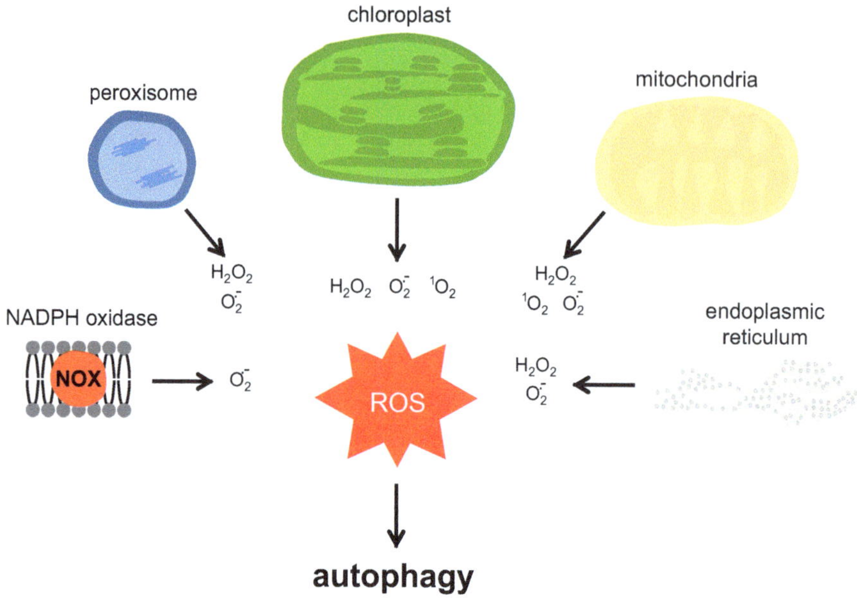

Fig. 1 ROS generated in different cellular compartments regulate autophagy in photosynthetic organisms. ROS is produced in different organelles such as peroxisome, mitochondria, chloroplast, and endoplasmic reticulum or in the plasma membrane as consequence of NOX activity. As a result of oxidative stress, autophagy is induced to remove damaged cellular components and to downregulate ROS production. Different forms of ROS produced in each compartment (H_2O_2, hydrogen peroxide; $^1O_2$, singlet oxygen; $O_2^{\bullet-}$, superoxide) are indicated

under light and dark conditions, whereas MV requires photosynthetic electron flow to produce ROS and to trigger this catabolic process (Perez-Perez et al. 2012b), suggesting that ROS generated in the chloroplast may signal to the activation of the cytoplasmic autophagic machinery. However, a lesser activation of autophagy has been detected in Chlamydomonas cells treated with MV in the dark, which might be due to the generation of ROS in the mitochondria (Perez-Perez et al. 2012b). In a recent study, the production of nitric oxide (NO) has been associated with the upregulation of some ATG proteins in response to salt stress in Chlamydomonas. Activation of autophagy in salt stressed cells might be due to oxidative damage caused by the inactivation of antioxidant enzymes and S-nitrosoglutathione reductase (Chen et al. 2016) although it is not clear how NO may signal to ATG proteins.

3 Activation of Autophagy by Photo-Oxidative Damage

The production of high levels of ROS in the chloroplast by different stresses has been connected to the activation of autophagy in algae. The analysis of Chlamydomonas carotenoid-deficient mutants revealed that the absence of

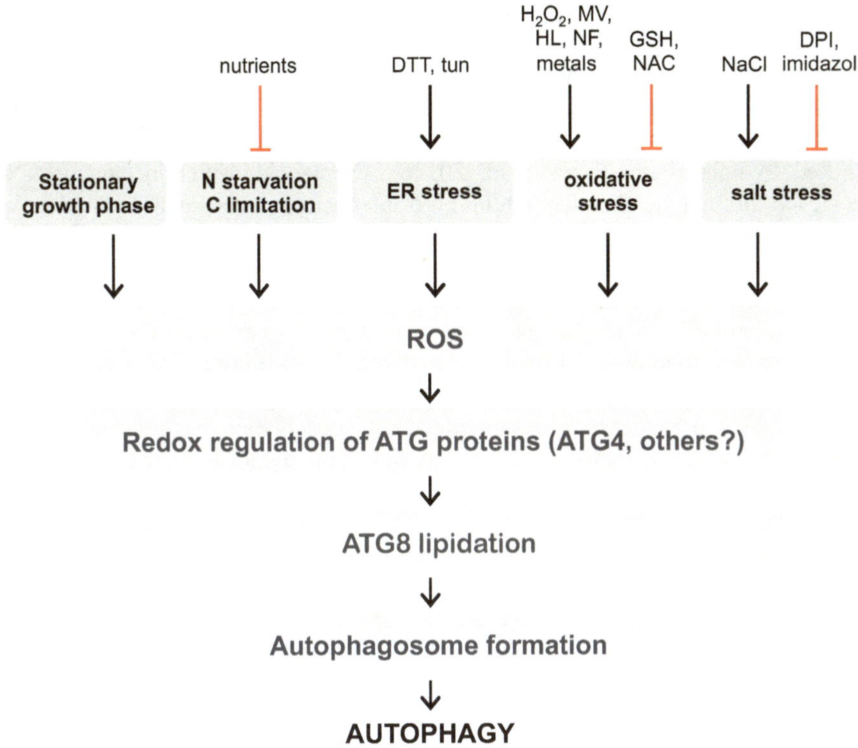

Fig. 2 Control of autophagy by stress signals in plants and algae. Stationary growth phase, nutrient (nitrogen or carbon) limitation, ER stress (caused by DTT or tunicamycin treatments), oxidative stress (triggered by hydrogen peroxide, methylviologen, norflurazon, metal toxicity, or high light), and salt stress result in ROS production. ROS regulate the activity of key ATG proteins such as the ATG4 protease, which modulates the lipidation of ATG8. Finally, the balance between modified and unmodified ATG8 regulates the formation of the autophagosome and thus autophagy progression. *DPI* diphenylene iodonium, *DTT* dithiothreitol, *GSH* glutathione, *HL* high light, *MV* methylviologen, *NAC* *N*-acetyl cysteine, *NF* norflurazon, *tun* tunicamycin

photoprotection leads to increased levels of ROS in the chloroplast and a pronounced increase in autophagic activity (Perez-Perez et al. 2012b). In plants and algae, carotenoids localize to the chloroplast, where they are synthesized, although extraplastidic carotenoids are frequently encountered among green algae under certain stress conditions (Grossman et al. 2004). Perhaps the most important function of carotenoids in photosynthetic organisms is to protect cells against excessive light and toxic ROS generated as by-products of photosynthesis. Carotenoids prevent photo-oxidative damage of the thylakoid membrane through quenching of excited intermediates like superoxide anion, singlet oxygen or triplet chlorophyll (^{3}Chl) molecules, and participate in energy dissipation when light absorption exceeds the photosynthetic capacity in a process known as nonphotochemical quenching (NPQ) (Li et al. 2009). The Chlamydomonas *lts1-204* mutant lacks the phytoene synthase gene, one of the first enzymes in the carotenoid biosynthetic pathway, and therefore cannot grow in the presence of light due to the total absence of protective carotenoids (Inwood et al. 2008). This mutant exhibits high levels of autophagy even in the dark, which are further induced when cells are shifted to light (Perez-Perez et al. 2012b). Carotenoid depletion can also be achieved through the chemical inhibition of phytoene synthase with the herbicide norflurazon (Sandmann and Albrecht 1990). It has been shown that norflurazon activates autophagy in Chlamydomonas cells (Fig. 2). Interestingly, norflurazon triggered autophagy in the light but not in the dark despite efficient carotenoid depletion under both conditions. This observation suggested that autophagy is associated with photo-oxidative damage caused by light-driven ROS production (Perez-Perez et al. 2012b). A role of light in the regulation of autophagy in photosynthetic organisms is also supported by the observation that high light stress, which generates ROS, activates autophagy in Chlamydomonas (Perez-Perez et al. 2012b). This is in agreement with the finding that the *npq1 lor1* Chlamydomonas mutant lacking specific photo-protective carotenoids (Niyogi et al. 1997) displays permanent high levels of autophagy when exposed to high light stress (Perez-Perez et al. 2012b).

A link between photo-oxidative damage of the chloroplast and autophagy has been recently proven in higher plants. In response to ultraviolet-B (UVB) damage or high light stress, entire photo-damaged chloroplasts from Arabidopsis leaves are transported to the vacuole for their degradation in a process termed chlorophagy (Izumi et al. 2017). This transport was not detected in autophagy-defective *atg* mutants, which exhibited UVB-sensitive phenotypes and accumulated collapsed chloroplasts. The production of ROS, likely superoxide, has been linked to the induction of chlorophagy in UVB damage since the use of ROS scavengers suppressed this degradative process (Izumi et al. 2017). How UVB-damaged chloroplasts are recognized by the autophagic machinery and the molecular mechanisms by which specific ROS regulate chlorophagy have not been established yet.

4 NADPH-Oxidases in the Control of Autophagy

ROS-generating enzymes such as NADPH-oxidases play a central role in the redox network in plants by integrating ROS with other signaling pathways (Suzuki et al. 2011). NADPH-oxidases, also known as respiratory burst oxidases (RBOs), are localized at the plasma membrane and catalyze the production of superoxide. These enzymes mediate many processes such as the hypersensitive response by pathogen infection or systemic signaling in response to biotic and abiotic stresses (Suzuki et al. 2011). NADPH-oxidases have been involved in autophagy regulation in plants. Pharmacological inhibition of NOX with the chemical inhibitor diphenylene iodonium (DPI) revealed that induction of autophagy by nutrient (nitrogen or carbon) limitation or salt stress requires NOX activity (Liu et al. 2009). This study revealed that ROS generated by NOX are required for the activation of autophagy in response to nutrient and salt stress, emphasizing a possible role of ROS in the control of autophagy in plants. However, it has also been reported that osmotic stress, which also triggers autophagy in plants, is not mediated by NOX (Liu et al. 2009). Therefore, ROS-dependent and ROS-independent pathways regulate autophagy in response to starvation, salt, and osmotic stresses.

NADPH-oxidases may also participate in the control of autophagy in response to pathogen infection in plants. Upon infection, ROS are generated at the plasma membrane by NOX, although other organelles such as mitochondria, chloroplasts, and/or peroxisomes maintain enhanced ROS production (Torres et al. 2005, 2006). Given the prominent role of ROS in the control of autophagy (Perez-Perez et al. 2012a), NOX might participate in the activation of this catabolic process during the initial stage of pathogen infection or at a subsequent phase associated with cell death programs. In this sense, it has been shown in Arabidopsis that NOX may play an important role in coupling intracellular ROS generated by these organelles to downstream changes in redox state and cell death associated with salicylic acid signaling (Chaouch et al. 2012). Further work is required to demonstrate the participation of NOX-generated ROS in the control of pathogen-induced autophagy.

5 ROS Participates in the Control of ER Stress-Induced Autophagy

Endoplasmic reticulum (ER) stress is activated by the accumulation of misfolded proteins in the ER under adverse environmental conditions. Under these conditions, the demand for protein folding exceeds the capacity of the system, which upregulates the expression of chaperones and other proteins required for the reestablishment of cell homeostasis, a signaling process known as the unfolded protein response (UPR) (Walter and Ron 2011). ER stress is also a strong inducer of autophagy, and this signal is conserved from yeast to mammals (Yorimitsu et al. 2006). More recently, induction of autophagy by ER stress has also been reported in algae and plants (Perez-Perez and Crespo 2010; Liu et al. 2012) (Fig. 2). How ER stress regulates autophagy in these organisms is not clear yet although it has been shown that IRE1b, a key protein in UPR signaling, is required for the formation of autophagosomes in response to ER stress (Liu et al. 2012). An association between ER stress and ROS production has been established in different organisms including plants and algae. The ER provides, through ER-resident oxidoreductases like ERO1, an oxidative environment that facilitates the oxidation of cysteines and thereby, the formation of disulfide bonds (Tu and Weissman 2004). Studies in plants and algae indicated that ER stress increased ERO1 expression and triggered ROS signaling (Ozgur et al. 2014; Perez-Martin et al. 2014). Given the role of ROS in the control of autophagy, it has been proposed that oxidative stress may contribute to the activation of this catabolic process by ER stress. Accordingly, the antioxidant glutathione partially suppressed ER stress-induced autophagy in Chlamydomonas (Perez-Martin et al. 2014) although additional studies will be required to unravel how ROS control autophagy in response to ER stress.

6 Metal Toxicity, Oxidative Stress, and Autophagy

Exposure to high concentrations of metals is toxic for most organisms. Metal toxicity has been associated with ROS production and oxidative stress signaling in plants and algae (Sharma and Dietz 2009; Glaesener et al. 2013; Jamers et al. 2013; Stoiber et al. 2013) although the connection to autophagy is largely unexplored in these organisms. Chlamydomonas has been widely used to investigate metal metabolism and the cellular response to metal excess and metal-limiting conditions in photosynthetic systems, and the development of genome-wide technologies has increased our current understanding about metal signaling in this alga (Blaby-Haas and Merchant 2012, 2013; Glaesener et al. 2013). A recent study has shown that high concentrations of nickel, cobalt, or copper trigger autophagy in Chlamydomonas (Perez-Martin et al. 2015). Transcriptome sequencing analysis of nickel-stressed cells revealed an increase in the abundance of the protein degradation machinery, including that responsible for autophagy, and a substantial overlap with the H_2O_2 response in cells treated with nickel ions. It has been proposed that excess nickel may cause oxidative damage, which in turn activates degradative pathways, including autophagy, to clear impaired components and recover cellular homeostasis (Perez-Martin et al. 2015). However, the molecular mechanisms by which metals trigger autophagy are currently unknown.

7 Redox Imbalance in Autophagy-Defective Mutants

A common feature of plant autophagy mutants is the early senescence phenotype and the hypersensitivity to starvation and oxidative stress conditions [for a review, see (Liu and Bassham 2012)]. Plant senescence is often associated with pronounced accumulation of ROS (Overmyer et al. 2003). Accordingly, it has been shown that the main reason why autophagy mutants senesce earlier than the wild type is the overaccumulation of ROS in these mutants (Yoshimoto et al. 2009). A recent transcriptomic and metabolomic analysis of rosette leaves of Arabidopsis autophagy mutants indicated that several metabolic pathways including the synthesis of glutathione are altered (Masclaux-Daubresse et al. 2014). This study revealed that autophagy-defective plants exhibited higher levels of both total glutathione and oxidized glutathione, which may reflect a redox imbalance in the absence of functional autophagy. Therefore, there seems to be a dual relation between autophagy and the redox status of the cell since multiple evidence indicate that ROS controls autophagy but at the same time this degradative process is necessary to maintain the redox balance of the cell in photosynthetic organisms.

Autophagy is also regulated by ROS-independent pathways in plants. In *Arabidopsis thaliana*, it has been demonstrated that mutant plants lacking L-cysteine desulfhydratase (*DES1*), which is involved in the degradation of cysteine and the production of hydrogen sulfide in the cytosol, exhibit increased levels of modified and unmodified ATG8 proteins. Treatment of *des1* mutant plants with hydrogen sulfide reversed ATG8 accumulation and lipidation, suggesting that cysteine-generated sulfide in the cytosol negatively regulates autophagy in Arabidopsis (Alvarez et al. 2012). Supporting this hypothesis, it has been reported that hydrogen sulfide negatively regulates autophagy induced in Arabidopsis roots under nitrogen deprivation (Laureano-Marin et al. 2016). Interestingly, the inhibitory effect of hydrogen sulfide on autophagy is unrelated with its antioxidant properties and seems to be via a redox independent pathway (Laureano-Marin et al. 2016). The molecular mechanism by which hydrogen sulfide regulates autophagy needs to be resolved.

8 Peroxisomes in the Redox Control of Autophagy

Autophagy is a primary defense mechanism to degrade oxidized molecules, and it also functions to remove ROS-generating organelles. Several studies have recently shown that peroxisomes are selectively degraded in plants via pexophagy, a specialized form of autophagy (Farmer et al. 2013; Kim et al. 2013; Shibata et al. 2013). Peroxisomes are small organelles present in most eukaryotes and essential for life in plants. These organelles harbor important metabolic activities such as fatty acid β-oxidation, hormone production, and photorespiration (Reumann and Bartel 2016). The peroxisome includes oxidative reactions that generate ROS and is predicted to be a major site of H_2O_2 production in plant cells. In addition, peroxisomes possess systems for scavenging H_2O_2 like ascorbate peroxidase and catalase, the enzyme responsible for the degradation of H_2O_2. Peroxisome-generated ROS has been shown to trigger pexophagy in Arabidopsis (Shibata et al. 2013). Autophagy mutant plants accumulate aggregated peroxisomes that contained high levels of inactive catalase. In addition, it has been shown that peroxisome aggregation can be induced in wild-type plants by exogenous application of H_2O_2, which results in peroxisome degradation (Shibata et al. 2013). Thus, pexophagy maintains peroxisome quality by degradation of highly oxidized, damaged peroxisomes. The demonstration of pexophagy in plants further strengthened the role of ROS in the control of autophagy but at the same time prompts additional questions like what is the signal for peroxisome degradation via autophagy or what is the pexophagy receptor in plants.

9 Molecular Mechanisms Underlying the Redox Control of Autophagy

Despite the extensive experimental evidence connecting autophagy and redox signaling, the precise molecular mechanisms by which ROS control this degradative process are largely unknown. In this regard, some progress has been recently done in this direction in plants and algae. In *Nicotiana benthamiana* plants, the cytosolic glyceraldehyde-3-phosphate dehydrogenases (GAPCs) binds and negatively regulates ATG3 (Han et al. 2015). This ATG protein is essential for autophagy and catalyzes the conjugation of ATG8 to phosphatidylethanolamine (PE), a process known as lipidation that is required for the formation of the autophagosome (Ichimura et al. 2000) (Fig. 3). Silencing of GAPCs in *N. benthamiana* plants resulted in autophagy activation whereas overexpression of GAPCs suppressed autophagy. Interestingly, treatment of tobacco plants with MV, which triggers oxidative stress and autophagy, inhibited the interaction of GAPCs with ATG3 (Han et al. 2015). This observation suggests that ROS may control autophagy through the regulation of GAPCs binding to ATG3 although how GAPCs modulate ATG3 activity is currently unknown.

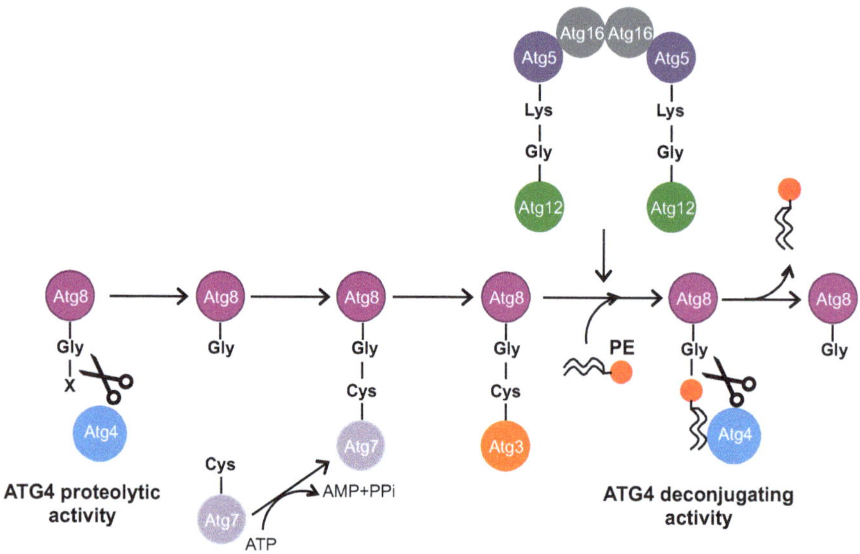

Fig. 3 Similar to protein ubiquitination, the C-terminal carboxyl group of the ubiquitin-like protein ATG8 is activated by the E1 enzyme ATG7, forming a thioester intermediate with its catalytic cysteine residue. Then, ATG8 is transferred to the catalytic cysteine residue of the E2 enzyme ATG3, to be finally conjugated to phosphatidylethanolamine (PE) via an amide bond. ATG4 mediates both C-terminal processing and deconjugation of ATG8 (proteolytic and deconjugating activities, respectively). The ATG12–ATG5-ATG16 conjugate enhances the E2 activity of ATG3 towards ATG8

A molecular mechanism linking ROS to the ATG4 protein has been identified in Chlamydomonas (Perez-Perez et al. 2016). ATG4 is a cysteine protease with an essential function in autophagosome biogenesis. ATG4 activity has a dual function on the lipidation process of ATG8 (Mizushima et al. 2011). On the one hand, ATG4 processes the C terminus of nascent ATG8 to expose the conserved glycine residue. This ATG4 activity is known as proteolytic activity and results in ATG8 lipidation. On the other hand, ATG4 can also cleave the amide bond between ATG8 and PE. This ATG4 activity is called deconjugating or delipidating activity and leads to ATG8 delipidation, releasing ATG8 from autophagosome membranes. This activity must be finely regulated otherwise autophagosome formation could not be completed if ATG8 is being constitutively delipidated. Thus, ATG4 plays an important role in the control of autophagy by allowing ATG8 recycling (Fig. 3).

At present, ATG4 is the only ATG protein that has been demonstrated to be redox regulated. The proteolytic activity of human ATG4A and ATG4B is reversibly inhibited by oxidation of Cys81, which might interfere with the activity of the nearby catalytic Cys77 (Scherz-Shouval et al. 2007). However, Cys81 is only conserved in tetrapods. The activity of ATG4 protease from *Saccharomyces cerevisiae* is also subject to redox regulation, although the underlying molecular mechanism differs from the one described in humans. It has been reported that yeast ATG4 is regulated by oxidoreduction of a single disulfide bond between Cys338 and Cys394 (Perez-Perez et al. 2014). This disulfide bond has a low redox potential (E_h: −289 mV at pH 7.0) and is very efficiently reduced by thioredoxin (Trx), suggesting that this oxidoreductase plays an important role in ATG4 regulation and therefore in autophagy (Perez-Perez et al. 2014). In higher plants, it has been shown that the activity of ATG4a and ATG4b from *Arabidopsis thaliana* is reversibly inhibited by oxidation with H_2O_2 (Woo et al. 2014), although the underlying molecular mechanism has not been described. In contrast, the molecular mechanism by which ATG4 is regulated has been recently reported in Chlamydomonas. Similarly to yeast ATG4, Chlamydomonas ATG4 is regulated by the formation of a single disulfide bond with a low redox potential (E_h: −278 mV at pH 7.0) that can be efficiently reduced by the NADPH/thioredoxin system (Perez-Perez et al. 2016). One of the cysteines forming this disulfide bond is Cys400 (homologous to yeast Cys338), and is conserved in other algae, yeasts, plants, and mammals. Both yeast and Chlamydomonas ATG4 proteins showed a second regulation level in vitro since the proteins formed oligomers under oxidizing conditions (Perez-Perez et al. 2016). Interestingly, regulation of ATG4 by oxidation and aggregation has been observed in vivo in Chlamydomonas cells subjected to stress conditions that generate ROS and activate autophagy. Depending on the redox potential, ATG4 can be found in Chlamydomonas in at least three conformational states ranging from a monomeric and active state to an oligomeric and inactive form (Perez-Perez et al. 2016). Oxidation and aggregation of ATG4 can be reversed by reducing agents, which maintain the equilibrium for the proper redox regulation of ATG4 activity. The current model for the redox control of autophagy in Chlamydomonas suggests that ATG4 is inactivated under stress conditions to prevent delipidation of

ATG8 at the site of autophagosome formation. Whether ATG4 proteins from other organisms have the same redox regulation needs to be further investigated.

Acknowledgments This work was supported in part by Ministerio de Economía y Competitividad grants BFU2015-68216-P and BIO2015-74432-JIN to JLC and MEPP, respectively.

References

Alvarez C, Garcia I, Moreno I, Perez-Perez ME, Crespo JL, Romero LC, Gotor C (2012) Cysteine-generated sulfide in the cytosol negatively regulates autophagy and modulates the transcriptional profile in Arabidopsis. Plant Cell 24:4621–4634

Apel K, Hirt H (2004) Reactive oxygen species: metabolism, oxidative stress, and signal transduction. Annu Rev Plant Biol 55:373–399

Bassham DC, Laporte M, Marty F, Moriyasu Y, Ohsumi Y, Olsen LJ, Yoshimoto K (2006) Autophagy in development and stress responses of plants. Autophagy 2:2–11

Blaby-Haas CE, Merchant SS (2012) The ins and outs of algal metal transport. Biochim Biophys Acta 1823:1531–1552

Blaby-Haas CE, Merchant SS (2013) Iron sparing and recycling in a compartmentalized cell. Curr Opin Microbiol 16:677–685

Chaouch S, Queval G, Noctor G (2012) AtRbohF is a crucial modulator of defence-associated metabolism and a key actor in the interplay between intracellular oxidative stress and pathogenesis responses in Arabidopsis. Plant J 69:613–627

Chen X, Tian D, Kong X, Chen Q, E F AA, Hu X, Jia A (2016) The role of nitric oxide signalling in response to salt stress in *Chlamydomonas reinhardtii*. Planta 244:651–669

Diaz-Troya S, Perez-Perez ME, Florencio FJ, Crespo JL (2008) The role of TOR in autophagy regulation from yeast to plants and mammals. Autophagy 4:851–865

Farmer LM, Rinaldi MA, Young PG, Danan CH, Burkhart SE, Bartel B (2013) Disrupting autophagy restores peroxisome function to an Arabidopsis lon2 mutant and reveals a role for the LON2 protease in peroxisomal matrix protein degradation. Plant Cell 25:4085–4100

Glaesener AG, Merchant SS, Blaby-Haas CE (2013) Iron economy in *Chlamydomonas reinhardtii*. Front Plant Sci 4:337

Grossman AR, Lohr M, Im CS (2004) *Chlamydomonas reinhardtii* in the landscape of pigments. Annu Rev Genet 38:119–173

Han S, Wang Y, Zheng X, Jia Q, Zhao J, Bai F, Hong Y, Liu Y (2015) Cytoplastic glyceraldehyde-3-phosphate dehydrogenases interact with ATG3 to negatively regulate autophagy and immunity in Nicotiana benthamiana. Plant Cell 27:1316–1331

He C, Klionsky DJ (2009) Regulation mechanisms and signaling pathways of autophagy. Annu Rev Genet 43:67–93

Ichimura Y, Kirisako T, Takao T, Satomi Y, Shimonishi Y, Ishihara N, Mizushima N, Tanida I, Kominami E, Ohsumi M, Noda T, Ohsumi Y (2000) A ubiquitin-like system mediates protein lipidation. Nature 408:488–492

Inwood W, Yoshihara C, Zalpuri R, Kim KS, Kustu S (2008) The ultrastructure of a *Chlamydomonas reinhardtii* mutant strain lacking phytoene synthase resembles that of a colorless alga. Mol Plant 1:925–937

Izumi M, Ishida H, Nakamura S, Hidema J (2017) Entire photodamaged chloroplasts are transported to the central vacuole by autophagy. Plant Cell. pii: tpc.00637.2016. doi: 10.1105/tpc.16.00637

Jamers A, Blust R, De Coen W, Griffin JL, Jones OA (2013) An omics based assessment of cadmium toxicity in the green alga *Chlamydomonas reinhardtii*. Aquat Toxicol 126:355–364

Kim J, Lee H, Lee HN, Kim SH, Shin KD, Chung T (2013) Autophagy-related proteins are required for degradation of peroxisomes in Arabidopsis hypocotyls during seedling growth. Plant Cell 25:4956–4966

Laureano-Marin AM, Moreno I, Romero LC, Gotor C (2016) Negative regulation of autophagy by sulfide is independent of reactive oxygen species. Plant Physiol 171:1378–1391

Li F, Vierstra RD (2012) Autophagy: a multifaceted intracellular system for bulk and selective recycling. Trends Plant Sci 17:526–537

Li Z, Wakao S, Fischer BB, Niyogi KK (2009) Sensing and responding to excess light. Annu Rev Plant Biol 60:239–260

Liu Y, Bassham DC (2012) Autophagy: pathways for self-eating in plant cells. Annu Rev Plant Biol 63:215–237

Liu Y, Xiong Y, Bassham DC (2009) Autophagy is required for tolerance of drought and salt stress in plants. Autophagy 5:954–963

Liu Y, Burgos JS, Deng Y, Srivastava R, Howell SH, Bassham DC (2012) Degradation of the endoplasmic reticulum by autophagy during endoplasmic reticulum stress in Arabidopsis. Plant Cell 24:4635–4651

Marshall RS, Li F, Gemperline DC, Book AJ, Vierstra RD (2015) Autophagic degradation of the 26S proteasome is mediated by the dual ATG8/ubiquitin receptor RPN10 in Arabidopsis. Mol Cell 58:1053–1066

Masclaux-Daubresse C, Clement G, Anne P, Routaboul JM, Guiboileau A, Soulay F, Shirasu K, Yoshimoto K (2014) Stitching together the multiple dimensions of autophagy using metabolomics and transcriptomics reveals impacts on metabolism, development, and plant responses to the environment in Arabidopsis. Plant Cell 26:1857–1877

Mizushima N, Yoshimori T, Ohsumi Y (2011) The role of Atg proteins in autophagosome formation. Annu Rev Cell Dev Biol 27:107–132

Niyogi KK, Bjorkman O, Grossman AR (1997) The roles of specific xanthophylls in photoprotection. Proc Natl Acad Sci U S A 94:14162–14167

Okamoto K, Kondo-Okamoto N, Ohsumi Y (2009) Mitochondria-anchored receptor Atg32 mediates degradation of mitochondria via selective autophagy. Dev Cell 17:87–97

Overmyer K, Brosche M, Kangasjarvi J (2003) Reactive oxygen species and hormonal control of cell death. Trends Plant Sci 8:335–342

Ozgur R, Turkan I, Uzilday B, Sekmen AH (2014) Endoplasmic reticulum stress triggers ROS signalling, changes the redox state, and regulates the antioxidant defence of *Arabidopsis thaliana*. J Exp Bot 65:1377–1390

Perez-Martin M, Perez-Perez ME, Lemaire SD, Crespo JL (2014) Oxidative stress contributes to autophagy induction in response to endoplasmic reticulum stress in *Chlamydomonas reinhardtii*. Plant Physiol 166:997–1008

Perez-Martin M, Blaby-Haas CE, Perez-Perez ME, Andres-Garrido A, Blaby IK, Merchant SS, Crespo JL (2015) Activation of autophagy by metals in *Chlamydomonas reinhardtii*. Eukaryot Cell 14:964–973

Perez-Perez ME, Crespo JL (2010) Autophagy in the model alga *Chlamydomonas reinhardtii*. Autophagy 6:562–563

Perez-Perez ME, Florencio FJ, Crespo JL (2010) Inhibition of target of rapamycin signaling and stress activate autophagy in *Chlamydomonas reinhardtii*. Plant Physiol 152:1874–1888

Perez-Perez ME, Lemaire SD, Crespo JL (2012a) Reactive oxygen species and autophagy in plants and algae. Plant Physiol 160:156–164

Perez-Perez ME, Couso I, Crespo JL (2012b) Carotenoid deficiency triggers autophagy in the model green alga *Chlamydomonas reinhardtii*. Autophagy 8:376–388

Perez-Perez ME, Zaffagnini M, Marchand CH, Crespo JL, Lemaire SD (2014) The yeast autophagy protease Atg4 is regulated by thioredoxin. Autophagy 10:1953–1964

Perez-Perez ME, Lemaire SD, Crespo JL (2016) Control of autophagy in Chlamydomonas is mediated through redox-dependent inactivation of the ATG4 protease. Plant Physiol 172:2219–2234

Reumann S, Bartel B (2016) Plant peroxisomes: recent discoveries in functional complexity, organelle homeostasis, and morphological dynamics. Curr Opin Plant Biol 34:17–26

Sandmann G, Albrecht M (1990) Accumulation of colorless carotenes and derivatives during interaction of bleaching herbicides with phytoene desaturation. Z Naturforsch C 45:487–491

Scherz-Shouval R, Shvets E, Fass E, Shorer H, Gil L, Elazar Z (2007) Reactive oxygen species are essential for autophagy and specifically regulate the activity of Atg4. EMBO J 26:1749–1760

Sharma SS, Dietz KJ (2009) The relationship between metal toxicity and cellular redox imbalance. Trends Plant Sci 14:43–50

Shemi A, Ben-Dor S, Vardi A (2015) Elucidating the composition and conservation of the autophagy pathway in photosynthetic eukaryotes. Autophagy 11:701–715

Shibata M, Oikawa K, Yoshimoto K, Kondo M, Mano S, Yamada K, Hayashi M, Sakamoto W, Ohsumi Y, Nishimura M (2013) Highly oxidized peroxisomes are selectively degraded via autophagy in Arabidopsis. Plant Cell 25:4967–4983

Stoiber TL, Shafer MM, Armstrong DE (2013) Induction of reactive oxygen species in *chlamydomonas reinhardtii* in response to contrasting trace metal exposures. Environ Toxicol 28:516–523

Suzuki N, Miller G, Morales J, Shulaev V, Torres MA, Mittler R (2011) Respiratory burst oxidases: the engines of ROS signaling. Curr Opin Plant Biol 14:691–699

Thompson AR, Vierstra RD (2005) Autophagic recycling: lessons from yeast help define the process in plants. Curr Opin Plant Biol 8:165–173

Torres MA, Jones JD, Dangl JL (2005) Pathogen-induced, NADPH oxidase-derived reactive oxygen intermediates suppress spread of cell death in *Arabidopsis thaliana*. Nat Genet 37:1130–1134

Torres MA, Jones JD, Dangl JL (2006) Reactive oxygen species signaling in response to pathogens. Plant Physiol 141:373–378

Tsukada M, Ohsumi Y (1993) Isolation and characterization of autophagy-defective mutants of *Saccharomyces cerevisiae*. FEBS Lett 333:169–174

Tu BP, Weissman JS (2004) Oxidative protein folding in eukaryotes: mechanisms and consequences. J Cell Biol 164:341–346

Walter P, Ron D (2011) The unfolded protein response: from stress pathway to homeostatic regulation. Science 334:1081–1086

Woo J, Park E, Dinesh-Kumar SP (2014) Differential processing of Arabidopsis ubiquitin-like Atg8 autophagy proteins by Atg4 cysteine proteases. Proc Natl Acad Sci U S A 111:863–868

Xiong Y, Contento AL, Nguyen PQ, Bassham DC (2007) Degradation of oxidized proteins by autophagy during oxidative stress in Arabidopsis. Plant Physiol 143:291–299

Yorimitsu T, Nair U, Yang Z, Klionsky DJ (2006) Endoplasmic reticulum stress triggers autophagy. J Biol Chem 281:30299–30304

Yoshimoto K, Jikumaru Y, Kamiya Y, Kusano M, Consonni C, Panstruga R, Ohsumi Y, Shirasu K (2009) Autophagy negatively regulates cell death by controlling NPR1-dependent salicylic acid signaling during senescence and the innate immune response in Arabidopsis. Plant Cell 21:2914–2927

Senescence-Associated Genes in Response to Abiotic/Biotic Stresses

Blanca Velasco-Arroyo, Mercedes Diaz-Mendoza, Maria Estrella Santamaria, Pablo Gonzalez-Melendi, Andrea Gomez-Sanchez, Ana Arnaiz, Manuel Martinez, and Isabel Diaz

Abstract Plant senescence is a complex physiological process consequence either of the natural lifespan or externally induced by abiotic and biotic factors. It comprises a coordinated sequence of molecular and biochemical events, phenotypically illustrated by changes in plant colour. Senescence is associated with alterations in chlorophyll and pigment content, reduction of photosynthesis, hydrolysis of macromolecules to produce more simple compounds and dismantling of cell organelles, to finally produce cell death. At the end, relocation of nutrients from the senescent tissues towards sink organs or growing tissues takes place to complete a recycling process. Consequently, the major part of the nitrogen is released as ammonium after being re-assimilated into amino acids to be exported via the phloem to the developing grains, fruits and tubers. During senescence, the reprograming of thousands of genes is triggered in response to specific senescence-promoting factors under a restricted regulatory control. The actual high-throughput omics technologies have led to the generation of integrative information, which has been used to understand the physiological changes during the onset and progression of senescence. This chapter covers an overview on plant senescence, particularly focussed on the senescence of the leaf, including the most recent findings about features, signalling, regulation and pathways involved in this natural or induced process.

B. Velasco-Arroyo, M. Diaz-Mendoza, M.E. Santamaria, P. Gonzalez-Melendi, A. Gomez-Sanchez, A. Arnaiz, M. Martinez, and I. Diaz (✉)
Centro de Biotecnologia y Genomica de Plantas, Universidad Politecnica de Madrid (UPM), Instituto Nacional de Investigacion y Tecnologia Agraria y Alimentaria (INIA), Campus Montegancedo, 28223 Pozuelo de Alarcon, Madrid, Spain
e-mail: blanca.velasco@upm.es; mercedes.diaz.mendoza@upm.es; me.santamaria@upm.es; pablo.melendi@upm.es; andrea.gomez@upm.es; ana.arnaiz.alonso@alumnos.upm.es; m.martinez@upm.es; i.diaz@upm.es

Progress in Botany (2018) 79: 89–110, DOI 10.1007/124_2017_1,
Published online: 1 March 2017

Contents

1 Introduction

Senescence is a natural process, which occurs in plants when the maturity phase is coming to its end leading either to the death or to the completion of a life cycle. Senescence-like processes occur in angiosperm and non-angiosperm land plants. This developmental phase is characterized by the striking changes in leaf colour observed during the autumn in trees and other perennial species in temperate regions. Similarly, in annual crops such as cereals, the green colour changes to golden as the grain ripens (Andersson et al. 2004; Lim et al. 2007; Munne-Bosch 2008). Phenotypical changes are the consequence of a set of coordinated sequential events including chlorophyll loss, reduction of photosynthesis, degradation of macromolecules, relocation of nutrients, dismantling of chloroplasts and, finally, cell death (Lim et al. 2007; Krupinska et al. 2012; Diaz-Mendoza et al. 2014; Schippers 2015). From a general perspective, this complex physiological process follows the simple concept of the three R's theory of the environment, reduce, reuse and recycle, with an extra R, remobilization. After *Reducing* the photosynthetic rate in response to the activation of a senescence program, a massive *Recycling* of nutrients begins that will be *Reutilized* as scaffolds for new macromolecules biosynthesis to insure the next generation survival. This implies an important *Remobilization* of nutrients through the phloem, from the source plant parts, such as the senescent leaves, towards sink organs such as emergent leaves, grains, tubers or fruits (Fig. 1). The strictly controlled program is an integral part of the flowering plant life cycle and is determined by endogenous developmental signals governed by the reproductive age (Podzimska-Sroka et al. 2015). In many monocarpic plants, the developing reproductive structures often govern the timing and onset of leaf senescence, thereby affecting all organs of a given plant (Munne-Bosch 2008). However, these intrinsic cues are continuously modulated by external abiotic environmental factors (drought or flooding, high irradiance or darkness, extreme temperatures, salinity, wounding or accumulation of pollutants) and biotic stresses (pathogens and pests) which modify, to some extent, the natural senescence programs of the plant. Both, endogenous and environmental signals perceived by a plant are integrated into the natural senescence program and subsequently

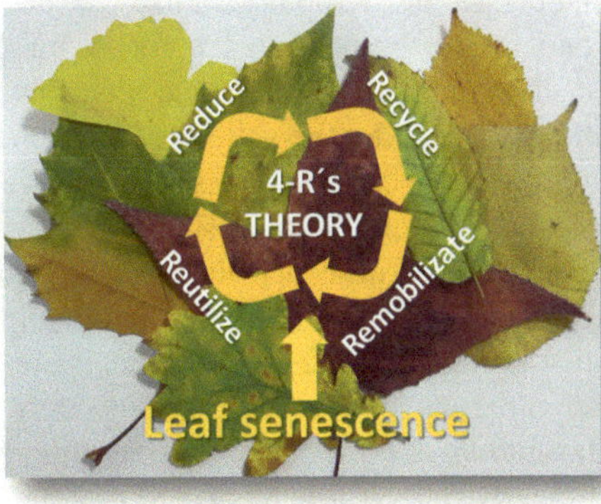

Fig. 1 Theory of the four R of leaf senescence. This physiological process follows the simple concept of the three R's theory of the environment (reduce, reuse and recycle) with an extra R, remobilization. After Reducing the photosynthetic rate in response to the activation of a senescence program, a massive Recycling of nutrients begins that will be Reutilized as scaffolds for new macromolecule biosynthesis and insurance of the next generation survival. This implies an important Remobilization of nutrients through the phloem, from the source plant parts, the senescent leaves, towards sink organs or emergent tissues

transmitted, forming complex interactions of regulatory pathways. Among them, hormonal routes and transcription factors (TFs), signalling transduction cascades of calcium, phosphatases, kinases and others, are involved in the control of the onset and progression of senescence. All the dramatic changes undergoing during senescence are finely controlled and do not constitute a mere chaotic event. These sophisticated networks transmit signals from the environment and determine multiple changes in gene expression patterns during senescence (Penfold and Buchanan-Wollaston 2014; Lin et al. 2015). The participation of hormones during the regulation of senescence is becoming evident through the characterization of mutants and global gene expression analysis. In general, senescence is accelerated by abscisic acid (ABA), brassinosteroids (BR), ethylene (ET), jasmonic acid (JA) and salicylic acid (SA), and slowed down by auxin, cytokinins (CK) and gibberellic acid (GA) (Podzimska-Sroka et al. 2015). Genetic and epigenetic mechanisms regulating phase change from juvenility to maturity directly influence the capacity for responding to senescence signals. Somehow, the degree of influence of such stresses will determine if it causes an impact on the yield (Gregersen et al. 2013; Thomas 2013).

Most of the studies on senescence are focused on the leaf, a crucial organ for the plant fitness. Considering that within the leaf the main source of nitrogen (N)-containing molecules is located inside the chloroplasts, it is not surprising that the earliest structural, biochemical and metabolic changes are observed here (Lim et al. 2007). All enzymes required for carbon fixation and N assimilation, such as the ribulose bisphosphate carboxylase-oxygenase (RuBisCo), as well as most of the proteins that plants can use for N recycling and mobilization, are inside this organelle (Masclaux-Daubresse and Krupinska 2014). A decrease in the overall protein anabolism is one of the best studied markers for the leaf senescence progress. Besides, the decline in photosynthesis and breakdown of the chlorophyll content is phenotypically evidenced by the yellowish of the leaf. The bulk macromolecule degradation mainly relies on proteolysis (Lim et al. 2007; Diaz-Mendoza et al. 2014). Among proteases, serine, and mostly cysteine proteases participate during important events related to senescence and stress (Martinez and Diaz 2008; Martinez et al. 2012; Roberts et al. 2012; Kidric et al. 2014). Furthermore, during leaf senescence, most of the fatty acids from membranes are oxidized to provide energy. An evident drop in the nuclei acid content, especially total RNA, has also been documented. As the amount of polysomes and ribosomes has been observed to decrease fairly early, it clearly reflects a reduction of protein synthesis (Lim et al. 2007). Leaf cells also require a certain energy status until late stages of senescence; thus, the nucleus and mitochondria, essential for gene expression and power generation are the last organelles to be degraded (Yoshida 2003; Lim et al. 2007). Together with the hormones, as main players in the regulation of leaf senescence, there are other signals involved in its onset. Among them, alterations in the cell sugar status as a consequence of the initial dismantling of the photosynthetic apparatus play an important role. There are some lines of research demonstrating that accumulation of sugars compromises the photosynthetic capacity and accelerates leaf senescence (Lim et al. 2007). In addition, the production and accumulation of reactive oxygen species (ROS), derived from alterations in the cell machineries, has also been proposed as an important promoting signal during natural and altered senescence. Albeit ROS production is known to have harmful effects upon diverse biomolecules, it has been proven that a given level of these molecules is required to trigger the activation of genetically programmed pathways of gene expression during leaf senescence (Khanna-Chopra 2012; Noctor et al. 2014, 2016). The degree of effectiveness in the response of the plant after the detection of a stressor factor will determine the degree of reversibility, delimiting a narrow border between degenerative cell death and senescence as a recycling process.

There exists a continuous effort in updating the resources related with leaf senescence information as demonstrate the last releases of the Leaf Senescence Database (LSD) (Liu et al. 2011; Li et al. 2014). A growing senescence community continuously sheds light on some relevant and particular aspects concerning senescence and stress. It is evidenced by the increased number of reports and reviews related to this fascinating topic, mainly focused on signalling and regulatory pathways, nutrient management and nitrogen use efficiency, chlorophyll and chloroplast degradation mechanisms, with a key participation of hydrolases and the regulators (Masclaux-Daubresse and Krupinska 2014; Diaz-Mendoza et al. 2014, 2016b). Besides, a huge number of mutants and/or transgenic plants have shown altered senescence phenotypes. To understand the intrinsic mechanistic principles of leaf senescence, the integration of multi-omics data and the analysis of senescence hallmarks network will provide information, and hopefully tools to detect and even control or prevent senescence events in leaves (Kim et al. 2016). Many investigations in the field of plant senescence and stress can be integrated into two different but complementary areas: research based upon the deciphering of the molecular basis underlying this crucial event at different layers, and translation of basic research to design tools through biotechnological approaches in combination with conventional breeding in order to manipulate senescence for agronomic advantages (Gan and Hörtensteiner 2013). Special emphasis is being undertaken in the maintenance or improvement of acceptable yields in important crops for feed and food such as cereals, in a context of an evident climate change scenario. It is then considered of pivotal importance to invest efforts to interpret the processes behind the decrease in productivity under adverse situations, which substantially relies on a deeper knowledge of chloroplast dismantling mechanisms in both model and crop species.

2 Leaf Senescence Events and Senescence-Associated Genes

A global picture of senescence networks, including feed-forward, feed-back, multiple signalling cascades and crosstalk mechanisms, is essential to understand the molecular programs inherent to these physiological processes. Identifying the team players is a first step, and to find regulatory networks based on transcript levels is a good starting point for controlling senescence key switches. Nowadays, accessibility to the massive whole-genome sequencing and the increasing availability of high-throughput technologies among different omics facilitate this knowledge and lead to the generation of integrative information (Kim et al. 2016; Moschen et al. 2016; Woo et al. 2016). During senescence, the reprograming of thousands of genes is triggered in response to specific senescence-promoting factors. Accordingly, down-regulated genes, mainly those related to photosynthesis anabolism, are usually referred as Senescence Down-regulated Genes (SDGs). On the opposite side, the orderly and orchestrated sequential changes in cellular physiology, biochemistry and metabolism are strongly triggered by a rapid alteration of the expression of an important battery of Senescence-Associated Genes (SAGs) (He et al. 2001; Breeze et al. 2008, 2011; Ay et al. 2014). These genes fall into different categories according to their function as they may be participating in protein, lipid and nucleic acid turnover, in transport of nutrients, in transcriptional regulation, in hormone response pathways and in defence mechanisms (Fig. 2). However, not all SAGs are induced by external cues and some stress-associated genes are not influenced by natural senescence, which evidences a complex crosstalk among routes involved in the developmental- or stress-induced senescence (He et al. 2001; Buchanan-Wollaston et al. 2005; Woo et al. 2016).

Leaf senescence has been intensively studied in model plants, mainly in *Arabidopsis thaliana* and *Populus* sp., with thousands of SAGs already identified in each species. This huge number of induced genes reflects the dramatic alteration in cellular physiology underlying this plant process. In the 1990s, Lohman et al. (1994)

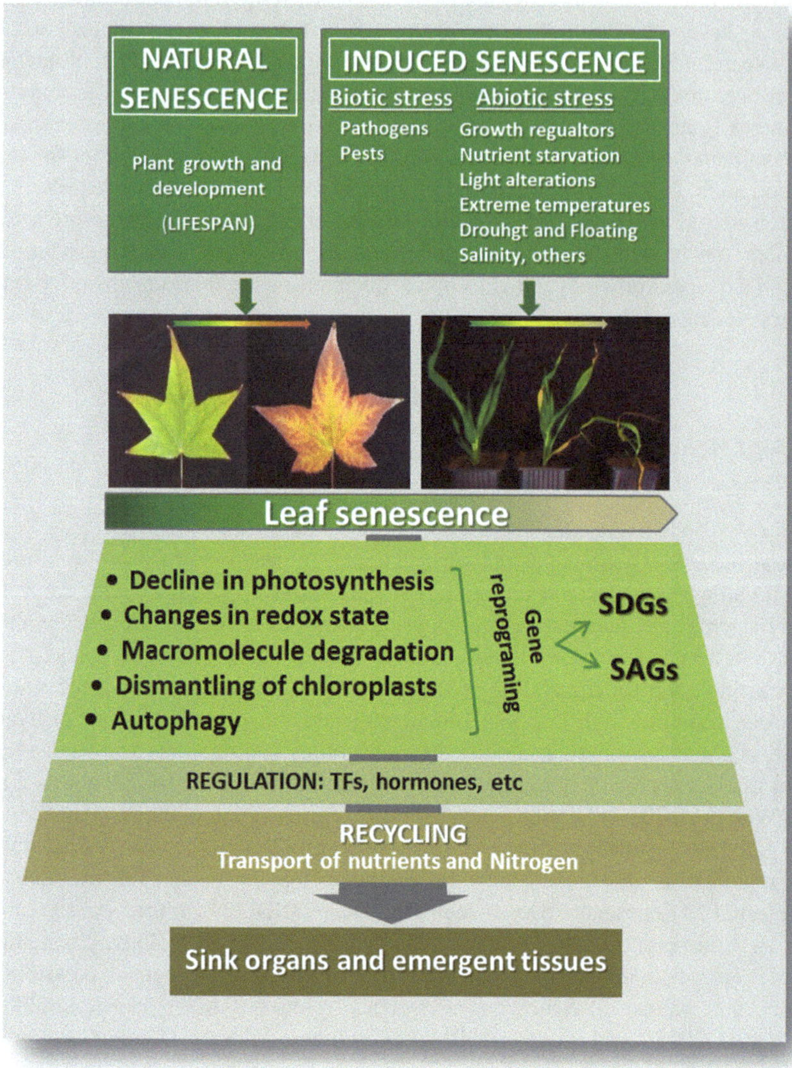

Fig. 2 Global picture of senescence networks. During senescence, the reprograming of thousands of genes is triggered in response to either natural or specific senescence-promoting factors. Accordingly, down-regulated genes, referred as Senescence Down-regulated Genes (SDGs), and an important battery of up-regulated genes, named Senescence-Associated Genes (SAGs), are expressed, under a strict regulation mediated mainly by TFs and hormones. Phenotypical changes are produced as consequence of a set of coordinated sequential events, including chlorophyll loss, reduction of photosynthesis, changes in the redox state, degradation of macromolecules, dismantling of chloroplasts and autophagy. Finally, relocation of nutrients from the senescent leaves towards sink organs and emergent tissues takes place completing a recycling process

constructed cDNA libraries representing mRNAs from *Arabidopsis* leaves at different senescence stages, and reported major changes in gene expression during natural leaf senescence. Among other SAGs, they identified the *SAG12* gene, probably the best-studied senescence-induced cathepsin L-like protease frequently used as senescence marker. Further studies with potential orthologues of *Arabidopsis SAG12* gene in other dicot and monocot species corroborated the important role of this gene in leaf senescence (Diaz-Mendoza et al. 2014, 2016b). The initial genetic screenings performed in *Arabidopsis* to identify mutants with modified senescence patterns revealed delayed senescence phenotypes in the three mutant termed *ore*-mutants: *ore1*, *ore4-1* and *ore9* (Oh et al. 1997). The *ore1* mutant encoded a modified NAC transcription factor that had impaired its function as positive regulator of senescence. Later on, several studies demonstrated that ORE1 binds and regulates some well-known SAGs players, such as *BFN1 (BIFUNCTIONAL NUCLEASE1)*, *SAG29/SWEET15* (involved in sugar transport) and *SINA1* (putative E3 ubiquitin ligase) (Matallana-Ramirez et al. 2013). Moreover, the abundance of *ORE1* transcripts is controlled by the *micro-RNA164* (*miR164*), which interacts with *ORE1* mRNA to trigger its degradation. Based on these and other data, Kim et al. (2009) suggested a trifurcate feed-forward regulatory pathway involving ORE1, *miR164* and the central protein of ethylene signalling EIN2 (ETHYLENE INSENSITIVE2). This regulatory complexity sustains a robust control of leaf senescence. The *ore4-1* mutant had an insertion in the promoter region of the plastid ribosomal small subunit protein 17 (PRPS17 gene). This mutation reduced the activity of the Photosystem I in the thylakoid membrane, and consequently, plants were deficient in chloroplast functions including photosynthesis (Woo et al. 2001). The *ore9* mutant consisted of a degenerate F-box motif that modified its natural function, by limiting the longevity of the leaf through ubiquitin-dependent proteolysis. Further studies with these mutants have provided experimental and/or genetic evidences that link the oxidative stress tolerance to the control of leaf longevity, and the reduction of metabolism and nucleic acid degradation to the extension of plant longevity (Woo et al. 2001, 2002). In 2004, Guo reported the construction of a database with 6,200 ESTs obtained from senescent rosette leaves from *Arabidopsis,* and categorized genes into clusters based on their potential function, including TFs, components of signal transduction pathways, protein turnover and transporters involved in nutrient recycling, among others. A similar experimental approach was used by Bhalerao et al. (2003) by generating an EST database from autumn *Populus* leaves, which allowed the identification of common SAG genes in both, *Arabidopsis* and *Populus* species. Moreover, Andersson et al. (2004) developed genomic tools to use *Populus sp.* (aspens and cottonwoods) as full-featured models for investigating fundamental aspects of the natural senescence of trees. They found a major shift in gene expression, coinciding with the massive chlorophyll degradation, that reflected a competence to energy generation by mitochondrial respiration, oxidation of fatty acids and nutrient mobilization. Multiple studies based on the microarray and RNA sequencing (RNAseq) technologies analysed gene expression profiles in *Arabidopsis* during leaf senescence, either natural or induced by abiotic stresses, throughout various time points (Zentgraf et al. 2004; Buchanan-Wollaston et al. 2005; van der Graff et al. 2006; Breeze et al. 2011). They highlighted the link between leaf senescence and the signalling pathways triggered by several hormones, including ABA, SA, JA and ET, and identified groups of active TFs at different stages. Li et al. (2012) built gene networks directed to identify common regulators of leaf senescence in *Arabidopsis* using the LSD. Their data revealed that CK, auxins, nitric oxide and other molecules and ions, such as Ca^{2+}, delayed leaf senescence, whereas ET, ABA, SA and JA promoted it. In addition, a functional analysis of candidate SAGs on LSD identified two TFs, a WRKY (WRKY75) and a Cys2/His2-type (AZF2), as positive regulators of leaf senescence. The loss of their function provided a delayed in the senescence. In contrast, the silencing of a protein phosphatase (AtMKP2) promoted early senescence. Very recently, a multidimensional transcriptome dataset of the *Arabidopsis* leaf has been obtained by analysing total- and small-RNA transcriptomes throughout its lifespan. These data described sophisticated regulatory networks comprising TFs, small regulatory RNA and temporal and specific inter-organelle coordination among transcriptomes (Woo et al. 2016). This excellent study concluded that the chloroplast transcriptome suffers major changes during leaf aging sharing expression patterns with nuclear transcripts that encode chloroplast-targeted proteins, whereas the mitochondrial transcriptome is not altered.

Transcriptional analyses to investigate the molecular mechanisms driving the leaf senescence in other plant species are not as extensive as in *Arabidopsis*, but have been considerably expanded to crop species. In particular, information concerning leaf senescence in cereals has broadened since it leads to massive remobilization of nitrogen and other nutrients to the developing grains with significant consequences on grain quality and yield. Microarrays and RNAseq data concerning this field have extended the information to barley, maize, wheat, rice and sorghum. For instance, Parrott et al. (2007) described high levels of plastidial and vacuolar gene encoding proteases in senescing barley leaves. Jukanti et al. (2008) found a new regulatory SAG in senescing primary barley leaves consisting on a transmembrane protein kinase. Ay et al. (2008) identified novel SAGs related to degradation or transport processes in senescent tissues. One of them encoded an ADP ribosylation factor 1-like protein (HvARF1) with four typical and well conserved ARF-like protein domains. More recently, Diaz-Mendoza et al. (2014) studied the transcriptional expression pattern of the whole C1A peptidase family after darkness-induced senescence in seedlings of barley and found the up-regulation of some cathepsin L-, B-, H- and F-like proteases. Hollmann et al. (2014) analysed the transcriptomes of senescing flag leaves collected from barley field plots with standard or high N supply and found that the NAC TF HvNAC026, the serine-protease SCPL51 and two autophagy factors, APG7 and ATG18F, might be major regulators and executors of N remobilization during barley leaf senescence in the field. Barley plants overexpressing the HvNAC005 NAC gene resulted in a strong phenotype with delayed development combined with precocious senescence (Christiansen et al. 2016). Notably, NAC- and WRKY-encoding genes were also up-regulated in senescence flag leaves of wheat (Gregersen and Holm 2007). Comparison of transcriptional data of natural leaf senescence in maize and *Arabidopsis* suggested that the mechanisms of leaf senescence were basically similar. In contrast, near 80% of differential expressed genes were altered when comparing natural and induced leaf senescence in maize, although many basic processes were common in both types of senescence (Zhang et al. 2014). Probably, a generic senescence program exists across monocot and dicot plants since homologous TFs have already been considered as SAGs in *Arabidopsis*, sorghum, rice, sunflower and many other species (Oh et al. 1997; Li et al. 2012; Liang et al. 2014; Moschen et al. 2016; Wu et al. 2016). Many advances have been also made to identify and characterize SAGs and senescence related mutants in dicot species other than *Arabidopsis*. There is information on senescence signals, metabolic pathways and regulation in cotton, soybean, winter oilseed rape, sunflower and sweet potato, among others (Huang et al. 2001; Brown and Hudson 2015; Koeslin-Findeklee et al. 2015; Lin et al. 2015; Moschen et al. 2015).

In addition to the large number of SAGs responding to the developmental progress, convergence and divergence in gene expression profiles induced by leaf senescence under different senescence-promoting abiotic and biotic stresses, or even under their combination, have also been analysed (Wingler and Roitsch 2008; Guo and Gan 2012; Machado-Assefh et al. 2014; Gupta et al. 2016b). From these data it may be concluded that just before the senescence program is initiated, treatment-induced and developmental senescence share very few expressed genes. On the other hand, when senescence starts, as evidence leaf yellowish, a great fraction of SAGs are common under natural and induced leaf senescence.

Transcripts are not the only part of the regulatory process, RNA stability, translation rates, protein processing and metabolite concentrations, among other factors, are also essential in the fine control of the sequential changes involved in leaf senescence. A great effort has been made to include proteomic analysis in leaf senescence research. Since the first proteomic study performed in green, early and late senescent leaves of white clover (Wilson et al. 2002) to the most recent innovative proteomic technologies carried out in field-grown maize or soybean (Wei et al. 2015; Gupta et al. 2016a), proteomic approaches have facilitated the identification of unknown proteins, revealed post-translational modification of proteins involved in senescence and provided tools for addressing biological questions about the topic process. The functional annotation of identified proteins has importantly increased over the last years, as well as the knowledge on altered metabolic pathways and biological processes affected by senescence. Proteomic approaches have demonstrated correlations between temporal patterns of protein abundance and the N mobilization in many species such as maize, oilseed rape, pea, poplar or *Arabidopsis* (Schiltz et al. 2004; Hebeler et al. 2008; Desclos et al. 2009; Islam et al. 2015; Wei et al. 2015). Proteomic data have also highlighted correlations between early senescence and up-regulated defence functions (Mason et al. 2016). Altogether, high-throughput proteomic analyses have identified proteins involved in different cellular responses and metabolic processes, including photosynthesis, photorespiration, glycolysis, cell defence, redox homeostasis, signal transduction, protein synthesis, degradation, folding and assembly.

As a consequence of the changes in genes expression throughout natural or induced-senescence, not only the protein content is altered but also other compounds and metabolites are modified. The metabolite profile analysed in leaves and developing seeds of *Arabidopsis* during senescence evidenced important changes in a variety of compounds, including amino acids, pigments, lipids, flavonols and sugars. This study provided a valuable catalogue of the spatio-temporal metabolome of the plant during developmental senescence (Watanabe et al. 2013). Integrative studies to explore global gene expression and metabolic changes during the onset and progression of induced senescence have been performed in *Arabidopsis*, maize and sunflower (Sekhon et al. 2012; Balazadeh et al. 2014; Moschen et al. 2016). An integrated multi-omics perspective, including genomic, epigenomic, transcriptomic, proteomic and metabolomics analyses, focused on the dynamic interactions within leaf senescence research has been recently reviewed by Kim et al. (2016). In a near future, comparative omic studies will open new insights to further understand this physiological process.

3 Overlaps, Similarities and Divergences Among Leaf Senescence, Abiotic and Biotic Stresses

Natural leaf senescence can be induced by endogenous and exogenous factors such as plant growth regulators, floating, nutrient starvation, dark, cold, heat, drought, salt and wound. Moreover, pathogen infection (bacteria, fungi, viruses) and pests, mainly phytophagous arthropods, can also promote or alter senescence. There are numerous reports demonstrating how abiotic stresses trigger leaf senescence by reprograming specific subsets of SAGs that are differentially expressed in distinct tissues. This has been widely reviewed in special recent issues focused on plant senescence (J Exp Bot vol. 65 and J Plant Growth Reg vol. 33 in 2014, Plants vol. 4 in 2015, and the special collection of New Phytol in Nov. 2016), as well as in other reviews from previous years (Quirino et al. 2000; Yoshida 2003; Guo and Gan 2005). In contrast, information about the interplay between leaf senescence and biotic stresses is still limited. The effects and interactions of biotic stress and senescence may be interpreted in two ways, either the presence of a biotic factor promotes senescence after surpassing plant defences through modification of common SAGs implied in primary metabolism; and conversely, it could happen that the fitness of the plant decreased because senescence was already activated in the plant, thus making the plant more prone for the establishment of pests and pathogens (Masclaux-Daubresse et al. 2010; Fagard et al. 2014; Ximenez-Embun et al. 2016). The first assumption would be the case, for instance, for the early senescence and premature cell death detected in rice after the inoculation with *Xanthomonas oryzae* of down-regulated *OsSAG12*-1 (Singh et al. 2013). In *Arabidopsis*, green peach aphid infestation accelerated senescence-like mechanisms by the elevated expression of several SAGs (Pegadaraju et al. 2005; Louis et al. 2010). The mechanisms that allow acclimation and adaptive responses to isolated biotic and abiotic stresses have been extensively characterized. Particularly, a good number of reports have appeared in the last years regarding the combination of both biotic and abiotic stresses and plant responses. A special issue on the interface between abiotic and biotic stress responses published in J Exp Bot vol. 67 in 2016 highlights the increasing interest for the subject. Deciphering the signalling pathways participating in the common crosstalk of biotic and abiotic stresses will allow the identification of new targets for increasing environmental resilience in crops (Foyer et al. 2016). The exposure to one type of stress confers a general increase in resistance to a range of different stresses, a phenomenon termed 'cross-tolerance'. This phenomenon relies on the synergistic co-activation of non-specific, stress-responsive pathways that cross biotic–abiotic stress boundaries, and which are usually related with altered redox and phytohormone signalling (Foyer et al. 2014). Cross-tolerance to different stresses triggered by an exposure to a single stress is widespread in plants. In potato plants exposed to dark-induced senescence, the aphid *Myzus persicae* showed a reduction in the pre-reproductive time (Machado-Assefh et al. 2014). As aforementioned, ABA is very important during the response to drought episodes as it binds to specific receptors that lead to changes in phosphatases/kinases cascades which induce genes related to the activation of ion channels and stomata closure. Through stomatal closure, the ABA-induced signalling pathway intersects with both abiotic and biotic stress factors (Lee and Luan 2012). In maize, differences at the proteome level were also detected depending on if two stresses, drought and the presence of the two-spotted spider mite, were applied individually or combined (Dworak et al. 2016). In conclusion, plant responses to confluent abiotic and biotic stresses are not only the addition of the responses to an independent stress. The presence of an abiotic stress may reduce or enhance susceptibility to a biotic pest or pathogen and vice versa. However, the significance of the differential plant responses and the precise roles of biotic/abiotic and senescence-responsive pathways remain still unknown.

4 Leaf Senescence and Nutrient Mobilization

Leaf senescence is strongly linked to a massive degradation of macromolecules to mobilize nutrients from leaves to sink tissues in order to sustain further growth and development. Protein breakdown is one of the most important hydrolytic processes in the senescent leaf with a crucial role in nutrient recycling, especially N. Many SAGs encode proteases which is consistent with increases in proteolytic activities and degradation, mainly, of chloroplastic proteins (Breeze et al. 2011; Roberts et al. 2012; Diaz-Mendoza et al. 2014; Velasco-Arroyo et al. 2016). Particularly Rubisco, which represents the major nitrogen investment in crops and the first source of transportable N, is the main target of proteases (Feller et al. 2008a, b; Masclaux-Daubresse et al. 2010). Changes in the proteolytic activities among different sub-cellular locations take place to cooperatively ensure efficient protein degradation into amino acids, amides and ammonium. As a result, a complex traffic of proteins, peptides and amino acids occurs among cell compartments involving chloroplasts, cytosol, special vesicles and lytic vacuoles (Roberts et al.

2012; Carrion et al. 2014; Avila-Ospina et al. 2014; Diaz-Mendoza et al. 2016b). Finally, a major part of the N is released as ammonium after being re-assimilated into amino acids to be exported via the phloem to the developing grains, fruits and tubers (Fig. 2). Therefore, the timing of leaf senescence is of pivotal importance for yield in crop species (Gregersen et al. 2013).

4.1 Coordinated Carbon and Nitrogen Assimilation During Remobilization Events

Senescence is predominantly controlled at the level of the individual leaf, and remobilization usually begins in older leaves towards the younger ones, making important contributions to the remobilization of the major part of photo-assimilates during seed filling, especially in cereals and legumes (Schiltz et al. 2004; Wiedemuth et al. 2005). More than 75% of the potentially remobilizable reduced N in plants is located inside the chloroplasts and mainly assembled into stromal components, such as the Rubisco and glutamine synthetase (GS) (Hörtensteiner and Feller 2002). Chlorophyll-apoprotein complexes from thylakoids represent the second major fraction. Likewise, it was estimated that around 70% of the N from senescing vegetative organs is exported during seed development in most annual crop plants. Although a part of ammonia is evaporated from leaves, the bulk ammonium content is exported from the senescing leaf and utilized to build new amino acids. An intense traffic of amino acids occurs along the phloem during grain developing and maturation stages. The major phloem-exported amino acid in barley and wheat is glutamate, followed by aspartate, glutamine, threonine and serine (Forde and Lea 2007). Two forms of GS have been identified in plants, the cytosolic GS1 and the chloroplastic/mitochondrial GS2. In non-senescing leaves, GS2 is the most abundant isoform in the mesophyll cells, where it assimilates ammonium originated from nitrate reduction and photorespiration. During leaf senescence GS1 fulfils a key function in the assimilation and recycling of ammonium generated from various catabolic processes (Masclaux-Daubresse et al. 2010). This role is particularly important in cereals after anthesis, and during grain development and filling, when nitrogen is remobilized to the reproductive sinks. In *Arabidopsis*, the total amino acid content is higher in shoots of plants grown under continuous N limitation in comparison to control conditions (Tschoep et al. 2009). Probably, this is due to a reduced utilization of amino acids for protein synthesis as a consequence of a slowdown in the tricarboxylic acid (TAC) cycle, which determined a general down-regulation of biosynthetic metabolism (Balazadeh et al. 2014). A rapid reversion in the cytosolic carbon to N (C/N) ratio is required to inverse leaf senescence. In both barley and *Arabidopsis*, senescence can be completely retreated when additional nitrate is resupplied after a nitrogen starvation period (Schildhauer et al. 2008). A situation of carbon fast (high CH levels) undergoing in source senescing organs may act as a first signal to start remobilization of nutrients; but a state of carbon starvation in the sink organs may also represent the initial stimulus for beginning the maintenance of molecules (Parrott et al. 2007). Importantly, a set of proteases were induced under these conditions. Given the complexity and the lack of precise descriptions on the events taking place during senescence, either developmental or stress-induced, there is a need to discern which is the main mechanism involved. Since amino acid and nutrient transport are usually the main hallmarks, it seems very likely that, in general, proteolysis represents the ruling process.

4.2 Nitrogen Economy in Plants: The 'Dilemma' of Senescence and the Stay-Green Trait

During leaf development, photosynthetic activity and cell growth processes decrease, whereas sucrose, fatty acid, nucleotide and amino acid metabolisms increase. The timing of senescence affects the length of the photosynthetic period, thus influencing grain filling in the case of cereals and, therefore, determining the yield and/or the quality. A delay or acceleration in leaf senescence has an important impact on yield, by maintaining or altering photosynthetic leaf area especially during the reproductive stage (Gregersen et al. 2013; Distelfeld et al. 2014; Have et al. 2016). Martinez and Diaz (2008) and Carrion et al. (2014) demonstrated that the accumulation of small 'senescence-associated vacuoles' (SAVs) with intense proteolytic activity in senescing leaves of soybean and *Arabidopsis* was associated with chloroplast protein breakdown. This finding indicates that autophagy is involved in the process since it is a complex vesicular pathway essential for cell longevity. Autophagic activity facilitates the removal of oxidized proteins, and contributes to the degradation of cell components to maintain cells alive long enough to assist nutrient recycling, mobilization and reallocation in other organs (Avila-Ospina et al. 2014; Masclaux-Daubresse 2016). *Arabidopsis* and maize autophagy mutants (*atg*) deficient in proteins of the core machinery displayed early leaf senescence and were hypersensitive to N limitation (Guiboileau et al. 2013; Li et al. 2015).

Senescence might reduce crop yield when it is prematurely induced under adverse environmental conditions. In 2008, Gregersen et al. raised the concept of 'Dilemma of senescence' questioning whether it is better to delay or to accelerate senescence. When late senescence occurs, looking at the grain, a higher CH content and a lower protein accumulation are observed, whereas when senescence starts earlier the opposite trend is appreciated. It is paramount to understand the molecular mechanisms behind senescence in order to improve these traits, depending on the end-product usage. For instance, applied to barley, this first situation would be desirable for malting purposes, since a higher amount of CH is necessary for fermentation. Conversely, the second situation would be ideal for manufacturing meals for animal feeding, since a high content in proteinaceous components is required (Fig. 3). Improvement of cereal cultivars requires a delicate balance

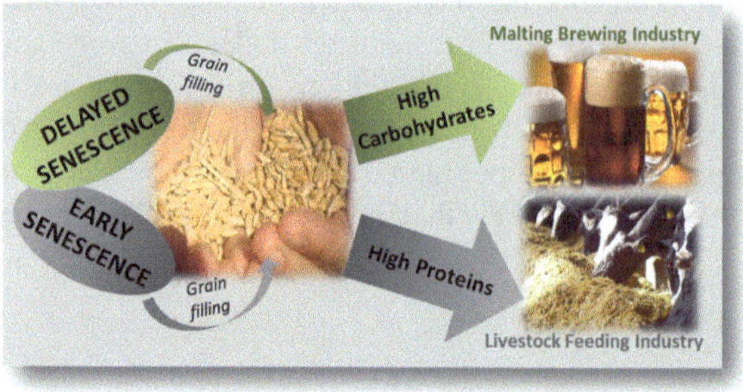

Fig. 3 Schematic hypothesis for the breeder's dilemma. Delayed senescence leads to an extended photosynthetic period resulting in an increased biomass accumulation, and in consequence, higher carbohydrate content in the grain. In contrast, accelerated senescence provokes a faster remobilization of nutrients translated into an increase in the grain number and size with higher protein content. Applied to barley and depending on the end-product usage, delayed senescence would be desirable for malting purposes, since a higher amount of carbohydrates is necessary for fermentation. Conversely, precocious senescence would be ideal for manufacturing meals for animal feeding, since a high content in proteinaceous components is required

among senescence timing, grain nutrient content, N use efficiency and yield. The physiological stage at which a plant faces a given stress will largely influence upon the activation and progression of the senescence programs, determining the efficiency in the remobilization of nutrients as a strategy for survival.

Stay-green mutants have been used to understand the correlation between senescence, yield and total N content of the grain in several crop species (Kichey et al. 2007). A wheat stay-green variety presented a reduced harvest index even after a prolonged grain filling period. It was postulated that remobilization of carbon was inefficient and that extra photo-assimilates remained in the vegetative parts instead of being translocated to the grain. N concentration in the straw of a stay-green line of wheat remained higher than in controls. Extended photosynthesis did not mean an increase in grain yield as expected; instead, these plants necessitated more N uptake to achieve a grain protein content comparable to that for wild-type (Chen et al. 2011). Another wheat mutant, *tasg1*, showing delayed leaf senescence, was identified as a functional stay-green (Hui et al. 2012). The explanation for the last examples in which extended photosynthesis did not result in higher harvest index relies on the fact that sink tissues may have a limitation in their capacity, which is in term influencing a major trait, the growth and size of the seed. One of the most common approaches to achieve stay-green varieties through biotechnology is based upon the expression of isopentenyltransferase (*IPT*), which codifies an enzyme that catalyses the rate-limiting step in CK synthesis (Gan and Amasino 1997), under the control of senescence-associated promoters. Binding of WRKY family members, among others, to the *cis*-elements on these promoters, is regulated by ABA. These experimental strategies determine an increased biomass in the crops in most of the cases, but this is not very commonly translated into improved seed yields. On the other hand, it was demonstrated a better performance of these transformed plants under certain adverse environmental stresses, such as drought (Gregersen et al. 2013). In a detailed transcriptome study performed in barley, in which late seed maturation and initial germination stage were analysed, the authors concluded that during maturation, the barley grain stores all required compounds and regulators, among them many TF, meaning that plant seeds prepare for germination already during seed maturation. This leads to conclude that maturation of the grain is a crucial developmental stage, and apparently alterations in source/sink communication influenced by modifications alongside senescence timing may have negative effects upon the accumulation of valuable elements required for a later and successful germination during next generation (Sreenivasulu et al. 2008). More recently, transgenic barley plants silencing or overexpressing the cathepsin F-like HvPap-1 cysteine protease have shown differential accumulation of storage molecules such as starch, proteins and free amino acids in the grain. Similarly, down-regulating cystatin *Icy-2* gene, one of the proteinaceous inhibitors of the cathepsin F-like protease, also has important effects on grain filling. Moreover, the HvPap-1 artificial microRNA lines revealed a drastic delay in the grain germination process (Diaz-Mendoza et al. 2016a).

Accelerated senescence was also achieved in several plant species by means of classical breeding, and in many cases, this was correlated with higher protein content in the seeds. The *Gpc-B1* locus was linked with accelerated flag leaf senescence and with a shorter grain filling period in wheat (Uauy et al. 2006). In barley, a similar locus was previously characterized (See et al. 2002). Gpc-B1 belongs to the NAC family of TF, which seems to be up-regulated in many expression studies in response to senescence in both cereals (Gregersen et al. 2013). These data pointed these members as ideal candidates to modulate leaf senescence (Gregersen et al. 2013; Distelfeld et al. 2014; Christiansen et al. 2016). In conclusion, plant senescence is a complex process with a great impact on agriculture closely linked to crop yield and quality, biomass production and bioenergy development. On the applied side, the biotechnological control of this process is a powerful tool to ameliorate the consequences of actual and coming stresses in the context of the climate change scenario. However, a wider picture of the global senescence mechanisms is still needed because many questions remain to be answered to ensure an efficient control of the plant lifespan history and longevity.

Acknowledgements This work was supported by the project from Ministerio de Economia y Competitividad e Industria of Spain (BIO2014-53508-R). B.V.-A. a and A.G.-S. had FPI contracts and M.E.S. Juan de la Cierva contract, all of them from the same Spanish Ministry.

References

Andersson A, Keskitalo J, Sjodin A, Bhalerao R, Sterky F, Wissel K, Tandre K, Aspeborg H, Moyle R, Ohmiya Y, Bhalerao R, Brunner A, Gustafsson P, Karlsson J, Lundeberg J, Nilsson O, Sandberg G, Strauss S, Sundberg B, Uhlen M, Jansson S, Nilsson P (2004) A transcriptional timetable of autumn senescence. Genome Biol 5:R24

Avila-Ospina L, Moison M, Yoshimoto K, Masclaux-Daubresse C (2014) Autophagy, plant senescence, and nutrient recycling. J Exp Bot 65:3799–3811
Ay N, Clauss K, Barth O, Humbeck K (2008) Identification and characterization of novel senescence-associated genes from barley (*Hordeum vulgare*) primary leaves. Plant Biol 10:121–135
Ay N, Janack B, Humbeck K (2014) Epigenetic control of plant senescence and linked processes. J Exp Bot 65:3875–3887
Balazadeh S, Schildhauer J, Araujo WL, Munne-Bosch S, Fernie AR, Proost S, Humbeck K, Mueller-Roeber B (2014) Reversal of senescence by N resupply to N-starved *Arabidopsis thaliana*: transcriptomic and metabolomic consequences. J Exp Bot 65:3975–3992
Bhalerao R, Keskitalo J, Sterky F, Erlandsson R, Bjorkbacka H, Birve SJ, Karlsson J, Gardestrom P, Gustafsson P, Lundeberg J, Jansson S (2003) Gene expression in autumn leaves. Plant Physiol 131:430–442
Breeze E, Harrison E, Page T, Warner N, Shen C, Zhang C, Buchanan-Wollaston V (2008) Transcriptional regulation of plant senescence: from functional genomics to systems biology. Plant Biol 10:99–109
Breeze E, Harrison E, McHattie S, Hughes L, Hickman R, Hill C, Kiddle S, Kim Y-S, Penfold CA, Jenkins D, Zhang C, Morris K, Jenner C, Jackson S, Thomas B, Tabrett A, Legaie R, Moore JD, Wild DL, Ott S, Rand D, Beynon J, Denby K, Mead A, Buchanan-Wollaston V (2011) High-resolution temporal profiling of transcripts during Arabidopsis leaf senescence reveals a distinct chronology of processes and regulation. Plant Cell 23:873–894
Brown AV, Hudson KA (2015) Developmental profiling of gene expression in soybean trifoliate leaves and cotyledons. BMC Plant Biol 15:169
Buchanan-Wollaston V, Page T, Harrison E, Breeze E, Lim PO, Nam HG, Lin JF, Wu SH, Swidzinski J, Ishizaki K, Leaver CJ (2005) Comparative transcriptome analysis reveals significant differences in gene expression and signalling pathways between developmental and dark/starvation-induced senescence in Arabidopsis. Plant J 42:567–585
Carrion C, Martinez D, Costa M, Guiamet J (2014) Senescence-associated vacuoles, as specific lytic compartment for degradation of chloroplast proteins? Plants 3:498–512
Chen CC, Han GQ, He HQ, Westcott M (2011) Yield, protein, and remobilization of water soluble carbohydrate and nitrogen of three spring wheat cultivars as influenced by nitrogen input. Agron J 103:786–795
Christiansen MW, Matthewman C, Podzimska-Sroka D, O'Shea C, Lindemose S, Møllegaard NE, Holme IB, Hebelstrup K, Skriver K, Gregersen PL (2016) Barley plants over-expressing the NAC transcription factor gene HvNAC005 show stunting and delay in development combined with early senescence. J Exp Bot 67:5259–5273
Desclos M, Etienne P, Coquet L, Jouenne T, Bonnefoy J, Segura R, Reze S, Ourry A, Avice JC (2009) A combined 15N tracing/proteomics study in *Brassica napus* reveals the chronology of proteomics events associated with N remobilization during leaf senescence induced by nitrate limitation or starvation. Proteomics 9:3580–3608
Diaz-Mendoza M, Velasco-Arroyo B, Gonzalez-Melendi P, Martinez M, Diaz I (2014) C1A cysteine protease-cystatin interactions in leaf senescence. J Exp Bot 65:3825–3833
Diaz-Mendoza M, Dominguez-Figueroa JD, Velasco-Arroyo B, Cambra I, Gonzalez-Melendi P, Lopez-Gonzalvez A, Garcia A, Hensel G, Kumlehn J, Diaz I (2016a) HvPap-1 C1A protease and HvCPI-2 cystatin contribute to barley grain filling and germination. Plant Physiol 170: 2511–2524
Diaz-Mendoza M, Velasco-Arroyo B, Santamaria ME, González-Melendi P, Martinez M, Diaz I (2016b) Plant senescence and proteolysis: two processes with one destiny. Genet Mol Biol 39: 329–338
Distelfeld A, Avni R, Fischer AM (2014) Senescence, nutrient remobilization, and yield in wheat and barley. J Exp Bot 65:37883–33798

Dworak A, Nykiel M, Walczak B, Miazek A, Szworst-Lupina D, Zagdanska B, Kiekiewicz M (2016) Maize proteomic responses to separate or overlapping soil drought and two-spotted spider mite stresses. Planta 244:939–960

Fagard M, Launay A, Clement G, Courtial J, Dellagi A, Farjad M, Krapp A, Soulie MC, Masclaux-Daubresse C (2014) Nitrogen metabolism meets phytopathology. J Exp Bot 65:5643–5656

Feller U, Anders I, Demirevska K (2008a) Degradation of Rubisco and other chloroplast proteins under abiotic stress. Plant Physiol 34:5–18

Feller U, Anders I, Mae T (2008b) Rubiscolytics: fate of Rubisco after its enzymatic function in a cell is terminated. J Exp Bot 59:1615–1624

Forde BG, Lea PJ (2007) Glutamate in plants: metabolism, regulation, and signalling. J Exp Bot 58:2339–2358

Foyer CH, Karpinska B, Krupinska K (2014) The functions of WHIRLY1 and REDOX-RESPONSIVE TRANSCRIPTION FACTOR 1 in cross tolerance responses in plants: a hypothesis. Philos Trans R Soc Lond B Biol Sci 369:2013–20226

Foyer CH, Rasool B, Davey JW, Hancock RD (2016) Cross-tolerance to biotic and abiotic stresses in plants: a focus on resistance to aphid infestation. J Exp Bot 67:2025–2037

Gan S, Amasino RM (1997) Making sense of senescence (molecular genetic regulation and manipulation of leaf senescence). Plant Physiol 3:313–319

Gan SS, Hörtensteiner S (2013) Frontiers in plant senescence research: from bench to bank. Plant Mol Biol 82:503–504

Gregersen PL, Holm PB (2007) Transcriptome analysis of senescence in the flag leaf of wheat (*Triticum aestivum* L). Plant Biotechnol J 5:192–206

Gregersen PL, Holm PB, Krupinska K (2008) Leaf senescence and nutrient remobilisation in barley and wheat. Plant Biol 10:37–49

Gregersen PL, Culetic A, Boschian L, Krupinska K (2013) Plant senescence and crop productivity. Plant Mol Biol 82:603622

Guiboileau A, Avila-Ospina L, Yoshimoto K, Soulay F, Azzopardi M, Marmagne A, Lothier J, Masclaux-Daubresse C (2013) Physiological and metabolic consequences of autophagy deficiency for the management of nitrogen and protein resources in Arabidopsis leaves depending on nitrate availability. New Phytol 199:683–694

Guo Y, Gan S (2005) Leaf senescence: signals, execution, and regulation. Curr Top Dev Biol 71: 83–112

Guo Y, Gan S (2012) Convergence and divergence in gene expression profiles induced by leaf senescence and 27 senescence-promoting hormonal, pathological and environmental stress treatments. Plant Cell Environ 35:644–655

Guo Y, Cai Z, Gan S (2004) Transcriptome of Arabidopsis leaf senescence. Plant Cell Environ 27:521–549

Gupta R, Lee SJ, Min CW, Kim SW, Park KH, Bae DW, Lee BW, Agrawal GK, Rakwal R, Kim ST (2016b) Coupling of gel-based 2-DE and 1-DE shotgun proteomics approaches to dig deep into the leaf senescence proteome of *Glycine max*. J Proteomics 148:65–74

Gupta A, Sarkar AK, Senthil-Kumar M (2016a) Global transcriptional analysis reveals unique and shared responses in *Arabidopsis thaliana* exposed to combined drought and pathogen stress. Front Plant Sci 7:686

Have M, Marmagne A, Chardon F, Masclaux-Daubresse C (2016) Nitrogen remobilisation during leaf senescence: lessons from Arabidopsis to crops. J Exp Bot. doi:10.1093/jxb/erw365 (in press)

He Y, Tang W, Swain JD, Green AL, Jack TP, Gan S (2001) Networking senescence-regulating pathways by using Arabidopsis enhancer trap lines. Plant Physiol 126:707–716

Hebeler R, Oeljeklaus S, Reidegeld KA, Eisenacher M, Stephan C, Sitek B, Stühler K, Meyer HE, Sturre MJ, Dijkwel PP, Warscheid B (2008) Study of early leaf senescence in *Arabidopsis thaliana* by quantitative proteomics using reciprocal 14N/15N labeling and difference gel electrophoresis. Mol Cell Proteomics 7:108–120

Hollmann J, Gregersen PL, Krupinska K (2014) Identification of predominant genes involved in regulation and execution of senescence-associated nitrogen remobilization in flag leaves of field grown barley. J Exp Bot 65:3963–3974
Hörtensteiner S, Feller U (2002) Nitrogen metabolism and remobilization during senescence. J Exp Bot 53:927937
Huang YJ, To KY, Yap MN, Chiang WJ, Suen DF, Chen SCG (2001) Cloning and characterization of leaf senescence up-regulated genes in sweet potato. Physiol Plant 113:384–339
Hui Z, Tian FX, Wang GK, Wang GP, Wang W (2012) The antioxidative defense system is involved in the delayed senescence in a wheat mutant tasg1. Plant Cell Rep 31:1073–1084
Islam N, Li G, Garrett WM, Lin R, Sriram G, Cooper B, Coleman GD (2015) Proteomics of nitrogen remobilization in poplar bark. J Proteome Res 14:1112–1126
Jukanti AK, Heidlebaugh NM, Parrott DL, Fischer IA, McInnerney K, Fischer AM (2008) Comparative transcriptome profiling of near-isogenic barley (*Hordeum vulgare*) lines differing in the allelic state of a major grain protein content locus identifies genes with possible roles in leaf senescence and nitrogen reallocation. New Phytol 177:333–349
Khanna-Chopra R (2012) Leaf senescence and abiotic stresses share reactive oxygen species-mediated chloroplast degradation. Protoplasma 249:469–481
Kichey T, Hirel B, Heumez E, Dubois F, Le Gouis J (2007) In winter wheat (*Triticum aestivum* L.), post-anthesis nitrogen uptake and remobilisation to the grain correlates with agronomic traits and nitrogen physiological markers. Field Crop Res 102:22–32
Kidric M, Kos J, Sabotic J (2014) Proteases and their endogenous inhibitors in the plant response to abiotic stress. Bot Serbica 38:139–158
Kim JH, Woo HR, Kim J, Lim PO, Lee IC, Choi SH, Hwang D, Nam HG (2009) Trifurcate feed-forward regulation of age-dependent cell death involving *miR164* in *Arabidopsis*. Science 323: 1053–1057
Kim J, Woo HR, Nam HG (2016) Toward systems to understand of leaf senescence: an integrated multi-omics perspective on leaf senescence research. Mol Plant 9:813–825
Koeslin-Findeklee F, Rizi VS, Becker MA, Parra-Londono S, Arif M, Balazadeh S, Mueller-Roeber B, Kunze R, Horst WJ (2015) Transcriptomic analysis of nitrogen starvation- and cultivar-specific leaf senescence in winter oilseed rape (*Brassica napus* L). Plant Sci 233: 174–185
Krupinska K, Mulisch M, Hollmann J, Tokarz K, Zschiesche W, Kage H, Humbeck K, Bilger W (2012) An alternative strategy of dismantling of the chloroplasts during leaf senescence observed in a high-yield variety of barley. Physiol Plant 144:189–200
Lee SC, Luan S (2012) ABA signal transduction at the crossroad of biotic and abiotic stress responses. Plant Cell Environ 35:53–60
Li Z, Peng J, Wen X, Guo H (2012) Gene network analysis and functional studies of senescence-associated genes reveal novel regulators of Arabidopsis leaf senescence. J Integr Plant Biol 54:526–539
Li Z, Zhao Y, Liu X, Peng J, Guo H, Luo J (2014) LSD 2.0: an update of the leaf senescence database. Nucleic Acids Res 42:1200–1205
Li F, Chung T, Pennington JC, Federico ML, Kaeppler HF, Kaeppler SM, Otegui MS, Vierstra RD (2015) Autophagic recycling plays a central role in maize nitrogen remobilization. Plant Cell 27:1389–1408
Liang C, Wang Y, Zhu Y, Tang J, Hu B, Liu L, Ou S, Wu H, Sun X, Chu J, Chu C (2014) OsNAP connects abscisic acid and leaf senescence by fine-tuning abscisic acid biosynthesis and directly targeting senescence-associated genes in rice. Proc Natl Acad Sci U S A 111:10013–10018
Lim PO, Kim HJ, Nam HG (2007) Leaf senescence. Annu Rev Plant Biol 58:115–136
Lin M, Pang C, Fan S, Song M, Wei H, Yu S (2015) Global analysis of the *Gossypium hirsutum* L. transcriptomic during leaf senescence by RNA-Seq. BMC Plant Biol 15:43
Liu X, Li Z, Jiang Z, Zhao Y, Peng J, Jin J, Guo H, Luo J (2011) LSD: a leaf senescence database. Nucleic Acids Res 39:1–5

Lohman K, Gan S, Manorama CJ, Amasino RM (1994) Molecular analysis of natural leaf senescence in *Arabidopsis thaliana*. Physiol Plant 92:322–328
Louis J, Lorenc-Kukula K, Singh V, Reese J, Jander G, Shah J (2010) Antibiosis against the green peach aphid requires the *Arabidopsis thaliana MYZUS PERSICAE*-INDUCED LIPASE1 gene. Plant J 64:800–811
Machado-Assefh CR, Lucatti AF, Alvarez AE (2014) Induced senescence promotes the feeding activities and nymph development of *Myzus persicae* (Hemiptera: Aphididae) on potato plants. J Insect Sci 14:155
Martinez M, Diaz I (2008) The origin and evolution of plant cystatins and their target cysteine proteinases indicate a complex functional relationship. BMC Evol Biol 8:198
Martinez M, Cambra I, Gonzalez-Melendi P, Santamarıa ME, Dıaz I (2012) C1A cysteine-proteases and their inhibitors in plants. Physiol Plant 145:85–94
Masclaux-Daubresse C (2016) Autophagy controls carbon, nitrogen, and redox homeostasis in plants. Autophagy 12:896–987
Masclaux-Daubresse C, Krupinska K (2014) Preface. J Exp Bot 65:3781–3782
Masclaux-Daubresse C, Daniel-Vedele F, Dechorgnat J, Chardon F, Gaufichon L, Suzuki A (2010) Nitrogen uptake, assimilation and remobilization in plants: challenges for sustainable and productive agriculture. Ann Bot 105:1141–1157
Mason KE, Hilmer JK, Maaty WS, Reeves BD, Grieco PA, Bothner B, Fischer AM (2016) Proteomic comparison of near-isogenic barley (*Hordeum vulgare* L.) germplasm differing in the allelic state of a major senescence QTL identifies numerous proteins involved in plant pathogen defense. Plant Physiol Biochem 109:114–127
Matallana-Ramirez LP, Rauf M, Farage-Barhom S, Dortay H, Xue GP, Dröge-Laser W, Lers A, Balazadeh S, Mueller-Roeber B (2013) NAC transcription factor ORE1 and senescence-induced *BIFUNCTIONAL NUCLEASE1* (*BFN1*) constitute a regulatory cascade in *Arabidopsis*. Mol Plant 6:1438–1452
Moschen S, Luoni SB, Di Rienzo JA, Caro MP, Tohge T, Watanabe M, Hollman J, Gonzalez S, Rivarola M, Garcia-Garcia F, Dopazo J, Hopp HE, Hoefgen R, Fernie AR, Paniego N, Fernandez P, Heinz RA (2015) Integrating transcriptomic and metabolomics analysis to understand natural leaf senescence in sunflower. Plant Biotechnol J 14:719–734
Moschen S, Higgins J, Di Rienzo JA, Heinz RA, Paniego N, Fernandez P (2016) Network and biosignature analysis for the integration of transcriptomic and metabolomic data to characterize leaf senescence process in sunflower. BMC Bioinf 17:174
Munne-Bosch S (2008) Do perennials really senesce? Trends Plant Sci 13:216–220
Noctor G, Mhamdi A, Foyer CH (2014) The roles of reactive oxygen metabolism in drought: not so cut and dried. Plant Physiol 164:1636–1648
Noctor G, Mhamdi A, Foyer CH (2016) Oxidative stress and antioxidative systems: recipes for successful data collection and interpretation. Plant Cell Environ:1140–1160
Oh SA, Park JH, Lee GI, Paek KH, Park SK, Nam HG (1997) Identification of three genetic loci controlling leaf senescence in *Arabidopsis thaliana*. Plant J 12:527–535
Parrott DL, McInnerney K, Feller U, Fischer AM (2007) Steam-girdling of barley (*Hordeum vulgare*) leaves leads to carbohydrate accumulation and accelerated leaf senescence, facilitating transcriptomic analysis of senescence-associated genes. New Phytol 176:56–69
Pegadaraju V, Knepper C, Reese J, Shah J (2005) Premature leaf senescence modulated by the Arabidopsis PHYTOALEXIN DEFICIENT4 gene is associated with defense against the phloem-feeding green peach aphid. Plant Physiol 139:1927–1934
Penfold CA, Buchanan-Wollaston V (2014) Modelling transcriptional networks in leaf senescence. J Exp Bot 65:3859–3873
Podzimska-Sroka D, O'shea C, Gregersen PL, Skriver K (2015) NAC transcription factors in senescence: from molecular structure to function in crops. Plants 4:412–448
Quirino BF, Noh YS, Himelblau E, Amasino RM (2000) Molecular aspects of leaf senescence. Trends Plant Sci 5:278–282

Roberts IN, Caputo C, Criado MV, Funk C (2012) Senescence-associated proteases in plants. Physiol Plant 145:130–139

Schildhauer J, Wiedemuth K, Humbeck K (2008) Supply of nitrogen can reverse senescence processes and affect expression of genes coding for plastidic glutamine synthetase and lysine-ketoglutarate reductase/saccharopine dehydrogenase. Plant Biol 10:76–84

Schiltz S, Gallardo K, Huart M, Negroni L, Sommerer N, Burstin J (2004) Proteome reference maps of vegetative tissues in pea. An investigation of nitrogen mobilization from leaves during seed filling. Plant Physiol 135:2241–2260

Schippers JHM (2015) Transcriptional networks in leaf senescence. Curr Opin Plant Biol 27: 77–83

See D, Kanazin V, Kephart K, Blake T (2002) Mapping genes controlling variation in barley grain protein concentration. Crop Sci 42:680685

Sekhon RS, Childs KL, Santoro N, Foster CE, Buell CR, de Leon N, Kaeppler SM (2012) Transcriptional and metabolic analysis of senescence induced by preventing pollination in maize. Plant Physiol 159:1730–1744

Singh S, Giri MK, Singh PK, Siddiqui A, Nandi AK (2013) Down-regulation of OsSAG12-1 results in enhanced senescence and pathogen-induced cell death in transgenic rice plants. J Biosci 38:583–592

Sreenivasulu N, Usadel B, Winter A, Radchuk V, Scholz U, Stein N, Weschke W, Strickert M, Close TJ, Stitt M, Graner A, Wobus U (2008) Barley grain maturation and germination: metabolic pathway and regulatory network commonalities and differences highlighted by new mapMan/PageMan profiling tools. Plant Physiol 146:1738–1758

Thomas H (2013) Senescence, ageing and death of the whole plant. New Phytol 197:696–711

Tschoep H, Gibon Y, Carillo P, Armengaud P, Szecowka M, Nunes-Nesi A, Fernie AR, Koehl K, Stitt M (2009) Adjustment of growth and central metabolism to a mild but sustained nitrogen-limitation in Arabidopsis. Plant Cell Environ 32:300–318

Uauy C, Distelfeld A, Fahima T, Blechl A, Dubcovsky J (2006) A NAC gene regulating senescence improves grain protein, zinc, and iron content in wheat. Science 314:1298–1301

Van der Graff E, Schwacke R, Schneider A, Desimone M, Fluggue UI, Kunze R (2006) Transcriptioon analyissi of Arabidopsis membrane tarnsporters and hormone pathways during developmental and induced leaf senescence. Plant Physiol 141:776–792

Velasco-Arroyo B, Diaz-Mendoza M, Gandullo J, Gonzalez-Melendi P, Santamaria ME, Dominguez-Figueroa JD, Hensel G, Martinez M, Kumlehn J, Diaz I (2016) HvPap-1 C1A protease actively participates in barley proteolysis mediated by abiotic stresses. J Exp Bot 67: 4297–4310

Watanabe M, Balazadeh S, Tohge T, Erban A, Giavalisco P, Kopka J, Mueller-Roeber B, Fernie AR, Hoefgen R (2013) Comprenhensive dissection of saptiotemporal metabolic shifts in primary, secondary, and lipid metabolism during developmental senescence in Arabidopsis. Plant Physiol 162:1290–1310

Wei S, Wang X, Zhang J, Liu P, Zhao B, Li G, Dong S (2015) The role of nitrogen in leaf senescence of summer maize and analysis of underlying mechanisms using comparative proteomics. Plant Sci 233:72–81

Wiedemuth K, Müller J, Kahlau A, Amme S, Mock H, Grzam A, Hell R, Egle K, Beschow H, Humbeck K (2005) Successive maturation and senescence of individual leaves during barley whole plant ontogeny reveals temporal and spatial regulation of photosynthetic function in conjunction with C and N metabolism. J Plant Physiol 162:1226–1236

Wilson KA, McManus MT, Gordon ME, Jordan TW (2002) The proteomics of senescence in leaves of white clover, *Trfolium repens* (L.) Proteomics 2:1114–1122

Wingler A, Roitsch T (2008) Metabolic regulation of leaf senescence: interactions of sugar signalling with biotic and abiotic stress responses. Plant Biol 10:50–62

Woo HR, Chung KM, Park JH, Oh SA, Ahn T, Hong SH, Jang SK, Nam HG (2001) ORE9, an F-box protein that regulates leaf senescence in Arabidopsis. Plant Cell 13:1779–1790

Woo HR, Goh CH, Park JH, Teyssendier de la Serve B, Kim JH, Park YL, Nam HG (2002) Extended leaf longevity in the ore4-1 mutant of Arabidopsis with a reduced expression of a plastid ribosomal protein gene. Plant J 31:331–340

Woo HR, Koo HJ, Kim J, Jeong H, Yang JO, Lee IW, Jun JH, Choi SH, Park SJ, Kang B, Kim YW, Phee BK, Kim JH, Seo C, Park C, Kim SC, Park S, Lee B, Lee S, Hwang D, Nam HG, Lim POK (2016) Programming of plant leaf senescence with temporal and inter-organellar coordination of transcriptome in Arabidopsis. Plant Physiol 171:452–467

Wu XY, Hu WJ, Luo H, Xia Y, Zhao Y, Wang LD, Zhan LM, Luo JC (2016) Transcriptome profiling of developmental leaf senescence in sorghum (*Sorghum bicolor*). Plant Mol Biol 92: 555–580

Ximenez-Embun MG, Ortego F, Castañera P (2016) Drought-stressed tomato plants trigger bottom–up effects on the invasive *Tetranychus evansi*. PLoS One 11:e0145275

Yoshida S (2003) Molecular regulation of leaf senescence. Curr Opin Plant Biol 6:7984

Zentgraf U, Jobst J, Kolb D, Rentsch D (2004) Senescence-related gene expression profiles of rosette leaves of *Arabidopsis thaliana*: leaf age versus plant age. Plant Mol 6:178–183

Zhang WY, Xu YC, Li WL, Yang L, Yue X, Zhang XS, Zhao XY (2014) Transcriptional analysis of natural leaf senescence in maize. PLoS One 9:e115617

Interactions Between Copper Homeostasis and Metabolism in Plants

Anna Schulten and Ute Krämer

Abstract Vascular plants require the micronutrient copper (Cu) at merely 6–8 μg g^{-1} leaf dry biomass. Yet insufficient bioavailable soil Cu is an immediate threat for growth, stress resistance and reproduction, because the flux through major metabolic pathways depends critically on Cu-containing metalloproteins, for example in the electron transport chains of both photosynthesis and respiration, in cell wall biosynthesis and in secondary metabolism. A first understanding is emerging of how a tightly knit Cu homeostasis network orchestrates Cu uptake, distribution, storage and remobilisation to meet a plant's nutritional demands while mitigating the unsurpassed toxicity potential of this nutrient element. In order to manage environmental, developmental and metabolic transitions, a plant must efficiently coordinate its metabolism with Cu homeostasis. In this review, we focus on interactions between Cu homeostasis and metabolism in plants, drawing also from insights obtained through the study of other biological systems. We describe evidence for possible roles of CuZn superoxide dismutases, convergence between Cu and sugar signalling, as well as the transcriptional remodelling predominantly of cell wall and secondary metabolism in response to Cu deficiency. The biological functions of most Cu-dependent proteins, which are often members of large protein families, remain to be demonstrated. Finally, we pinpoint future research needs towards an improved understanding of the interactions between Cu homeostasis and metabolism, which will provide avenues for improving not only crop quality, but importantly also crop resilience and yields.

Communicated by Ulrich Lüttge

A. Schulten and U. Krämer (✉)
Molecular Genetics and Physiology of Plants, Ruhr University Bochum, Universitätsstr. 150, ND3/30, 44801 Bochum, Germany
e-mail: Ute.Kraemer@ruhr-uni-bochum.de

Progress in Botany (2018) 79: 111–146, DOI 10.1007/124_2017_7,
Published online: 22 April 2017

Contents

1 Introduction

As sessile organisms, land plants are subject to a multitude of abiotic and biotic environmental challenges. Their perpetual acclimation is required for survival and reproductive success in non-optimal, fluctuating and spatially heterogeneous environments. Moreover, as primary producers, plants generate photosynthates and a rich variety of other organic compounds, and they are capable of acquiring all inorganic nutrients to sustain their own lifecycle, as well as the ecosystems depending on them.

Many soils are either globally or locally deficient in bioavailable mineral nutrients, or exhibit mineral imbalances that can lead to a hazardous excess of a specific mineral. Plants respond to such challenging edaphic conditions by activating nutrient acquisition through mobilisation and uptake, by managing nutrient distribution and utilisation and by implementing storage or detoxification of excess ions, in a coordinated manner at the plant, cellular and subcellular levels (Marschner and Marschner 2012; Jeong and Guerinot 2009; Plaxton and Tran 2011; Masclaux-Daubresse et al. 2010; Hawkesford and De Kok 2006). Besides these very well studied processes, plant mineral nutrient status or exposure to toxic levels of minerals can have profound effects on plant metabolism, growth and development (Salome et al. 2013; Forde and Lorenzo 2001; Barberon et al. 2016; Giehl et al. 2014). Vice versa, environmental conditions can change periodically following predictable circadian and seasonal rhythms, or stochastically, thus generating vast fluctuations in parameters like temperature, light and water supply, which can have enormous effects on a plant's metabolic state. In the face of the needs of dynamic metabolic activity, plants must efficiently manage the resources of available mineral nutrients in the soil in order to ensure their biochemical functionality following environmentally triggered metabolic adjustments and growth, as well as exogenously or endogenously triggered developmental transitions. This aspect remains less explored by researchers to date (Hong et al. 2013; Perea-Garcia et al. 2016; Haydon et al. 2015).

Transition metal nutrients can serve as a paradigm for examining the interactions between inorganic nutrient homeostasis and plant metabolism for three reasons. First, cellular biochemistry depends vitally on protein cofactors that contain or consist of transition metals. Second, transition metal availability in the environment is often sub-optimal. And third, transition metals are severely toxic when present in excess.

The dynamic range between deficiency and toxicity is particularly narrow for copper (Cu), which among all transition metal nutrients is positioned at the maximum end of the Irving–Williams series of ligand binding affinities (Fraústo da Silva and Williams 2001) (a list of all abbreviations is given in Table 1). As a redox-active metal, Cu can alternate between the oxidised (+II) and reduced (+I) oxidation states under physiologically relevant conditions, thus allowing the catalysis of single electron transfer reactions. Indeed, among the enzymes with known three-dimensional structures, 93% of Cu-containing enzymes were found to belong to the class of oxidoreductases (Waldron et al. 2009). Redox cycling of Cu can lead to the production of toxic hydroxyl radicals via the participation of Cu ions in Fenton Chemistry (Halliwell and Gutteridge 1984), which is employed to fight microbial pathogens in animal innate immunity (Hood and Skaar 2012). Furthermore, Cu cations can displace other metal cations in metalloproteins, which can result in loss of enzymatic activity or erroneous protein structures (Festa and Thiele 2011; Fraústo da Silva and Williams 2001). Therefore, regulation and control of Cu homeostasis must be extraordinarily precise in order to ensure the functionality of proteins requiring Cu as a cofactor while preventing the accumulation of a harmful excess of Cu.

During the history of life on Earth, Cu is among a few metals which became more bioavailable as a result of increasing atmospheric O_2 levels following the evolution of oxygenic photosynthesis (Fraústo da Silva and Williams 2001). Iron (Fe), on the other hand, is poorly soluble in the presence of O_2, leading to a strong decrease in Fe solubility as a consequence of the oxygenation of Earth's atmosphere. A comparative genomics study revealed that most aerobic prokaryotes employ Cu-containing proteins, while most anaerobic prokaryotes do not, indicating a link between oxygenated environments and the shift to a more prominent use of Cu as a micronutrient (Ridge et al. 2008). At least to some degree, biological metal utilisation is thus driven by metal availability. The genome of the model plant *Arabidopsis thaliana* encodes around 200 Cu-dependent proteins according to estimates from both experimental work and the occurrence of conserved Cu binding motifs in the proteome (Andreini et al. 2008; Bernal et al. 2012).

2 The Quest for Cu Homeostasis

The demands on Cu homeostasis are governed primarily by abundant Cu-dependent proteins of central importance in plant metabolism. Cu is a cofactor in the respiratory chain enzyme cytochrome *c* oxidase (COX), in which the main catalytic subunits and their metal-containing redox centres are conserved among prokaryotes and eukaryotes, including plants (Kadenbach et al. 2000). Another widely found Cu protein is CuZn superoxide dismutase (CSD), which is highly abundant in plants and other eukaryotes. Three isoforms are known in *A. thaliana*, and these are localised in the cytosol (CSD1), chloroplasts (CSD2) and peroxisomes (CSD3), respectively (Kliebenstein

Table 1 List of abbreviations

Abbreviation	Description
4CL	4-Coumarate-CoA Ligase
ALDH2C4	Aldehyde Dehydrogenase 2C4
AOX	Alternative Oxidase
ARPN	Plantacyanin
ATX1	Cu chaperone Anti-oxidant1 (also *Saccharomyces cerevisiae*)
C3H	*p*-Coumarate-3-Hydroxylase
C4H	Cinnamate 4-Hydroxylase
CAD8	Cinnamyl Alcohol Dehydrogenase8
CCC2	Cu-transporting P-type ATPase2 (*Saccharomyces cerevisiae*)
CCH	Cu Chaperone
CCOAMT	Caffeoyl-CoA 3-*O*-Methyltransferase
CCS	Cu Chaperone for Superoxide dismutase
CERK1	Chitin Elicitor Receptor Kinase1
CK	Casein Kinase
CKL	Casein Kinase 1-Like
CLP	Caseinolytic Protease
COPT (COPT1, COPT2, COPT5, COPT6)	Copper transporter (members of protein family)
COX (COX1, COX2, COX3, COX5b.1)	Cytochrome *c* Oxidase (subunits)
COX (COX11, COX17)	Cu chaperone for Cytochrome *c* Oxidase
CRR1	Copper Response Regulator1 (*Chlamydomonas reinhardtii*)
CSD (CDS1, CDS2, CDS3)	CuZn Superoxide Dismutase (isoforms)
CTR3	Copper Transport Protein3 (*Podospora anserina*)
Cu	Copper
CuREs	Cu Responsive *cis*-regulatory Elements
CYC6	Cytochrome c_6
DOPA	L-3,4-dihydroxyphenylalanine
DCL1	Dicer-Like1
EDTA	Ethylenediaminetetraacetic acid
F5H	Ferulate 5-Hydroxylase
FA	Fatty acid
Fe	Iron
FET3	Ferrous Transport3 (*Saccharomyces cerevisiae*)
FRO (FRO4, FRO5)	Ferric Reductase Oxidase (members of protein family)
FSD (FSD1, FSD2, FSD3)	Fe superoxide dismutase (isoforms)
FTR1	Fe Transporter1 (*Saccharomyces cerevisiae*)
HBED	N,N′-di-(2-hydroxybenzoyl)-ethylenediamine-N,N′-diacetic acid
HCT	Hydroxycinnamoyl-CoA shikimate/quinate Hydroxycinnamoyl Transferase
HMA (HMA5, HMA6, HMA7, HMA8)	Heavy-Metal ATPase (members of protein family)

(continued)

Table 1 (continued)

Abbreviation	Description
HXK1	Hexokinase1
HY5	Elongated Hypocotyl5
IMS	Mitochondrial intermembrane space
K	Potassium
LAC (LAC4, LAC17)	Laccase (members of protein family)
LMCO	Laccase-like multi-copper oxidase
LPR (LPR1, LPR2)	Low Phosphate Root (isoforms)
MAC1	Metal-binding Activator1 (*Saccharomyces cerevisiae*)
MCO	Multi-copper oxidase
Metab.	Metabolism
miRNA	MicroRNA
Mito.	Mitochondrial
Mn	Manganese
MS	Murashige and Skoog
MSD1	Mn Superoxide Dismutase1
Na	Sodium
Ni	Nickel
OMT1	Caffeate *O*-Methyltransferase1
OPP	Oxidative pentose phosphate
PAA (PAA1/HMA6, PAA2/HMA8)	Cu transporting P-type ATPase of Arabidopsis (members of protein family)
PAL3	Phenylalanine Ammonia Lyase3
PC (PC1, PC2)	Plastocyanin
PCH1	Plastid Chaperone1
PET	Photosynthetic electron transport
Pi	Phosphate
PPO	Polyphenol oxidase
pri-miR398	Primary transcript of miR398
RACK1A	Receptor for Activated C Kinase1A
RAN1 (HMA7)	Responsive to Antagonist1
RNAi	RNA interference
RNA-Seq	Sequencing-based transcriptome analysis
ROS	Reactive oxygen species
SBP	Squamosa promoter Binding Protein
SCO1	Synthesis of Cytochrome *c* Oxidase1
SHAM	Salicylhydroxamic acid
SKS	SKU5 Similar
SKU5	Skewed5
SOD (SOD1, SOD2)	Superoxide dismutase (*Saccharomyces cerevisiae*, mammals)
SPL7	Squamosa Promoter-binding protein Like7
TCA	Tricarboxylic acid cycle
UBQ3	Polyubiquitin3
WNK8	With No lysine Kinase8

(continued)

Table 1 (continued)

Abbreviation	Description
Wnt signalling	Wingless/integration1 signalling pathway originally identified in *Drosophila melanogaster*
XET	Xyloglucan endo-transglycosylase
YCK	Casein Kinase1-gamma homolog (*Saccharomyces cerevisiae*)
YSL2	Yellow Stripe-Like2
Zn	Zinc

Gene/protein names refer to *Arabidopsis thaliana* (and other plants), unless indicated otherwise in the description

et al. 1998). A number of Cu-containing proteins have key roles in biochemical processes characteristic of plants. Among these is plastocyanin (PC), the most abundant Cu protein in vascular plants (Marschner and Marschner 2012). It acts as a soluble electron carrier between the cytochrome b_6f complex and photosystem I of the photosynthetic electron transport chain in the thylakoid lumen (Redinbo et al. 1994). In addition, receptor proteins for the plant hormone ethylene require the association of a Cu^+ ion for ethylene binding activity (Rodriguez et al. 1999). The implications of this for a crosstalk between ethylene signalling and Cu homeostasis have not been addressed by targeted studies yet. Amine oxidases and laccase (LAC)-like proteins, the latter belonging to the large group of multi-copper oxidases (MCOs), are Cu-dependent proteins that have been linked to functions in cell wall differentiation and modelling (McCaig et al. 2005; Paschalidis and Roubelakis-Angelakis 2005). An involvement of MCOs in cellular Fe uptake or export, as known in yeast and human, respectively, was thought to be absent in land plants. Recent publications, however, strongly support an involvement of MCOs in specific aspects of plant Fe homeostasis (Bernal et al. 2012; Müller et al. 2015).

A dedicated segment of the plant metal homeostasis network ensures a functional complement of Cu metalloproteins while preventing excessive accumulation of Cu in any location within the plant. Cu is acquired from the soil via high-affinity Cu^+ transporters of the COPT family of transmembrane proteins (Sancenon et al. 2003, 2004; Kampfenkel et al. 1995). Prior to cellular import by COPTs, Cu present in the +II oxidation state in an oxygenated environment must thus be reduced to Cu^+. In plants, this role is accomplished by specific plasma membrane proteins of gp91phox-related NADPH oxidases, the root surface Cu(II) chelate reductases (annotated as ferric reductase oxidase, FRO) FRO4 and FRO5 (Bernal et al. 2012). In *A. thaliana*, the COPT family has six members. COPT1 is plasma membrane-localised, and the gene encoding it is highly expressed in root tips. Antisense-silencing of *COPT1* results in a decrease in Cu levels by 40–60%, so that COPT1 is thought to make the predominant contribution to root Cu uptake (Sancenon et al. 2004). COPT2 was found in both green tissues and in some parts of the roots. In addition to constituting a secondary pathway for root Cu uptake especially under Cu deficiency, roles in phosphate sensing and Fe homeostasis have been suggested for COPT2 based on an upregulation of *COPT2* transcript levels in response to deficiencies of the corresponding nutrients (Perea-Garcia et al. 2013). *COPT6* is primarily expressed in shoots and may have a role in Cu distribution from green tissues to reproductive organs under Cu-deficient growth conditions (Garcia-Molina et al. 2013; Jung et al. 2012). The vacuolar membrane was described as the localisation of COPT5, which probably functions in the mobilisation of stored Cu from the vacuole into the cytoplasm (Garcia-Molina et al. 2011; Klaumann et al. 2011). Interestingly, COPT5 was one of the few transition metal transporters among the 30 most abundant tonoplast membrane proteins identified in a proteomics study of cultured Arabidopsis cells (Jaquinod et al. 2007). Given that a substantial proportion of cellular Cu localises to chloroplasts and no transporter mediating vacuolar Cu sequestration has been identified in Arabidopsis to date, the autophagy of chloroplast components followed by nutrient remobilisation from the vacuole appears to contribute to cellular cycling of Cu (Wang and Blumwald 2014; Lanquar et al. 2010; Ishida et al. 2008; Izumi et al. 2010). Cu^+ entering the cytosol via COPT transporters is thought to be transferred directly onto Cu delivery proteins, the so-called Cu metallochaperones, which mediate the trafficking of Cu^+ to target proteins through direct protein–protein interactions (O'Halloran and Culotta 2000). The underlying metallochaperone model for intracellular Cu trafficking was developed based on work in yeast and mammals, and it differs markedly from our general concept of how other monovalent cations, for example potassium (K^+) or sodium (Na^+), are distributed within the cytosol. Essentially, the metallochaperone model implies the absence of free aqueous Cu^+ ions, and thus the absence of Cu^+ diffusion, in the cytosol under virtually all physiological conditions (Rae et al. 1999). This concept is compelling given the chemical properties and pre-eminent toxicity of Cu^+, as well as the propensity of Cu^+ to disproportionate in aqueous solution ($Cu^+ + Cu^+ \rightarrow Cu^0 + Cu^{2+}$).

Homologs of the most extensively studied Cu metallochaperone, Cu Chaperone for Superoxide dismutase (CCS) from yeast, have been identified in several plant species (Burkhead et al. 2009). In *A. thaliana*, there is one homolog of CCS which is active in both the cytosol and plastids (Cohu et al. 2009). Another Cu chaperone involved in Cu delivery to chloroplasts is the recently identified Plastid Chaperone1 (PCH1), which delivers Cu to the Cu transporting P-type ATPase of Arabidopsis1/Heavy-Metal ATPase6 (PAA1/HMA6) located in the chloroplast inner envelope membrane (Blaby-Haas et al. 2014). Furthermore, there are two homologs of the soluble Cu chaperone Anti-oxidant1 (ATX1) from yeast, namely ATX1 and Cu Chaperone (CCH) in *A. thaliana* (Puig et al. 2007b; Himelblau et al. 1998). CCH differs from *At*ATX1 and *Sc*ATX1 by an extended C-terminus (Puig et al. 2007b). It was hypothesised that this C-terminal domain is involved in symplastic CCH protein transport through plasmodesmata and that CCH may mediate long-distance transport of Cu via the phloem (Mira et al. 2001; Puig et al. 2007b). *At*ATX1 was shown to interact with both HMA5 and HMA7, and it thus likely acts to deliver Cu to these Cu transporting P-type ATPases (Puig et al. 2007b; Andres-Colas et al. 2006). HMA7 (also known as Responsive to Antagonist1, RAN1) is required for the biogenesis of functional ethylene receptors and is thought to localise to the endomembrane system due to its ability to complement the yeast *ccc2* mutant (Hirayama et al. 1999; Binder et al. 2010). *ScCCC2* encodes a Cu-transporting P-type ATPase found to localise to a post-Golgi department for Cu delivery to the secretory pathway (Yuan et al. 1997). HMA5 was hypothesised to mediate Cu export from the root symplast for root-to-shoot transport via the apoplastic xylem. This hypothesis is based on the phenotype of a *hma5* mutant, which exhibited Cu sensitivity and accumulation of Cu in roots, but the intracellular location of HMA5 has not been determined yet (Andres-Colas et al. 2006; Burkhead et al. 2009).

A prerequisite for the adjustment of metabolic processes dependent on Cu availability is the existence of Cu sensing mechanisms. In the green alga *Chlamydomonas reinhardtii*, Copper Response Regulator1 (CRR1) is a Squamosa promoter Binding Protein (SBP) domain-containing transcription factor acting as a master regulator for Cu deficiency-responsive gene expression (Kropat et al. 2005). The SBP domain binds to Cu Responsive *cis*-regulatory Elements (CuREs) that contain the core sequence motif GTAC. In vitro experiments showed that the SBP domain does not bind to DNA in the presence of Cu^+ ions, suggesting that CRR1 activity can be altered by its direct interaction with Cu(I) (Sommer et al. 2010). It was hypothesised that the two zinc (Zn)-finger domains of SBP are occupied by Zn(II) in Cu-deficient conditions, whereas Zn (II) is substituted by Cu(I) under Cu-replete conditions, thus abrogating DNA binding of CRR1 and the transcriptional activation of its target genes (Sommer et al. 2010). This could provide a model for the inactivation of the Cu deficiency response. However, it remains unknown whether the substitution of Zn(II) by Cu(I) also occurs in vivo in the cytosol or in the nucleus. Experimental studies suggested the absence of free Cu^+ ions in the cytosol (Rae et al. 1999), so that in vivo the delivery of Cu(I) to CRR1 would most likely be mediated by a Cu metallochaperone. In relation to the open question of how Cu is sensed in photosynthetic eukaryotes, it is interesting to note that *C. reinhardtii* overaccumulates Cu under Zn limitation in a CRR1-dependent manner in association with physiological Cu deficiency despite a high cellular Cu content (Hong-Hermesdorf et al. 2014). Cu was found to accumulate in structures which might be lysosome-related organelles. The sequestration of Cu was suggested to prevent mismetallation of proteins with Zn^{2+} binding sites under Zn-deficient growth conditions. In Arabidopsis, the closest homolog of CRR1 is the transcription factor Squamosa Promoter-binding protein Like7 (SPL7). Like CRR1, SPL7 binds to a GTAC core motif via the conserved SBP domain and was shown to be a central regulator of Cu deficiency responses (Yamasaki et al. 2009; Bernal et al. 2012). Although *SPL7* mRNA levels are unresponsive to plant Cu status, SPL7 activity relating to the Cu deficiency regulon is observed only under low Cu conditions. Protein stability or localisation was suggested to regulate SPL7 activation dependent on Cu availability (Garcia-Molina et al. 2014; Yamasaki et al. 2009). SPL7 contains a putative transmembrane helix in the vicinity of its C-terminus, which was proposed to anchor SPL7 to a membrane, probably of the endoplasmic reticulum (Garcia-Molina et al. 2014). As this conflicts with a nuclear localisation of SPL7, which is required for its function as a transcription factor, a proteolytic cleavage event was proposed to release and thus activate SPL7 following a Cu deficiency-triggered ER stress response. In line with this suggestion, the SPL7 protein truncated behind the DNA binding SBP domain was able to complement *spl7* mutant phenotypes under Cu-deficient growth conditions (Garcia-Molina et al. 2014). Moreover, several SPL7 target genes displayed constitutive activation in these transgenic lines irrespective of plant Cu status, indicating that the abilities of this truncated SPL7 protein variant to bind DNA and activate transcription are not influenced by cellular Cu levels. It remains to be determined, however, how exactly the proteolytic processing of

SPL7 is regulated by Cu availability and what the identities are of the proteases involved in this process.

Although CRR1 and SPL7 both bind to CuREs and activate transcriptional responses to Cu deficiency, the two organisms *C. reinhardtii* and *A. thaliana* differ fundamentally in the metabolic strategies by which they adjust. In *C. reinhardtii*, CRR1 activates Cytochrome c_6 (CYC6) which acts as a heme-containing substitute for PC (Eriksson et al. 2004; Quinn and Merchant 1995). The concurrent degradation of PC is mediated by a protease which also is under the transcriptional control of CRR1 (Castruita et al. 2011). In the seed plant Arabidopsis, however, PC is essential for photosynthesis (Weigel et al. 2003). Instead, the abundant CSD proteins are replaced by Fe superoxide dismutase (FSD) in a process that is thought to promote the allocation of available Cu to essential Cu metalloproteins like PC for maintenance of their function even under Cu deficiency (Yamasaki et al. 2007; Abdel-Ghany et al. 2005). This mechanism economising on Cu is mediated by SPL7, which activates both *FSD1* and a set of microRNAs (miRNAs) effecting the cleavage of transcripts encoding Cu-dependent proteins (Fig. 1), for example *CSD1* and *CSD2*

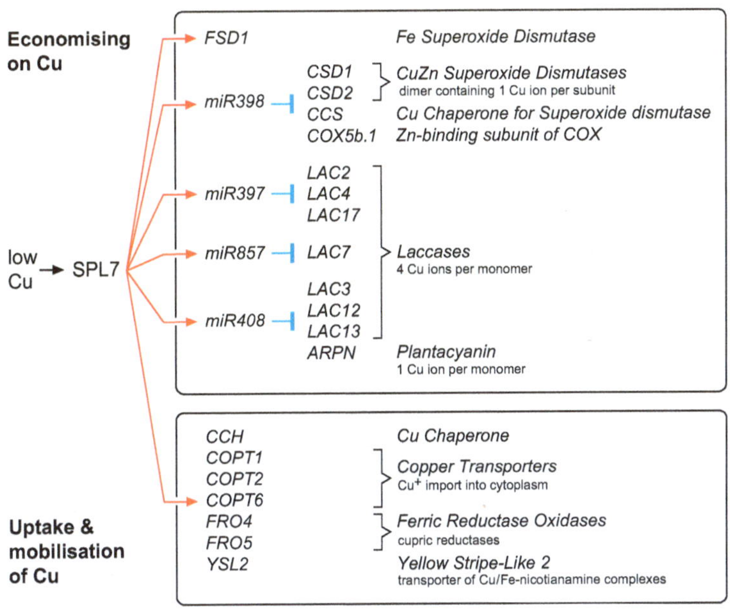

Fig. 1 The Cu deficiency response in *Arabidopsis thaliana* dependent on the transcription factor Squamosa Promoter binding protein-Like 7 (SPL7). SPL7 acts in the transcriptional activation of Cu acquisition systems under Cu-deficient growth conditions. Additionally, SPL7 mediates the microRNA-dependent downregulation of transcripts encoding several abundant Cu metalloproteins in Cu-deficient plants. Most importantly, CuZn superoxide dismutases (CSD) are replaced by Fe superoxide dismutases (FSD), which is thought to economise on Cu for maintaining essential functions of other Cu-dependent proteins. See main text for details. *Red colour* indicates transcriptional activation; *blue colour* indicates post-transcriptional downregulation

(Yamasaki et al. 2009). The miRNAs involved in the post-transcriptional downregulation of mRNAs encoding Cu metalloproteins are the so-called Cu miRNAs miR397, miR398, miR408 and miR857. Along with the transcripts of *CSD1* and *CDS2*, miR398 targets the *CCS* mRNA as well as *COX5b.1*, which encodes the Zn binding subunit of COX (Yamasaki et al. 2007). The target transcripts of miR397, miR408 and miR857 encode several laccase-like multi-copper oxidases (LCMOs), whereby miR408 additionally targets plantacyanin (ARPN) (Abdel-Ghany and Pilon 2008). SPL7 is also required for enhancing the transcription of genes encoding proteins implicated in Cu uptake and mobilisation under Cu deficiency (Fig. 1) (Yamasaki et al. 2009; Bernal et al. 2012). *FRO4* and *FRO5* are among the transcripts that are most strongly upregulated under Cu deficiency, in an SPL7-dependent manner (Bernal et al. 2012). In addition, the expression of *COPT1*, *COPT2* and *COPT6* increased in response to Cu deficiency, which was not observed in *spl7* mutants (Bernal et al. 2012; Yamasaki et al. 2009; Jung et al. 2012). Similarly, under Cu-deficient growth conditions the increase in transcript levels of *Yellow Stripe-Like2* (*YSL2*), which encodes a putative transporter of metal-nicotianamine complexes, also depends on SPL7 (Yamasaki et al. 2009; Bernal et al. 2012). YSL2 was suggested to act in lateral metal movement of metals in the vasculature (DiDonato et al. 2004). Studies using different approaches like microarray (Yamasaki et al. 2009), RNA sequencing (RNA-Seq) (Bernal et al. 2012), or chromatin-immunoprecipitation followed by sequencing (Zhang et al. 2014) identified a large number of putative SPL7 targets, and only a small proportion thereof has been characterised with respect to their functions in Cu homeostasis. While the network of proteins directly mediating Cu homeostasis is conserved across eukaryotes of different kingdoms, the current state of our knowledge suggests that the regulatory proteins and mechanisms maintaining Cu homeostasis are not conserved: CRR1 and SPL7 both belong to a transcription factor family which is characterised by the SBP domain that is uniquely found in green algae and plants (Birkenbihl et al. 2005).

A number of key proteins acting in plant Cu homeostasis and its regulation have been identified and characterised in some detail, and excellent reviews have summarised this work (Burkhead et al. 2009; Aguirre and Pilon 2015; Puig et al. 2007a; Curie et al. 2009). However, much less is known about how their regulation may be tied to metabolic signals and how plant Cu status feeds back to adjust metabolism, growth and development. In this review, we describe the current status of our understanding of interactions between Cu homeostasis and metabolism, also drawing from a variety of non-plant model systems in order to provide a basis for future research in this area.

3 Copper in Chloroplast Photosynthesis

Plants depend on photosynthesis for their energy supply and the biosynthesis of organic compounds. Glyceraldehyde 3-phosphate, one of the primary products of the Calvin Cycle, can serve as the initial substrate in multiple biosynthetic pathways of plants, including the synthesis of hexose sugars. The biochemical reaction series required for photosynthesis take place almost exclusively in chloroplasts. Chloroplasts demand the highest cellular Cu quota in plants: about one third of leaf Cu (Shikanai et al. 2003). Cu is primarily required in PC as one of the most abundant proteins in the thylakoid lumen (Schubert et al. 2002), which is involved in all photosynthetic electron transport. PC is a nuclear-encoded protein and thus imported into the chloroplast and the thylakoid lumen subsequent to its synthesis (Weisbeek et al. 1989). Because of the strong unfolding capacity of the chloroplast import machinery, the association of the PC apoprotein with its Cu cofactor presumably occurs in the same compartment where the mature protein serves its function (Merchant and Dreyfuss 1998). Therefore, Cu must be separately delivered to the thylakoid lumen for PC maturation to take place. The Cu transporters involved in this delivery have been identified as PAA1/HMA6 and PAA2/HMA8, two Cu^+ pumps of the P-type ATPase superfamily (Shikanai et al. 2003; Abdel-Ghany et al. 2005). PAA1 is localised in the inner chloroplast envelope membrane to transport Cu into the stroma (Shikanai et al. 2003). PAA2 transports Cu from the stroma across the thylakoid membrane into the thylakoid lumen (Abdel-Ghany et al. 2005). Whereas PAA1 protein abundance was not modulated in response to Cu status, PAA2 protein stability was observed to increase in plants grown under low Cu conditions (Tapken et al. 2012). PAA2 stabilisation was proposed to facilitate Cu flux into the thylakoids when Cu is limited. Under Cu-replete conditions, lower abundance of PAA2 protein might act to retain more Cu in the stroma or the cytosol for distribution to Cu-dependent proteins other than PC (Tapken et al. 2012). The chloroplast Caseinolytic Protease (CLP) was identified as responsible for the Cu-dependent increase in PAA2 turnover (Tapken et al. 2015). The mechanism of how this is triggered in a Cu-dependent manner has not been fully resolved, but it was suggested to involve CLP substrate recognition based on Cu-dependent conformational changes of PAA2.

Cu deprivation was reported to cause a strong reduction in photosynthetic electron transport (PET) capacity in both poplar (*Populus trichocarpa)* and hybrid poplar (*P. tremula* × *P. alba*), especially in young leaves of the latter (Ravet et al. 2011; Shahbaz et al. 2015). In line with this observation, PC protein abundance was severely decreased under these conditions, which is the likely cause for the observed decrease in PET. Non-specific or secondary effects on PET were judged minimal on the basis that the maximum quantum yield of photosystem II (F_v/F_m), which operates upstream of PC in the PET, was only slightly affected even by long-term Cu depletion in hybrid poplar (Shahbaz et al. 2015). Like several other angiosperms, including poplar, rice and tobacco, *A. thaliana* expresses two PC isoforms called PC1 and PC2. Studies performed with *A. thaliana pc1* and *pc2* mutants revealed that the isoforms are differentially regulated in response to Cu (Abdel-Ghany 2009). PC2 protein is generally much more abundant than PC1, and only PC2 protein levels decrease in response to Cu deficiency. PC1 was thus proposed to drive basal PET under Cu deficiency. As PC activity is essential for PET (Weigel et al. 2003), the differential regulation of PC isoforms may constitute an adaptive mechanism to cope with fluctuating Cu availability (Abdel-Ghany 2009). The spatiotemporal analysis of the effects of Cu resupply to Cu-depleted *P. trichocarpa* plants showed that Cu is preferentially allocated to PC, particularly in young leaves, whereas CSD and polyphenol oxidase (PPO) protein levels exhibited delayed recovery (Ravet et al. 2011). As expected, PET activity recovered proportionally with increasing PC abundance. The effects of Cu deficiency are reflected in plant biomass production, which was reduced by about half in Cu-depleted compared to Cu-replete *P. trichocarpa* plants (Ravet et al. 2011). This emphasises the critical importance of sufficient Cu supply for optimal photosynthesis rates and net primary production in land plants.

4 Copper in Mitochondrial Respiration

Mitochondria are at the centre of respiratory metabolism. The oxygen-consuming redox reactions of the mitochondrial electron transport chain are employed to generate a proton gradient across the inner membrane of mitochondria which drives the synthesis of ATP as a carrier molecule to provide energy for a variety of plant metabolic processes. Cu-containing COX is the terminal enzymatic complex IV of the mitochondrial respiratory chain and is composed of 12–14 subunits in plants (Millar et al. 2004). Except for the three subunits COX1, COX2 and COX3, all subunits are encoded in the nuclear genome (Unseld et al. 1997), necessitating a coordinated assembly of subunits for COX biogenesis. The number of different metal co-factors found in COX, including two hemes, three Cu ions and Zn, magnesium and Na ions, further complicates the assembly (Carr and Winge 2003). The Cu ions are coordinated by COX1 in a mononuclear Cu site designated Cu_B, and by COX2 in the binuclear Cu_A site. To this date, there is little plant-specific information about the biogenesis of COX, but the process is well characterised in yeast. In a simple model, the Cytochrome *c* Oxidase Cu chaperone COX17 acquires Cu in the mitochondrial intermembrane space (IMS) and delivers it to COX11 and Synthesis of Cytochrome *c* Oxidase1 (SCO1), which are associated with the inner mitochondrial membrane (Robinson and Winge 2010). COX11 then donates Cu to the Cu_B site in COX1, whereas SCO1 delivers Cu to the Cu_A site in COX2. How Cu is transported to the IMS in the first place, however, remains to be determined. Plant homologs of COX17 and SCO1 were identified in *A. thaliana* and demonstrated to complement the corresponding yeast mutants, suggesting that they encode proteins required for COX activity (Garcia et al. 2016; Attallah et al. 2011). *At*COX11 was not able to complement the respiratory deficiency phenotype of the yeast *Δcox11* strain, but its localisation and the characterisation of COX11 knockdown lines support a role of *At*COX11 in COX assembly (Radin et al. 2015). COX activity was determined in pea plants (*Pisum sativum*) grown hydroponically under different Cu regimens and found to be moderately decreased (by about 26%) in Cu-deficient plants (Ayala and Sandmann 1988). In *C. reinhardtii*, COX-dependent respiration was strongly impaired in Cu-deficient cells (by about 75% compared to Cu-replete cultures), despite the operating Cu sparing mechanisms (Kropat et al. 2015). In vascular plants, protein levels of COX2 were reported to be unaltered under Cu deficiency (Yamasaki et al. 2007; Ravet et al. 2011), which was proposed to indicate that Cu delivery to COX might have an even higher priority than delivery to PC (Ravet et al. 2011). However, it was not examined whether Cu was bound to COX2 protein even under Cu-deficient conditions. COX5b contains a Zn cofactor, and the mRNA of its isoform *COX5b.1* is a target of miR398 and is mildly downregulated under Cu-deficient growth conditions (Yamasaki et al. 2007; Bernal et al. 2012). Whether COX5b protein levels decrease under Cu deficiency in plants is not known. Unlike animals, plants contain a second terminal oxidase of the respiratory electron transport chain in addition to COX, the cyanide-insensitive Alternative Oxidase (AOX) (Millenaar and Lambers 2003). AOX couples the oxidation of ubiquinol with the reduction of O_2 to H_2O, which introduces a branch away from the canonical mitochondrial electron transport pathway. Electron flow via AOX results in a reduced ATP yield of respiration because it bypasses the proton-pumping complexes cytochrome *c* reductase (complex III) and the oxygen-reducing COX (complex IV) of the mitochondrial electron transport chain. AOX activity in plants has generally been implicated in metabolic homeostasis to confer increased tolerance against several abiotic and biotic stresses, for example low temperatures or salt stress (Vanlerberghe 2013). AOX is thought to prevent the production of reactive oxygen species (ROS) and to ensure continuous oxidation of the NADH pool when the cytochrome pathway is restricted (Millenaar and Lambers 2003). This allows the continuation of the citric acid cycle to generate reducing molecules for the mitochondrial electron transport chain and precursors for amino acid metabolism. The mRNA levels of *AOX1D*, one of five AOX isoforms in *A. thaliana*, were upregulated in roots in an SPL7-dependent manner under Cu deficiency (Bernal et al. 2012). Strong upregulation of the mRNA abundance of the AOX isoform *AOX2* was also observed in Cu-depleted cultures of *C. reinhardtii* (Kropat et al. 2015; Castruita et al. 2011). AOX protein levels were not determined in these studies. Inhibitors such as cyanide, which inhibits the cytochrome pathway, or the AOX inhibitor salicylhydroxamic acid (SHAM), can be used to study AOX capacity, the maximum possible flux of electrons to AOX. AOX capacity, however, gives no indication of AOX activity, which is the actual flux of electrons to AOX and can be measured with oxygen isotopes, based on the observation that AOX and COX have different discrimination ratios for heavy oxygen ($^{18}O^{16}O$) (Guy et al. 1989). To our knowledge, the oxygen discrimination technique has not been employed so far to study the effects of Cu deficiency on AOX respiration in plants. Despite relying on the use of inhibitors to distinguish between COX and AOX contributions to respiration, studies on the fungal model system *Podospora anserina* provide interesting clues for the possible role of AOX respiration under Cu deficiency. The long-living *grisea* mutant of *P. anserina* has a mutation leading to a splice defect in the gene *GRISEA* (Osiewacz and Nuber 1996), which encodes a transcription factor involved in the regulation of Cu homeostasis (Borghouts and Osiewacz 1998). The mutant suffers from cellular Cu deficiency because it is unable to enhance expression of the gene encoding the high-affinity Cu transporter *Pa*CTR3 (Borghouts et al. 2002). Both transcript and protein abundance of *Pa*AOX were increased in *grisea* in response to reduced COX activity (Borghouts et al. 2001). The same response, combined with a similar longevity phenotype as *grisea*, was observed in two other *P. anserina* mutants called *ex1* and *cox5:BLE*, which are COX-deficient but not affected in Cu homeostasis (Schulte et al. 1988; Dufour et al. 2000). This indicates that disruptions in mitochondrial electron transport, instead of or in addition to cellular Cu deficiency, can lead to increased expression of *Pa*AOX, a mechanism which resembles the so-called retrograde response reported in yeast (Liao et al. 1991; Liao and Butow 1993). Respiration via AOX results in reduced generation of ROS, which was observed in both *P. anserina* and plants (Dufour et al. 2000; Wagner and Moore 1997). In *P. anserina*, the reduction in ROS levels was suggested to contribute to the extended lifespan of the *grisea* mutant, in conjunction with the stabilisation of mitochondrial DNA in the absence of GRISEA (Borghouts et al. 1997; Borghouts et al. 2001). By affecting the generation of signalling molecules such as superoxide, AOX may play a role in mitochondrial signalling pathways of Cu-deficient plants. More experiments are needed to understand how respiratory metabolism is affected by Cu homeostasis and how the underlying regulatory pathways are organised.

5 Copper-Zinc Superoxide Dismutases: Additional Roles Beyond Oxidative Stress Tolerance?

Because of their ability to catalyse the dismutation of superoxide to hydrogen peroxide, superoxide dismutases (SOD) have been associated with functions in oxidative stress tolerance of plants and other organisms living in oxygen- or ROS-rich environments (McCord and Fridovich 1969; Alscher et al. 2002). Four types of SODs have been described so far, classified by their active site containing both Cu and Zn, or Fe, manganese (Mn) or nickel (Ni) as metal cofactors. The three Arabidopsis CuZnSOD isoforms were named CSD1 to CSD3 (Kliebenstein et al. 1998), whereas the predominantly cytosolic CuZnSOD of yeast and mammals is designated Superoxide Dismutase1 (SOD1) (Culotta et al. 2006). Yeast and mammalian SOD2 contains Mn as a cofactor and localises to the mitochondrial matrix (Culotta et al. 2006), in analogy to Arabidopsis Mn Superoxide Dismutase1 (MSD1) (Kliebenstein et al. 1998; Morgan et al. 2008). FeSOD seems to be absent from fungi and animals (Miller 2012). In Arabidopsis, there are three FeSOD isoforms called FSD1, FSD2 and FSD3, with FSD1 as the most abundantly expressed isoform (Pilon et al. 2011; Kliebenstein et al. 1998). FSD2 and FSD3 are plastid-localised. The localisation remains unresolved for FSD1, because evidence from different studies indicated contrasting plastid or cytosolic localisation (Pilon et al. 2011).

Despite the general association of SOD function with ROS scavenging, the efforts to elucidate their biological roles have often remained inconclusive (Alscher et al. 2002; Kliebenstein et al. 1998). In dependence on plant Cu and Fe status, *CSD1*, *CDS2* and *FSD1* mRNA levels were shown to be reciprocally regulated, which was reflected in protein abundance, as well (Yamasaki et al. 2007; Abdel-Ghany et al. 2005; Cohu and Pilon 2007; Waters et al. 2012). This could be interpreted as an outcome of the importance of maintaining superoxide scavenging capacity under all growth conditions. However, this interpretation is questioned by the results of a study on *At*CCS, a metallochaperone that traffics the Cu cofactor to CSD1 and CSD2 isoforms (Cohu et al. 2009; Huang et al. 2012). Under Cu-sufficient growth conditions, a *ccs* mutant thus lacks both FSD1 and CSD activity in the cytosol and the chloroplast. Surprisingly, the *ccs* mutant showed no visible phenotype, no measurable photosynthetic defects and no change in superoxide levels, not even after exposure to high-light (Cohu et al.

2009). This suggests that CSDs are dispensable for the metabolism of superoxide generated during photosynthesis or elsewhere in the cell. A more recent publication reported CCS-independent residual activity of *At*CSD1, which was concededly much lower than *At*CSD1 activity in the presence of CCS (about 36% of wild-type levels) (Huang et al. 2012). Given that there was no evident phenotype of the Arabidopsis *ccs* mutant, the authors reasoned that the CCS-independent CSD activity must be sufficient for normal plant growth even under high light conditions. The feasibility of this hypothesis is supported by a study in yeast, which demonstrated that only about 2% of normal SOD1 protein levels is required to protect cells against superoxide toxicity (Corson et al. 1998).

The biological function of cellular amounts of CSD proteins in excess of its required enzymatic activity has not been resolved. It was hypothesised that their high abundance is linked to an additional role in buffering cellular Cu levels based on their ability to sequester Cu ions (Cohu et al. 2009). This has also been suggested to occur in yeast (Culotta et al. 1997). In addition, another role for CSD, i.e. outside its enzymatic function in the conversion of superoxide to hydrogen peroxide, was proposed based on a study in yeast. The authors showed that CCS and SOD1 are necessary for the activation of Metal-binding Activator1 (MAC1), a Cu binding transcription factor that activates Cu deficiency responses (Wood and Thiele 2009). Diminishing cellular oxidative stress levels did not rescue the activation of MAC1 in *sod1Δ* cells. Instead, the catalytic activity of SOD1 promoted the ability of MAC1 to bind to CuRE upstream of its target genes under conditions of Cu deficiency. Both CCS and SOD1 proteins partially localised to the nucleus of yeast cells. While nuclear localisation of CCS and SOD1 has also been observed in mammalian cells (Casareno et al. 1998; Crapo et al. 1992), it has not been reported in plant cells so far (Huang et al. 2012; Kliebenstein et al. 1998; Pilon et al. 2011). Moreover, the Arabidopsis *ccs* mutant, in which the authors could not detect any CSD activity, grew as well as wild-type plants on Cu-deficient media (Cohu et al. 2009). By comparison, the growth of *spl7* mutants is severely impaired under low-Cu growth conditions (Yamasaki et al. 2009; Bernal et al. 2012), suggesting that catalytic activity of CSD is not required for activation of the Cu deficiency response in Arabidopsis.

In yeast, SOD1 was reported to act as an integrator of metabolic signals from oxygen and glucose (Reddi and Culotta 2013). Upon the metabolic switch to aerobic fermentation in the presence of glucose, the repression of mitochondrial respiration requires two Casein Kinase1-gamma homologs (YCK1, YCK2). The stabilisation of YCK1 and YCK2 proteins requires interaction with catalytically active *Sc*SOD1, which was demonstrated to bind to the C-terminal region of the YCK1 protein downstream of the catalytic region. Three lysine residues (K383, K386, K390) present in the C-terminal region of both YCK proteins were shown to mediate YCK1 protein turnover, but not SOD1 binding. Yeast SOD1 was also able to stabilise a bovine CK1γ homolog, and *Sc*YCK1 in turn was stabilised by SOD1 from nematodes and humans (Reddi and Culotta 2013), suggesting the conservation of SOD1-mediated regulation of CK1 isoforms across fungal and animal kingdoms. In mammalian cells CK1γ isoforms play a role in the regulation of fermentative metabolism *via* the Wnt (Wingless/integration1 originally identified in Drosophila) signalling pathway (Sethi and Vidal-Puig 2010).

To date, there is no evidence for Wnt-like genes in plants (Holstein 2012), but plant homologs were identified of proteins that form a complex acting as a negative regulator of Wnt signalling in animals (Sharma et al. 2014), suggesting a possible existence of Wnt signalling-related pathways. The roles of plant CK1 family members are not well understood. The Arabidopsis genome contains 13 annotated genes encoding 14 Casein Kinase 1-Like (CKL) isoforms (Lee et al. 2005). Compared to their homologs in yeast, these proteins are highly conserved within the catalytic domain, but display nearly no similarity to YCK1 in their C-terminus. CKL13 is the only isoform in which two out of the three lysine residues important for SOD1-dependent regulation of *Sc*YCK1/*Sc*YCK2 are conserved, similar to bovine CK1. To date, publicly available protein–protein interaction data do not support any interactions between SODs and CK1-like proteins of Arabidopsis. However, multiple proteins were found to interact with Arabidopsis CSD1, CSD2 and CSD3 proteins. Interestingly, Receptor for Activated C Kinase1A (RACK1A), which was proposed to act as a MAPK cascade scaffolding protein, is among the proteins reported to interact with CSD1 (Su et al. 2015; Kundu et al. 2013), alongside Polyubiquitin3 (UBQ3) and Chitin Elicitor Receptor Kinase1 (CERK1). RACK1A is a phosphorylation target of With No lysine Kinase8 (WNK8), which has a role in glucose sensing via a plasma membrane heterotrimeric G-protein pathway (Urano et al. 2012, 2015). Epistasis analysis suggested that RACK1A acts downstream of WNK8 in the same glucose response pathway (Urano et al. 2015). Whether this pathway involves interaction between RACK1A and CSD1 remains unconfirmed. Notably, other Cu-containing proteins, namely PC2 and plantacyanin, were identified as potential interaction partners of RACK1A, as well (Kundu et al. 2013). All in all, it remains to be established whether CSDs have regulatory functions in plants, and even their precise roles in antioxidant defence remain unclear. It will be interesting to study the integration of these functions with the Cu-dependent regulation of CSDs.

6 Links Between Carbohydrate Metabolism and Copper Homeostasis in Plants

It is well established that sugar molecules can act as signals in plants (Smeekens 2000; Rolland et al. 2002; Price et al. 2004; Sheen 2014; Ruan 2014). Regulatory effects of sugars have been reported at the levels of transcription, transcript stability, translation and protein stability (Rolland et al. 2006). It is now thought that transcript levels of many genes are regulated coordinately by light and carbohydrate signalling (Thum et al. 2004), with a predominance of genes involved in metabolism among the targets of regulation. Despite the comparably low abundance of Cu in both the soil and the plant ionome, central pathways in photosynthesis and respiration depend on Cu-requiring proteins. Therefore, it is reasonable to hypothesise on the existence of regulatory crosstalk between Cu homeostasis and carbohydrate metabolism. Beyond this, with carbohydrates exerting a major influence on the regulation of plant growth and development (Smeekens et al. 2010; Gibson 2005),

Cu availability might constitute an additional factor in these regulatory networks. Individual links between sugar signalling and elements of Cu homeostasis have indeed been reported, but an in-depth understanding of the underlying regulatory networks remains to be established.

The abundance of the Cu miR398 was reported to respond not only to Cu deficiency, but also to a variety of abiotic stresses, including high salinity (Jagadeeswaran et al. 2009), temperature (Guan et al. 2013) and nitrogen starvation (Liang et al. 2012). In addition, the levels of miR398 were reported to increase in response to elevated sucrose concentrations in growth media. The increase in miR398 abundance in response to sucrose resulted in a decrease in both mRNA and protein levels of CSD1 and CSD2. This was proposed to reflect a decreased need for CSD activity upon reduced ROS production as a result of sugar-mediated inhibition of photosynthesis (Dugas and Bartel 2008). Even though the *COX5b.1* transcript, which encodes the Zn binding subunit of COX, is a validated target of miR398 (Jones-Rhoades and Bartel 2004), the abundance of *COX5b.1* mRNA was not affected by sucrose treatment (Dugas and Bartel 2008). The activity of the *COX5b.1* promoter responds positively to sucrose (Welchen et al. 2004), and this might account for the constant mRNA levels observed. Accordingly, *COX5b.1* mRNA abundance is expected to respond differentially to sucrose between conditions of Cu deficiency and Cu sufficiency.

In Arabidopsis, there are three miR398 isoforms (miR398a, miR398b, miR398c). The mature sequences of these three isoforms differ in only a single nucleotide and are thus challenging to distinguish. Differences in promoter sequences of these isoforms suggest differential gene regulation: a promoter analysis for miR398 in *P. trichocarpa* revealed that only a single CuRE is present within the 1 kb-region upstream of the stem-loop structure of the *MIR398a* gene, in contrast to eight and ten CuREs in this region for the *MIR398b* and *MIR398c* genes, respectively (Ravet et al. 2011). Accordingly, in Arabidopsis primary transcript levels of miR398b and miR398c (pri-miR398b and pri-miR398c) increased under Cu-deficient conditions in dependence on SPL7, whereas pri-miR398a was apparently not SPL7-dependent and much less abundant than the other two isoforms under both Cu deficiency and sufficiency (Bernal et al. 2012).

Dugas and Bartel (2008) concluded that miR398 levels are independently regulated in response to Cu and sucrose. In 12-day-old wild-type seedlings, the increase in mature miR398 abundance in response to sucrose was not influenced by Cu supply, and the decrease in miR398 abundance with increasing exogenous Cu occurred both in the presence and absence of sucrose supplementation. The use of *mir398* mutants should allow to determine which isoforms are sucrose-responsive. In a *mir398a/c* double mutant, the residual miR398 abundance was not increased under high-sucrose conditions, suggesting that miR398b is not sucrose-responsive (Dugas and Bartel 2008). It was puzzling, however, that no sucrose-dependent increase in residual miR398 was observed in *mir398a* and *mir398c* single mutants, either.

In a more recent study, sucrose-dependent increases in transcript abundance of both mature miR398 and miR408 were reported in 21-day old wild-type seedlings of *A. thaliana* (Ren and Tang 2012). In contrast to the findings of Dugas and Bartel (2008), there was no response to sucrose under Cu-sufficient growth conditions.

Additionally, mature miR398 and miR408 were undetectable in transgenic *SPL7*-artificial-miRNA lines, in which the relative transcript levels of *SPL7* were decreased to 50% of the wild type, under any sucrose or Cu condition. The authors thus concluded the increase in the levels of these miRNAs in response to sucrose to be dependent on SPL7 activity. Contrastingly, they also reported a decrease in *FSD1* and *COPT1* transcript abundance in response to sucrose. These genes are known to be transcriptionally upregulated under Cu deficiency and belong to the SPL7 regulon along with *MIR398b/c* (Abdel-Ghany et al. 2005; Yamasaki et al. 2009; Bernal et al. 2012). Earlier, Dugas and Bartel (2008) showed discrepant results of both lower and higher FSD1 protein levels in Arabidopsis wild-type seedlings grown on 0% sucrose compared to 1% sucrose in two independent experiments conducted under what they describe as identical growth conditions. The discrepancy in FSD1 protein levels was not discussed by the authors. Interestingly, *FSD1* mRNA levels showed circadian oscillations and were light-responsive according to a study by Kliebenstein et al. (1998). *FSD1* transcript abundance peaked during the day, but decreased with increasing light intensities. The latter observation might correspond to the observed decrease in *FSD1* mRNA levels under high sugar observed by Ren and Tang (2012). If there is indeed a sucrose-dependent uncoupling of two subsets of the SPL7 regulon, the underlying molecular mechanisms remain to be elucidated. It is also unknown how a sugar signal might be integrated with the regulation of SPL7 transcription factor activity.

In a study on miRNAs involved in energy signalling, pri-miR398c was identified among a set of miRNAs with reduced transcript abundance after exposure of 5-week-old detached Arabidopsis leaves to darkness (Confraria et al. 2013). This repression was observed to a similar degree in the miRNA biogenesis mutant *dcl1-9*. Indeed, not all miRNAs are processed by Dicer-Like1 (DCL1), but due to its major role in biogenesis of mature miRNAs (Rajagopalan et al. 2006) it is possible that the reduction of pri-miR398c transcript levels is caused by a change in the rates of its transcription rather than altered processing. Interestingly, the primary transcript levels observed in the light could not be restored by supplying external glucose to dark-grown seedlings (Confraria et al. 2013). Considering earlier work on miR398c (Dugas and Bartel 2008), this may indicate that miR398c is regulated in a sucrose-specific way. However, repression of pri-miR398c after exposure to darkness is also consistent with light-dependent regulation of *MIR398c* transcription, suggesting that the regulation of SPL7 activity could be light-dependent as well. Crosstalk between Cu and light signalling was also proposed in a publication demonstrating the co-regulation of numerous genes by the transcription factors SPL7 and Elongated Hypocotyl5 (HY5) (Zhang et al. 2014), the latter acting downstream of the light receptor network in mediating the transcription of light-induced genes. In addition to the interaction of HY5 and SPL7 proteins, HY5 was found to bind to the promoter of *SPL7*. Increased transcript levels of *SPL7* in the *hy5* mutant compared to wild type indicate that HY5 binding acts to suppress *SPL7* transcription.

In Arabidopsis seedlings, *PC* gene transcription is repressed by sucrose, which was interpreted to be part of the feedback regulation of photosynthesis by the availability of carbohydrate metabolites (Dijkwel et al. 1996). At the time of the publication, only one of two Arabidopsis *PC* genes was known, and later sequence analysis revealed that the studied isoform was *PC1*. So far, there is no data on whether *PC2* transcription is repressed by sucrose, as well. Unlike PC1, the protein abundance of PC2 is controlled by Cu availability (Abdel-Ghany 2009). Since PC2 is the predominant isoform and considering the generally high PC abundance, PC2 is thought to be the major Cu sink in green tissues. In a *pc2* mutant, CSD1 and CSD2 accumulate even under Cu deficiency conditions, presumably to buffer cellular Cu levels in the absence of the preferred Cu sink PC2 (Abdel-Ghany 2009). In addition, levels of pri-miR398b were downregulated in *pc2*. However, whether these changes result from increased cellular Cu availability or the downregulation of this miRNA as a consequence of decreased photosynthetic activity and thus lower sugar status in the *pc2* mutant is not known at this point. These observations also raise the question of how Cu may be buffered upon sucrose-mediated downregulation of PC, given that CSD1 and CSD2 protein levels were reported to decrease with sucrose supplementation as a result of increased abundance of miR398 (Dugas and Bartel 2008). However, these considerations are only relevant if *PC2* promoter activity is indeed influenced by carbohydrate status as is known only for *PC1* promoter activity to date.

Another link between sugar signalling and Cu homeostasis comes from the identification of a glucose-insensitive mutant called *gig*, which was found to be allelic to *paa1* (Lee et al. 2012). As described above, PAA1/HMA6 is located in the inner chloroplast envelope membrane where it mediates the transport of Cu^+ into the chloroplast stroma. Whereas the photosynthetic phenotype of the *gig*/*paa1* mutant was recovered by supplementation with exogenous Cu, the addition of Cu could not restore glucose sensitivity in *gig*/*paa1*. Additionally, a *gig hxk1* double mutant showed the same growth phenotype as the *gig* single mutant in the presence of high glucose, suggesting that *gig*/*paa1* is epistatic to *hxk1*. *Hexokinase1* (*HXK1*) encodes a protein with dual roles in glucose metabolism and signalling (Moore et al. 2003).

All in all, while different lines of evidence support crosstalk between sugar signalling and Cu homeostasis, there are some discrepancies between published datasets, and many questions remain unanswered. For example, what is the physiological relevance of the sugar-dependent downregulation of CSD1 and CSD2? What constitutes the mechanistic basis of the glucose-insensitivity observed in the *paa1* mutant? How are signals relaying Cu and sugar status sensed and integrated for subsequent metabolic adjustments? We are thus still at the very beginning of understanding how Cu might fulfil regulatory roles in carbohydrate metabolism.

7 Roles of Copper in the Metabolism of Phenolic Compounds in Plants

Despite the central roles of Cu-dependent proteins in photosynthesis and mitochondrial respiration, at the transcript level Cu deficiency has little effect on these processes (Fig. 2). Instead, Arabidopsis responds to Cu deficiency primarily by modulating transcript abundance of genes acting in cell wall and secondary metabolism of Arabidopsis. Limpness and wilting have long been established as symptoms of Cu deficiency in

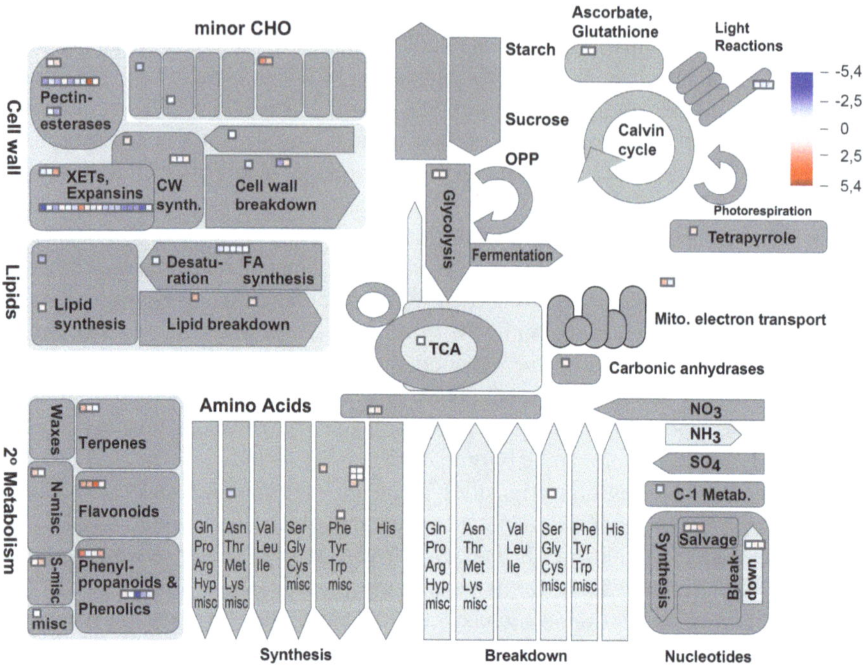

Fig. 2 Metabolism chart summarising the transcriptional response of *Arabidopsis thaliana* to Cu deficiency. Shown is a MapMan metabolic pathway-based visualisation of RNA-Seq data from shoots of wild-type *A. thaliana* grown under Cu-deficient and Cu-sufficient conditions (Bernal et al. 2012; Thimm et al. 2004). The *colour scale* represents Log_2 fold change (Cu-deficient *vs.* Cu-sufficient): *red-coloured squares* denote transcripts upregulated under Cu deficiency, and *blue-coloured squares* represent transcripts downregulated under Cu deficiency. Functions prominently upregulated under Cu deficiency are aromatic amino acid biosynthesis in Amino Acids, and flavonoid biosynthesis in Secondary Metabolism. Functions prominently downregulated under Cu deficiency are related to cell surface proteoglycans and cell wall modifications in Cell Wall, and fatty acid synthesis and elongation in Lipids. Abbreviations: *CW synth.* cell wall synthesis (cellulose and hemicellulose synthesis), *FA* fatty acid, *Mito.* mitochondrial, *Metab.* metabolism, *OPP* oxidative pentose phosphate, *TCA* tricarboxylic acid cycle, *XET* xyloglucan endo-transglycosylase. Amino acids are abbreviated by their three letter codes

plants. Mechanical weakness can account for such symptoms. Changes in cell wall composition, and particularly impaired lignification, have been associated with Cu deficiency in plants (Marschner and Marschner 2012). Consistent with these observations, Cu limitation led to severe decreases in lignin content of both wheat leaves and stem tissues (Marschner and Marschner 2012). Several Cu-dependent enzymes have been implicated in functions relating to lignin biosynthesis. Among these enzymes are Cu-containing amine oxidases, which catalyse the degradation of polyamines, mainly putrescine and cadaverine (Bagni and Tassoni 2001). Ammonia, hydrogen peroxide and an aminoaldehyde are formed as reaction products, and hydrogen peroxide in turn can act as a substrate for peroxidases catalysing lignin polymerisation in the apoplast

(Angelini et al. 1990). The enzyme activity of Cu amine oxidases was reported to decrease under Cu deficiency (Delhaize et al. 1985). In addition, some members of the MCO superfamily were proposed to have roles in lignification. The MCO superfamily includes laccases, ascorbate oxidases, ferroxidases and caeruloplasmin, which catalyse the four-electron reduction of O_2 to H_2O coupled with the one-electron oxidation of a substrate in a variety of organisms (Hoegger et al. 2006). The sequence homology of MCOs is overall low, but their structures and the three different types of Cu binding sites are conserved also in plant MCOs (Giardina et al. 2010). Members of the plant *Skewed5 (SKU5) Similar (SKS)* gene family are structurally related to MCOs, but the Cu binding sites are not fully conserved so that they likely lack oxidase activity (Sedbrook et al. 2002). Despite a wide distribution and large size of MCO and MCO-like protein families in plants, their biochemical and physiological functions are not well understood to date (Turlapati et al. 2011).

Laccases make up the largest family among the MCOs, comprising 17 genes in *A. thaliana* (McCaig et al. 2005). Given the complexity of the gene family, the term Laccase-like MCOs (LMCOs) was introduced to account for the likely multitude of biological functions of MCO enzymes (Hoopes and Dean 2004). LMCOs have co-ordination centres containing four Cu atoms and are generally able to catalyse the oxidation of a wide range of substrates, for example phenols and amines. Some LMCOs have been identified as targets of Cu miRNAs, i.e. their transcripts are thought to be post-transcriptionally downregulated under Cu deficiency (Fig. 1) (Abdel-Ghany and Pilon 2008). Two targets of the Cu miRNA miR397, namely *LAC4* and *LAC17*, were proposed to mediate lignin biosynthesis in Arabidopsis (Berthet et al. 2011). Whereas single mutants showed only a moderate decrease in lignin content, respectively, lignin levels in stems of *lac4 lac17* double mutants were reduced by up to 40% in comparison to the wild type. Both *lac17* and *lac4 lac17* accumulated more sinapoyl malate, which the authors hypothesised to reflect the metabolic redirection of phenolic intermediates the mutants were unable to use in lignification (Berthet et al. 2011). Sinapoyl malate was identified as a powerful UV-B protective agent in plants (Dean et al. 2014). Given that both *LAC4* and *LAC17* transcript levels are post-transcriptionally downregulated by miR397 under Cu deficiency (Abdel-Ghany and Pilon 2008), lignification might be altered similarly as observed in the *lac4 lac17* double mutants. This suggests the possibility of a Cu-dependent modulation of phenylpropanoid metabolism (Fig. 3), which serves as a starting point for the production not only of lignins but also of a variety of additional secondary metabolites like flavonoids or coumarins (Fraser and Chapple 2011). Several coumarins were identified as plant-derived chelators important in the mobilisation of Fe in soils (Clemens and Weber 2016; Schmidt et al. 2014; Schmid et al. 2014; Fourcroy et al. 2014), and this family of compounds may well comprise yet unidentified chelators of Cu that could assist in Cu acquisition or its distribution within the plant. Lignin biosynthesis and deposition increase rapidly in response to biotic stress (Tronchet et al. 2010; Menden et al. 2007). With the role of lignification in plant resistance against pathogens, it will be interesting to further explore whether defects in Cu homeostasis influence the susceptibility to pathogens. Finally, several authors have provided evidence for specific ferroxidase activities of a subset of plant MCOs, in analogy to well-characterised human and yeast ferroxidases that act to oxidise Fe(II) to Fe(III). In yeast, the MCO Ferrous Transport3 (FET3) acts in

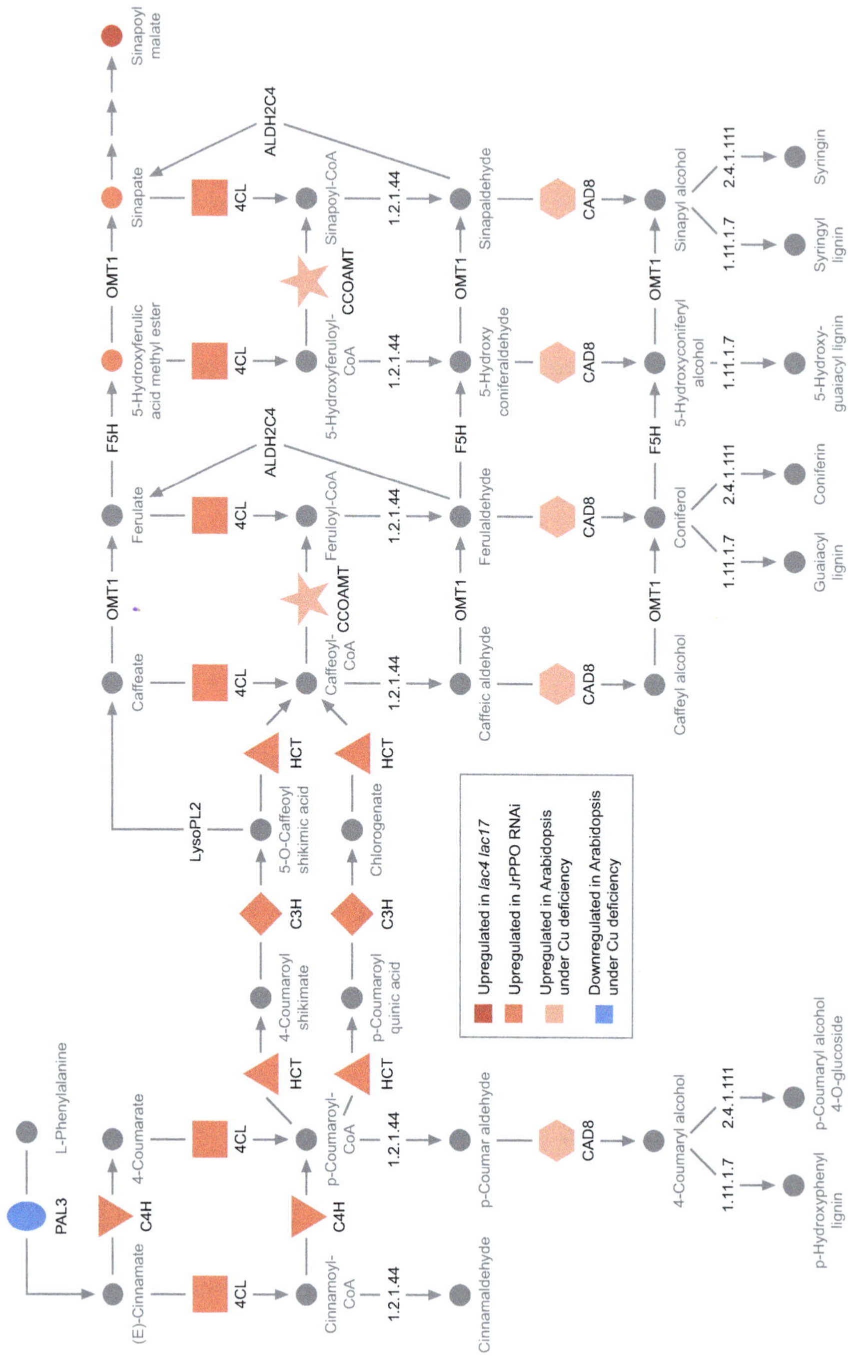
L-Phenylalanine
PAL3
(E)-Cinnamate
C4H
4-Coumarate
4CL
Cinnamoyl-CoA
p-Coumaroyl-CoA
1.2.1.44
Cinnamaldehyde
p-Coumar aldehyde
CAD8
4-Coumaryl alcohol
1.11.1.7
2.4.1.111
p-Hydroxyphenyl lignin
p-Coumaryl alcohol 4-O-glucoside
HCT
4-Coumaroyl shikimate
p-Coumaroyl quinic acid
C3H
5-O-Caffeoyl shikimic acid
Chlorogenate
LysoPL2
Caffeate
Caffeoyl-CoA
CCOAMT
Caffeic aldehyde
Caffeyl alcohol
OMT1
Ferulate
Feruloyl-CoA
Ferulaldehyde
Coniferol
Guaiacyl lignin
Coniferin
ALDH2C4
F5H
5-Hydroxyferulic acid methyl ester
5-Hydroxyferuloyl-CoA
5-Hydroxy coniferaldehyde
5-Hydroxyconiferyl alcohol
5-Hydroxy-guaiacyl lignin
Sinapate
Sinapoyl-CoA
Sinapaldehyde
Sinapyl alcohol
Syringyl lignin
Syringin
Sinapoyl malate
Upregulated in lac4 lac17
Upregulated in JrPPO RNAi
Upregulated in Arabidopsis under Cu deficiency
Downregulated in Arabidopsis under Cu deficiency

Fig. 3 Putative effects of Cu deficiency on phenylpropanoid metabolism. RNA-Seq data from shoots of wild-type *Arabidopsis thaliana* grown under Cu-deficient and -sufficient conditions (Bernal et al. 2012) are shown together with RNA-Seq and metabolite data from walnut (*Juglans regia*) Jr*PPO* RNAi lines (Araji et al. 2014) and with metabolite data from *A. thaliana lac4 lac17* double mutants (Berthet et al. 2011). This is based on the hypothesis that phenylpropanoid metabolism is similarly altered under both Cu deficiency and in the shown transgenic and mutant plants defective in specific Cu-dependent metalloproteins (see main text for details). Enzymatic steps are labelled using EC numbers. Enzyme names are given instead when a step is either catalysed by an enzyme encoded by a single genetic locus in Arabidopsis or regulated at the transcript level in the experiments shown here. *Multiple consecutive arrows* summarise intermediate enzymatic steps that are not detailed here. The pathway shown is based on the KEGG pathway map 00940 for *A. thaliana* (Kanehisa and Goto 2000). Abbreviations: *PAL3* Phenylalanine Ammonia Lyase3 (At5g04230), *C4H* Cinnamate 4-Hydroxylase (At2g30490, CYP73A5), *4CL* 4-Coumarate-CoA Ligase, *HCT* Hydroxycinnamoyl-CoA shikimate/quinate Hydroxycinnamoyl Transferase (At5g48930), *C3H* *p*-Coumarate-3-Hydroxylase (At2g40890, CYP98A3), *CCOAMT* Caffeoyl-CoA 3-*O*-Methyltransferase (At5g54160, COMT), *ALDH2C4* Aldehyde Dehydrogenase 2C4 (At3g24503), *OMT1* Caffeate *O*-Methyltransferase1 (At5g54160), *F5H* Ferulate 5-Hydroxylase, *CAD8* Cinnamyl Alcohol Dehydrogenase8 (At4g37990)

high affinity Fe acquisition by generating Fe^{3+} as the substrate for subsequent cellular import by Fe Transporter1 (FTR1) (De Silva et al. 1995; Stearman et al. 1996). In humans, effective export of Fe(II) from various cell types requires the subsequent extracytoplasmic oxidation by MCOs, for example caeruloplasmin (Vashchenko and MacGillivray 2013). In plants, the first LMCO with ferroxidase activity was detected in tulip poplar (*Liriodendron tulipifera*) (Hoopes and Dean 2004). Additionally, impaired root-to-shoot Fe movement in Cu-deficient *spl7* mutant plants was associated with a decrease in net ferroxidase activity, which might be mediated by one or several MCOs under Cu-sufficient growth conditions (Bernal et al. 2012). Recently, evidence was provided that the MCOs Low Phosphate Root1 (LPR1) and LPR2, originally identified based on their role in phosphate (Pi) sensing, exhibit ferroxidase activity (Müller et al. 2015). The attenuation of primary root growth under Pi deficiency was suggested to depend on extracellular redox cycling of Fe that arrests the division of meristematic cells, initiated by LPR-dependent Fe^{3+} production in the apoplast. Notably, LPR-mediated ferroxidase activity was also implicated in lignin polymerisation in roots during Pi deficiency (Ziegler et al. 2016).

Another connection between Cu homeostasis and phenylpropanoid metabolism might be constituted by polyphenol oxidases (PPOs), which contain a dinuclear Cu centre. These enzymes catalyse the conversion of monophenols to *o*-diphenols or *o*-diphenols to *o*-quinones and are broadly distributed in plants, but missing in *A. thaliana* (Mayer 2006; Schubert et al. 2002). The novel miR1444 was shown to target the transcripts of plastid-localised PPOs in *P. trichocarpa* and was strongly upregulated under Cu deficiency (Ravet et al. 2011). Correspondingly, PPO enzyme activity was decreased under Cu deficiency in *P. trichocarpa* (Ravet et al. 2011), and this had previously also been reported for chrysanthemum (*Chrysanthemum morifolium*) (Marschner and Marschner 2012). So far, miR1444 has only been identified in *P. trichocarpa*. Reduced PPO enzyme activity under Cu limitation may alternatively be caused by lack of Cu required as cofactor. Classically, PPOs have been associated with roles in plant defence. Loss of cellular compartmentalisation, e.g. due to wounding by pathogens or feeding herbivores, is thought to bring together the plastid-localised PPOs with their phenolic substrates that are accumulated in vacuoles and cell walls (Vaughn et al. 1988). The molecular mechanisms of defences against biotic stress involving PPO activity, however, have been difficult to pinpoint (Mayer 2006). PPO function has recently been studied in walnut (*Juglans regia*), which is a suitable model because it has only a single *PPO* gene (Escobar et al. 2008). Silencing of *JrPPO* via RNAi led to major alterations in the metabolism of phenolic compounds, and in particular in the expression of genes of the phenylpropanoid pathway (Fig. 3) and the levels of metabolites derived from this pathway (Araji et al. 2014). A decrease in L-3,4-dihydroxyphenylalanine (DOPA) and an increase in tyramine levels in the *PPO*-RNAi walnut lines were proposed to be caused by lack of *Jr*PPO, which would normally catalyse the *o*-hydroxylation of tyrosine to DOPA and of tyramine to dopamine. This is supported by the observation that the necrotic phenotype of PPO-silenced walnut plants corresponds to the phenotype observed after exogenous application of tyramine. The effects of different Cu regimes on *Jr*PPO activity have not been studied yet. Considering the many open questions regarding PPO function and the indication for different or multiple roles of PPO in different plant species (Boeckx et al. 2015), the relationship between PPO function and Cu homeostasis requires further study (see also Fig. 3).

In recent years, there has been important progress in our understanding of the roles of some MCOs in plants. Roles of Cu and its homeostasis in aromatic amino acid metabolism have emerged and require further study. New insights into Cu homeostasis might arise from the future elucidation of additional functions of MCOs in plants.

8 Experimental Design Across Research Fields

In experimental design, emphasis is usually placed on a selection of aspects that are specific to each individual research field, i.e. carbohydrate metabolism, development, diurnal/circadian rhythms or metal homeostasis. In order to reliably advance our knowledge in combining these fields, we must carefully consider, design and also document in publications all of these aspects, including day length and temperatures, carbohydrate supply, light quality and intensity, plant age, time of day of harvest and all aspects of the composition of growth media. Controlled and appropriate Cu supply to plants has been a major bottleneck to date (Fig. 4). Very importantly, the most

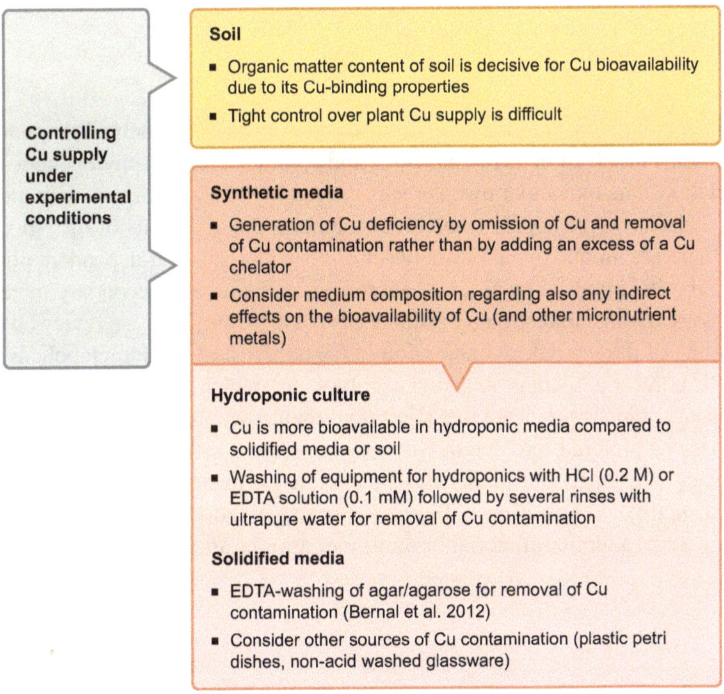

Fig. 4 Key considerations for controlling plant Cu supply under experimental conditions. Controlled and appropriate Cu conditions are of paramount importance in research aiming to advance our understanding of the interactions between Cu homeostasis and plant metabolism. See main text for details

common plant growth medium or fertiliser used in Arabidopsis research, Murashige and Skoog (MS) or half-strength MS medium, is acutely Cu-deficient. This is a consequence of its excessive Fe(III)-ethylenediaminetetraacetic acid (EDTA) content and the strong preference of EDTA for the chelation of Cu(II) over Fe(III) (Chaney 1988). Controlling Cu supply in nutrient solutions requires the use of physiologically appropriate concentrations of Fe bound to alternative, better suited synthetic Fe chelators, such as N,N′-di-(2-hydroxybenzoyl)-ethylenediamine-N,N′-diacetic acid (HBED). Moreover, Cu contamination relevant in deficiency experiments arises from agars and agaroses, as well as from plastics, for example petri dishes, and from non-acid washed glassware (Salome et al. 2014). We strongly advise against the use of chelators to achieve deficiencies of Cu or other micronutrient metals (Fig. 4). This is based on the observation that the addition of chelators caused symptoms in hydroponically grown plants differing from those generated by omission of metals (Salome et al. 2014). In soil, the bioavailability of Cu differs substantially depending on soil composition. Tight control of Cu supply to soil-grown plants is thus difficult to implement. Differences in Cu supply in soils can be achieved by watering plants without or with a Cu solution (e.g. 2 mM $CuSO_4$ for *A. thaliana*), especially on Cu-deficient soils. The phenotype of Cu deficiency sensitive mutants such as *spl7* can serve as a rough proxy for bioavailable Cu in experimental soils.

9 Conclusions

The evolutionary conservation of cellular Cu homeostasis has helped to identify key plant proteins involved in its maintenance and – within a shorter phylogenetic range – its regulation. The molecular mechanisms integrating Cu homeostasis with metabolic signals and vice versa appear to be diverse among the kingdoms of life. In vascular plants, there are unique dynamics between Cu homeostasis, Cu-dependent photosynthesis and mitochondrial respiration. Moreover, cell wall and secondary metabolism of plants are distinct and strongly influenced by Cu status. Here we reviewed our understanding of the regulatory integration of Cu homeostasis with metabolism and the metabolic control of plant growth and development, which is still in its infancy. Studies in other organisms indicate that additional regulatory roles of plant CSD enzymes should be explored in more detail. Not only soil-dependent nutritional state, but also the efficacy of endogenous homeostasis of Cu are of major agricultural relevance for the yield, quality, composition and stress resistance of crops, and this will require considerably more attention in agricultural research in the future.

References

Abdel-Ghany SE (2009) Contribution of plastocyanin isoforms to photosynthesis and copper homeostasis in *Arabidopsis thaliana* grown at different copper regimes. Planta 229(4):767–779. doi:10.1007/s00425-008-0869-z

Abdel-Ghany SE, Pilon M (2008) MicroRNA-mediated systemic down-regulation of copper protein expression in response to low copper availability in Arabidopsis. J Biol Chem 283 (23):15932–15945. doi:10.1074/jbc.M801406200

Abdel-Ghany SE, Muller-Moule P, Niyogi KK, Pilon M, Shikanai T (2005) Two P-type ATPases are required for copper delivery in *Arabidopsis thaliana* chloroplasts. Plant Cell 17(4):1233–1251. doi:tpc.104.030452 [pii]

Aguirre G, Pilon M (2015) Copper delivery to chloroplast proteins and its regulation. Front Plant Sci 6:1250. doi:10.3389/fpls.2015.01250

Alscher RG, Erturk N, Heath LS (2002) Role of superoxide dismutases (SODs) in controlling oxidative stress in plants. J Exp Bot 53(372):1331–1341

Andreini C, Banci L, Bertini I, Rosato A (2008) Occurrence of copper proteins through the three domains of life: a bioinformatic approach. J Proteome Res 7(1):209–216. doi:10.1021/pr070480u

Andres-Colas N, Sancenon V, Rodriguez-Navarro S, Mayo S, Thiele DJ, Ecker JR, Puig S, Peñarrubia L (2006) The Arabidopsis heavy metal P-type ATPase HMA5 interacts with metallochaperones and functions in copper detoxification of roots. Plant J 45(2):225–236. doi:10.1111/j.1365-313X.2005.02601.x

Angelini R, Manes F, Federico R (1990) Spatial and functional correlation between diamine-oxidase and peroxidase activities and their dependence upon de-etiolation and wounding in chick-pea stems. Planta 182(1):89–96. doi:10.1007/BF00239989

Araji S, Grammer TA, Gertzen R, Anderson SD, Mikulic-Petkovsek M, Veberic R, Phu ML, Solar A, Leslie CA, Dandekar AM, Escobar MA (2014) Novel roles for the polyphenol oxidase enzyme in secondary metabolism and the regulation of cell death in walnut. Plant Physiol 164 (3):1191–1203. doi:10.1104/pp.113.228593

Attallah CV, Welchen E, Martin AP, Spinelli SV, Bonnard G, Palatnik JF, Gonzalez DH (2011) Plants contain two SCO proteins that are differentially involved in cytochrome *c* oxidase function and copper and redox homeostasis. J Exp Bot 62(12):4281–4294. doi:10.1093/jxb/err138

Ayala MB, Sandmann G (1988) The role of Cu in respiration of pea plants and heterotrophically growing Scenedesmus cells. Z Naturforsch 43c:438–442

Bagni N, Tassoni A (2001) Biosynthesis, oxidation and conjugation of aliphatic polyamines in higher plants. Amino Acids 20(3):301–317

Barberon M, Vermeer JE, De Bellis D, Wang P, Naseer S, Andersen TG, Humbel BM, Nawrath C, Takano J, Salt DE, Geldner N (2016) Adaptation of root function by nutrient-induced plasticity of endodermal differentiation. Cell 164(3):447–459. doi:10.1016/j.cell.2015.12.021

Bernal M, Casero D, Singh V, Wilson GT, Grande A, Yang H, Dodani SC, Pellegrini M, Huijser P, Connolly EL, Merchant SS, Krämer U (2012) Transcriptome sequencing identifies *SPL7*-regulated copper acquisition genes *FRO4*/*FRO5* and the copper dependence of iron homeostasis in Arabidopsis. Plant Cell 24(2):738–761. doi:10.1105/tpc.111.090431

Berthet S, Demont-Caulet N, Pollet B, Bidzinski P, Cezard L, Le Bris P, Borrega N, Herve J, Blondet E, Balzergue S, Lapierre C, Jouanin L (2011) Disruption of *LACCASE4* and *17* results in tissue-specific alterations to lignification of *Arabidopsis thaliana* stems. Plant Cell 23(3):1124–1137. doi:10.1105/tpc.110.082792

Binder BM, Rodriguez FI, Bleecker AB (2010) The copper transporter RAN1 is essential for biogenesis of ethylene receptors in Arabidopsis. J Biol Chem 285(48):37263–37270. doi:10.1074/jbc.M110.170027

Birkenbihl RP, Jach G, Saedler H, Huijser P (2005) Functional dissection of the plant-specific SBP-domain: overlap of the DNA-binding and nuclear localization domains. J Mol Biol 352 (3):585–596. doi:10.1016/j.jmb.2005.07.013

Blaby-Haas CE, Padilla-Benavides T, Stube R, Arguello JM, Merchant SS (2014) Evolution of a plant-specific copper chaperone family for chloroplast copper homeostasis. Proc Natl Acad Sci U S A 111(50):E5480–E5487. doi:10.1073/pnas.1421545111

Boeckx T, Winters AL, Webb KJ, Kingston-Smith AH (2015) Polyphenol oxidase in leaves: is there any significance to the chloroplastic localization? J Exp Bot 66(12):3571–3579. doi:10.1093/jxb/erv141

Borghouts C, Osiewacz HD (1998) GRISEA, a copper-modulated transcription factor from *Podospora anserina* involved in senescence and morphogenesis, is an ortholog of MAC1 in *Saccharomyces cerevisiae*. Mol Gen Genet 260(5):492–502

Borghouts C, Kimpel E, Osiewacz HD (1997) Mitochondrial DNA rearrangements of *Podospora anserina* are under the control of the nuclear gene *grisea*. Proc Natl Acad Sci U S A 94 (20):10768–10773

Borghouts C, Werner A, Elthon T, Osiewacz HD (2001) Copper-modulated gene expression and senescence in the filamentous fungus *Podospora anserina*. Mol Cell Biol 21(2):390–399. doi:10.1128/MCB.21.2.390-399.2001

Borghouts C, Scheckhuber CQ, Stephan O, Osiewacz HD (2002) Copper homeostasis and aging in the fungal model system *Podospora anserina*: differential expression of *PaCtr3* encoding a copper transporter. Int J Biochem Cell Biol 34(11):1355–1371. doi:S135727250200078X [pii]

Burkhead JL, Reynolds KA, Abdel-Ghany SE, Cohu CM, Pilon M (2009) Copper homeostasis. New Phytol 182(4):799–816. doi:10.1111/j.1469-8137.2009.02846.x

Carr HS, Winge DR (2003) Assembly of cytochrome *c* oxidase within the mitochondrion. Acc Chem Res 36(5):309–316. doi:10.1021/ar0200807

Casareno RL, Waggoner D, Gitlin JD (1998) The copper chaperone CCS directly interacts with copper/zinc superoxide dismutase. J Biol Chem 273(37):23625–23628

Castruita M, Casero D, Karpowicz SJ, Kropat J, Vieler A, Hsieh SI, Yan W, Cokus S, Loo JA, Benning C, Pellegrini M, Merchant SS (2011) Systems biology approach in Chlamydomonas reveals connections between copper nutrition and multiple metabolic steps. Plant Cell 23 (4):1273–1292. doi:10.1105/tpc.111.084400

Chaney RL (1988) Plants can utilize iron form Fe-N,N′-di-(2-hydroxybenzoyl)-ethylenediamine-N, N′-diacetic acid, a ferric chelate with 10^6 greater formation constant than Fe-EDDHA. J Plant Nutr 11(6–11):1033–1050. doi:10.1080/01904168809363867

Clemens S, Weber M (2016) The essential role of coumarin secretion for Fe acquisition from alkaline soil. Plant Signal Behav 11(2):e1114197. doi:10.1080/15592324.2015.1114197

Cohu CM, Pilon M (2007) Regulation of superoxide dismutase expression by copper availability. Physiol Plant 129(4):747–755. doi:10.1111/j.1399-3054.2007.00879.x

Cohu CM, Abdel-Ghany SE, Gogolin Reynolds KA, Onofrio AM, Bodecker JR, Kimbrel JA, Niyogi KK, Pilon M (2009) Copper delivery by the copper chaperone for chloroplast and cytosolic copper/zinc-superoxide dismutases: regulation and unexpected phenotypes in an Arabidopsis mutant. Mol Plant 2(6):1336–1350. doi:10.1093/mp/ssp084

Confraria A, Martinho C, Elias A, Rubio-Somoza I, Baena-Gonzalez E (2013) miRNAs mediate SnRK1-dependent energy signaling in Arabidopsis. Front Plant Sci 4:197. doi:10.3389/fpls.2013.00197

Corson LB, Strain JJ, Culotta VC, Cleveland DW (1998) Chaperone-facilitated copper binding is a property common to several classes of familial amyotrophic lateral sclerosis-linked superoxide dismutase mutants. Proc Natl Acad Sci U S A 95(11):6361–6366

Crapo JD, Oury T, Rabouille C, Slot JW, Chang LY (1992) Copper,zinc superoxide dismutase is primarily a cytosolic protein in human cells. Proc Natl Acad Sci U S A 89(21):10405–10409

Culotta VC, Klomp LW, Strain J, Casareno RL, Krems B, Gitlin JD (1997) The copper chaperone for superoxide dismutase. J Biol Chem 272(38):23469–23472

Culotta VC, Yang M, O'Halloran TV (2006) Activation of superoxide dismutases: putting the metal to the pedal. Biochim Biophys Acta 1763(7):747–758. doi:10.1016/j.bbamcr.2006.05.003

Curie C, Cassin G, Couch D, Divol F, Higuchi K, Le Jean M, Misson J, Schikora A, Czernic P, Mari S (2009) Metal movement within the plant: contribution of nicotianamine and yellow stripe 1-like transporters. Ann Bot 103(1):1–11. doi:10.1093/aob/mcn207

De Silva DM, Askwith CC, Eide D, Kaplan J (1995) The FET3 gene product required for high affinity iron transport in yeast is a cell surface ferroxidase. J Biol Chem 270(3):1098–1101

Dean JC, Kusaka R, Walsh PS, Allais F, Zwier TS (2014) Plant sunscreens in the UV-B: ultraviolet spectroscopy of jet-cooled sinapoyl malate, sinapic acid, and sinapate ester derivatives. J Am Chem Soc 136(42):14780–14795. doi:10.1021/ja5059026

Delhaize E, Loneragan JF, Webb J (1985) Development of three copper metalloenzymes in clover leaves. Plant Physiol 78(1):4–7

DiDonato RJ Jr, Roberts LA, Sanderson T, Eisley RB, Walker EL (2004) Arabidopsis yellow stripe-Like2 (YSL2): a metal-regulated gene encoding a plasma membrane transporter of nicotianamine-metal complexes. Plant J 39(3):403–414. doi:10.1111/j.1365-313X.2004.02128.x

Dijkwel PP, Kock P, Bezemer R, Weisbeek PJ, Smeekens S (1996) Sucrose represses the developmentally controlled transient activation of the plastocyanin gene in *Arabidopsis thaliana* seedlings. Plant Physiol 110(2):455–463.

Dufour E, Boulay J, Rincheval V, Sainsard-Chanet A (2000) A causal link between respiration and senescence in *Podospora anserina*. Proc Natl Acad Sci U S A 97(8):4138–4143. doi:10.1073/pnas.070501997

Dugas DV, Bartel B (2008) Sucrose induction of Arabidopsis miR398 represses two Cu/Zn superoxide dismutases. Plant Mol Biol 67(4):403–417. doi:10.1007/s11103-008-9329-1

Eriksson M, Moseley JL, Tottey S, Del Campo JA, Quinn J, Kim Y, Merchant S (2004) Genetic dissection of nutritional copper signaling in chlamydomonas distinguishes regulatory and target genes. Genetics 168(2):795–807. doi:10.1534/genetics.104.030460

Escobar MA, Shilling A, Higgins P, Uratsu SL, Dandekar AM (2008) Characterization of polyphenol oxidase from walnut. J Am Soc Hortic Sci 133(6):852–858

Festa RA, Thiele DJ (2011) Copper: an essential metal in biology. Curr Biol 21(21):R877–R883. doi:10.1016/j.cub.2011.09.040

Forde B, Lorenzo H (2001) The nutritional control of root development. Plant and Soil 232 (1):51–68. doi:10.1023/a:1010329902165

Fourcroy P, Siso-Terraza P, Sudre D, Saviron M, Reyt G, Gaymard F, Abadia A, Abadia J, Alvarez-Fernandez A, Briat JF (2014) Involvement of the ABCG37 transporter in secretion of scopoletin and derivatives by Arabidopsis roots in response to iron deficiency. New Phytol 201 (1):155–167. doi:10.1111/nph.12471

Fraser CM, Chapple C (2011) The phenylpropanoid pathway in Arabidopsis. Arabidopsis Book 9: e0152. doi:10.1199/tab.0152

Fraústo da Silva JJR, Williams RJP (2001) The biological chemistry of the elements. Oxford University Press, Oxford

Garcia L, Welchen E, Gey U, Arce AL, Steinebrunner I, Gonzalez DH (2016) The cytochrome *c* oxidase biogenesis factor AtCOX17 modulates stress responses in Arabidopsis. Plant Cell Environ 39(3):628–644. doi:10.1111/pce.12647

Garcia-Molina A, Andres-Colas N, Perea-Garcia A, Del Valle-Tascon S, Peñarrubia L, Puig S (2011) The intracellular Arabidopsis COPT5 transport protein is required for photosynthetic electron transport under severe copper deficiency. Plant J 65(6):848–860. doi:10.1111/j.1365-313X.2010.04472.x

Garcia-Molina A, Andres-Colas N, Perea-Garcia A, Neumann U, Dodani SC, Huijser P, Peñarrubia L, Puig S (2013) The Arabidopsis COPT6 transport protein functions in copper distribution under copper-deficient conditions. Plant Cell Physiol 54(8):1378–1390. doi:10.1093/pcp/pct088

Garcia-Molina A, Xing S, Huijser P (2014) Functional characterisation of Arabidopsis SPL7 conserved protein domains suggests novel regulatory mechanisms in the Cu deficiency response. BMC Plant Biol 14:231. doi:10.1186/s12870-014-0231-5

Giardina P, Faraco V, Pezzella C, Piscitelli A, Vanhulle S, Sannia G (2010) Laccases: a never-ending story. Cell Mol Life Sci 67(3):369–385. doi:10.1007/s00018-009-0169-1

Gibson SI (2005) Control of plant development and gene expression by sugar signaling. Curr Opin Plant Biol 8(1):93–102. doi:10.1016/j.pbi.2004.11.003
Giehl RF, Gruber BD, von Wiren N (2014) It's time to make changes: modulation of root system architecture by nutrient signals. J Exp Bot 65(3):769–778. doi:10.1093/jxb/ert421
Guan Q, Lu X, Zeng H, Zhang Y, Zhu J (2013) Heat stress induction of miR398 triggers a regulatory loop that is critical for thermotolerance in Arabidopsis. Plant J 74(5):840–851. doi:10.1111/tpj.12169
Guy RD, Berry JA, Fogel ML, Hoering TC (1989) Differential fractionation of oxygen isotopes by cyanide-resistant and cyanide-sensitive respiration in plants. Planta 177(4):483–491. doi:10.1007/BF00392616
Halliwell B, Gutteridge JM (1984) Oxygen toxicity, oxygen radicals, transition metals and disease. Biochem J 219(1):1–14
Hawkesford MJ, De Kok LJ (2006) Managing sulphur metabolism in plants. Plant Cell Environ 29 (3):382–395
Haydon MJ, Roman A, Arshad W (2015) Nutrient homeostasis within the plant circadian network. Front Plant Sci 6:299. doi:10.3389/fpls.2015.00299
Himelblau E, Mira H, Lin SJ, Culotta VC, Peñarrubia L, Amasino RM (1998) Identification of a functional homolog of the yeast copper homeostasis gene ATX1 from Arabidopsis. Plant Physiol 117(4):1227–1234
Hirayama T, Kieber JJ, Hirayama N, Kogan M, Guzman P, Nourizadeh S, Alonso JM, Dailey WP, Dancis A, Ecker JR (1999) RESPONSIVE-TO-ANTAGONIST1, a Menkes/Wilson disease-related copper transporter, is required for ethylene signaling in Arabidopsis. Cell 97 (3):383–393. doi:S0092-8674(00)80747-3 [pii]
Hoegger PJ, Kilaru S, James TY, Thacker JR, Kues U (2006) Phylogenetic comparison and classification of laccase and related multicopper oxidase protein sequences. FEBS J 273(10):2308–2326. doi:10.1111/j.1742-4658.2006.05247.x
Holstein TW (2012) The evolution of the Wnt pathway. Cold Spring Harb Perspect Biol 4(7):a007922. doi:10.1101/cshperspect.a007922
Hong S, Kim SA, Guerinot ML, McClung CR (2013) Reciprocal interaction of the circadian clock with the iron homeostasis network in Arabidopsis. Plant Physiol 161(2):893–903. doi:10.1104/pp.112.208603
Hong-Hermesdorf A, Miethke M, Gallaher SD, Kropat J, Dodani SC, Chan J, Barupala D, Domaille DW, Shirasaki DI, Loo JA, Weber PK, Pett-Ridge J, Stemmler TL, Chang CJ, Merchant SS (2014) Subcellular metal imaging identifies dynamic sites of Cu accumulation in Chlamydomonas. Nat Chem Biol 10(12):1034–1042. doi:10.1038/nchembio.1662
Hood MI, Skaar EP (2012) Nutritional immunity: transition metals at the pathogen-host interface. Nat Rev Microbiol 10(8):525–537. doi:10.1038/nrmicro2836
Hoopes JT, Dean JF (2004) Ferroxidase activity in a laccase-like multicopper oxidase from *Liriodendron tulipifera*. Plant Physiol Biochem 42(1):27–33. doi:10.1016/j.plaphy.2003.10.011
Huang CH, Kuo WY, Weiss C, Jinn TL (2012) Copper chaperone-dependent and -independent activation of three copper-zinc superoxide dismutase homologs localized in different cellular compartments in Arabidopsis. Plant Physiol 158(2):737–746. doi:10.1104/pp.111.190223
Ishida H, Yoshimoto K, Izumi M, Reisen D, Yano Y, Makino A, Ohsumi Y, Hanson MR, Mae T (2008) Mobilization of rubisco and stroma-localized fluorescent proteins of chloroplasts to the vacuole by an ATG gene-dependent autophagic process. Plant Physiol 148(1):142–155. doi:10.1104/pp.108.122770
Izumi M, Wada S, Makino A, Ishida H (2010) The autophagic degradation of chloroplasts via rubisco-containing bodies is specifically linked to leaf carbon status but not nitrogen status in Arabidopsis. Plant Physiol 154(3):1196–1209. doi:10.1104/pp.110.158519
Jagadeeswaran G, Saini A, Sunkar R (2009) Biotic and abiotic stress down-regulate miR398 expression in Arabidopsis. Planta 229(4):1009–1014. doi:10.1007/s00425-009-0889-3

Jaquinod M, Villiers F, Kieffer-Jaquinod S, Hugouvieux V, Bruley C, Garin J, Bourguignon J (2007) A proteomics dissection of *Arabidopsis thaliana* vacuoles isolated from cell culture. Mol Cell Proteomics 6(3):394–412. doi:10.1074/mcp.M600250-MCP200

Jeong J, Guerinot ML (2009) Homing in on iron homeostasis in plants. Trends Plant Sci 14(5):280–285. doi:10.1016/j.tplants.2009.02.006

Jones-Rhoades MW, Bartel DP (2004) Computational identification of plant microRNAs and their targets, including a stress-induced miRNA. Mol Cell 14(6):787–799. doi:10.1016/j.molcel.2004.05.027

Jung HI, Gayomba SR, Rutzke MA, Craft E, Kochian LV, Vatamaniuk OK (2012) COPT6 is a plasma membrane transporter that functions in copper homeostasis in Arabidopsis and is a novel target of SQUAMOSA promoter-binding protein-like 7. J Biol Chem 287(40):33252–33267. doi:10.1074/jbc.M112.397810

Kadenbach B, Huttemann M, Arnold S, Lee I, Bender E (2000) Mitochondrial energy metabolism is regulated via nuclear-coded subunits of cytochrome c oxidase. Free Radic Biol Med 29 (3–4):211–221. doi:S0891-5849(00)00305-1 [pii]

Kampfenkel K, Kushnir S, Babiychuk E, Inze D, Van Montagu M (1995) Molecular characterization of a putative *Arabidopsis thaliana* copper transporter and its yeast homologue. J Biol Chem 270(47):28479–28486

Kanehisa M, Goto S (2000) KEGG: kyoto encyclopedia of genes and genomes. Nucleic Acids Res 28(1):27–30. doi:gkd027 [pii]

Klaumann S, Nickolaus SD, Furst SH, Starck S, Schneider S, Ekkehard Neuhaus H, Trentmann O (2011) The tonoplast copper transporter COPT5 acts as an exporter and is required for interorgan allocation of copper in *Arabidopsis thaliana*. New Phytol 192(2):393–404. doi:10.1111/j.1469-8137.2011.03798.x

Kliebenstein DJ, Monde RA, Last RL (1998) Superoxide dismutase in Arabidopsis: an eclectic enzyme family with disparate regulation and protein localization. Plant Physiol 118(2):637–650

Kropat J, Tottey S, Birkenbihl RP, Depege N, Huijser P, Merchant S (2005) A regulator of nutritional copper signaling in Chlamydomonas is an SBP domain protein that recognizes the GTAC core of copper response element. Proc Natl Acad Sci U S A 102(51):18730–18735. doi:10.1073/pnas.0507693102

Kropat J, Gallaher SD, Urzica EI, Nakamoto SS, Strenkert D, Tottey S, Mason AZ, Merchant SS (2015) Copper economy in Chlamydomonas: prioritized allocation and reallocation of copper to respiration vs. photosynthesis. Proc Natl Acad Sci U S A 112(9):2644–2651. doi:10.1073/pnas.1422492112

Kundu N, Dozier U, Deslandes L, Somssich IE, Ullah H (2013) Arabidopsis scaffold protein RACK1A interacts with diverse environmental stress and photosynthesis related proteins. Plant Signal Behav 8(5):e24012. doi:10.4161/psb.24012

Lanquar V, Ramos MS, Lelievre F, Barbier-Brygoo H, Krieger-Liszkay A, Krämer U, Thomine S (2010) Export of vacuolar manganese by AtNRAMP3 and AtNRAMP4 is required for optimal photosynthesis and growth under manganese deficiency. Plant Physiol 152(4):1986–1999. doi:10.1104/pp.109.150946

Lee JY, Taoka K, Yoo BC, Ben-Nissan G, Kim DJ, Lucas WJ (2005) Plasmodesmal-associated protein kinase in tobacco and Arabidopsis recognizes a subset of non-cell-autonomous proteins. Plant Cell 17(10):2817–2831. doi:10.1105/tpc.105.034330

Lee SA, Yoon EK, Heo JO, Lee MH, Hwang I, Cheong H, Lee WS, Hwang YS, Lim J (2012) Analysis of Arabidopsis glucose insensitive growth mutants reveals the involvement of the plastidial copper transporter PAA1 in glucose-induced intracellular signaling. Plant Physiol 159(3):1001–1012. doi:10.1104/pp.111.191726

Liang G, He H, Yu D (2012) Identification of nitrogen starvation-responsive microRNAs in *Arabidopsis thaliana*. PLoS One 7(11):e48951. doi:10.1371/journal.pone.0048951

Liao X, Butow RA (1993) RTG1 and RTG2: two yeast genes required for a novel path of communication from mitochondria to the nucleus. Cell 72(1):61–71. doi:0092-8674(93)90050-Z [pii]

Liao XS, Small WC, Srere PA, Butow RA (1991) Intramitochondrial functions regulate nonmitochondrial citrate synthase (CIT2) expression in *Saccharomyces cerevisiae*. Mol Cell Biol 11 (1):38–46

Marschner H, Marschner P (2012) Marschner's mineral nutrition of higher plants, 3rd edn. Elsevier/Academic Press, London/Waltham, MA

Masclaux-Daubresse C, Daniel-Vedele F, Dechorgnat J, Chardon F, Gaufichon L, Suzuki A (2010) Nitrogen uptake, assimilation and remobilization in plants: challenges for sustainable and productive agriculture. Ann Bot 105(7):1141–1157. doi:10.1093/aob/mcq028

Mayer AM (2006) Polyphenol oxidases in plants and fungi: going places? A review. Phytochemistry 67(21):2318–2331. doi:10.1016/j.phytochem.2006.08.006

McCaig BC, Meagher RB, Dean JF (2005) Gene structure and molecular analysis of the laccase-like multicopper oxidase (LMCO) gene family in *Arabidopsis thaliana*. Planta 221(5):619–636. doi:10.1007/s00425-004-1472-6

McCord JM, Fridovich I (1969) Superoxide dismutase. An enzymic function for erythrocuprein (hemocuprein). J Biol Chem 244(22):6049–6055

Menden B, Kohlhoff M, Moerschbacher BM (2007) Wheat cells accumulate a syringyl-rich lignin during the hypersensitive resistance response. Phytochemistry 68(4):513–520. doi:10.1016/j.phytochem.2006.11.011

Merchant S, Dreyfuss BW (1998) Posttranslational assembly of photosynthetic metalloproteins. Annu Rev Plant Physiol Plant Mol Biol 49:25–51. doi:10.1146/annurev.arplant.49.1.25

Millar AH, Eubel H, Jansch L, Kruft V, Heazlewood JL, Braun HP (2004) Mitochondrial cytochrome *c* oxidase and succinate dehydrogenase complexes contain plant specific subunits. Plant Mol Biol 56(1):77–90. doi:10.1007/s11103-004-2316-2

Millenaar FF, Lambers H (2003) The alternative oxidase: *in vivo* regulation and function. Plant Biol 5(1):2–15. doi:10.1055/s-2003-37974

Miller AF (2012) Superoxide dismutases: ancient enzymes and new insights. FEBS Lett 586 (5):585–595. doi:10.1016/j.febslet.2011.10.048

Mira H, Martinez-Garcia F, Peñarrubia L (2001) Evidence for the plant-specific intercellular transport of the Arabidopsis copper chaperone CCH. Plant J 25(5):521–528. doi:tpj985 [pii]

Moore B, Zhou L, Rolland F, Hall Q, Cheng WH, Liu YX, Hwang I, Jones T, Sheen J (2003) Role of the Arabidopsis glucose sensor HXK1 in nutrient, light, and hormonal signaling. Science 300(5617):332–336. doi:10.1126/science.1080585

Morgan MJ, Lehmann M, Schwarzlander M, Baxter CJ, Sienkiewicz-Porzucek A, Williams TC, Schauer N, Fernie AR, Fricker MD, Ratcliffe RG, Sweetlove LJ, Finkemeier I (2008) Decrease in manganese superoxide dismutase leads to reduced root growth and affects tricarboxylic acid cycle flux and mitochondrial redox homeostasis. Plant Physiol 147(1):101–114. doi:10.1104/pp.107.113613

Müller J, Toev T, Heisters M, Teller J, Moore KL, Hause G, Dinesh DC, Burstenbinder K, Abel S (2015) Iron-dependent callose deposition adjusts root meristem maintenance to phosphate availability. Dev Cell 33(2):216–230. doi:10.1016/j.devcel.2015.02.007

O'Halloran TV, Culotta VC (2000) Metallochaperones, an intracellular shuttle service for metal ions. J Biol Chem 275(33):25057–25060. doi:10.1074/jbc.R000006200

Osiewacz HD, Nuber U (1996) GRISEA, a putative copper-activated transcription factor from *Podospora anserina* involved in differentiation and senescence. Mol Gen Genet 252(1–2):115–124

Paschalidis KA, Roubelakis-Angelakis KA (2005) Sites and regulation of polyamine catabolism in the tobacco plant. Correlations with cell division/expansion, cell cycle progression, and vascular development. Plant Physiol 138(4):2174–2184. doi:10.1104/pp.105.063941

Perea-Garcia A, Garcia-Molina A, Andres-Colas N, Vera-Sirera F, Perez-Amador MA, Puig S, Peñarrubia L (2013) Arabidopsis copper transport protein COPT2 participates in the cross talk between iron deficiency responses and low-phosphate signaling. Plant Physiol 162(1):180–194. doi:10.1104/pp.112.212407

Perea-Garcia A, Andres-Borderia A, Mayo de Andres S, Sanz A, Davis AM, Davis SJ, Huijser P, Peñarrubia L (2016) Modulation of copper deficiency responses by diurnal and circadian rhythms in *Arabidopsis thaliana*. J Exp Bot 67(1):391–403. doi:10.1093/jxb/erv474

Pilon M, Ravet K, Tapken W (2011) The biogenesis and physiological function of chloroplast superoxide dismutases. Biochim Biophys Acta 1807(8):989–998. doi:10.1016/j.bbabio.2010.11.002

Plaxton WC, Tran HT (2011) Metabolic adaptations of phosphate-starved plants. Plant Physiol 156(3):1006–1015. doi:10.1104/pp.111.175281

Price J, Laxmi A, St Martin SK, Jang JC (2004) Global transcription profiling reveals multiple sugar signal transduction mechanisms in Arabidopsis. Plant Cell 16(8):2128–2150. doi:10.1105/tpc.104.022616

Puig S, Andres-Colas N, Garcia-Molina A, Peñarrubia L (2007a) Copper and iron homeostasis in Arabidopsis: responses to metal deficiencies, interactions and biotechnological applications. Plant Cell Environ 30(3):271–290. doi:10.1111/j.1365-3040.2007.01642.x

Puig S, Mira H, Dorcey E, Sancenon V, Andres-Colas N, Garcia-Molina A, Burkhead JL, Gogolin KA, Abdel-Ghany SE, Thiele DJ, Ecker JR, Pilon M, Peñarrubia L (2007b) Higher plants possess two different types of ATX1-like copper chaperones. Biochem Biophys Res Commun 354(2):385–390. doi:10.1016/j.bbrc.2006.12.215

Quinn JM, Merchant S (1995) Two copper-responsive elements associated with the Chlamydomonas *Cyc6* gene function as targets for transcriptional activators. Plant Cell 7(5):623–628. doi:10.1105/tpc.7.5.623

Radin I, Mansilla N, Rodel G, Steinebrunner I (2015) The Arabidopsis COX11 homolog is essential for cytochrome *c* oxidase activity. Front Plant Sci 6:1091. doi:10.3389/fpls.2015.01091

Rae TD, Schmidt PJ, Pufahl RA, Culotta VC, O'Halloran TV (1999) Undetectable intracellular free copper: the requirement of a copper chaperone for superoxide dismutase. Science 284(5415):805–808

Rajagopalan R, Vaucheret H, Trejo J, Bartel DP (2006) A diverse and evolutionarily fluid set of microRNAs in *Arabidopsis thaliana*. Genes Dev 20(24):3407–3425. doi:10.1101/gad.1476406

Ravet K, Danford FL, Dihle A, Pittarello M, Pilon M (2011) Spatiotemporal analysis of copper homeostasis in *Populus trichocarpa* reveals an integrated molecular remodeling for a preferential allocation of copper to plastocyanin in the chloroplasts of developing leaves. Plant Physiol 157(3):1300–1312. doi:10.1104/pp.111.183350

Reddi AR, Culotta VC (2013) SOD1 integrates signals from oxygen and glucose to repress respiration. Cell 152(1–2):224–235. doi:10.1016/j.cell.2012.11.046

Redinbo MR, Yeates TO, Merchant S (1994) Plastocyanin: structural and functional analysis. J Bioenerg Biomembr 26(1):49–66

Ren L, Tang G (2012) Identification of sucrose-responsive microRNAs reveals sucrose-regulated copper accumulations in an SPL7-dependent and independent manner in *Arabidopsis thaliana*. Plant Sci 187:59–68. doi:10.1016/j.plantsci.2012.01.014

Ridge PG, Zhang Y, Gladyshev VN (2008) Comparative genomic analyses of copper transporters and cuproproteomes reveal evolutionary dynamics of copper utilization and its link to oxygen. PLoS One 3(1):e1378. doi:10.1371/journal.pone.0001378

Robinson NJ, Winge DR (2010) Copper metallochaperones. Annu Rev Biochem 79:537–562. doi:10.1146/annurev-biochem-030409-143539

Rodriguez FI, Esch JJ, Hall AE, Binder BM, Schaller GE, Bleecker AB (1999) A copper cofactor for the ethylene receptor ETR1 from Arabidopsis. Science 283(5404):996–998

Rolland F, Moore B, Sheen J (2002) Sugar sensing and signaling in plants. Plant Cell 14(Suppl):S185–S205

Rolland F, Baena-Gonzalez E, Sheen J (2006) Sugar sensing and signaling in plants: conserved and novel mechanisms. Annu Rev Plant Biol 57:675–709. doi:10.1146/annurev.arplant.57.032905.105441

Ruan YL (2014) Sucrose metabolism: gateway to diverse carbon use and sugar signaling. Annu Rev Plant Biol 65:33–67. doi:10.1146/annurev-arplant-050213-040251

Salome PA, Oliva M, Weigel D, Krämer U (2013) Circadian clock adjustment to plant iron status depends on chloroplast and phytochrome function. EMBO J 32(4):511–523. doi:10.1038/emboj.2012.330

Salome PA, Bernal M, Krämer U (2014) Circadian life without micronutrients: effects of altered micronutrient supply on clock function in Arabidopsis. Methods Mol Biol 1158:227–238. doi:10.1007/978-1-4939-0700-7_16

Sancenon V, Puig S, Mira H, Thiele DJ, Peñarrubia L (2003) Identification of a copper transporter family in *Arabidopsis thaliana*. Plant Mol Biol 51(4):577–587

Sancenon V, Puig S, Mateu-Andres I, Dorcey E, Thiele DJ, Peñarrubia L (2004) The Arabidopsis copper transporter COPT1 functions in root elongation and pollen development. J Biol Chem 279(15):15348–15355. doi:10.1074/jbc.M313321200

Schmid NB, Giehl RF, Doll S, Mock HP, Strehmel N, Scheel D, Kong X, Hider RC, von Wiren N (2014) Feruloyl-CoA 6′-Hydroxylase1-dependent coumarins mediate iron acquisition from alkaline substrates in Arabidopsis. Plant Physiol 164(1):160–172. doi:10.1104/pp.113.228544

Schmidt H, Günther C, Weber M, Sporlein C, Loscher S, Böttcher C, Schobert R, Clemens S (2014) Metabolome analysis of *Arabidopsis thaliana* roots identifies a key metabolic pathway for iron acquisition. PLoS One 9(7):e102444. doi:10.1371/journal.pone.0102444

Schubert M, Petersson UA, Haas BJ, Funk C, Schröder WP, Kieselbach T (2002) Proteome map of the chloroplast lumen of *Arabidopsis thaliana*. J Biol Chem 277(10):8354–8365. doi:10.1074/jbc.M108575200

Schulte E, Kück U, Esser K (1988) Extrachromosomal mutants from *Podospora anserina*: permanent vegetative growth in spite of multiple recombination events in the mitochondrial genome. Mol Gen Genet 211(2):342–349. doi:10.1007/bf00330614

Sedbrook JC, Carroll KL, Hung KF, Masson PH, Somerville CR (2002) The Arabidopsis *SKU5* gene encodes an extracellular glycosyl phosphatidylinositol-anchored glycoprotein involved in directional root growth. Plant Cell 14(7):1635–1648

Sethi JK, Vidal-Puig A (2010) Wnt signalling and the control of cellular metabolism. Biochem J 427(1):1–17. doi:10.1042/BJ20091866

Shahbaz M, Ravet K, Peers G, Pilon M (2015) Prioritization of copper for the use in photosynthetic electron transport in developing leaves of hybrid poplar. Front Plant Sci 6:407. doi:10.3389/fpls.2015.00407

Sharma M, Pandey A, Pandey GK (2014) beta-catenin in plants and animals: common players but different pathways. Front Plant Sci 5:143. doi:10.3389/fpls.2014.00143

Sheen J (2014) Master regulators in plant glucose signaling networks. J Plant Biol 57(2):67–79. doi:10.1007/s12374-014-0902-7

Shikanai T, Müller-Moule P, Munekage Y, Niyogi KK, Pilon M (2003) PAA1, a P-type ATPase of Arabidopsis, functions in copper transport in chloroplasts. Plant Cell 15(6):1333–1346

Smeekens S (2000) Sugar-induced signal transduction in plants. Annu Rev Plant Physiol Plant Mol Biol 51:49–81. doi:10.1146/annurev.arplant.51.1.49

Smeekens S, Ma J, Hanson J, Rolland F (2010) Sugar signals and molecular networks controlling plant growth. Curr Opin Plant Biol 13(3):274–279. doi:10.1016/j.pbi.2009.12.002

Sommer F, Kropat J, Malasarn D, Grossoehme NE, Chen X, Giedroc DP, Merchant SS (2010) The CRR1 nutritional copper sensor in Chlamydomonas contains two distinct metal-responsive domains. Plant Cell 22(12):4098–4113. doi:10.1105/tpc.110.080069

Stearman R, Yuan DS, Yamaguchi-Iwai Y, Klausner RD, Dancis A (1996) A permease-oxidase complex involved in high-affinity iron uptake in yeast. Science 271(5255):1552–1557

Su J, Xu J, Zhang S (2015) RACK1, scaffolding a heterotrimeric G protein and a MAPK cascade. Trends Plant Sci 20(7):405–407. doi:10.1016/j.tplants.2015.05.002

Tapken W, Ravet K, Pilon M (2012) Plastocyanin controls the stabilization of the thylakoid Cu-transporting P-type ATPase PAA2/HMA8 in response to low copper in Arabidopsis. J Biol Chem 287(22):18544–18550. doi:10.1074/jbc.M111.318204

Tapken W, Kim J, Nishimura K, van Wijk KJ, Pilon M (2015) The Clp protease system is required for copper ion-dependent turnover of the PAA2/HMA8 copper transporter in chloroplasts. New Phytol 205(2):511–517. doi:10.1111/nph.13093

Thimm O, Bläsing O, Gibon Y, Nagel A, Meyer S, Krüger P, Selbig J, Müller LA, Rhee SY, Stitt M (2004) MAPMAN: a user-driven tool to display genomics data sets onto diagrams of metabolic pathways and other biological processes. Plant J 37(6):914–939. doi:10.1111/j.1365-313X.2004.02016.x

Thum KE, Shin MJ, Palenchar PM, Kouranov A, Coruzzi GM (2004) Genome-wide investigation of light and carbon signaling interactions in Arabidopsis. Genome Biol 5(2):R10. doi:10.1186/gb-2004-5-2-r10

Tronchet M, Balague C, Kroj T, Jouanin L, Roby D (2010) Cinnamyl alcohol dehydrogenases-C and D, key enzymes in lignin biosynthesis, play an essential role in disease resistance in Arabidopsis. Mol Plant Pathol 11(1):83–92. doi:10.1111/j.1364-3703.2009.00578.x

Turlapati PV, Kim KW, Davin LB, Lewis NG (2011) The laccase multigene family in *Arabidopsis thaliana*: towards addressing the mystery of their gene function(s). Planta 233(3):439–470. doi:10.1007/s00425-010-1298-3

Unseld M, Marienfeld JR, Brandt P, Brennicke A (1997) The mitochondrial genome of *Arabidopsis thaliana* contains 57 genes in 366,924 nucleotides. Nat Genet 15(1):57–61. doi:10.1038/ng0197-57

Urano D, Phan N, Jones JC, Yang J, Huang J, Grigston J, Taylor JP, Jones AM (2012) Endocytosis of the seven-transmembrane RGS1 protein activates G-protein-coupled signalling in Arabidopsis. Nat Cell Biol 14(10):1079–1088. doi:10.1038/ncb2568

Urano D, Czarnecki O, Wang X, Jones AM, Chen JG (2015) Arabidopsis receptor of activated C kinase1 phosphorylation by WITH NO LYSINE8 KINASE. Plant Physiol 167(2):507–516. doi:10.1104/pp.114.247460

Vanlerberghe GC (2013) Alternative oxidase: a mitochondrial respiratory pathway to maintain metabolic and signaling homeostasis during abiotic and biotic stress in plants. Int J Mol Sci 14 (4):6805–6847. doi:10.3390/ijms14046805

Vashchenko G, MacGillivray RT (2013) Multi-copper oxidases and human iron metabolism. Nutrients 5(7):2289–2313. doi:10.3390/nu5072289

Vaughn KC, Lax AR, Duke SO (1988) Polyphenol oxidase: the chloroplast oxidase with no established function. Physiol Plant 72(3):659–665. doi:10.1111/j.1399-3054.1988.tb09180.x

Wagner AM, Moore AL (1997) Structure and function of the plant alternative oxidase: its putative role in the oxygen defence mechanism. Biosci Rep 17(3):319–333

Waldron KJ, Rutherford JC, Ford D, Robinson NJ (2009) Metalloproteins and metal sensing. Nature 460(7257):823–830. doi:10.1038/nature08300

Wang S, Blumwald E (2014) Stress-induced chloroplast degradation in Arabidopsis is regulated via a process independent of autophagy and senescence-associated vacuoles. Plant Cell 26 (12):4875–4888. doi:10.1105/tpc.114.133116

Waters BM, McInturf SA, Stein RJ (2012) Rosette iron deficiency transcript and microRNA profiling reveals links between copper and iron homeostasis in *Arabidopsis thaliana*. J Exp Bot 63(16):5903–5918. doi:10.1093/jxb/ers239

Weigel M, Varotto C, Pesaresi P, Finazzi G, Rappaport F, Salamini F, Leister D (2003) Plastocyanin is indispensable for photosynthetic electron flow in *Arabidopsis thaliana*. J Biol Chem 278 (33):31286–31289. doi:10.1074/jbc.M302876200

Weisbeek P, Hageman J, de Boer D, Pilon R, Smeekens S (1989) Import of proteins into the chloroplast lumen. J Cell Sci Suppl 11:199–223

Welchen E, Chan RL, Gonzalez DH (2004) The promoter of the Arabidopsis nuclear gene *COX5b-1*, encoding subunit 5b of the mitochondrial cytochrome c oxidase, directs tissue-specific expression by a combination of positive and negative regulatory elements. J Exp Bot 55(405):1997–2004. doi:10.1093/jxb/erh223

Wood LK, Thiele DJ (2009) Transcriptional activation in yeast in response to copper deficiency involves copper-zinc superoxide dismutase. J Biol Chem 284(1):404–413. doi:10.1074/jbc.M807027200

Yamasaki H, Abdel-Ghany SE, Cohu CM, Kobayashi Y, Shikanai T, Pilon M (2007) Regulation of copper homeostasis by micro-RNA in Arabidopsis. J Biol Chem 282(22):16369–16378. doi:10.1074/jbc.M700138200

Yamasaki H, Hayashi M, Fukazawa M, Kobayashi Y, Shikanai T (2009) SQUAMOSA promoter binding protein-Like7 is a central regulator for copper homeostasis in Arabidopsis. Plant Cell 21(1):347–361. doi:10.1105/tpc.108.060137

Yuan DS, Dancis A, Klausner RD (1997) Restriction of copper export in *Saccharomyces cerevisiae* to a late Golgi or post-Golgi compartment in the secretory pathway. J Biol Chem 272(41):25787–25793

Zhang H, Zhao X, Li J, Cai H, Deng XW, Li L (2014) MicroRNA408 is critical for the HY5-SPL7 gene network that mediates the coordinated response to light and copper. Plant Cell 26 (12):4933–4953. doi:10.1105/tpc.114.127340

Ziegler J, Schmidt S, Chutia R, Müller J, Böttcher C, Strehmel N, Scheel D, Abel S (2016) Non-targeted profiling of semi-polar metabolites in Arabidopsis root exudates uncovers a role for coumarin secretion and lignification during the local response to phosphate limitation. J Exp Bot 67(5):1421–1432. doi:10.1093/jxb/erv539

Interplay of Circadian Rhythms and Light in the Regulation of Photosynthesis-Derived Metabolism

Matthew A. Jones

Abstract Alternating periods of day and night confer an environmental rhythm upon terrestrial plants. Seasonal changes in light intensity and duration (as well as integrals of temperature) inform developmental decisions that directly impact upon plant growth. In response to the selective pressure of these daily rhythms, plants have evolved an endogenous, biological oscillator that coincides with these patterns. These circadian rhythms allow plants to anticipate daily transitions and consequently allocate specific metabolic functions to certain times of day. The circadian system also has a dramatic effect upon development, with the external coincidence model describing how plants measure day length to induce flowering under inductive conditions. Plants' responses to environmental change are therefore a distillation of direct responses to abiotic factors and moderating factors derived from endogenous biological rhythms. This review summarizes our understanding of how metabolic processes are governed by these interactions, with particular attention to carbon and redox metabolism, two processes derived from photosynthesis.

Contents

The original version of this chapter was revised: Figure 1 has been updated. The erratum to this chapter is available at DOI 10.1007/124_2017_12.

Communicated by Ulrich Lüttge

M.A. Jones (✉)
School of Biological Sciences, University of Essex, Colchester, UK
e-mail: matthew.jones@essex.ac.uk

Progress in Botany (2018) 79: 147–172, DOI 10.1007/124_2017_2,
Published online: 1 March 2017

1 The Circadian System Is an Endogenous Timer that Modulates Responses to Light and Temperature

Plants' photosynthetic nature emphasizes the importance of marrying energy harvested from the sun with growth and reproductive development. Light availability varies not only with the season but also across the course of the day, and plants adjust their accumulation and use of metabolites to prevent starvation during the night while also providing resources for growth. Given the relevance of the photoperiodic cycle to plant growth and development, it is unsurprising that 90% of the transcriptome responds to changes in light, temperature, or a combination of these two signals (Michael et al. 2008). The selective pressure conferred by these day/night transitions has driven the evolution of a complex relationship between metabolism, photoreceptors, and an endogenous timing mechanism known as the circadian clock (from the Latin *circa diem*: about a day).

The circadian clock is defined by three primary characteristics. Circadian systems display "gating" of responses (i.e., cellular responses to extrinsic factors are modulated dependent upon the state of the internal oscillator), continue to oscillate for several days in constant conditions, and are temperature compensated (McClung 2006). Such qualities enable the circadian system to be buffered against minor irregularities in day length or light irradiance and so enable this timekeeper to reliably serve both as an internal reference and as a modulator of cellular responses. A subset of the diurnally rhythmic "diurnal" genes are directly controlled by the circadian system. Fully one third of transcripts expressed in *Arabidopsis* continue to oscillate under constant light and temperature and so are defined as being regulated by the circadian system, thereby highlighting the importance of this pervasive endogenous oscillator in regulating plant behavior (Covington et al. 2008).

The importance of the circadian system has led to the identification of over 30 genetic loci that contribute to the maintenance of these rhythms (Hsu and Harmer 2014; Millar 2016). Importantly, the regular oscillations of the circadian system also lend themselves to mathematical abstraction, enabling models of reciprocal transcriptional control to be generated and tested (Fogelmark and Troein 2014; Locke et al. 2005). The most recent models of the circadian transcriptional oscillator describe a cascade of interlocking negative feedback loops that successively repress gene expression, coupled with a smaller number of transcriptional activators

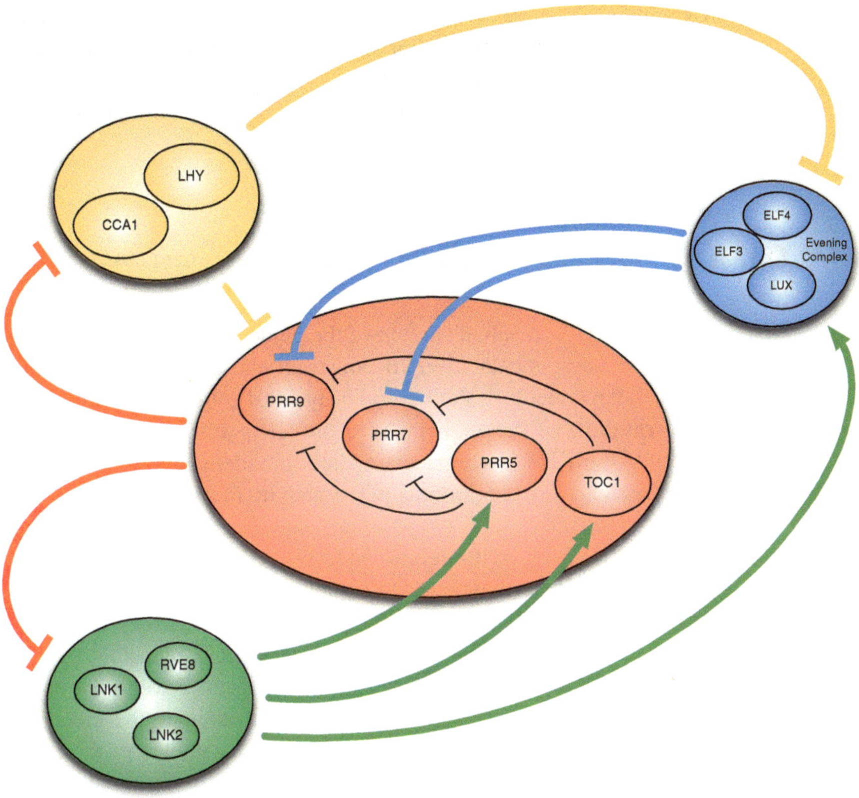

Fig. 1 A simplified schematic of the transcriptional feedback loops within the circadian system. Groups of proteins indicate shared regulation and/or effect upon targets' transcription as indicated by the *matching colored lines*. REVEILLE8 (RVE8) acts with LNK1 and LNK2 to activate transcription of evening-phased genes including *EARLY FLOWERING3* (*ELF3*), *ELF4*, *LUX ARRHYTHMO* (*LUX*), *TIMING OF CAB EXPRESSION1* (*TOC1*), and *PSEUDORESPONSE REGULATOR5* (*PRR5*; Nusinow et al. 2011; Rawat et al. 2011; Xie et al. 2014). ELF3, ELF4, and LUX together comprise the Evening Complex that represses expression of *PRR9* and *PRR7* (Helfer et al. 2011; Mizuno et al. 2014). TOC1 and PRR5 act in turn with PRR7 and PRR9 to repress expression of *CCA1* and *LHY* until shortly before dawn (Gendron et al. 2012; Huang et al. 2012; Nakamichi et al. 2010, 2012). CCA1 and LHY subsequently close the loop by repressing expression of evening-phased genes (Alabadi et al. 2001). Numerous defined proteins and their interactions have been removed for clarity – readers are directed to recent comprehensive reviews of the circadian system for a more thorough discussion (Hsu and Harmer 2014; Millar 2016)

(Fig. 1; Fogelmark and Troein 2014; Hsu and Harmer 2014). To enhance their utility, these transcription/translation feedback loops (TTFL) are entrained to environmental rhythms of light and temperature. Morning-phased clock components including *CIRCADIAN CLOCK ASSOCIATED1* (*CCA1*), *LATE ELONGATED HYPOCOTYL* (*LHY*), *PSEUDORESPONSE REGULATOR9* (*PRR9*), *NIGHT LIGHT-INDUCIBLE AND CLOCK-REGULATED1* (*LNK1*), and *LNK2* are all induced by light (Ito et al. 2003; Rugnone et al. 2013; Wang and Tobin 1998), whereas PRR9 and PRR7 are important for entrainment to temperature, possibly via temperature-derived modulation of light signaling pathways or the action of the Evening Complex (Fig. 1; Gould et al. 2013; Mizuno et al. 2014; Salomé and McClung 2005).

1.1 Transcriptional and Metabolic Rhythms Are Interlinked

Although comparable TTFL mechanisms have been described in many other species, the individual components used to construct these systems appear to have been recruited independently (Young and Kay 2001). Instead, more recent work suggests that rhythms of metabolism were prevalent in the last eukaryotic common ancestor, with the TTFL systems arising later in independent lineages. For example, many species demonstrate rhythms in the activity of metabolically linked enzymes including casein kinase and enzymes involved in histone modification (Hastings et al. 2008; Jones et al. 2010; Perales and Más 2007). Similarly, circadian rhythms of peroxiredoxin (Prx) oxidation are apparent across a diverse range of species (Edgar et al. 2012; O'Neill and Reddy 2011; O'Neill et al. 2011). As Prx are scavengers of reactive oxygen species (ROS; Dietz 2011), it is plausible that the rhythmic accumulation of metabolites such as ROS and sugars drove the development of the ancestral circadian system to anticipate and accentuate the metabolic changes that are a consequence of the diel cycle. What remains unclear is the extent to which the TTFL and metabolic clocks are interlinked – it is apparent that metabolic clocks can continue to oscillate in the absence of transcriptional rhythms but it is unlikely that these systems operate independently of one another. For example, the transcription of many rate-limiting enzymes is controlled by the circadian system (Covington et al. 2008).

Although the circadian control of transcript abundance has been intensively studied, plants also contend with dramatic changes in the concentrations of numerous metabolites during the day. Sugars, amino acids, and organic acids all vary over the course of a day, with a subset of these oscillating with a circadian rhythm (Espinoza et al. 2010; Figueroa et al. 2016; Lu et al. 2005; Sulpice et al. 2014). Similarly, concentrations of micronutrients (that serve as enzymatic cofactors) often display a diurnal rhythm, although circadian rhythms in these are not necessarily apparent. The phototrophic nature of plants emphasizes the importance of micronutrients that contribute towards photosynthesis, with magnesium, copper, and iron being particularly critical. Perturbation of Mg^{2+} transport is sufficient to alter both protein metabolism and the circadian system, demonstrating how these two systems are interlinked (Feeney et al. 2016; Hermans et al. 2010; Wilkinson et al. 1990). In comparison, excess copper accumulation impairs circadian amplitude in constant light, although copper deficiency does not appear to alter the clock (Andrés-Colás et al. 2010; Salomé et al. 2013). The effect of iron deficiency upon the clock is light dependent and requires protein translation in plastids. Plants lacking either *GIGANTEA* or *ZEITLUPE* (which regulate the stability and subcellular localization of TOC1 and PRR5 in a light-dependent manner, Kim et al. 2013, 2007; Más et al. 2003) are insensitive to iron-dependent circadian effects. Such data suggest a role for these proteins in this signaling cascade, although phytochrome signaling also influences these rhythms (Chen et al. 2013; Hong et al. 2013; Salomé et al. 2013). In addition, the mis-expression of *FERRITIN* (which stores intracellular iron) in clock mutants suggests that the clock reciprocally contributes to iron homeostasis (Ravet et al. 2009). PRR7 represses expression of *FERRITIN*s while plants lacking *TIME FOR COFFEE* demonstrate increased accumulation of *FERRITIN1* (Duc et al. 2009; Liu et al. 2013). The circadian accumulation of micronutrients is not restricted to those directly required for photochemistry, however. Rhythms of intracellular potassium, lithium, sodium, cobalt, zinc, and selenium have all been reported in the photosynthetic alga *Ostreococcus tauri* (Feeney et al. 2016), while daily and circadian fluctuations of cytosolic calcium concentration are also observed (Dodd et al. 2007; Johnson et al. 1995; Love et al. 2004). As many of these elements serve as cofactors for enzymatic activity, it is likely that these rhythmic variations directly influence metabolism, although the ubiquitous role of these ions has precluded elucidation of specific signaling pathways (Haydon et al. 2015).

1.2 Photosynthetic Parameters Vary with a Circadian Rhythm

Light is the predominant environmental factor that governs plant growth and development, being a source of both photosynthetic energy and environmental information. Most plants fix CO_2 directly from the atmosphere via the Calvin–Benson cycle, although derivations to maximize water-use efficiency or increase productivity have arisen to adapt to specific habitats (Hartwell et al. 2016; Sage et al. 2012). The photosynthetic machinery makes up a large proportion of the protein in the cell, and the maintenance and repair of this apparatus consequently requires significant resources and management. Since light is the driver of photochemistry, rates of photosynthesis typically correlate with photon fluence rates over the course of a day (Lawson and Blatt 2014). However, circadian regulation of photosynthesis is also apparent. Rhythms of carbon assimilation are observed when plants are grown under constant light (Dodd et al. 2004; Hennessey and Field 1991; Salome et al. 2002), and both photochemical quenching and the maximum efficiency of PhotoSystem II (PSII) oscillate over circadian time (Litthauer et al. 2015; Rascher et al. 2001). The cumulative effect of this circadian modulation is an increased fitness (as measured by biomass; Dodd et al. 2005), although apportion of the effect for each of these physiological consequences in the ultimate increase in yield has yet to be satisfactorily completed.

Nuclear transcripts encoding components of both PhotoSystem I (PSI) and PSII increase before subjective dawn (Harmer et al. 2000), and up to 70% of chloroplast encoded, protein-coding transcripts can be regulated by the circadian clock (Noordally et al. 2013). However, this variation in transcript does not necessarily demonstrate that the quantity of the photosynthetic machinery varies over the course of a day. The abundance of the most prevalent proteins in a cell does not differ at dusk compared to dawn (Baerenfaller et al. 2012), despite changes in transcript level being apparent in three quarters of these genes. This discrepancy probably arises from the metabolic cost of protein metabolism; it is likely more efficient to retain functional protein or alter its activity via posttranslational modification rather than to perform catabolism at the end of each day. Instead, it is thought that the retention of fluctuating transcript levels in these cases either facilitates improved responses to environmental stress relative to stable transcript accumulation or reflects a differential rate of replacement over the course of a day. This would enable replacement of damaged proteins while maintaining steady-state levels (Baerenfaller et al. 2012).

One convenient measure of circadian rhythms is achieved through monitoring PSII operating efficiency, which estimates the efficiency at which light absorbed by PSII is used for reduction of the primary quinone electron acceptor (Baker 2008; Litthauer et al. 2015; Rascher et al. 2001). Circadian rhythms monitored via this method are disrupted by mutations known to impair nuclear TTFL oscillations, again highlighting the interplay between TTFL and metabolic rhythms (Litthauer et al. 2015). Although the mechanism underlying rhythmic fluctuation of PSII operating efficiency remains unclear, the rate of consumption of both NADPH and ATP are major factors known to alter PSII operating efficiency (Baker 2008). Availability of NADPH and ATP is influenced by photosynthetic parameters such as the efficiency of carboxylation and the regeneration rate of ribulose 1,5-bisphosphate, as well as by photorespiration and the rate of transport of carbohydrates out of the cell (Baker 2008).

One of the primary physiological mechanisms that alters rates of photosynthesis is stomatal conductance, which influences the concentration of CO_2 within the leaf, and consequently within the chloroplast (Farquhar and Sharkey 1982; Lawson and Blatt 2014). While intrinsically linked to light quality and quantity (Farquhar and Sharkey 1982; McAusland et al. 2016), oscillations of stomatal aperture are apparent under constantly lit conditions, demonstrating a contribution from the circadian system (Dodd et al. 2004; Hennessey and Field 1991; Litthauer et al. 2015; Somers et al. 1998). One possible mechanism underlying these circadian changes in stomatal aperture is the modulation of abscisic acid (ABA) signaling. ABA induces stomatal closure, and sensitivity to ABA varies over the course of a day (Covington et al. 2008; Mizuno and Yamashino 2008; Robertson et al. 2008). This sensitivity is conferred in part by circadian regulation of the magnesium protoporphyrin IX chelatase large subunit (ABAR/CHLH/GUN5, which acts within the ABA signaling network; Legnaioli et al. 2009; Pokhilko et al. 2013). TOC1 (Fig. 1) represses the accumulation of *ABAR*, leading to reduced ABA sensitivity and an increased stomatal aperture (Legnaioli et al. 2009). Such changes in stomatal aperture could subsequently alter the rate of photosynthesis and PSII operating efficiency.

1.3 Circadian Regulation of Crassulacean Acid Metabolism

The opening of stomata during the day to enable CO_2 diffusion has the consequence of increasing water loss via transpiration (Lawson and Blatt 2014). Crassulacean Acid Metabolism (CAM) is an adaptation of photosynthetic metabolism that enables improved water-use efficiency by temporally separating CO_2 fixation from stomatal opening (Hartwell et al. 2016). Rather than atmospheric CO_2 fixation occurring directly through the activity of Rubisco, an additional metabolic step has evolved that allows CAM plants to close their stomata during the day, subsequently limiting water loss. The benefits of this adaptation have led to the independent evolution of CAM on several occasions in response to arid and semiarid climates (Smith and Winter 1996).

This temporal separation of atmospheric carbon fixation demonstrates a clear circadian pattern of metabolic regulation that has been well documented in the literature, with circadian rhythms of net CO_2 exchange (anti-phasic to those observed in C3 plants) being reported in the model CAM genus *Kalanchoë* since the 1960s (Lüttge and Ball 1978; Warren and Wilkins 1961). During the night, CAM plants open their stomata and fix CO_2 via the activity of phospho*enol*pyruvate carboxylase (PPC) and malate dehydrogenase. PPC activity is regulated by a combination of circadian-regulated expression of *PPC*, combined with phosphorylation of PPC which modulates its carboxylase activity (Hartwell et al. 1999, 2016; Nimmo 2000; Wilkins 1992). The rhythmic activity of PPC generates malate, which is subsequently stored as malic acid in the vacuole (Hartwell et al. 2016). At dawn, malate moves out of the vacuole into the cytosol where it is decarboxylated by either NAD-malic enzyme (NAD-ME) or phosphopyruvate carboxykinase (PCK) depending on the variant of CAM performed (Dittrich 1976; Holtum et al. 2005). This causes CO_2 levels within the leaf to increase (Hartwell 2008) and induces closure of stomata, enabling Rubisco to refix the CO_2 into the Calvin–Benson cycle in the presence of light.

Although distinct from the Calvin–Benson cycle, the metabolic processes underlying CAM are linked to a comparable transcriptional circadian oscillator. Several clock genes are present and oscillate with a circadian rhythm in CAM species such as *Kalanchoë fedtschenkoi* and *Mesembryanthemum crystallinum* (Boxall et al. 2005; Dever et al. 2015). Such genetic similarities and physiological discrepancies suggest a degree of plasticity in the relationship between the transcriptional circadian oscillator and metabolic processes, with the endogenous clock providing a timing reference that guides metabolism to optimize the timing of physiological processes.

2 Balancing Carbon Metabolism to Optimize Growth

Once CO_2 has been fixed by the Calvin–Benson cycle, a choice arises whether to immediately utilize the resultant sugars for growth, respiration, or defense, or to synthesize starch as a transient energy store. This is a complex decision as although the energetic cost of growth is greater at night than during the day, sufficient starch needs to be accumulated during the day to prevent starvation at night (Graf and Smith 2011; Sharkey 2015). The proportion of photoassimilates used to synthesize starch or complete respiration is influenced by the life history of the plant. Under conditions providing either low irradiance or short photoperiods, only limited photosynthesis is possible. This leads to a "source-limited" phenotype – i.e., insufficient amounts of carbon, which is typically associated with reduced growth (Smith and Stitt 2007). Conversely, plants can be "sink-limited" in longer photoperiods (or high irradiance) when more carbon is fixed from the atmosphere than can be used for growth. These latter conditions are particularly prevalent in nutrient-limited conditions (Smith and Stitt 2007). It is therefore beneficial for plants to control both the synthesis and degradation of starch to optimize carbohydrate availability during the diel cycle, with reference to both day length and available photosynthetic capacity. This balance is achieved by monitoring carbon assimilation during the day while also accounting for starvation the previous night. The circadian system contributes to this biochemical calculation by providing a timing reference that indicates the expected time until dawn (Graf et al. 2010). This combination of signaling ensures efficient utilization of photoassimilates during the day while also preventing starvation during the night.

2.1 *Trehalose 6-Phosphate Acts as a Signal to Modulate Sucrose Metabolism*

Plants determine their intracellular sucrose levels (and hence their carbon assimilation) by monitoring the concentration of trehalose 6-phosphate (Tre6P), a relationship that is encapsulated by the sucrose-Tre6P nexus hypothesis (Figueroa and Lunn 2016). Concentrations of both of these metabolites increase over the course of a day, decreasing at night in line with the diurnal cycle (Lunn et al. 2006; Martins et al. 2013). Sucrose and Tre6P form a negative feedback loop, with sucrose inducing Tre6P production and Tre6P conversely limiting sucrose accumulation (Fig. 2; Figueroa and Lunn 2016).

Although the mechanism by which sucrose induces Tre6P production remains to be clarified, it does appear to require protein synthesis, as blocking translation prevents accumulation of Tre6P (Lunn et al. 2006; Yadav et al. 2014). In comparison, much more is known about the repression of sucrose accumulation by Tre6P. Experimental induction of Tre6P accumulation induces a reduction in sucrose levels by diverting photoassimilates away from sucrose, instead promoting nitrate assimilation and organic acid production via the posttranslational regulation of PPC and NITRATE REDUCTASE (Figueroa et al. 2016). This mechanism has obvious benefits in the context of source/sink relationships. If sucrose synthesis exceeds demand in sink tissues, then the diversion of photoassimilates into nitrogen metabolism generates additional amino acids that may subsequently alleviate the sink limitation (Figueroa and Lunn 2016). Such a mechanism ensures that plants optimize growth dependent upon sink tissue requirements.

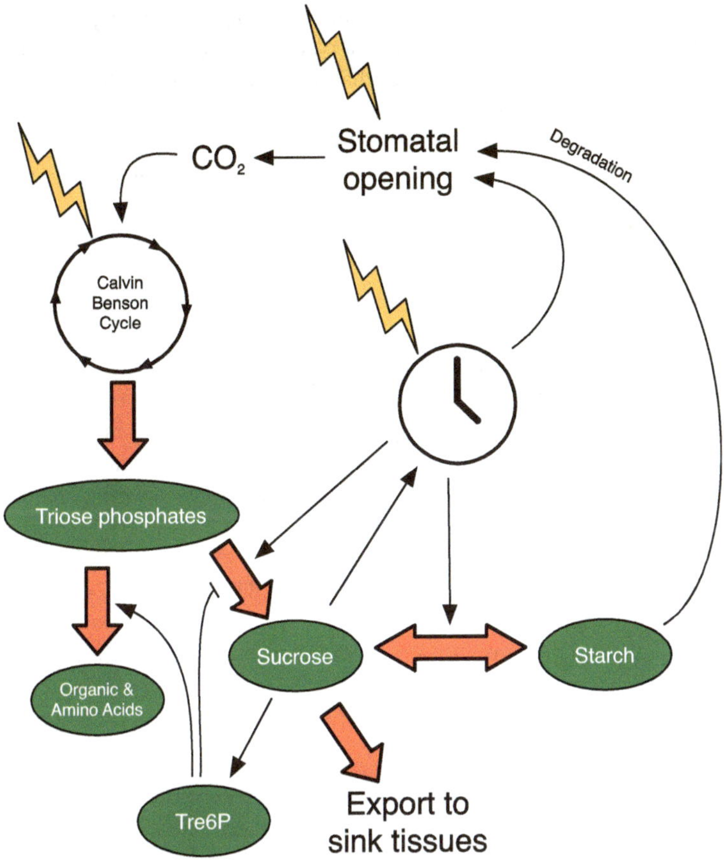

Fig. 2 Interplay of carbon metabolism and the circadian system. Light induces photosynthesis, generating triose phosphates that are used in metabolism. The circadian system acts to regulate sucrose anabolism and starch metabolism, although the former may be via an indirect mechanism related to the starvation phenotype induced by premature exhaustion of starch reserves (see text). The circadian system itself is entrained by light and by sucrose-initiated signals

2.2 *Stocking Up: Control of Starch Synthesis*

Plants precisely control sucrose and starch metabolism so as to balance carbon supply throughout the day/night cycle. Plants typically use starch as a transitory store of carbon, with this starch being catabolized during the night to maintain cellular energy levels (Graf and Smith 2011). It has long been believed that starch accumulated in a linear fashion during the day, but it is becoming apparent that the partitioning of fixed carbon is more dynamic than originally thought – at least in source tissues (Kölling et al. 2015). The proportion of fixed carbon converted into starch is controlled by a number of factors including the preexisting energy status of the cell, the current rate of photosynthesis, and the length of the photoperiod (Sharkey 2015). In particular, carbon starvation has a dramatic effect upon carbon partitioning and utilization. Plants that have experienced starvation during the preceding night will prioritize the production of starch rather than utilizing newly assimilated carbon for respiration and growth (Gibon et al. 2004). Conversely, in sink-limited conditions (when plants assimilate an excess of carbon) starch reserves are not depleted at the end of the night and there is a delay before starch begins to accumulate (Hädrich et al. 2012; Sulpice et al. 2014).

Starch synthesis is regulated by ADP glucose pyrophosphorylase (AGPase), which is activated by a rising ratio of glycerate-3-phosphate to inorganic phosphate (Stitt et al. 2010; Zeeman et al. 2010). Immediately after dawn, most of the fixed carbon is exported from source leaves, with only 15% of fixed carbon being converted into starch (Kölling et al. 2015). Instead, much of the fixed carbon is diverted to structural compounds such as proteins, cell walls, lipids, pigments, and waxes (Kölling et al. 2015). Later in the day, this relationship changes, with a decrease in fixed carbon export correlated with an increase in starch conversion (to almost a third; Kölling et al. 2015). This leads to an accumulation of starch in source leaves by the end of the day that can be utilized during the night (Sulpice et al. 2014). However, these proportions are highly plastic dependent upon the prevalent photoperiod – *Arabidopsis* will survive in extremely short photoperiods with as little as 3 h light every 24 h (Gibon et al. 2009). To achieve this, the proportion of fixed carbon diverted to starch synthesis at the beginning of the diel period is increased (Gibon et al. 2009; Sulpice et al. 2014). Such a strategy, accompanied by a reduced rate of starch degradation, emphasizes maintenance over growth but allows plants to tolerate shortened photoperiods when induced by shading or changing seasons.

2.3 Raiding the Store: Degradation of Starch

The regulation of starch degradation is dependent upon whether plant growth is sink- or source-limited. In sink-limited conditions, Tre6P acts to inhibit starch degradation by limiting the accumulation of sucrose, leading to the retention of starch reserves at the end of the night (Martins et al. 2013). Conversely, starch reserves are almost entirely exhausted by the end of the night in source-limited conditions (Gibon et al. 2009; Graf and Smith 2011; Smith and Stitt 2007). This efficient utilization of starch ensures that sugars are available throughout the night for respiration while maximizing potential growth (Stitt and Zeeman 2012). Sugar levels and plant growth rates are dramatically reduced upon exhaustion of starch (Graf et al. 2010; Usadel et al. 2008), demonstrating the importance of regulating carbohydrate availability during the night. Starch is degraded in a near linear fashion during the night (Graf et al. 2010; Lu et al. 2005), with the rate of degradation varying according to day length and the amount of accumulated starch (Graf et al. 2010; Lu et al. 2005). This rate is primarily influenced by the time until the subsequent dawn, with this information being provided by the circadian system (Graf et al. 2010). Plants therefore adjust starch degradation rates so that starch is utilized by dawn as anticipated by the endogenous oscillator (Graf et al. 2010).

Although the involvement of the circadian system in starch degradation is clear, the biological mechanisms that underlie this observed behavior are less obvious. The appropriate degradation rate is set by integrating information reporting the time until dawn (derived from the circadian system) and the amount of starch available. As starch is stored as granules in the chloroplast, it is not immediately obvious how cells are able to determine this latter component. It has been proposed that a soluble molecule (known as "S"), whose accumulation is correlated with cellular starch levels, may serve this role (Scialdone et al. 2013). Molecular interactions between S and a circadian component would then derive the necessary degradation rate. Alternatively, it is possible that the rate of starch degradation is directly influenced by the circadian system (Feugier and Satake 2012). In this latter case, sugar sensing within the cell would alter circadian phase, leading to continuous changes in the starch degradation rate during the night (Webb and Satake 2015). While the ability of the circadian system to be reset by acute levels of sucrose has been documented (Dalchau et al. 2011; Haydon et al. 2013), it remains unclear how minutely degradation rates of starch can be regulated. Equally, plants grown in source-limiting conditions continue to degrade starch in anticipation of dawn rather than increasing starch degradation at the start of the night to compensate for this limitation, suggesting that acute sugar levels do not set the rate of starch degradation (Pilkington et al. 2015). In either case, as the protein levels of starch degrading enzymes are relatively stable in diel conditions it is apparent that regulation of these enzymes occurs at a posttranscriptional level (Gibon et al. 2006; Lu et al. 2005; Skeffington et al. 2014; Smith et al. 2004).

One possible integrator of information derived from starch accumulation and the circadian system is PHOSPHOGLUCAN, WATER DIKINASE (PWD). This enzyme contributes to a cycle of phosphorylation reactions that are essential for normal starch degradation, and as a consequence retains starch at dawn (Smith 2012). Plants lacking PWD do not decrease their rate of starch degradation in response to an unexpected early dusk, but it remains unclear whether this reflects an upper limit upon starch degradation in these plants rather than representing a decoupling of starch degradation from the circadian system. Despite this caveat, the amount of granule-bound phosphate is correlated with cellular starch levels over diurnal time, suggesting that a phosphate-derived signal would be able to signal cellular starch content (Scialdone et al. 2013). Another useful mutant to dissect this mechanism may be *time for coffee*, as these have an excess starch phenotype despite having a shortened circadian period (Hall et al. 2003; Sanchez-Villarreal et al. 2013). Future developments in this field will be awaited with interest.

2.4 Interplay Between Carbon Metabolism and the Circadian System

The utility of light as an entraining signal and as the driving force of the Calvin–Benson cycle confounds attempts to separate their roles in carbon metabolism, although current hypotheses state that the clock acts indirectly to influence starch synthesis through its regulation of starch degradation (Fig. 2; Sulpice et al. 2014). In source-limited conditions (when plant growth is constrained by insufficient carbon assimilation), starch reserves are exhausted by the end of the night and starch synthesis occurs almost immediately after dawn (Sulpice et al. 2014). Analogous changes in carbon partitioning are observed in plants with accelerated circadian systems that prematurely degrade starch reserves before dawn (Graf et al. 2010). In plants with a shortened circadian period, the proportion of carbon fixed into starch in the morning was higher than in wild type (Kölling et al. 2015). However, it is not yet clear whether this phenotype is a direct result of the increased pace of the circadian system or if this altered partitioning is a consequence of the end of night starvation response observed in these mutants (Graf et al. 2010; Kölling et al. 2015). It seems likely, however, that the daily availability of light is the primary driver of photosynthesis, with plants referring to the circadian system overnight to prepare for dawn (Dodd et al. 2014).

The accumulation of photosynthate also informs progression of the circadian system (Bläsing et al. 2005; Haydon et al. 2013, Müller et al. 2014). Leaf glucose and sucrose levels peak approximately 4 h after dawn and subsequently fall to a baseline level (Haydon et al. 2013). As exogenous sucrose is sufficient to reset the circadian system, it has been proposed that a "metabolic dawn" acts as an additional entrainment mechanism of the clock (Dalchau et al. 2011; Dodd et al. 2015). Interestingly, disruption of CAM metabolism in *K. fedtschenkoi* via RNA silencing of NAD-ME not only abolishes CAM but also dampens circadian rhythms of gene expression (Dever et al. 2015). PPC KINASE activity is inhibited by malate (Carter et al. 1991), raising the possibility that malate acts as a feedback signal to modulate the activity of PPC through regulation of PPC phosphorylation (Nimmo 2000). Interestingly, circadian rhythms of CO_2 output can be maintained even in the absence of rhythmic PPC phosphorylation, suggesting that malate oscillations are sufficient to maintain a subset of circadian outputs in CAM plants (Borland et al. 1999, Carter et al. 1995). Such data suggest that feedback from carbon metabolism enhances the amplitude of rhythms in both C3 and CAM species. This influence from carbon metabolism would add to the robustness of light entrainment in natural photoperiods where initial sensing of dawn by photoreceptors can be substantiated by the metabolic products of photosynthesis, and may be of particular value under fluctuating light conditions (caused by shade or cloud cover). This additional factor linking the clock with assimilated carbon would also be of use in the changing seasons, when fluence rates can vary substantially.

The modulation of nuclear gene transcription via sugar signaling appears to occur via a pathway incorporating PRR7 and CCA1, as mutants lacking either of these proteins do not reset their circadian rhythms in response to exogenous sucrose (Haydon et al. 2013). In addition, *PRR7* accumulates when photosynthesis is inhibited whereas *CCA1* is repressed (Haydon et al. 2013). Mathematical models have suggested that a subunit of SNF-RELATED KINASE1 (SnRK1) acts as a clock-regulated sensor of carbon deficit (Pokhilko et al. 2014), although a role for Tre6P within the circadian system has yet to be examined.

3 Changes in Redox Status Contribute to Diel Rhythmicity

The biochemical reactions that fix carbon from the atmosphere simultaneously generate ROS and reactive nitrogen species (RNS) as by-products. As a consequence, the daily variation in light availability confers a rhythm upon the generation of ROS/RNS from photosynthesis (Del Río 2015). Plants integrate this environmental information via adapted ROS scavenging pathways that optimize responses to these metabolites, despite their potentially damaging nature (Baxter et al. 2014). ROS are also harnessed as a defense against pathogens (Spoel and Van Ooijen 2014). Complex regulatory pathways exist to balance the accumulation of ROS as signaling and catalytic components while also limiting potential damage (Baxter et al. 2014). In a signaling context, most research has examined the role of H_2O_2 although increasing evidence documents a role for redox-derived posttranslational modifications that modulate protein behavior (Sevilla et al. 2015). Levels of H_2O_2 and other indicators of redox status vary with a diurnal rhythm, peaking at midday (Lai et al. 2012). These variations are correlated with both the formation of stromules and the accumulation of ROS-responsive genes over diurnal time (Brunkard et al. 2015; Lai et al. 2012) suggesting that the redox status of cells varies over the course of a day and that this has consequences for the metabolic functioning of the cell.

ROS and RNS accumulation and the resultant oxidation are regulated by a network of systems including those based upon ascorbate peroxidases (APX), glutathione (GSH), and thioredoxin (Trx)/Prx (Fig. 3; Dietz 2016; Sevilla et al. 2015). ROS/RNS often oxidize thiol groups of specific cysteine residues, and the enzymatic reduction of this oxidization enables the maintenance of biologically relevant signaling. Thiol reduction is controlled by Prx, Trx, and sulfiredoxins (Srx) (Sevilla et al. 2015; Yu et al. 2014). The *Arabidopsis* genome expresses ten different Prx genes, which are localized to different compartments including the nucleus, cytosol, mitochondria, and the chloroplast (Sevilla et al. 2015). In unstressed conditions, Prx homodimers scavenge H_2O_2 by generating a sulfenic acid group upon the peroxidatic cysteine (Fig. 3; Sevilla et al. 2015). This acidic group then reacts with a resolving cysteine within the homodimer to generate a disulfide bridge that is subsequently converted back to a reduced state by biological thiols such as Trx or glutaredoxin (Sevilla et al. 2015; Spoel and Van Ooijen 2014). In addition to this reversible oxidation, Prx can also become hyperoxidized, forming either a sulfinic ($Cys\text{-}SO_2H$) or sulfonic ($Cys\text{-}SO_3H$) species when in contact with higher concentrations of H_2O_2 (Sevilla et al. 2015). The sulfinic species inhibits Prx enzymatic function, but can be reduced by Srx. By contrast, reduction of the sulfonic form has yet to be reported (Sevilla et al. 2015).

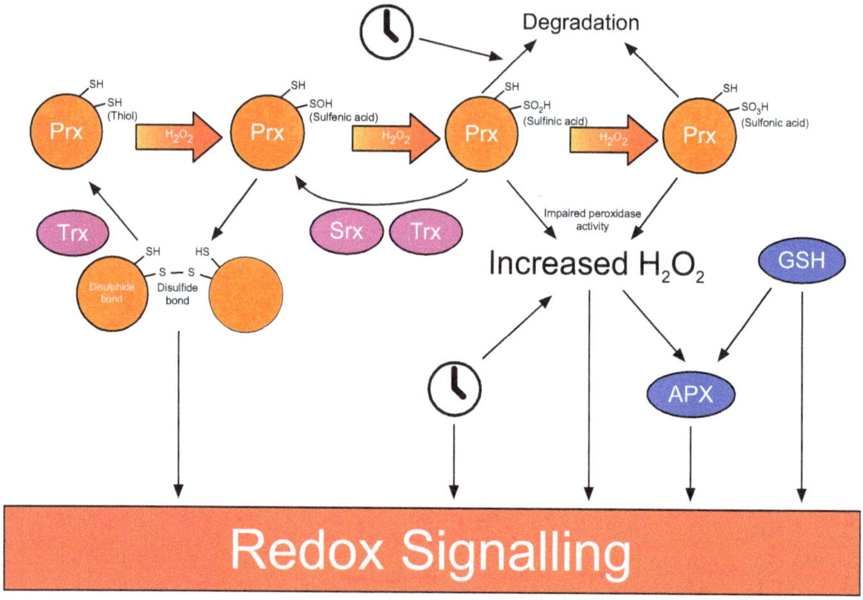

Fig. 3 Interplay between ROS scavenging and signaling. Oxidation of thiol groups upon peroxiredoxins (Prx) alters protein interactions, protein stability, and Prx peroxidase activity. A subset of these posttranslational modifications can be reversed either by Thioredoxins (Trx) or Sulfiredoxins (Srx). Increased H_2O_2 can either be detoxified by the parallel Ascorbate peroxidase (APX)/ Glutathione (GSH) pathway or induce redox-dependent signaling. The circadian system regulates the rhythmic accumulation of hyperoxidized Prx and contributes to the intracellular accumulation of H_2O_2 while also modulating redox signaling

The biochemical changes induced by these oxidation events perform a dual role by modulating protein–protein interactions in addition to acting as ROS and RNS scavengers (Sevilla et al. 2015). Protein cysteine residues are highly sensitive to thiol oxidation and it is through this posttranslational modification that changes in protein multimerization, stability, localization, and activity can be precipitated (Fig. 3; Sevilla et al. 2015; Spoel and Van Ooijen 2014). In particular, the Cys sulfenic species can subsequently form intra- or intermolecular disulfide bonds (Sevilla et al. 2015). Although solvent-exposed cysteine residues with the appropriate biochemical properties to act as physiologically relevant redox sensors are comparatively rare (Spadaro et al. 2010), the enzymatic reversal of this process enables the use of these disulfide bridges as a biologically relevant signaling state (Sevilla et al. 2015; Spoel and Van Ooijen 2014; Wouters et al. 2010). One candidate is Prx itself (Hall et al. 2009). A Prx signaling state could either manifest as the formation of a disulfide bridge with a signaling partner, multimerization of Prx (altering Prxs' role as chaperone proteins), or as inactivation of Prx peroxidase activity (inducing localized accumulation of H_2O_2; Hall et al. 2009; Lázaro et al. 2013).

Given this daily fluctuation in redox status, it is natural to assume that the circadian system regulates plants' responses to ROS production. In agreement with this hypothesis, endogenous rhythms of H_2O_2 accumulation persist in constant light (Lai et al. 2012). The role of the clock is sufficient to mitigate some ROS damage as circadian clock mutants demonstrate increased sensitivity to ROS stress (Lai et al. 2012). The clock's contribution to ROS quenching may arise from regulation of ROS scavengers, as circadian control of protein oxidation have been observed. The oxidation status of Prx and the activity of catalases, as well as the ratio of the reduction–oxidation coenzymes NADPH and $NADP^+$, continue to oscillate with a circadian rhythm, in phase with rhythmic accumulation of H_2O_2 (Edgar et al. 2012; Lai et al. 2012; Zhong and McClung 1996; Zhou et al. 2015). Interestingly, rhythms of Prx oxidation persist in the absence of transcription in red blood cells and the unicellular alga *O. tauri* (O'Neill and Reddy 2011; O'Neill et al. 2011), and also continue in plants with an arrhythmic transcriptional oscillator (Edgar et al. 2012). At least in red blood cells, these rhythms are not dependent upon the reductive capacity of Srx and Trx (Cho et al. 2014). Instead, it appears that increased degradation of the hyperoxidized enzyme occurs via the proteasome. As rhythms of Prx oxidation are also observed in species that lack Srx (Edgar et al. 2012; Stangherlin and Reddy 2013), it is possible that selective protein degradation of oxidized Prx is an evolutionarily conserved mechanism that underlies these rhythms rather than being representative of rhythmic oxidation and reduction of a comparatively stable pool of protein.

3.1 Linking Redox Metabolism to Changes in Gene Expression

Beyond their effects upon the enzymatic activity and multimerization of ROS/RNS scavenging systems, changes in redox status also precipitate changes in transcription. ROS-induced gene expression is modulated by the circadian system, with ROS-responsive genes accumulating mainly during the subjective day in plants grown under constant conditions (Lai et al. 2012). It remains unclear whether changes in cellular redox status are communicated to the nucleus directly (via ROS/RNS) or through a signaling cascade initiated by localized ROS/RNS production; both mechanisms have been reported (Foyer and Noctor 2016). In addition, the effects of redox status on translation have been documented (Benina et al. 2015).

Although still a developing field in plant biology, redox-dependent regulation of transcription factors has been well described in bacterial and mammalian fields (Dietz 2014). Redox status has numerous consequences on transcription factor activity in these latter systems, altering nuclear localization, protein conformation, oligomerization, co-factor coordination, posttranslational modification, and protein stability (Dietz 2014). Many plant-specific examples arise from stress-induced pathways (for which readers are referred to comprehensive reviews – Dietz 2014; Lázaro et al. 2013; Sevilla et al. 2015), but the importance of redox signaling in unstressed conditions is also beginning to emerge. Microarray analysis has revealed several cis-regulatory elements that are associated with H_2O_2 stress (Allu et al. 2014). In addition, the accumulation pattern of

ROS determines root development (Dunand et al. 2007), and the mutant *root meristemless* has a defect in GSH synthesis (Cheng et al. 1995; Vernoux et al. 2000). A common pattern of transcriptional regulation by ROS is their contribution to feedback loops that regulate the accumulation of ROS. Transcription factors such as UPBEAT1 feedback to control H_2O_2 levels in the root (Tsukagoshi et al. 2010), while genes encoding enzymes such as catalase are expressed with a circadian rhythm (Lai et al. 2012); CCA1 has subsequently been found bound to the promoter of *TRX* genes (Barajas-López et al. 2011).

Cellular redox status also influences gene expression by altering the activity of enzymes responsible for epigenetic marks upon chromatin. Epigenetic marks (most commonly histone acetylation, methylation, and phosphorylation) are highly dynamic and are correlated with transcriptional activity (Young et al. 2010). For example, repression of *CCA1* and *LHY* by PRR proteins requires histone acetylase activity (Fig. 1; Wang et al. 2013). Many of the enzymes necessary to add or remove these epigenetic markers require metabolic cofactors for their activity. NAD^+ is necessary for the enzymatic activity of SIRTUIN histone deacetylases, whereas Fe^{2+} is a cofactor for JUMONJI histone demethylases (Klose et al. 2006; Sauve et al. 2006). The majority of the work examining the interplay between the availability of these metabolites and enzymatic function has been completed in mammalian systems although the application of the SIRTUIN inhibitor nicotinamide reduces the accumulation of histone methylation marks associated with gene transcription and lengthens circadian period in plants (Malapeira et al. 2012). Additionally, ROS stress induces a global increase in histone methylation in mammalian systems, potentially via the inactivation of histone demethylases and deacetylases (Niu et al. 2015). It will therefore be of interest to determine whether accumulation of NAD^+ and Fe^{2+} *in planta* is dependent upon cellular redox status, and whether modulation of these levels via redox stress is sufficient to regulate gene expression and the circadian system.

4 Perspectives

Light is a prerequisite for photosynthesis, but it is equally apparent that the rhythms conferred by the diel cycle have induced the evolution of the circadian system to facilitate anticipation of dawn and dusk, as well as enabling measurement of day length and the tracking of seasons. Our understanding of metabolic, circadian, and light signaling is now advancing such that we are beginning to understand the complex relationship between these interdependent pathways. Future work will describe how oscillating levels of metabolites and posttranslational protein modification relay information pertaining to the metabolic status of the cell to the nucleus. This information – coupled with knowledge of how transcription factors within the circadian system modulate these signals – will enable the manipulation of plant responses' to daylight and temperature to optimize growth, yield, and cellular responses to biotic or abiotic stress.

Acknowledgments For Theo, whose impact on nycthemeral rhythms is profound and persistent. Thanks to Uli Bechtold, Lauren Headland, and Tracy Lawson for critical reading of the manuscript.

References

Alabadi D, Oyama T, Yanovsky M, Harmon F, Mas P, Kay S (2001) Reciprocal regulation between TOC1 and LHY/CCA1 within the Arabidopsis circadian clock. Science 293(5531):880–883

Allu AD, Soja AM, Wu A, Szymanski J, Balazadeh S (2014) Salt stress and senescence: identification of cross-talk regulatory components. J Exp Bot 65(14):3993–4008. doi:10.1093/jxb/eru173

Andrés-Colás N, Perea-García A, Puig S, Peñarrubia L (2010) Deregulated copper transport affects Arabidopsis development especially in the absence of environmental cycles. Plant Physiol 153 (1):170–184. doi:10.1104/pp.110.153676

Baerenfaller K, Massonnet C, Walsh S, Baginsky S, Bühlmann P, Hennig L, Hirsch-Hoffmann M, Howell KA, Kahlau S, Radziejwoski A, Russenberger D, Rutishauser D, Small I, Stekhoven D, Sulpice ER, Svozil J, Wuyts N, Stitt M, Hilson P, Granier C, Gruissem W (2012) Systems-based analysis of Arabidopsis leaf growth reveals adaptation to water deficit. Mol Syst Biol 8:606. doi:10.1038/msb.2012.39

Baker NR (2008) Chlorophyll fluorescence: a probe of photosynthesis *in vivo*. Annu Rev Plant Biol 59:89–113. doi:10.1146/annurev.arplant.59.032607.092759

Barajas-López JD, Serrato AJ, Cazalis R, Meyer Y, Chueca A, Reichheld J-P, Sahrawy M (2011) Circadian regulation of chloroplastic f and m thioredoxins through control of the CCA1 transcription factor. J Exp Bot 62(6):2039–2051. doi:10.1093/jxb/erq394

Baxter A, Mittler R, Suzuki N (2014) ROS as key players in plant stress signalling. J Exp Bot 65 (5):1229–1240. doi:10.1093/jxb/ert375

Benina M, Ribeiro DM, Gechev TS, Mueller-Roeber B, Schippers JHM (2015) A cell type-specific view on the translation of mRNAs from ROS-responsive genes upon paraquat treatment of *Arabidopsis thaliana* leaves. Plant Cell Environ 38(2):349–363. doi:10.1111/pce.12355

Bläsing OE, Gibon Y, Günther M, Höhne M, Morcuende R, Osuna D, Thimm O, Usadel B, Scheible W-R, Stitt M (2005) Sugars and circadian regulation make major contributions to the global regulation of diurnal gene expression in Arabidopsis. Plant Cell 17(12):3257–3281. doi:10.1105/tpc.105.035261

Borland A, Hartwell J, Jenkins GI, Wilkins M, Nimmo HG (1999) Metabolite control overrides circadian regulation of phosphoenolpyruvate carboxylase kinase and CO_2 fixation in crassulacean acid metabolism. Plant Physiol 121(3):889–896

Boxall S, Foster J, Bohnert H, Cushman J, Nimmo H, Hartwell J (2005) Conservation and divergence of circadian clock operation in a stress-inducible crassulacean acid metabolism species reveals clock compensation against stress. Plant Physiol 137(3):969–982. doi:10.1104/pp.104.054577

Brunkard JO, Runkel AM, Zambryski PC (2015) Chloroplasts extend stromules independently and in response to internal redox signals. Proc Natl Acad Sci U S A 112(32):10044–10049. doi:10.1073/pnas.1511570112

Carter PJ, Nimmo HG, Fewson CA, Wilkins MB (1991) Circadian rhythms in the activity of a plant protein kinase. EMBO J 10(8):2063–2068

Carter PJ, Wilkins MB, Nimmo HG, Fewson C (1995) The role of temperature in the regulation of the circadian rhythm of CO_2 fixation in *Bryophyllum fedtschenkoi*. Planta 196(2). doi:10.1007/BF00201399

Chen Y-Y, Wang Y, Shin L-J, Wu J-F, Shanmugam V, Tsednee M, Lo J-C, Chen C-C, Wu S-H, Yeh K-C (2013) Iron is involved in the maintenance of circadian period length in Arabidopsis. Plant Physiol 161(3):1409–1420. doi:10.1104/pp.112.212068

Cheng JC, Seeley KA, Sung ZR (1995) RML1 and RML2, Arabidopsis genes required for cell proliferation at the root tip. Plant Physiol 107(2):365–376

Cho C-S, Yoon HJ, Kim JY, Woo HA, Rhee SG (2014) Circadian rhythm of hyperoxidized peroxiredoxin II is determined by hemoglobin autoxidation and the 20S proteasome in red blood cells. Proc Natl Acad Sci U S A 111(33):12043–12048. doi:10.1073/pnas.1401100111

Covington MF, Maloof JN, Straume M, Kay SA, Harmer SL (2008) Global transcriptome analysis reveals circadian regulation of key pathways in plant growth and development. Genome Biol 9 (8):R130. doi:10.1186/gb-2008-9-8-r130

Dalchau N, Baek SJ, Briggs HM, Robertson FC, Dodd AN, Gardner MJ, Stancombe MA, Haydon MJ, Stan G-B, Gonçalves JM, Webb AAR (2011) The circadian oscillator gene GIGANTEA mediates a long-term response of the *Arabidopsis thaliana* circadian clock to sucrose. Proc Natl Acad Sci U S A 108(12):5104–5109. doi:10.1073/pnas.1015452108

Del Río LA (2015) ROS and RNS in plant physiology: an overview. J Exp Bot 66(10):2827–2837. doi:10.1093/jxb/erv099

Dever LV, Boxall SF, Kneřová J, Hartwell J (2015) Transgenic perturbation of the decarboxylation phase of Crassulacean acid metabolism alters physiology and metabolism but has only a small effect on growth. Plant Physiol 167(1):44–59. doi:10.1104/pp.114.251827

Dietz K-J (2011) Peroxiredoxins in plants and cyanobacteria. Antioxid Redox Signal 15(4):1129–1159. doi:10.1089/ars.2010.3657

Dietz K-J (2014) Redox regulation of transcription factors in plant stress acclimation and development. Antioxid Redox Signal 21(9):1356–1372. doi:10.1089/ars.2013.5672

Dietz K-J (2016) Thiol-based peroxidases and ascorbate peroxidases: why plants rely on multiple peroxidase systems in the photosynthesizing chloroplast? Mol Cells 39(1):20–25. doi:10.14348/molcells.2016.2324

Dittrich P (1976) Nicotinamide adenine dinucleotide-specific "malic" enzyme in Kalanchoë daigremontiana and other plants exhibiting crassulacean acid metabolism. Plant Physiol 57 (2):310–314

Dodd AN, Parkinson K, Webb AAR (2004) Independent circadian regulation of assimilation and stomatal conductance in the *ztl-1* mutant of Arabidopsis. New Phytol 162(1):63–70. doi:10.1111/j.1469-8137.2004.01005.x

Dodd AN, Salathia N, Hall A, Kevei E, Tóth R, Nagy F, Hibberd JM, Millar AJ, Webb AAR (2005) Plant circadian clocks increase photosynthesis, growth, survival, and competitive advantage. Science 309(5734):630–633. doi:10.1126/science.1115581

Dodd AN, Gardner MJ, Hotta CT, Hubbard KE, Dalchau N, Love J, Assie J-M, Robertson FC, Jakobsen MK, Gonçalves J, Sanders D, Webb AAR (2007) The Arabidopsis circadian clock incorporates a cADPR-based feedback loop. Science 318(5857):1789–1792

Dodd AN, Dalchau N, Gardner MJ, Baek S-J, Webb AAR (2014) The circadian clock has transient plasticity of period and is required for timing of nocturnal processes in Arabidopsis. New Phytol 201(1):168–179. doi:10.1111/nph.12489

Dodd AN, Belbin FE, Frank A, Webb AAR (2015) Interactions between circadian clocks and photosynthesis for the temporal and spatial coordination of metabolism. Front Plant Sci 6:245. doi:10.3389/fpls.2015.00245

Duc C, Cellier F, Lobréaux S, Briat J-F, Gaymard F (2009) Regulation of iron homeostasis in *Arabidopsis thaliana* by the clock regulator time for coffee. J Biol Chem 284(52):36271–36281. doi:10.1074/jbc.M109.059873

Dunand C, Crèvecoeur M, Penel C (2007) Distribution of superoxide and hydrogen peroxide in Arabidopsis root and their influence on root development: possible interaction with peroxidases. New Phytol 174(2):332–341. doi:10.1111/j.1469-8137.2007.01995.x

Edgar RS, Green EW, Zhao Y, Van Ooijen G, Olmedo M, Qin X, Xu Y, Pan M, Valekunja UK, Feeney KA, Maywood ES, Hastings MH, Baliga NS, Merrow M, Millar AJ, Johnson CH, Kyriacou CP, O'Neill JS, Reddy AB (2012) Peroxiredoxins are conserved markers of circadian rhythms. Nature 485(7399):459–464. doi:10.1038/nature11088

Espinoza C, Degenkolbe T, Caldana C, Zuther E, Leisse A, Willmitzer L, Hincha DK, Hannah MA (2010) Interaction with diurnal and circadian regulation results in dynamic metabolic and transcriptional changes during cold acclimation in Arabidopsis. PLoS One 5(11):e14101. doi:10.1371/journal.pone.0014101

Farquhar GD, Sharkey TD (1982) Stomatal conductance and photosynthesis. Annu Rev Plant Physiol 33:317–345

Feeney KA, Hansen LL, Putker M, Olivares-Yañez C, Day J, Eades LJ, Larrondo LF, Hoyle NP, O'Neill JS, Van Ooijen G (2016) Daily magnesium fluxes regulate cellular timekeeping and energy balance. Nature 532(7599):375–379. doi:10.1038/nature17407

Feugier FG, Satake A (2012) Dynamical feedback between circadian clock and sucrose availability explains adaptive response of starch metabolism to various photoperiods. Front Plant Sci 3:305. doi:10.3389/fpls.2012.00305

Figueroa CM, Lunn JE (2016) A tale of two sugars: trehalose 6-phosphate and sucrose. Plant Physiol 172(1):7–27. doi:10.1104/pp.16.00417

Figueroa CM, Feil R, Ishihara H, Watanabe M, Kölling K, Krause U, Höhne M, Encke B, Plaxton WC, Zeeman SC, Li Z, Schulze WX, Hoefgen R, Stitt M, Lunn JE (2016) Trehalose 6-phosphate coordinates organic and amino acid metabolism with carbon availability. Plant J 85(3):410–423. doi:10.1111/tpj.13114

Fogelmark K, Troein C (2014) Rethinking transcriptional activation in the Arabidopsis circadian clock. PLoS Comput Biol 10(7):e1003705. doi:10.1371/journal.pcbi.1003705

Foyer CH, Noctor G (2016) Stress-triggered redox signalling: what's in pROSpect? Plant Cell Environ 39(5):951–964. doi:10.1111/pce.12621

Gendron JM, Pruneda-Paz JL, Doherty CJ, Gross AM, Kang SE, Kay SA (2012) Arabidopsis circadian clock protein, TOC1, is a DNA-binding transcription factor. Proc Natl Acad Sci U S A 109(8):3167–3172. doi:10.1073/pnas.1200355109

Gibon Y, Bläsing OE, Palacios-Rojas N, Pankovic D, Hendriks JHM, Fisahn J, Höhne M, Günther M, Stitt M (2004) Adjustment of diurnal starch turnover to short days: depletion of sugar during the night leads to a temporary inhibition of carbohydrate utilization, accumulation of sugars and post-translational activation of ADP-glucose pyrophosphorylase in the following light period. Plant J 39(6):847–862. doi:10.1111/j.1365-313X.2004.02173.x

Gibon Y, Usadel B, Blaesing OE, Kamlage B, Hoehne M, Trethewey R, Stitt M (2006) Integration of metabolite with transcript and enzyme activity profiling during diurnal cycles in Arabidopsis rosettes. Genome Biol 7(8):R76. doi:10.1186/gb-2006-7-8-R76

Gibon Y, Pyl E-T, Sulpice R, Lunn JE, Höhne M, Günther M, Stitt M (2009) Adjustment of growth, starch turnover, protein content and central metabolism to a decrease of the carbon supply when Arabidopsis is grown in very short photoperiods. Plant Cell Environ 32(7):859–874. doi:10.1111/j.1365-3040.2009.01965.x

Gould PD, Ugarte N, Domijan M, Costa M, Foreman J, Macgregor D, Rose K, Griffiths J, Millar AJ, Finkenstädt B, Penfield S, Rand DA, Halliday KJ, Hall AJW (2013) Network balance via CRY signalling controls the Arabidopsis circadian clock over ambient temperatures. Mol Syst Biol 9:650. doi:10.1038/msb.2013.7

Graf A, Smith AM (2011) Starch and the clock: the dark side of plant productivity. Trends Plant Sci 16(3):169–175. doi:10.1016/j.tplants.2010.12.003

Graf A, Schlereth A, Stitt M, Smith AM (2010) Circadian control of carbohydrate availability for growth in Arabidopsis plants at night. Proc Natl Acad Sci U S A 107(20):9458–9463. doi:10.1073/pnas.0914299107

Hädrich N, Hendriks JHM, Kötting O, Arrivault S, Feil R, Zeeman SC, Gibon Y, Schulze WX, Stitt M, Lunn JE (2012) Mutagenesis of cysteine 81 prevents dimerization of the APS1 subunit of ADP-glucose pyrophosphorylase and alters diurnal starch turnover in *Arabidopsis thaliana* leaves. Plant J 70(2):231–242. doi:10.1111/j.1365-313X.2011.04860.x

Hall A, Bastow RM, Davis SJ, Hanano S, McWatters HG, Hibberd V, Doyle MR, Sung S, Halliday KJ, Amasino RM, Millar AJ (2003) The TIME FOR COFFEE gene maintains the amplitude and timing of Arabidopsis circadian clocks. Plant Cell 15(11):2719–2729. doi:10.1105/tpc.013730. tpc.013730 [pii]

Hall A, Karplus PA, Poole LB (2009) Typical 2-Cys peroxiredoxins – structures, mechanisms and functions. FEBS J 276(9):2469–2477. doi:10.1111/j.1742-4658.2009.06985.x

Harmer SL, Hogenesch JB, Straume M, Chang HS, Han B, Zhu T, Wang X, Kreps JA, Kay SA (2000) Orchestrated transcription of key pathways in Arabidopsis by the circadian clock. Science 290:2110–2113

Hartwell J (2008) The circadian clock in CAM plants. In: Annual plant reviews. doi:10.1016/j.ejcts.2009.02.013

Hartwell J, Gill A, Nimmo GA, Wilkins MB, Jenkins GI, Nimmo HG (1999) Phosphoenolpyruvate carboxylase kinase is a novel protein kinase regulated at the level of expression. Plant J 20(3):333–342. tpj609 [pii]

Hartwell J, Dever LV, Boxall SF (2016) Emerging model systems for functional genomics analysis of crassulacean acid metabolism. Curr Opin Plant Biol 31:100–108. doi:10.1016/j.pbi.2016.03.019

Hastings MH, Maywood ES, O'Neill JS (2008) Cellular circadian pacemaking and the role of cytosolic rhythms. Curr Biol 18(17):R805–R815. doi:10.1016/j.cub.2008.07.021

Haydon MJ, Mielczarek O, Robertson FC, Hubbard KE, Webb AAR (2013) Photosynthetic entrainment of the *Arabidopsis thaliana* circadian clock. Nature 502(7473):689–692. doi:10.1038/nature12603

Haydon MJ, Román Á, Arshad W (2015) Nutrient homeostasis within the plant circadian network. Front Plant Sci 6:299. doi:10.3389/fpls.2015.00299

Helfer A, Nusinow DA, Chow BY, Gehrke AR, Bulyk ML, Kay SA (2011) LUX ARRHYTHMO encodes a nighttime repressor of circadian gene expression in the Arabidopsis core clock. Curr Biol 21(2):126–133. doi:10.1016/j.cub.2010.12.021

Hennessey TL, Field CB (1991) Circadian rhythms in photosynthesis: oscillations in carbon assimilation and stomatal conductance under constant conditions. Plant Physiol 96(3):831–836. doi:10.1023/A:1011807229154

Hermans C, Vuylsteke M, Coppens F, Craciun A, Inzé D, Verbruggen N (2010) Early transcriptomic changes induced by magnesium deficiency in *Arabidopsis thaliana* reveal the alteration of circadian clock gene expression in roots and the triggering of abscisic acid-responsive genes. New Phytol 187(1):119–131. doi:10.1111/j.1469-8137.2010.03258.x

Holtum JAM, Smith JAC, Neuhaus HE (2005) Intracellular transport and pathways of carbon flow in plants with crassulacean acid metabolism. Funct Plant Biol 32(5):429. doi:10.1071/FP04189

Hong S, Kim SA, Guerinot ML, McClung CR (2013) Reciprocal interaction of the circadian clock with the iron homeostasis network in Arabidopsis. Plant Physiol 161(2):893–903. doi:10.1104/pp.112.208603

Hsu PY, Harmer SL (2014) Wheels within wheels: the plant circadian system. Trends Plant Sci 19(4):240–249. doi:10.1016/j.tplants.2013.11.007

Huang W, Perez-Garcia P, Pokhilko A, Millar AJ, Antoshechkin I, Riechmann JL, Mas P (2012) Mapping the core of the Arabidopsis circadian clock defines the network structure of the oscillator. Science 336(6077):75–79. doi:10.1126/science.1219075

Ito S, Matsushika A, Yamada H, Sato S, Kato T, Tabata S, Yamashino T, Mizuno T (2003) Characterization of the APRR9 pseudo-response regulator belonging to the APRR1/TOC1 quintet in *Arabidopsis thaliana*. Plant Cell Physiol 44(11):1237–1245

Johnson CH, Knight MR, Kondo T, Masson P, Sedbrook J, Haley A, Trewavas A (1995) Circadian oscillations of cytosolic and chloroplastic free calcium in plants. Science 269(5232):1863–1865

Jones MA, Covington MF, DiTacchio L, Vollmers C, Panda S, Harmer SL (2010) Jumonji domain protein JMJD5 functions in both the plant and human circadian systems. Proc Natl Acad Sci U S A 107(50):21623–21628. doi:10.1073/pnas.1014204108

Kim W, Fujiwara S, Suh S, Kim J, Kim Y, Han L, David K, Putterill J, Nam H, Somers D (2007) ZEITLUPE is a circadian photoreceptor stabilized by GIGANTEA in blue light. Nature 449(7160):356–360

Kim J, Geng R, Gallenstein RA, Somers DE (2013) The F-box protein ZEITLUPE controls stability and nucleocytoplasmic partitioning of GIGANTEA. Development 140(19):4060–4069. doi:10.1242/dev.096651

Klose R, Kallin E, Zhang Y (2006) JmjC-domain-containing proteins and histone demethylation. Nat Rev Genet 7(9):715–727

Kölling K, Thalmann M, Müller A, Jenny C, Zeeman SC (2015) Carbon partitioning in *Arabidopsis thaliana* is a dynamic process controlled by the plants metabolic status and its circadian clock. Plant Cell Environ 38(10):1965–1979. doi:10.1111/pce.12512

Lai AG, Doherty CJ, Mueller-Roeber B, Kay SA, Schippers JHM, Dijkwel PP (2012) CIRCADIAN CLOCK-ASSOCIATED 1 regulates ROS homeostasis and oxidative stress responses. Proc Natl Acad Sci U S A 109(42):17129–17134. doi:10.1073/pnas.1209148109

Lawson T, Blatt MR (2014) Stomatal size, speed, and responsiveness impact on photosynthesis and water use efficiency. Plant Physiol 164(4):1556–1570. doi:10.1104/pp.114.237107

Lázaro JJ, Jimenez A, Camejo D, Iglesias-Baena I, Martí MDC, Lázaro-Payo A, Barranco-Medina S, Sevilla F (2013) Dissecting the integrative antioxidant and redox systems in plant mitochondria. Effect of stress and S-nitrosylation. Front Plant Sci 4:460. doi:10.3389/fpls.2013.00460

Legnaioli T, Cuevas J, Más P (2009) TOC1 functions as a molecular switch connecting the circadian clock with plant responses to drought. EMBO J 28(23):3745–3757. doi:10.1038/emboj.2009.297

Litthauer S, Battle MW, Lawson T, Jones MA (2015) Phototropins maintain robust circadian oscillation of PSII operating efficiency under blue light. Plant J 83(6):1034–1045. doi:10.1111/tpj.12947

Liu T, Carlsson J, Takeuchi T, Newton L, Farré EM (2013) Direct regulation of abiotic responses by the Arabidopsiscircadian clock component PRR7. Plant J. doi:10.1111/tpj.12276

Locke J, Southern M, Kozma-Bognar L, Hibberd V, Brown P, Turner M, Millar A (2005) Extension of a genetic network model by iterative experimentation and mathematical analysis. Mol Syst Biol 1(2005):0013

Love J, Dodd AN, Webb AAR (2004) Circadian and diurnal calcium oscillations encode photoperiodic information in Arabidopsis. Plant Cell 16(4):956–966. doi:10.1105/tpc.020214

Lu Y, Gehan JP, Sharkey TD (2005) Daylength and circadian effects on starch degradation and maltose metabolism. Plant Physiol 138(4):2280–2291. doi:10.1104/pp.105.061903

Lunn JE, Feil R, Hendriks JHM, Gibon Y, Morcuende R, Osuna D, Scheible W-R, Carillo P, Hajirezaei M-R, Stitt M (2006) Sugar-induced increases in trehalose 6-phosphate are correlated with redox activation of ADPglucose pyrophosphorylase and higher rates of starch synthesis in *Arabidopsis thaliana*. Biochem J 397(1):139–148. doi:10.1042/BJ20060083

Lüttge U, Ball E (1978) Free running oscillations of transpiration and CO_2 exchange in CAM plants without a concomitant rhythm of malate levels. Z Pflanzenphysiol 90(1):69–77. doi:10.1016/S0044-328X(78)80226-8

Malapeira J, Khaitova LC, Más P (2012) Ordered changes in histone modifications at the core of the Arabidopsis circadian clock. Proc Natl Acad Sci U S A 109(52):21540–21545. doi:10.1073/pnas.1217022110

Martins MCM, Hejazi M, Fettke J, Steup M, Feil R, Krause U, Arrivault S, Vosloh D, Figueroa CM, Ivakov A, Yadav UP, Piques M, Metzner D, Stitt M, Lunn JE (2013) Feedback inhibition of starch degradation in Arabidopsis leaves mediated by trehalose 6-phosphate. Plant Physiol 163(3):1142–1163. doi:10.1104/pp.113.226787

Más P, Kim W-Y, Somers DE, Kay SA (2003) Targeted degradation of TOC1 by ZTL modulates circadian function in *Arabidopsis thaliana*. Nature 426(6966):567–570. doi:10.1038/nature02163

McAusland L, Vialet-Chabrand S, Davey P, Baker NR, Brendel O, Lawson T (2016) Effects of kinetics of light-induced stomatal responses on photosynthesis and water-use efficiency. New Phytol 211(4):1209–1220. doi:10.1111/nph.14000

McClung CR (2006) Plant circadian rhythms. Plant Cell 18(4):792–803. doi:10.1105/tpc.106.040980

Michael T, Mockler T, Breton G, McEntee C, Byer A, Trout J, Hazen S, Shen R, Priest H, Sullivan C, Givan S, Yanovsky M, Hong F, Kay S, Chory J (2008) Network discovery pipeline elucidates conserved time-of-day-specific cis-regulatory modules. PLoS Genet 4(2):e14. doi:10.1371/journal.pgen.0040014

Millar AJ (2016) The intracellular dynamics of circadian clocks reach for the light of ecology and evolution. Annu Rev Plant Biol 67:595–618. doi:10.1146/annurev-arplant-043014-115619

Mizuno T, Yamashino T (2008) Comparative transcriptome of diurnally oscillating genes and hormone-responsive genes in *Arabidopsis thaliana*: insight into circadian clock-controlled daily responses to common ambient stresses in plants. Plant Cell Physiol 49(3):481–487. doi:10.1093/pcp/pcn008

Mizuno T, Nomoto Y, Oka H, Kitayama M, Takeuchi A, Tsubouchi M, Yamashino T (2014) Ambient temperature signal feeds into the circadian clock transcriptional circuitry through the EC night-time repressor in *Arabidopsis thaliana*. Plant Cell Physiol 55(5):958–976. doi:10.1093/pcp/pcu030

Müller LM, von Korff M, Davis SJ (2014) Connections between circadian clocks and carbon metabolism reveal species-specific effects on growth control. J Exp Bot 65(11):2915–2923. doi:10.1093/jxb/eru117

Nakamichi N, Kiba T, Henriques R, Mizuno T, Chua N-H, Sakakibara H (2010) PSEUDO-RESPONSE REGULATORS 9, 7, and 5 are transcriptional repressors in the Arabidopsis circadian clock. Plant Cell 22(3):594–605. doi:10.1105/tpc.109.072892

Nakamichi N, Kiba T, Kamioka M, Suzuki T, Yamashino T, Higashiyama T et al (2012) Transcriptional repressor PRR5 directly regulates clock-output pathways. Proc Natl Acad Sci U S A 109(42):17123–17128. doi:10.1073/pnas.1205156109

Nimmo HG (2000) The regulation of phosphoenolpyruvate carboxylase in CAM plants. Trends Plant Sci 5(2):75–80

Niu Y, DesMarais TL, Tong Z, Yao Y, Costa M (2015) Oxidative stress alters global histone modification and DNA methylation. Free Radic Biol Med 82:22–28. doi:10.1016/j.freeradbiomed.2015.01.028

Noordally ZB, Ishii K, Atkins KA, Wetherill SJ, Kusakina J, Walton EJ, Kato M, Azuma M, Tanaka K, Hanaoka M, Dodd AN (2013) Circadian control of chloroplast transcription by a nuclear-encoded timing signal. Science 339(6125):1316–1319. doi:10.1126/science.1230397

Nusinow DA, Helfer A, Hamilton EE, King JJ, Imaizumi T, Schultz TF, Farré EM, Kay SA (2011) The ELF4-ELF3-LUX complex links the circadian clock to diurnal control of hypocotyl growth. Nature 475:398–402. doi:10.1038/nature10182

O'Neill JS, Reddy AB (2011) Circadian clocks in human red blood cells. Nature 469(7331):498–503. doi:10.1038/nature09702

O'Neill JS, Van Ooijen G, Dixon LE, Troein C, Corellou F, Bouget F-Y, Reddy AB, Millar AJ (2011) Circadian rhythms persist without transcription in a eukaryote. Nature 469(7331):554–558. doi:10.1038/nature09654

Perales M, Más P (2007) A functional link between rhythmic changes in chromatin structure and the Arabidopsis biological clock. Plant Cell 19(7):2111–2123. doi:10.1105/tpc.107.050807

Pilkington SM, Encke B, Krohn N, Höhne M, Stitt M, Pyl E-T (2015) Relationship between starch degradation and carbon demand for maintenance and growth in *Arabidopsis thaliana* in different irradiance and temperature regimes. Plant Cell Environ 38(1):157–171. doi:10.1111/pce.12381

Pokhilko A, Más P, Millar AJ (2013) Modelling the widespread effects of TOC1 signalling on the plant circadian clock and its outputs. BMC Syst Biol 7:23. doi:10.1186/1752-0509-7-23

Pokhilko A, Flis A, Sulpice R, Stitt M, Ebenhöh O (2014) Adjustment of carbon fluxes to light conditions regulates the daily turnover of starch in plants: a computational model. Mol Biosyst 10(3):613–627. doi:10.1039/c3mb70459a

Rascher U, Hütt MT, Siebke K, Osmond B, Beck F, Lüttge U (2001) Spatiotemporal variation of metabolism in a plant circadian rhythm: the biological clock as an assembly of coupled individual oscillators. Proc Natl Acad Sci U S A 98(20):11801–11805. doi:10.1073/pnas.191169598

Ravet K, Touraine B, Boucherez J, Briat J-F, Gaymard F, Cellier F (2009) Ferritins control interaction between iron homeostasis and oxidative stress in Arabidopsis. Plant J 57(3):400–412. doi:10.1111/j.1365-313X.2008.03698.x

Rawat R, Takahashi N, Hsu PY, Jones MA, Schwartz J, Salemi MR, Phinney BS, Harmer SL (2011) REVEILLE8 and PSEUDO-REPONSE REGULATOR5 form a negative feedback loop within the Arabidopsis circadian clock. PLoS Genet 7(3):e1001350. doi:10.1371/journal.pgen.1001350

Robertson FC, Skeffington AW, Gardner MJ, Webb AAR (2008) Interactions between circadian and hormonal signalling in plants. Plant Mol Biol 69(4):419–427. doi:10.1007/s11103-008-9407-4

Rugnone ML, Faigón Soverna A, Sanchez SE, Schlaen RG, Hernando CE, Seymour DK, Mancini AC, Weigel D, Más P, Yanovsky MJ (2013) LNK genes integrate light and clock signaling networks at the core of the Arabidopsis oscillator. Proc Natl Acad Sci U S A 110(29):12120–12125. doi:10.1073/pnas.1302170110

Sage RF, Sage TL, Kocacinar F (2012) Photorespiration and the evolution of C4 photosynthesis. Annu Rev Plant Biol 63:19–47. doi:10.1146/annurev-arplant-042811-105511

Salomé PA, McClung CR (2005) PSEUDO-RESPONSE REGULATOR 7 and 9 are partially redundant genes essential for the temperature responsiveness of the Arabidopsis circadian clock. Plant Cell 17(3):791–803. doi:10.1105/tpc.104.029504

Salome PA, Michael TP, Kearns EV, Fett-Neto AG, Sharrock RA, McClung CR (2002) The out of phase 1 mutant defines a role for PHYB in circadian phase control in Arabidopsis. Plant Physiol 129(4):1674–1685

Salomé PA, Oliva M, Weigel D, Krämer U (2013) Circadian clock adjustment to plant iron status depends on chloroplast and phytochrome function. EMBO J 32(4):511–523. doi:10.1038/emboj.2012.330

Sanchez-Villarreal A, Shin J, Bujdoso N, Obata T, Neumann U, Du S-X, Ding Z, Davis AM, Shindo T, Schmelzer E, Sulpice R, Nunes-Nesi A, Stitt M, Fernie AR, Davis SJ (2013) TIME FOR COFFEE is an essential component in the maintenance of metabolic homeostasis in *Arabidopsis thaliana*. Plant J 76(2):188–200. doi:10.1111/tpj.12292

Sauve AA, Wolberger C, Schramm VL, Boeke JD (2006) The biochemistry of sirtuins. Annu Rev Biochem 75:435–465. doi:10.1146/annurev.biochem.74.082803.133500

Scialdone A, Mugford ST, Feike D, Skeffington A, Borrill P, Graf A, Smith AM, Howard M (2013) Arabidopsis plants perform arithmetic division to prevent starvation at night. Elife 2: e00669. doi:10.7554/eLife.00669

Sevilla F, Camejo D, Ortiz-Espin A, Calderon A, Lazaro JJ, Jimenez A (2015) The thioredoxin/peroxiredoxin/sulfiredoxin system: current overview on its redox function in plants and regulation by reactive oxygen and nitrogen species. J Exp Bot 66(10):2945–2955. doi:10.1093/jxb/erv146

Sharkey TD (2015) Understanding carbon partitioning and its role in determining plant growth. Plant Cell Environ 38(10):1963–1964. doi:10.1111/pce.12543

Skeffington AW, Graf A, Duxbury Z, Gruissem W, Smith AM (2014) Glucan, water dikinase exerts little control over starch degradation in Arabidopsis leaves at night. Plant Physiol 165 (2):866–879. doi:10.1104/pp.114.237016

Smith AM (2012) Starch in the Arabidopsis plant. Starch Stärke. doi:10.1002/star.201100163

Smith AM, Stitt M (2007) Coordination of carbon supply and plant growth. Plant Cell Environ 30 (9):1126–1149. doi:10.1111/j.1365-3040.2007.01708.x

Smith JAC, Winter K (1996) Taxonomic distribution of crassulacean acid metabolism. In: Chapter 27, vol 114. Springer, Berlin; Heidelberg, pp 427–436. doi:10.1007/978-3-642-79060-7_27

Smith SM, Fulton DC, Chia T, Thorneycroft D, Chapple A, Dunstan H, Hylton C, Zeeman SC, Smith AM (2004) Diurnal changes in the transcriptome encoding enzymes of starch metabolism provide evidence for both transcriptional and posttranscriptional regulation of starch metabolism in Arabidopsis leaves. Plant Physiol 136(1):2687–2699. doi:10.1104/pp.104.044347

Somers D, Webb A, Pearson M, Kay S (1998) The short-period mutant, toc1-1, alters circadian clock regulation of multiple outputs throughout development in *Arabidopsis thaliana*. Development 125(3):485–494

Spadaro D, Yun B-W, Spoel SH, Chu C, Wang Y-Q, Loake GJ (2010) The redox switch: dynamic regulation of protein function by cysteine modifications. Physiol Plant 138(4):360–371. doi:10.1111/j.1399-3054.2009.01307.x

Spoel SH, Van Ooijen G (2014) Circadian redox signaling in plant immunity and abiotic stress. Antioxid Redox Signal 20(18):3024–3039. doi:10.1089/ars.2013.5530

Stangherlin A, Reddy AB (2013) Regulation of circadian clocks by redox homeostasis. J Biol Chem 288(37):26505–26511. doi:10.1074/jbc.R113.457564

Stitt M, Zeeman SC (2012) Starch turnover: pathways, regulation and role in growth. Curr Opin Plant Biol 15(3):282–292. doi:10.1016/j.pbi.2012.03.016

Stitt M, Lunn J, Usadel B (2010) Arabidopsis and primary photosynthetic metabolism – more than the icing on the cake. Plant J 61(6):1067–1091. doi:10.1111/j.1365-313X.2010.04142.x

Sulpice R, Flis A, Ivakov AA, Apelt F, Krohn N, Encke B, Abel C, Feil R, Lunn JE, Stitt M (2014) Arabidopsis coordinates the diurnal regulation of carbon allocation and growth across a wide range of photoperiods. Mol Plant 7(1):137–155. doi:10.1093/mp/sst127

Tsukagoshi H, Busch W, Benfey PN (2010) Transcriptional regulation of ROS controls transition from proliferation to differentiation in the root. Cell 143(4):606–616. doi:10.1016/j.cell.2010.10.020

Usadel B, Bläsing OE, Gibon Y, Retzlaff K, Höhne M, Günther M, Stitt M (2008) Global transcript levels respond to small changes of the carbon status during progressive exhaustion of carbohydrates in Arabidopsis rosettes. Plant Physiol 146(4):1834–1861. doi:10.1104/pp.107.115592

Vernoux T, Wilson RC, Seeley KA, Reichheld JP, Muroy S, Brown S, Maughan SC, Cobbett CS, Van Montagu M, Inze D, May MJ, Sung ZR (2000) The ROOT MERISTEMLESS1/CADMIUM SENSITIVE2 gene defines a glutathione-dependent pathway involved in initiation and maintenance of cell division during postembryonic root development. Plant Cell 12(1):97–110

Wang Z, Tobin E (1998) Constitutive expression of the CIRCADIAN CLOCK ASSOCIATED 1 (CCA1) gene disrupts circadian rhythms and suppresses its own expression. Cell 93(7):1207–1217

Wang L, Kim J, Somers DE (2013) Transcriptional corepressor TOPLESS complexes with pseudoresponse regulator proteins and histone deacetylases to regulate circadian transcription. Proc Natl Acad Sci U S A 110(2):761–766. doi:10.1073/pnas.1215010110

Warren D, Wilkins M (1961) An endogenous rhythm in the rate of dark-fixation of carbon dioxide in leaves of *Bryophyllum fedtschenkoi*. Nature 191(4789):686–688. doi:10.1038/191686a0

Webb AAR, Satake A (2015) Understanding circadian regulation of carbohydrate metabolism in Arabidopsis using mathematical models. Plant Cell Physiol 56(4):586–593. doi:10.1093/pcp/pcv033

Wilkins MB (1992) Tansley review No. 37 circadian rhythms: their origin and control. New Phytol 121:347–375

Wilkinson SR, Welch RM, Mayland HF, Grunes DL (1990) Magnesium in plants: uptake, distribution, function and utilization by man and animals. In: Metal ions in biological systems, vol 26, p 33

Wouters MA, Fan SW, Haworth NL (2010) Disulfides as redox switches: from molecular mechanisms to functional significance. Antioxid Redox Signal 12(1):53–91. doi:10.1089/ars.2009.2510

Xie Q, Wang P, Liu X, Yuan L, Wang L, Zhang C, Li Y, Xing H, Zhi L, Yue Z, Zhao C, McClung CR, Xu X (2014) LNK1 and LNK2 are transcriptional coactivators in the Arabidopsis circadian oscillator. Plant Cell 26(7):2843–2857. doi:10.1105/tpc.114.126573

Yadav UP, Ivakov A, Feil R, Duan GY, Walther D, Giavalisco P, Piques M, Carillo P, Hubberten H-M, Stitt M, Lunn JE (2014) The sucrose-trehalose 6-phosphate (Tre6P) nexus: specificity and mechanisms of sucrose signalling by Tre6P. J Exp Bot 65(4):1051–1068. doi:10.1093/jxb/ert457

Young MW, Kay SA (2001) Time zones: a comparative genetics of circadian clocks. Nat Rev Genet 2(9):702–715. doi:10.1038/35088576

Young NL, Dimaggio PA, Garcia BA (2010) The significance, development and progress of high-throughput combinatorial histone code analysis. Cell Mol Life Sci 67(23):3983–4000. doi:10.1007/s00018-010-0475-7

Yu M, Lamattina L, Spoel SH, Loake GJ (2014) Nitric oxide function in plant biology: a redox cue in deconvolution. New Phytol 202(4):1142–1156. doi:10.1111/nph.12739

Zeeman SC, Kossmann J, Smith AM (2010) Starch: its metabolism, evolution, and biotechnological modification in plants. Annu Rev Plant Biol 61:209–234. doi:10.1146/annurev-arplant-042809-112301

Zhong HH, McClung CR (1996) The circadian clock gates expression of two Arabidopsis catalase genes to distinct and opposite circadian phases. Mol Gen Genet 251(2):196–203

Zhou M, Wang W, Karapetyan S, Mwimba M, Marqués J, Buchler NE, Dong X (2015) Redox rhythm reinforces the circadian clock to gate immune response. Nature 523:472–476. doi:10.1038/nature14449

Photosynthesis-Related Functions of Vasculature-Associated Chlorenchymatous Cells

Zbigniew Miszalski, Andrzej Kornaś, and Elżbieta Kuźniak

Abstract In most biochemical, molecular, and genetic studies, a leaf is regarded as a uniformly responding unit, however leaves are not homogeneous in structure and function. Leaf venation is in continuity with the vascular system within leaf petiols and stems. Leaf veins are typically encircled by bundle sheath (BS) cells containing chloroplasts and photosynthetic cells adjacent to the vasculature are also found in petiols and stems. In C_3 plants, BS cells have been shown to be preadapted for the role in C_4 photosynthesis and this may explain the polyphyletic evolution of C_4 photosynthesis. The photosynthetically active radiation (400–700 nm) reaching the chloroplast-containing cells adjacent to the vasculature in leaves, petiols, and stems is of lower intensity and enriched with longer wavelengths (~500–700 nm) when compared with that absorbed by mesophyll cells. The CO_2 diffusion from the air to the vasculature-adjacent chlorenchymatous cells is also expected to be slow in comparison to mesophyll cells. However, the vasculature can be supplied with malate which releases CO_2 after decarboxylation and with respiratory CO_2 from heterotrophic tissues transported in the xylem. It could be expected that high CO_2 concentration at the green cells around the vasculature supports carboxylation and photosynthesis. However, CO_2-rich

---

Communicated by Ulrich Lüttge

Z. Miszalski (✉)
Institute of Plant Physiology, Polish Academy of Sciences, Niezapominajek 21, Kraków 30-239, Poland

Małopolska Centre of Biotechnology, Jagiellonian University, Gronostajowa 7a, Kraków 30-387, Poland
e-mail: miszalski@ifr-pan.edu.pl

A. Kornaś
Institute of Biology, Pedagogical University, Podchorążych 2, Kraków 30-084, Poland

E. Kuźniak
Department of Plant Physiology and Biochemistry, Faculty of Biology and Environmental Protection, University of Lodz, Banacha 12/16, Łódź 90-237, Poland

Progress in Botany (2018) 79: 173–196, DOI 10.1007/124_2017_5,
Published online: 29 March 2017

environment in stems impedes the photochemical activity of the photosynthetic vascular cells possibly through acidification of protoplasm and impairment of the pH-dependent excess energy quenching followed by reduction in the efficiency of heat dissipation. Light-dependent reduction in CO_2 release, as shown in experiments on stems can predominantly be attributed to corticular refixation. All these can affect chloroplast ultrastructure, the composition of photosynthetic electron transport chain components, and the photosynthetic enzymatic machinery in these cells.

In BS cells, a higher heterotrophic/autotrophic ratio than in mesophyll cells causes a decrease in O_2 level, intensity of photorespiration, and Mehler reaction. All these result in a decrease in PSII efficiency and stimulation of cyclic phosphorylation providing energetic support for the metabolism of BS cells.

One can conclude that vasculature-associated chlorenchymatous cells which are not able to fix atmospheric CO_2 with the help of RubisCO can be supplied with carbon originating from β-carboxylating processes, which is reflected in the ^{13}C discrimination value, and this process takes place in tissues growing at low light intensities enriched with longer wavelengths.

Contents

1 Introduction

In higher plants, the green leaves make up the majority of photosynthesizing organs and the physiology of leaves is an important field of plant physiology. In most biochemical, molecular, and genetic studies, a leaf is regarded as a uniformly responding unit, however, it is worthy to remember that leaves are not homogeneous in structure and function. Leaves are composed of many cell types of diverse physiology and structure forming epidermis containing stomata, mesophyll, and vascular tissues. Leaf veins, composed of highly specialized phloem and xylem tissues which are usually associated with parenchyma and reinforced with collenchyma or sclerenchyma, form a dense venation network. In the dicotyledonous plants, leaf venation consists of minor and major veins, including the midrib veins.

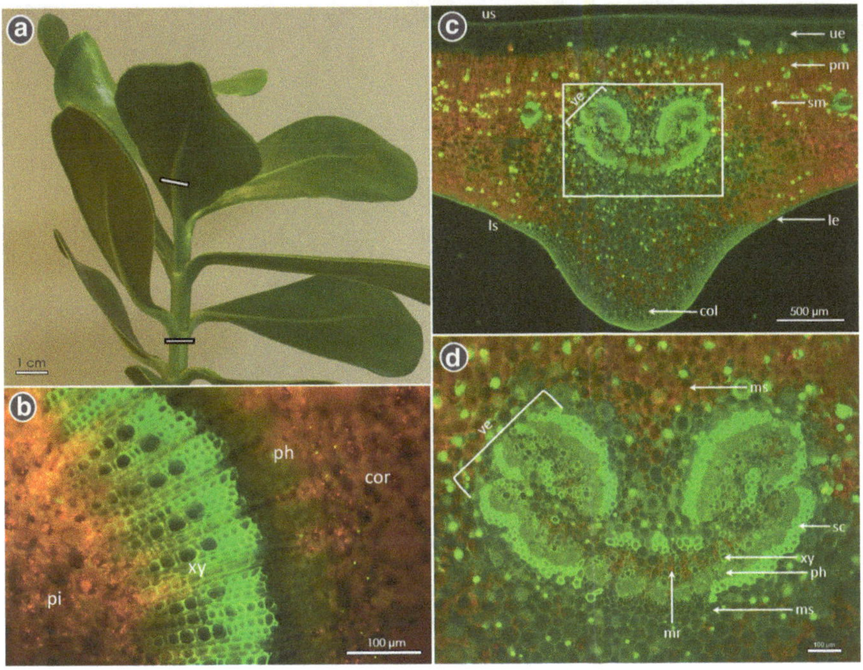

Fig. 1 Morphology (**a**) and anatomy (**b–d**) of *Clusia fluminensis*. Cross section of stem (**b**). Cross section of midrib observed using epifluorescence microscopy (**c**). Magnified image of leaf vascular bundle (**d**). *Red,* and *green* colors correspond to autofluorescence of chlorophyll and cell walls, respectively. Cutting areas were indicated by bands in (**a**). *Cor* cortex, *col* collenchyma, *le* lower epidermis, *ls* lower surface, *mr* mesophyll ray, *ms* mesophyll cells surrounding vein, *ph* phloem, *pi* pith, *pm* palisade mesophyll cells, *sc* sclerenchyma, *sm* spongy parenchyma cells, *ue* upper epidermis, *us* upper surface, *ve* vein (vascular bundle), *xy* xylem

Leaf venation is in continuity with the vascular system within leaf stalks (petiols), stems, and roots and is responsible not only for transport of water and solutes but also plays specific functions in communication between organs. The vascular tissue can also conduct light (especially far-red light) between plant organs (Sun et al. 2003). During senescence, leaf veins and the tissues around the vascular bundle stay active longer than mesophyll cells (Peoples et al. 1980; Niewiadomska et al. 2009).

Leaf veins are typically encircled by bundle sheath (BS) cells containing chloroplasts. Although the function of the vascular system is traditionally realized to be related to the mechanical support, transport, and signalling, it also contains green cells (Figs. 1 and 2) and their potential photosynthesis-related role has recently emerged (Leegood 2008; Kuźniak et al. 2016).

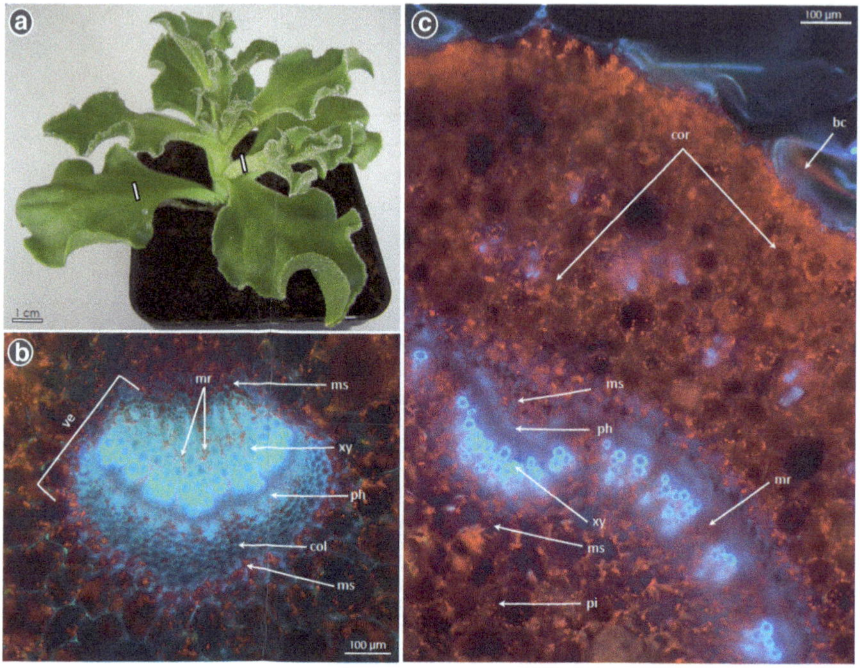

Fig. 2 Morphology (**a**) and anatomy (**b–c**) of *Mesembryanthemum crystallinum*. Cross section of the central leaf part observed using epifluorescence microscopy (**b**). Cross section of stem (**c**). *Red*, and *blue* colors correspond to autofluorescence of chlorophyll and cell walls, respectively. Cutting areas were indicated by bands in (**a**). *bc* bladdes cell, *cor* cortex, *col* collenchyma, *mr* mesophyll ray, *ms* mesophyll cells surrounding vein, *ph* phloem, *pi* pith, *ve* vein (vascular bundle), *xy* xylem

2 Bundle Sheath Cells

As suggested by Haberlandt (1914) over 100 years ago, the structure and localization of green bundle sheath cells in plants allow to suspect that in addition to photosynthesis they perform also other undiscovered functions. However, until now many details of their diverse functions are not elucidated. In leaves of *Arabidopsis*, bundle sheath cells constitute about 15% of the chloroplast-containing cells within the leaf (Kinsman and Pyke 1998). The bundle sheath usually consists of thin-walled cells which are elongated along the length of the bundle (Crookston and Moss 1974; Esau 1977). In leaves of C_4 plants, BS cells form a layer surrounding the vascular bundles and this arrangement is known as "Kranz anatomy" or "Kranz structure" (after the German Kranz, wreath). BS contains chloroplasts and is a site of the Calvin cycle. According to Slewinski (2013), the Kranz-type C_4 system in photosynthesizing organs such as leaves conceptually corresponds to the endodermis/starch sheath found in the roots and stems. Many grasses have two layers of cells surrounding the BSs of leaves (Eastman et al. 1988). In the inner layer, the cells are typically nonchlorochymatous and form the mestome sheath. The outer layer of chlorenchymatous cells is called the parenchymatous BS. In barley, a C_3 plant, S-type, and L-type cells surrounding BS have been distinguished. S-type cells contain small chloroplasts (one third of the volume of mesophyll chloroplasts) and the chloroplasts in L-type cells are similar to those in mesophyll cells (Williams et al. 1989).

In some plants, leaves referred to as heterobaric, have BS extensions consisting of cells with thickened walls which prevent lateral gas diffusion within the mesophyll. In legume species, BS extends laterally beneath the cells of palisade mesophyll forming the paraveinal mesophyll which transfers photoassimilates to the phloem (Leegood 2008).

In C_4 plants, BS forms a specialized compartment in which CO_2 can be concentrated around ribulose-1,5-bisphosphate Carboxylase/Oxygenase (RubisCO), thus suppressing photorespiration. BS in C_4 plants show clear structural and ultrastructural differences when compared with C_3 or C_2 (also known as C_3-C_4) plants (Voznesenskaya et al. 2010). In *Mesembryanthemum crystallinum*, a facultative C_3-CAM (Crassulacean Acid Metabolism) species, no clear differences in the midrib anatomy in leaves of C_3- and CAM-performing plants were found (Kuźniak et al. 2016).

3 General Functions of Bundle Sheath Cells

Besides the well-known specific metabolic functions of BS in the photosynthetic/photorespiratory carbon metabolism in C_4 plants, its roles include mechanical support as well as hydraulic, storage, and signaling functions (von Caemmerer and Furbank 2003; Griffiths et al. 2013). In C_3 plants, parenchymal cells containing chloroplasts within BS have diverse functions and are considered highly metabolically active (van Bel and Knoblauch 2000; van Bel 2003). They are able to perform some photosynthetic, transport, antioxidant, and storage roles (Kinsman and Pyke 1998; Leegood 2008). Moreover, the analysis of reticulate leaf mutants of *Arabidopsis* highlighted the role of chloroplasts of the BS cells in supplying the mesophyll cells with metabolites during leaf development (Lundquist et al. 2014). Phloem parenchyma is involved in the transport of sugars to the phloem known as "phloem loading" and in carbohydrate metabolism. Xylem parenchyma cells store water, mineral nutrients, and carbohydrates and also respond to wounding (Myburg and Sederoff 2001). Their photosynthetic properties in C_3 plants are not known in details. According to Pfanz et al. (2002), their contribution to the total photosynthetic area and productivity is rather low. It is also questionable whether their photosynthetic activity goes beyond CO_2 fixation. It was shown that starch accumulates in both mesophyll and BS cells in C_4 plants (Lunn and Furbank 1997) and C_3 plants (Miyake and Maeda 1976; Williams et al. 1989; Miyake 2016). In C_3 plants, the accumulation and dissipation of starch in BS chloroplasts occurs in the dark. Therefore it has recently been proposed that this starch is not an assimilatory but a storage type originating from carbohydrates in other organs (Miyake 2016; Tsutsumi et al. 2008).

Similar to C_4 plants, vascular tissues of C_3 plants, e.g. *Nicotiana tabacum, Betula pendula,* and leaf veins of the C_3-CAM intermediate plants, *M. crystallinum* show high activities of decarboxylating enzymes (Hibberd and Quick 2002; Wittmann et al. 2006; Ślesak et al. 2008b). It seems that BS of C_3 plants can be regarded as preadapted for the role in C_4 photosynthesis and it is suggested that the decarboxylation mechanisms developed in C_3 plants were recruited to C_4 metabolism (Hibberd and Quick 2002). All these observations could help to understand why C_4 metabolism evolved at least 66 times in different plant families within the past 35 million years, parallel to lowering CO_2 concentration in the atmosphere (Sage 2013).

4 RubisCO Activity and CO_2 and O_2 Concentrations in Stems and Bundle Sheath-Adjacent Chlorenchymatous Cells

Photosynthesizing plants fix CO_2 in the reaction catalyzed by RubisCO. This enzyme developed 3.5 billion years ago, prior to oxygenation of the atmosphere, when CO_2 pressure was $0.5–1.0 \times 10^6$ Pa in contrast to the current 38–40 Pa, and the O_2 concentration was lower than 0.2%. During evolution it has maintained the affinity to O_2, which in the atmosphere with high O_2/CO_2 ratio causes the photorespiration process. Along with the rise of O_2 in the Earth's atmosphere, the photosynthesizing organisms have evolved effective inorganic carbon concentrating mechanisms limiting photorespiration by reducing the O_2/CO_2 ratio at RubisCO, such as C_4 or CAM metabolism or systems developed in cyanobacteria (Badger and Price 2003; Lüttge 2011). In C_4 plants, photorespiration is strongly reduced in BS cells and in CAM plants it is suppressed in Phase III (Niewiadomska and Borland 2008), when the decarboxylation of nocturnally stored malate leads to 2- to 60-fold increase in internal CO_2 partial pressure when compared to C_3 plants. However, this mechanism does not eliminate photorespiration in CAM plants because vigorous CO_2 assimilation accompanied by photosynthetic oxygen evolution also leads to O_2 concentrating of up to 40% (Lüttge 2011). Other cellular CO_2-concentrating mechanisms have been identified in the glycine-shuttling species with no close evolutionary relationship to C_4 plants (Sage et al. 2011, 2014). These plants representing the so-called C_2 photosynthesis or C_3-C_4 intermediacy use a photorespiratory glycine shuttle to concentrate CO_2 into the BS. To date, this mechanism has been identified in *Brassicaceae, Asteraceae, Chenopodiaceae, Boraginaceae,* and *Poaceae*. It cannot be excluded that a similar pathway occurs in other plant families. The presence of this process, at least in some plants, could support the suggestion that higher CO_2 concentration limiting photorespiration is important for the physiology of vascular tissues. Hot conditions, by reducing the solubility of CO_2 to greater degree than O_2, increase O_2/CO_2, and favor the

RubisCO-catalyzed oxygenation of ribulose-1,5-bisphosphate triggering photorespiration which produces an abundance of glycine for shuttling into an interior compartment, usually the BS tissue. Intensive photosynthetically active radiation (PAR) can indirectly increase temperature of the leaf and thereby enhance the O_2/CO_2 ratio. Larger enhancement in net photosynthesis due to glycine shuttling is observed in atmospheres of reduced CO_2 and at warmer temperatures (Sage 2013). C_2 photosynthesis is also described as the evolutionary bridge from C_3 to C_4 photosynthesis (Sage et al. 2014). Reassimilation of the photorespired CO_2 operates also in C_3 species, e.g. rice and wheat (Busch et al. 2013).

The O_2/CO_2 gas-exchange and photosynthesis in the chloroplast-containing cells adjacent to BS in C_3 plants were not described, however, they are likely to resemble those in the photosynthetically competent cells of the young twigs described by Wittmann et al. (2006). Leaf veins as well as leaf stalks and stems are often covered with epidermis lacking stomata or showing a severe reduction in their number. Thus, the vertical CO_2 diffusion from the air to the veinal and vasculature-adjacent chlorenchymatous cells is limited (Duarte et al. 2005; Morison et al. 2005). As to the lateral diffusion of CO_2 inside homobaric leaves of dicotyledonous plants, its contribution to photosynthesis still needs elucidation. Under natural conditions, it was shown to be insufficient to support photosynthesis in veins because of the high assimilatory activity of mesophyll cells along the pathway of CO_2 diffusion (Morison et al. 2005). However, at low assimilation rates, when stomata are closed, the lateral CO_2 diffusion may to some extent enhance photosynthesis (Pieruschka et al. 2006).

It is worthy to note that due to the limited possibility to diffuse out of the plant, the respiratory and/or photorespiratory CO_2 can be supplied to the vascular tissues where its concentration can be kept at high level. High CO_2 concentration can also result, at least partly, from high ratio of heterotrophic/autotrophic cells and low light intensity reaching these tissues in comparison to leaf mesophyll (Yiotis et al. 2009). Many authors have reported high CO_2 concentration within stems (Cernusak and Marshall 2000; Yiotis and Manetas 2010). In addition, in C_3 leaves the cells surrounding the vascular system can be supplied with CO_2 derived from malate transported by the xylem vessels (Hibberd and Quick 2002; Janacek et al. 2009). According to Carrodus and Triffett (1975) CO_2 concentration in the xylem sap ranged from 1 to 26%. CO_2 released due to decarboxylation of malate can modify the rates of photosynthetic CO_2 fixation and photorespiration. Chlorophyll-containing cells within the stems, in addition to other functions, are able to assimilate the stem internal CO_2 in the light and to produce carbohydrates contributing to biomass building (Pfanz et al. 2002). For example, in *Clusia* and *M. crystallinum* plants, cells located in the midribs and stems may use CO_2 for photosynthesis and are equipped with carboxylating enzymes, as shown by the expression levels of the respective genes (Kocurek et al. 2015; Nosek et al. 2015b).

Involvement of BS cells in photosynthesis in the CO_2-enriched internal environment can be further supported by results from laboratory experiments with stem segments equilibrated under various CO_2 levels (0.036% to the very high 20%) which showed progressive decrease in photochemical yield, with values obtained at

20% being similar to those from the field studies (at 0.04% CO_2) on intact stems (twigs) (Manetas 2004). According to Manetas (2004), high CO_2 concentration in combination with low light intensity reaching these tissues can suppress the photoprotective nonphotochemical quenching (NPQ) mechanism increasing the risk of photoinhibition. It is expected that high CO_2 concentration will support carboxylation and photosynthesis, however, CO_2-rich environment in stems impede the photochemical activity possibly through acidification of protoplasm and impairment of the pH-dependent excess energy quenching followed by reduction in the efficiency of heat dissipation. Plants exposed for long time periods to elevated CO_2 decrease their photosynthetic activity and this phenomenon is described as acclimation (Paul and Foyer 2001).

One can expect that in leaves, the increased CO_2 availability as well as lower light intensity penetrating to the chlorenchymatous cells adjacent to BS, in comparison to the palisade and spongy parenchyma, can cause similar changes as described above for the green cells in stems. Although there are no experimental data confirming directly this hypothesis, it was supported by the results showing that in *M. crystallinum* plants, regardless of photosynthetic type, the expression of the large (rbcL) RubisCO subunit in the midribs is lower than in mesophyll cells (Nosek et al. 2015b). Tsutsumi et al. (2008) also found a differential expression pattern of both RubisCO subunits in BS and mesophyll cells, although it was not reflected by enzyme activity. Comparable levels of protein and mRNA of RubisCO subunits detected in BS and mesophyll chloroplasts of rice (Yamane et al. 2012) suggest that Calvin cycle operates in both locations. However, in petiols and stems only a very low amount of RubisCO protein was detected (Kocurek et al. 2015). This can be the effect of acclimation to low light and high CO_2 concentration (Foyer et al. 2012). Although high CO_2 level in chlorenchymatous cells adjacent to BS and their photochemical activity have been shown, net photosynthetic CO_2 uptake in these cells was not found (to our knowledge). Stem internal re-fixation (not net uptake) of CO_2 in young stems and branches may compensate for 60–90% of the potential respiratory carbon loss (Pfanz et al. 2002). While analyzing the photosynthetic/photochemical activity of stems it is necessary to distinguish between the chlorenchymatous cells adjacent to BS and the corticular/chlorenchymal tissues just below peridermis. Isolated chlorenchymal tissues showed high rates of net photosynthesis being up to 75% of those for leaf photosynthesis (Pfanz et al. 2002) and the corticular photosynthesis appears to be an effective mechanism for recapturing CO_2 before it diffuses out of the stem. Using isotopic tracer methods, total leaf refixation of photorespired CO_2 has been estimated to range from 15 to 80%. However, these studies did not provide separate quantification of the refixation of CO_2 photorespired directly within the cell and of that returning to the cell from intercellular spaces, nor were they able to evaluate any potential for chloroplast CO_2 enrichment (Sage 2013).

Light-dependent reduction in CO_2 release, as shown in experiments on stems of young birch (*B. pendula*), can predominantly be attributed to corticular refixation. Optimal corticular photosynthesis measured as electron transport rate (ETR) and gas exchange occurs at higher temperatures and is able to refix up to 97% of CO_2

produced by respiration. In illuminated stems and branches exposed to respiratory and non-respiratory conditions, photorespiration does not play a predominant role in carbon exchange due to high CO_2 concentration. The photosynthetic capacity of the cortex is high, however, it is masked by peridermal resistances as proven by O_2 gas exchange measurements in peeled stems (Wittmann et al. 2006). In the cortex, photorespiratory activity can be inhibited also by the high corticular C_i (internal CO_2 concentration). Furthermore, chloroplasts of the proper wood or pith fraction take also part in stem internal photosynthesis. Although no direct experimental evidence has been provided, it was suggested that the oxygen evolved during wood or pith photosynthesis may be decisive in avoiding/reducing stem internal anaerobiosis during the day (Pfanz et al. 2002), but its high concentration could also promote photorespiration. In stems, the diffusive resistance to gas exchange, in conjunction with high heterotrophic/autotrophic cell ratio, plays a more important role than in leaves. Thus, in stems CO_2 partial pressure is very high while that of O_2 may fall to hypoxic levels due to respiratory activity of heterotrophic as well as autotrophic cells (Pfanz et al. 2002; Borisjuk and Rolletschek 2009).

Oxygen produced during photosynthesis can be transported in the xylem sap to heterotrophic tissues, such as sapwood or roots (Gansert 2003). According to Sage (2013), the removal of O_2 excess from the tissue is not practical due to high energy costs. Excess of O_2 cannot be utilized in respiration. In addition, as shown in studies on whole *Glycine* and *Plantago* sp. leaves, the rate of mitochondrial respiration in the light is from 16 to 77% less than in darkness (Krömer 1995; Atkin et al. 2007). Even if increased O_2 concentration significantly affects oxygenation/carboxylation ratio, high CO_2 concentration in BS and adjacent cells can hinder photorespiration. Non-photorespiratory conditions (1% O_2 which allowed establishment of non-photorespiratory conditions for 20 min) can reduce PSII activity in both leaf veins and lamina. This can be the effect of particular high energy demand for photorespiration and/or inhibition of Mehler reaction and overreduction of NADP (Lüttge 2008). In general, in the deeply laying tissues in stems the intensity of photorespiration is expected to be low. According to Lüttge (2011), the anticipated suppression of photorespiration attributed to the high CO_2 concentration is not observed in light-exposed tissues showing vigorous O_2 production. In BS, the increased CO_2 concentration together with low O_2 content would help to keep photorespiration at lower level than in mesophyll cells (Lüttge 2008). It was found that high oxygenating activity of RubisCO in the mesophyll is required to suppress ribulose-1,5-bisphosphate oxygenation in the BS (von Caemmerer 1989; Monson and Rawsthorne 2000). We may suggest that low rate of oxygenation reaction results from low PSII activity and intensive oxygen consumption in heterotrophic cells within vascular tissues. In CAM plants during phase III of CAM metabolism because of vigorous oxygen production and lack of CO_2 sources (Miszalski et al. 1998, 2001) photorespiration within BS can be extremely intensive. It was shown that in stems and leaf veins photorespiration is not fully suppressed and this process uses some part of energy from the photochemical activity (Gansert 2003; Lüttge 2008).

The early Devonian true vascular terrestrial leafless land plants, *Cooksonia* and *Rhynia* (Willis and McElwain 2002) grew in an atmosphere probably containing ca. tenfold higher CO_2 but 25% lower O_2 partial pressure in comparison to their concentrations in the present Earth's atmosphere. Moreover, their stomatal densities were roughly two orders of magnitude lower compared to extant plants (Beerling and Woodward 1997). Thus, it is tempting to speculate that in these plants photosynthesis was carried out within the axial organs (green cells in the stems close to their vasculature) and their ability to fix atmospheric CO_2 was somehow similar to the one expected in BS adjacent cells.

5 PEPC and ^{13}C Discrimination

Plants fix CO_2 mostly with the help of two carboxylating enzymes: RubisCO and phospho*enol*pyruvate carboxylase (PEPC). PEPC has a 60-fold higher affinity to its substrate (HCO_3^-) than RubisCO to CO_2. Using photosynthetic carbon isotope discrimination, it is possible to distinguish between plants that use PEPC for CO_2 prefixation (β-carboxylating C_4 and CAM plants) and plants which most of CO_2 fix with RubisCO (C_3 plants). Non-photosynthetic or heterotrophic tissues tend to be enriched in ^{13}C compared with leaves. This can be explained by differences in biochemical composition of heterotrophic tissues compared with leaves, seasonal separation of growth of leaves and heterotrophic tissues or differential use of day versus night sucrose (Rut et al. 2008; Cernusak et al. 2009; Kocurek et al. 2015). CAM metabolism enabling CO_2 prefixation with PEPC also takes place in darkness and allows successful acclimation to the limiting low light conditions (Ceusters et al. 2011). Stems of deciduous and coniferous trees were characterized by lower ^{13}C discrimination values, with PEPC being 10–23 times more abundant than in leaves (Berveiller and Damesin 2008). This could indicate that in stems, PEPC is more strongly involved in biomass building than in leaves. Similarly, veins of *Plantago media* and *M. crystallinum* are enriched in ^{13}C in comparison to the lamina (Nosek et al. 2015b; Miszalski et al. 2016). It was also shown that the enzymatic machinery of β-carboxylating plants can work at photosynthesis-limiting low light (deep shade) conditions (Ceusters et al. 2011). These experiments were done on the CAM bromeliad *Aechmea* "Maya" exposed to severe light limitation. Just the opposite was noticed for biomass of wood originating from corticular photosynthesis. About 11% of wood biomass was constructed from corticular photosynthate with a mean δ^{13}C of −34.8% (indicating RubisCO activity) and the mean proportional refixation rate during corticular photosynthesis was 0.71 (Cernusak and Hutley 2011). These observations suggest that some veinal and stem tissues which are not able to fix atmospheric CO_2 with the help of RubisCO can be supported with carbon originating from β-carboxylating processes. This can also happen at low light intensities in vasculature-associated green cells. When analyzing intensity of carboxylation we may speak about pre-fixation of CO_2 molecules originating from atmosphere and not only about carbon originating from refixation of respired CO_2. Because of high CO_2 concentration in stem and veinal tissues, even low amount of PEPC can ensure effective carbon assimilation. We may suggest that in deciduous plants this process can take place also before or after appearing of leaves or during winter. The same can be true also for vascular tissues of senescing leaves, which stay green longer than the interveinal mesophyll.

6 Chloroplast Ultrastructure and Photochemical Activity

Many plants utilize a specific acclimation strategy involving molecular rearrangements at the level of chloroplast ultrastructure. The BS chloroplasts of rice, a C_3 plant, are smaller than mesophyll chloroplasts and their profile is lens-shaped while mesophyll chloroplasts are somewhat amorphous with extended stromal regions (Kinsman and Pyke 1998; Sage and Sage 2009; Yamane et al. 2012).

Acclimation strategies include adjustments in reaction center stoichiometry and in antenna size. This can be partly monitored through consequent changes in chlorophyll content and chl*a*/chl*b* ratio. All these are long-term or short-term responses (Karpinski et al. 1999; Scheibe et al. 2005). The most important factor affecting chloroplast ultrastructure is light. Low light is usually enriched with longer wavelengths (500–700 nm) reaching the deep-laying cells within the leaf or stem and is more effectively absorbed by PSI (Pilarski et al. 2008; Wientjes et al. 2013). The C_4-performing plants, *Miscanthus* and maize, have been shown to coordinate the activities of mesophyll and bundle sheath cells when exposed to blue light and this appeared to be related to differences in the response of cyclic electron flux (PSI) and chloroplast rearrangement (Sun et al. 2014). Blue light is a key determinant in regulating decarboxylation and the relationships between starch and organic acids turnover (Ceusters et al. 2014). Due to the poor penetration of blue light (short wavelength) into the BS, the subsequent production of ATP was insufficient. Moreover, lower efficiency of PSII under blue light could suggest insufficient production of NADPH (Sun et al. 2014). The responses of plants exposed to very low irradiance include large increase in the levels of PSI and complex changes in the levels of minor LHCII and LHCI components (Bailey et al. 2004). Additionally, the temperature of dense-packed cells within vascular and bark tissues which absorb significant amounts of red-infrared wavelength could increase stimulating oxygenase activity of RubisCO and photorespiration (Sage 2013).

Chlorophyll fluorometers enable fast and reliable assessment of photosynthetic electron transport (PET) chain activity which expresses the long-term and short-term acclimation responses of PSII and PSI to variable stimuli sensed by the photosynthetic apparatus. In general, high photochemical efficiency of PSII is found in leaves of high-light acclimated plants while high PSI content is characteristic for plants grown at low photosynthetically active radiation (PAR), i.e. 35 μmol photons m^{-2} s^{-1} (Bailey et al. 2001). Many C_4 plants show agranal, PSII-deficient chloroplasts in BS and well-developed granal chloroplasts with high content of PSII in mesophyll cells (Hofer et al. 1992; Voznesenskaya et al. 2010). Clear day/night changes in chloroplast ultrastructure were described for *M. crystallinum* (Niewiadomska et al. 2011). In maize, the agranal BS chloroplasts show low or no activity to photoreduce NADP and both BS and mesophyll chloroplasts are active in cyclic photophosphorylation (Andersen et al. 1972). In the C_4 plant *Flaveria trinervia,* BS cells maintained up to 23% of the PSII capacity shown by mesophyll chloroplasts and this was accompanied by lower oxygen production and depletion of three protein components of the water splitting system of PSII. It was speculated that this mechanism is not an acquisition of C_4 plants but is based on an evolutionary principle which was used by C_3 plants to control the activity of PET from water to plastoquinone (Höfer et al. 1992). Thus, a similar machinery could be expected in C_3 plants, however, most studies on leaf photosynthesis do not distinguish between photochemical activity of veinal and mesophyll tissues. Unlike C_4 plants, in leaves of C_3 or CAM-performing plants it is usually difficult to separate the green cells adjacent to BS from mesophyll. Therefore, in these plants the interference of mesophyll cells in fluorometric, biochemical, and molecular analyses of midribs cannot be excluded. Therefore, one can expect that the differences between the BS-adjacent cells and mesophyll cells are in fact more distinct than those between midribs and mesophyll.

In leaf veins of C_3 and CAM *M. crystallinum* plants, PSII efficiency was lower than in the interveinal areas (Kuźniak et al. 2016). This was also observed in *Clusia* leaves, however, only in some parts of the day and during midday PSII efficiency in veins was even higher than in the mesophyll. This goes parallel to O_2 concentration (Lüttge 2008). Similar observations have been made for petiols and pedicels of *Zantedeschia aethiopica* (Yiotis and Manetas 2010). It was also shown that PSI efficiency in veins, in both C_3 and CAM plants, is slightly increased in comparison to mesophyll (Nosek et al. 2015b; Kuźniak et al. 2016). In *M. crystallinum* showing C_3 or CAM metabolism, smaller grana in BS in comparison to mesophyll were noted suggesting also lower PSII activity. One can expect that lower PSII activity and O_2 production would reduce the number of mitochondria and ATP would be delivered by PSI from cyclic phosphorylation. Such interpretation can be supported by similar amount of grana in BS and mesophyll cells observed in the C_2 plant, *Portulaca cryptopetala*. In contrast to C_3 plants, BSs in *P. cryptopetala* are also rich in mitochondria containing glycine decarboxylase responsible for CO_2 production (Tobin et al. 1991). In *P. media* leaves (C_3 species), no significant differences in maximum quantum yield of PS II and PSII vitality index between midribs and lamina were found in leaves grown in high light but the blockage of electron flow between Q_A and Q_B was less visible in midribs than in mesophyll. Moreover, midribs showed lower dissipation of quantum energy in comparison to mesophyll (Miszalski et al. 2016). These results indicate that BS-adjacent cells can make use of absorbed light energy and they are somehow better protected from high light stress. However, in *Arabidopsis* leaves, BS cells showed lower PSII efficiency and they were more susceptible to photoinhibition than mesophyll cells, as manifested by the decrease in operating efficiency of non-cyclic electron transport (Fryer et al. 2002; Gorecka et al. 2014).

Similarly to differences within the leaf blade, the photochemical activity in vein parenchymal tissues in stems of several tree species differed significantly from that in leaves. When compared to leaves, stem tissues of *Eleagnus angustifolia, Platanus orientalis, Nerium oleander,* and *Quercus coccifera* showed lower chlorophyll levels and lower PSII efficiency, with a steep decreasing gradient towards the stem center. In the xylem rays, PSII/PSI ratios were lower than those in the photosynthetically active corticular layer. The results indicate an increasing, inherent, and light-independent inactivation of PSII along the twig depth. In the deep-located cell layers, there was an efficient electron flow to PSI from stromal donors in the absence of PSII activity (Yiotis et al. 2009). Twig cortices of *E. angustifolius* have also been shown to have low linear electron transport rates when illuminated. In addition, their PSII population was characterized by a high proportion of inactive non-QB-reducing centers and an incomplete fluorescence quenching during slow induction phase (Kotakis et al. 2006). Thus, in stems, PSII is an inefficient electron donor to PSI and/or reductive pentose phosphate cycle. In spite of PSII deficiency, pools of intermediate electron carriers and potential PSI activity were more than sufficient to support the observed rates of linear PET.

In the corticular tissue of *Pinus sylvestris*, fluorescence quenching analysis revealed higher reduction states of Q_A and of the plastoquinone (PQ) pool than in leaves. An adaptive significance was ascribed to PSII inactivation. The redox state of the PQ-pool significantly affects the activity pattern of main antioxidants (Nosek et al. 2015a). Data on differences in adenylate pool sizes and PSI activity between stems and needles of *P. sylvestris* corroborated reduction states of the PQ-pool. These differences were even more clear during winter (Ivanov et al. 2006). *Populus tremula* also showed lowest values of PSII yield during the winter period when plants grown under field conditions were exposed to strong light. This could be the result of photoinhibition (Solhaug and Haugen 1998).

Differences in the photochemical activity are parallel to CO_2 fixation and photorespiratory intensity. As discussed earlier, in stem segments of deciduous trees light-adapted PSII photochemical yield was decreased (Manetas 2004), probably due to high CO_2 level. As normalized ETR in leaves at 0% O_2 is lower than at 5–21% O_2 (Yiotis and Manetas 2010) and with decreasing O_2 concentration (non photorespiratory conditions) PSII efficiency is lowering in both central veins and lamina (Duarte and Lüttge 2007), it can be expected that low photochemical activity of PSII results from anaerobiosis and less effective Mehler reaction which is responsible for dissipating excess of light energy. Low light intensity and limited activity of PSII found in BS adjacent cells in tissues rich in heterotrophic cells would also suggest low O_2 production and following this, hypoxia (Pfanz et al. 2002; Borisjuk and Rolletschek 2009). All these relationships could point to the possibility that tissues with low PSII activity operate under non-photorespiratory conditions. In BS cells, the protection against photorespiration is of low importance and a lower PSII/PSI activity ratio intensifies cyclic ATP phosphorylation keeping NADP reduction at low level. In leaves, with increasing CO_2 concentration, the high ATP/NADPH ratio necessary for the combined activity of Calvin cycle and photorespiration decreased (Foyer 2012). This could be the result of very high ATP demand for purposes other than photosynthesis. As photorespiration plays a prominent role in defense against pathogens (Sørhagen et al. 2013), this could also affect the response of BS cells to pathogen attack. It was shown that the type of photosynthesis can affect pathogen development (Kuźniak et al. 2010; Gabara et al. 2012). Photorespiration together with the so-called malate valve are important components establishing the ATP/NADPH ratio, simultaneously protecting against oxidative stress. It was found that CO_2 fixation, respiration, and metabolism of crucial metabolites malate and citrate depend on light intensity in *Clusia* (Kornas et al. 2009, 2010a; Miszalski et al. 2013). Theoretical calculations for leaf mesophylls allow to expect that PET can produce more reducing equivalents relative to ATP than required for photosynthesis (Stitt 1986). For *P. sylvestris* stems, an adaptive significance was ascribed to PSII inactivation (in comparison to needles) as a way of adjusting the ATP/NADPH ratio and replenishing the ATP pool. Data on adenylate pool sizes and PSI activity of stems and needles of *P. sylvestris* corroborated this hypothesis (Ivanov et al. 2006).

The differences in photochemical activity of green cells within stems paralleled those in the level of photosynthetic gases and PAR light availability. Some small differences in light intensity as well CO_2 levels within leaf tissues are also observed. Higher oxygenation/carboxylation ratio and the resulting increase in photorespiration rates in petioles were accompanied by enhanced cyclic electron flow around PSI (Yiotis and Manetas 2010).

7 The Involvement of Veins and Cells Near to the Veins in Signaling Pathways

The involvement of the vascular system in the intraorganismic signaling in plants has been reported, but it still needs elucidation. The vascular continuity is fundamental for the long-distance transport of hormones and electrical and hydraulic signals (Lacombe and Achard 2016). Mobile molecules such as secreted peptides in the xylem as well as proteins and RNAs in the phloem can also serve as systemic signals transmitting information through the entire plant organism (Vlot et al. 2008; Notaguchi and Okamoto 2015). Moreover, many proteins identified in the phloem sap of different plant species have predicted functions in stress and defense responses (Kehr 2006). Transmission of signals between mesophyll cells and phloem is mediated by BS cells (Szechynska-Hebda et al. 2010). The occurrence of stress-related phytohormones, namely jasmonic acid (JA) and its derivatives, abscisic acid (ABA) as well as salicylic acid (SA) and methyl salicylate in phloem and xylem exudates suggests that the vasculature plays a role in plant systemic stress response (Furch et al. 2014). Hormonal signals transported through vasculature are amplified via biosynthesis gene induction in the vascular tissues. For example, enzymes of JA biosynthesis have been detected in vascular parenchyma cells (Hause et al. 2000). Recently it has been demonstrated that signals stimulating jasmonate biosynthesis in response to wounding are propagated via parenchyma cells adjacent to mature xylem vessels which are sensitive to the xylem hydrostatic pressure changes upon wounding (Farmer et al. 2014). In *Arabidopsis* exposed to dehydration conditions, ABA biosynthetic enzymes and the drought-responsive genes were specifically expressed in the vascular parenchyma cells (Endo et al. 2008). Under high light stress, ABA biosynthesis in *Arabidopsis* is initiated in vascular parenchyma cells and activates a signaling network in BS cells involving OPEN STOMATA1 (OST1) kinase, heterotrimeric G proteins, glutathione-based redox signal and H_2O_2 sourced from chloroplasts which induces ASCORBATE PEROXIDASE2 (APX2) specifically expressed in BS cells (Galvez-Valdivieso et al. 2009; Gorecka et al. 2014). This pattern of regulation observed for *APX2*, encoding cytosolic APX, supports the concept that retrograde signals from chloroplasts, which regulate the expression of nuclear stress-responsive genes, act by merging with a signal originating in an external tissue (Pfannschmidt et al. 2009), i.e. vascular parenchyma. Moreover, the correlation between ABA secreted from the vascular parenchyma cells and effective physiological response of leaves to excess light mediated by reactive oxygen species (ROS) and associated with the rapid induction of *APX2* in the BS cells demonstrates a functional organization of the leaf (Karpinski et al. 1999; Fryer et al. 2003; Gorecka et al. 2014). In plants, APX which interacts tightly with ascorbic acid and the ascorbate-glutathione cycle plays a pivotal role in the regulation of redox-based signaling (Kuźniak and Skłodowska 2005; Foyer and Noctor 2011). The significance of the signaling pathway regulating *APX2* is further supported by the results indicating that cytosolic APX is more important for stress protection, including combination stress, than APX isoenzymes located in other organelles (Davletova et al. 2005; Koussevitzky et al. 2008).

In *Arabidopsis*, 25% of the high light-responsive genes examined in the study was preferentially expressed in petioles containing a higher percentage in BS cells than in the leaf blades. These data suggest that BS is more responsive to fluctuating light conditions than leaf blade (Bechtold et al. 2008). In the moderately heterobaric leaves of *Arabidopsis*, in which the lateral CO_2 diffusion is limited, under high light conditions the BS cell chloroplasts are more susceptible to photoinhibition than the mesophyll ones due to increased singlet oxygen production in PSII antennae and the reduction in PSII operating efficiency. ABA signaling in high light-exposed BS cells also determines their susceptibility to photoinhibition by influencing non-photochemical quenching, suggesting that in these cells ABA plays a complex regulatory role (Gorecka et al. 2014).

In *Arabidopsis* plants, in which chlorophyll content in veins and cells close to veins was reduced, the veinal fixation of CO_2 supplied to the xylem, and therefore the photosynthetic ability and carbon skeleton supply to the shikimate pathway were also reduced (Janacek et al. 2009). As the shikimate pathway provides carbon skeletons for biosynthesis of phenolic compounds involved in signaling and defense against abiotic and biotic stresses, in these plants the biosynthesis of stress-signaling metabolites, e.g. SA could be impaired. In *Catharanthus roseus* phloem parenchyma supplies carbon skeletons to the MEP (methylerythritol-phosphate) pathway, using glyceraldehyde 3-phosphate and pyruvate produced by photosynthesis as substrates and providing isoprenoid units for ABA biosynthesis (Burlat et al. 2004).

Redox reactions in cells have developed in strict connection to molecular oxygen and its derivatives, i.e. reactive oxygen species (ROS) (Kornas et al. 2010b). Histochemical methods localizing ROS production in leaves usually show that superoxide anion radical ($O_2^{\bullet-}$) and H_2O_2 preferentially accumulate in the vicinity of veins (Ślesak et al. 2008a; Tanou et al. 2012; Kuźniak et al. 2016) arguing for the systemic nature of ROS signaling. As in the photosynthesizing cells, PET is the major source of ROS and they are predominantly produced when CO_2 is limited (Voss et al. 2013). The vasculature-associated parenchyma cells can contribute to ROS accumulation in the veins' vicinity. The vascular-driven long-distance ROS signaling has been confirmed in many studies and its relationship with light signaling has recently emerged. For example, in leaves of *Pelargonium zonale*, NADPH-oxidase-dependent accumulation of H_2O_2 was related to photosynthetic activity and occurred in the apoplast of main veins extending from the petiole towards the adjacent photosynthetically active cells in the green sectors of leaves exposed to high light (Vidović et al. 2016). Similar localization of the light-induced H_2O_2 was found in *M. crystallinum* (Kuźniak et al. 2016), and this compound can affect type of photosynthesis (Ślesak et al. 2007, 2008a; Surówka et al. 2016) and tissue formation (Libik-Konieczny et al. 2012; Konieczny et al. 2014) and programmed cell dead (PCD) process (Kuźniak et al. 2013). Hydrogen peroxide produced in the apoplast of BS cells was also shown to be involved in intracellular signal transmission during acclimation to high light associated with *APX2* induction (Szechynska-Hebda et al. 2010). The preferential accumulation of ROS in the vicinity of veins could imply that the cells near the vascular bundles are more exposed to ROS which diffuse out of the veins and therefore are more likely to react to the ROS-mediated systemic signal, as suggested by Tanou et al. (2012).

ROS-mediated signaling is controlled by a delicate balance between their production and scavenging. The outcome of ROS signaling depends on the ROS production site, signal intensity and duration (ROS signature) as well as their interaction with other signaling molecules (Gilroy et al. 2016). While in C_4 plants, the differential partitioning of antioxidants between BS and mesophyll cells determining the level of ROS at these locations is known (Pastori et al. 2000), in C_3 and CAM plants the data are limited and refer mainly to the regulation of *APX2*. In maize, the post-transcriptional regulation of glutathione reductase (GR) prevented the accumulation of GR protein in BS cells, and the decreased ability of these cells to regenerate the reduced glutathione was suggested to contribute to the chilling sensitivity of leaves (Pastori et al. 2000). The studies on the distribution of H_2O_2 and its relationship with APX and catalase (CAT) activities in *M. crystallinum* revealed that the midribs in leaves of CAM performing plants were more prone to produce H_2O_2 and exhibited higher level of APX activity than the leaf midribs in C_3 performing plants and the lamina cells of both photosynthetic types. These spatial and photosynthetic type-related differences in APX activity were not accompanied by CAT. The APX–H_2O_2 interaction could be important for the proposed mediatory

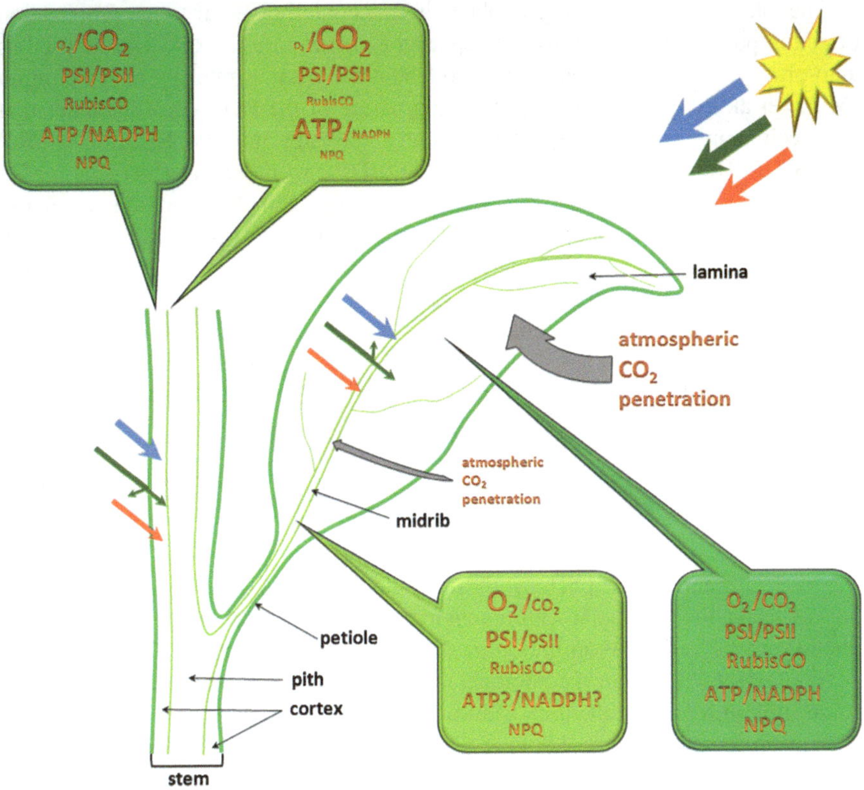

Fig. 3 Schematic characteristics of vasculature-associated photosynthetic cells located in leaves and stems. Font size correspond to the intensity of changes in comparison to leaf mesophyll. The width and color of arrows correspond to the absorptivity of wavelengths in the visible sunlight spectrum reaching the indicated areas: *blue arrow – violet-blue* wavelengths in the range of 400–500 nm; *green arrow – green-yellow* wavelengths in the range of 500–600 nm; *red arrow – red* wavelengths in the range of 600–700 nm. *NPQ* non-photochemical quenching, *PSI* photosystem I, *PS II* photosystem II, *RubisCO* ribulose-1,5-bisphosphate Carboxylase-Oxygenase

role of midrib chloroplasts of CAM plants in H_2O_2 signaling regulated by the malate shuttle (Kuźniak et al. 2016).

In plants, H_2O_2 signaling is closely linked to nitric oxide (NO) and the regulatory effects of NO are mediated through protein modifications, including S-nitrosylation (Astier et al. 2012; Saxena et al. 2016). In the context of systemic acquired resistance (SAR) signaling, it has been found that S-nitrosoglutathione reductase specifically expressed in xylem parenchyma cells is involved in SAR signal transmission through the vasculature (Rustérucci et al. 2007), linking the NO-induced protein S-nitrosylation to SAR. The vascular-driven long-distance ROS and also NO signaling was involved in the acclimation response of *Citrus aurantium* plants to salinity (Tanou et al. 2012).

The long-distance signaling in plants is conceptually similar to the propagation of action potentials in neurons. As reported for *Arabidopsis*, ROS can be used as auto-propagating wave transmitted throughout the plant at rates up to 8.4 cm/min (Miller et al. 2009). The auto-propagating nature of the ROS wave indicates that each cell along the pathway independently activates its own RESPIRATORY BURST OXIDASE HOMOLOG D (RBOHD) enzyme and generates ($O_2^{\bullet -}$) which dismutases to H_2O_2 spontaneously or by the action of superoxide dismutase. The ROS wave is integrated with other signaling pathways mediated by calcium, ABA as well as stress-specific metabolites and electric signals to generate highly coordinated systemic response to stress (Baxter et al. 2014; Gilroy et al. 2016). As it was demonstrated that BS cells are important for the long-distance electrical signaling in plants (Szechynska-Hebda et al. 2010), they are likely involved in the integration of systemic signaling in plants.

8 Concluding Remarks

Figure 3 summarizes the photosynthesis-related characteristics of the vasculature-adjacent and the corticular chlorenchymatous cells. Irrespective of the type of photosynthetic carbon assimilation metabolism, these chloroplast-containing cells are competent to perform the photochemical reactions of photosynthesis and share some structural and biochemical similarities with the BS cells in C_4 plants. However, they operate under conditions which differ significantly from those experienced by the mesophyll cells. Accordingly, it has been suggested that their role in plants is related rather to specific biological processes than to the photosynthetic CO_2 assimilation contributing to biomass yield. Taken together, we have evidence that these chlorenchymatous cells could be involved in long-distance signaling mediated by phytohormones as well as NO and H_2O_2, and in processes which allow plants to acclimatize to changes in the environment, especially to fluctuating light conditions, at the whole-organism level. However, understanding the specific physiological functions they exert at their destinations remains a challenge for the future.

Acknowledgments This work was supported by the Polish National Science Center (research grants N N304 156440 and 2011/01/B/NZ9/02540 and UMO-2016/21/B/NZ9/00813) and Alexander von Humboldt Foundation (AvH).

References

Andersen KS, Bain JM, Bishop DG, Smillie RM (1972) Photosystem II activity in agranal bundle sheath chloroplasts from *Zea mays*. Plant Physiol 49:461–466

Astier J, Kulik A, Koen E, Besson-Bard A, Bourque S, Jeandroz S et al (2012) Protein S-nitrosylation: what's going on in plants? Free Radic Biol Med 53:1101–1110

Atkin OK, Scheurwater I, Pons TL (2007) Respiration as a percentage of daily photosynthesis in whole plants is homeostatic at moderate, but not high, growth temperatures. New Phytol 174: 367–380

Badger MR, Price GD (2003) CO_2 concentrating mechanisms in cyanobacteria: molecular components, their diversity and evolution. J Exp Bot 54:609–622

Bailey S, Horton P, Walters RG (2004) Acclimation of *Arabidopsis thaliana* to the light environment: the relationship between photosynthetic function and chloroplast composition. Planta 218:793–802

Bailey S, Walters RG, Jansson S, Horton P (2001) Acclimation of *Arabidopsis thaliana* to the light environment: the existence of separate low light and high light responses. Planta 213:794–801

Baxter A, Mittler R, Suzuki N (2014) ROS as key players in plant stress signaling. J Exp Bot 65:1229–1240

Bechtold U, Richard O, Zamboni A, Gapper C, Geisler M, Pogson B, Karpinski S, Mullineaux PM (2008) Impact of chloroplastic- and extracellular-sourced ROS on high light-responsive gene expression in *Arabidopsis*. J Exp Bot 59:121–133

Beerling DJ, Woodward FI (1997) Changes in land plant function over the Phanerozoic: reconstructions based on the fossil record. Bot J Linn Soc 124:137–153

van Bel AJE (2003) The phloem, a miracle of ingenuity. Plant Cell Environ 26:125–149

van Bel AJE, Knoblauch M (2000) Sieve element and companion cell: the story of the comatose patient and the hyperactive nurse. Aust J Plant Physiol 27:477–487

Berveiller D, Damesin C (2008) Carbon assimilation by tree stems: potential involvement of phospho*enol*pyruvate carboxylase. Trees Struct Funct 22:149–157

Borisjuk L, Rolletschek H (2009) The oxygen status of the developing seed. New Phytol 182: 17–30

Burlat V, Oudin A, Courtois M, Rideau M, St-Pierre B (2004) Co-expression of three MEP pathway genes and geraniol 10-hydroxylase in internal phloem parenchyma of *Catharanthus roseus* implicates multicellular translocation of intermediates during the biosynthesis of monoterpene indole alkaloids and isoprenoid-derived primary metabolites. Plant J 38:131–141

Busch FA, Sage TL, Cousins AB, Sage RF (2013) C_3 plants enhance rates of photosynthesis by reassimilating photorespired and respired CO_2. Plant Cell Environ 36:200–212

von Caemmerer S (1989) A model of photosynthetic CO_2 assimilation and carbon-isotope discrimination in leaves of certain C_3–C_4 intermediates. Planta 178:463–474

von Caemmerer S, Furbank RT (2003) The C_4 pathway: an efficient CO_2 pump. Photosynth Res 77:191–207

Carrodus BB, Triffett ACK (1975) Analysis of respiratory gases in woody stems by mass spectrometry. New Phytol 74:243–246

Cernusak LA, Hutley LB (2011) Stable isotopes reveal the contribution of corticular photosynthesis to growth in branches of *Eucalyptus miniata*. Plant Physiol 155:515–523

Cernusak LA, Marshall JD (2000) Photosynthetic refixation in branches of Western White Pine. Funct Ecol 14:300–311

Cernusak LA, Tcherkez G, Keitel C, Cornwell WK, Santiago LS, Knohl A, Barbour MM, Williams DG, Reich PB, Ellsworth DS, Dawson TE, Griffiths HG, Farquhar GD, Wright IJ (2009) Why are non-photosynthetic tissues generally ^{13}C enriched compared with leaves in C_3 plants? Review and synthesis of current hypotheses. Funct Plant Biol 36:199–213

Ceusters J, Borland AM, Godts C, Londers E, Croonenborghs S, Van Goethem D, De Proft MP (2011) Crassulacean acid metabolism under severe light limitation: a matter of plasticity in the shadows? J Exp Bot 62:283–291

Ceusters J, Borland AM, Taybi T, Frans M, Godts C, De Proft MP (2014) Light quality modulates metabolic synchronization over the diel phases of crassulacean acid metabolism. J Exp Bot 65:3705–3714

Crookston RK, Moss DN (1974) Interveinal distance for carbohydrate transport in leaves of C_3 and C_4 grasses. Crop Sci 14:123–125

Davletova S, Rizhsky L, Liang H, Shengqiang Z, Oliver DJ, Coutu J, Shulaev V, Schlauch K, Mittler R (2005) Cytosolic ascorbate peroxidase 1 is a central component of the reactive oxygen gene network of *Arabidopsis*. Plant Cell 17:268–281

Duarte HM, Lüttge U (2007) Correlation between photorespiration, CO_2-assimilation and spatio-temporal dynamics of photosynthesis in leaves of C_3-photosynthesis/crassulacean acid metabolism-intermediate species *Clusia minor* L. (Clusiaceae). Trees 21:531–540

Duarte HM, Jakovljevic I, Kaiser F, Lüttge U (2005) Lateral diffusion of CO_2 in leaves of the crassulacean acid metabolism plant *Kalanchoë daigremontiana* Hamet et Perrier. Planta 220: 809–816

Eastman PAK, Peterson CA, Dengler NG (1988) Suberized bundle sheaths in grasses (*Poaceae*) of different photosynthetic types. II. Apoplastic permeability. Protoplasma 142:112–126

Endo A, Sawada Y, Takahashi H, Okamoto M, Ikegami K, Koiwai H, Seo M, Toyomasu T, Mitsuhashi W, Shinozaki K, Nakazono M, Kamiya Y, Koshiba T, Nambara E (2008) Drought induction of Arabidopsis 9-cis-epoxycarotenoid dioxygenase occurs in vascular parenchyma cells. Plant Physiol 147:1984–1993

Esau K (1977) Anatomy of seed plants, 2nd edn. Wiley, New York

Farmer EE, Gasperini D, Acosta IF (2014) The squeeze cell hypothesis for the activation of jasmonate synthesis in response to wounding. New Phytol 204:282–288

Foyer CH, Noctor G (2011) Ascorbate and glutathione: the heart of the redox hub. Plant Physiol 155:2–18

Foyer CH, Neukermans J, Queval G, Noctor G, Harbinson J (2012) Photosynthetic control of electron transport and the regulation of gene expression. J Exp Bot 63:1637–1661

Fryer MJ, Ball L, Oxborough K, Karpinski S, Mullineaux PM, Baker NR (2003) Control of *Ascorbate Peroxidase 2* expression by hydrogen peroxide and leaf water status during excess light stress reveals a functional organisation of *Arabidopsis* leaves. Plant J 33:691–705

Fryer MJ, Oxborough K, Mullineaux PM, Baker NR (2002) Imaging photo-oxidative stress responses in leaves. J Exp Bot 53:1249–1254

Furch ACU, Zimmermann MR, Kogel K-H, Reichelt M, Mithöfer A (2014) Direct and individual analysis of stress-related phytohormone dispersion in the vascular system of *Cucurbita maxima* after flagellin 22 treatment. New Phytol 201:1176–1182

Gabara B, Kuźniak E, Skłodowska M, Miszalski Z (2012) Ultrastructural and metabolic modifications at the plant–pathogen interface in *Mesembryanthemum crystallinum* leaves infected by *Botrytis cinerea*. Environ Exp Bot 77:33–43

Galvez-Valdivieso G, Fryer MJ, Lawson T, Slattery K, Truman W, Smirnoff N, Asami T, Davies WJ, Jones AM, Baker NR, Mullineaux PM (2009) The high light response in *Arabidopsis* involves ABA signaling between vascular and bundle sheath cells. Plant Cell 21:2143–2162

Gansert D (2003) Xylem sap flow as a major pathway for oxygen supply to the sapwood of birch (*Betula pubescens* Ehr.) Plant Cell Environ 26:1803–1814

Gilroy S, Białasek M, Suzuki N, Gorecka M, Devireddy AR, Karpiński S, Mittler R (2016) ROS, calcium, and electric signals: key mediators of rapid systemic signaling in plants. Plant Physiol 171:1606–1615

Gorecka M, Alvarez-Fernandez R, Slattery K, McAusland L, DaveyPA KS, Lawson T, Mullineaux PM (2014) Abscisic acid signalling determines susceptibility of bundle sheath cells to photoinhibition in high light-exposed *Arabidopsis* leaves. Phil Trans R Soc B 369: 20130234

Griffiths H, Weller G, Toy LFM, Dennis RJ (2013) You're so vein: bundle sheath physiology, phylogeny and evolution in C3 and C4 plants. Plant Cell Environ 36:249–261

Haberlandt G (1914) Physiological Plant Anatomy (trans: Drummond M). Macmillan and Co., London

Hause B, Stenzel I, Miersch O, Maucher H, Kramel R, Ziegler J, Wasternack C (2000) Tissue-specific oxylipin signature of tomato flowers: allene oxide cyclase is highly expressed in distinct flower organs and vascular bundles. Plant J 24:113–126

Hibberd JM, Quick WP (2002) Characteristics of C_4 photosynthesis in stems and petioles of C_3 flowering plants. Nature 415:451–454

Hofer MU, Santore UJ, Westhoff P (1992) Differential accumulation of the 10-,16- and 23-kDa peripheral components of the water-splitting complex of photosystem II in mesophyll and bundle sheath chloroplasts of the dicotyledonous C_4 plant *Flaveria trinervia* (Spreng.) C.Mohr. Planta 186:304–312

Ivanov AG, Krol M, Sveshnikov D, Malmberg G, Gardeström P, Hurry V, Öquist G, Huner NPA (2006) Characterization of the photosynthetic apparatus in cortical bark chlorenchyma of Scots pine. Planta 223:1165–1177

Janacek SH, Trenkamp S, Palmer B, Brown NJ, Parsley K, Stanley S, Astley HM, Rolfe SA, Quick WP, Fernie AR, Hibberd JM (2009) Photosynthesis in cells around veins of the C_3 plant *Arabidopsis thaliana* is important for both the shikimate pathway and leaf senescence as well as contributing to plant fitness. Plant J 59:329–343

Karpinski S, Reynolds H, Karpinska B, Wingsle G, Creissen G, Mullineaux P (1999) Systemic signaling and acclimation in response to excess excitation energy in *Arabidopsis*. Science 284: 654–657

Kehr J (2006) Phloem sap proteins: their identities and potential roles in the interaction between plants and phloem-feeding insects. J Exp Bot 57:767–774

Kinsman EA, Pyke KA (1998) Bundle sheath cells and cell-specific plastid development in *Arabidopsis* leaves. Development 125:1815–1822

Kocurek M, Kornas A, Pilarski J, Tokarz K, Lüttge U, Miszalski Z (2015) Photosynthetic activity of stems in two *Clusia* species. Trees Struct Funct 29:1029–1040

Konieczny R, Banaś AK, Surówka E, Michalec Ż, Miszalski Z, Libik-Konieczny M (2014) Pattern of antioxidant enzyme activities and hydrogen peroxide content during developmental stages of rhizogenesis from hypocotyls explants of *Mesembryanthemum crystallinum* L. Plant Cell Rep 33:165–177

Kornas A, Miszalski Z, Surówka E, Fischer-Schliebs E, Lüttge U (2010a) Light stress is not effective to enhanced Crassulacean acid metabolism (CAM). Z Naturforsch 65c:79–86

Kornas A, Kuźniak E, Ślesak I, Miszalski Z (2010b) The key role of the redox status in regulation of metabolism in photosynthesizing organisms. Acta Biochim Pol 57:143–151

Kornas A, Fischer-Schliebs E, Lüttge U, Miszalski Z (2009) Adaptation of the obligate CAM plant *Clusia alata* to light stress: metabolic responses. J Plant Physiol 166:1914–1922

Kotakis C, Petropoulou Y, Stamatakis K, Yiotis C, Manetas Y (2006) Evidence for active cyclic electron flow in twig chlorenchyma in the presence of an extremely deficient linear electron transport activity. Planta 225:45–253

Koussevitzky S, Suzuki N, Huntington S, Armijo L, Sha W, Cortes D, Shulaev V, Mittler R (2008) Ascorbate peroxidase 1 plays a key role in the response of *Arabidopsis thaliana* to stress combination. J Biol Chem 283:34197–34203

Krömer S (1995) Respiration during photosynthesis. Annu Rev Plant Physiol Plant Mol Biol 46: 45–70

Kuźniak E, Skłodowska M (2005) Compartment-specific role of the ascorbate-glutathione cycle in tomato leaf cells response to *Botrytis cinerea* infection. J Exp Bot 56:921–933

Kuźniak E, Kaźmierczak A, Wielanek M, Głowacki R, Kornas A (2013) Involvement of salicylic acid, glutathione and protein S-thiolation in plant cell-death mediated defence response of *Mesembryanthemum crystallinum* against *Botrytis cinerea*. Plant Physiol Biochem 63:30–38

Kuźniak E, Kornas A, Gabara B, Ullrich C, Skłodowska M, Miszalski Z (2010) Interaction of *Botrytis cinerea* with the intermediate C_3-CAM plant *Mesembryanthemum crystallinum*. Environ Exp Bot 69:137–147

Kuźniak E, Kornas A, Kaźmierczak A, Rozpądek P, Nosek M, Kocurek M, Zellnig G, Müller M, Miszalski Z (2016) Photosynthesis-related characteristics of the midrib and the interveinal lamina in leaves of the C_3–CAM intermediate plant *Mesembryanthemum crystallinum*. Ann Bot 117:1141–1151

Lacombe B, Achard P (2016) Long-distance transport of phytohormones through the plant vascular system. Curr Opin Plant Biol 34:1–8

Leegood RC (2008) Roles of the bundle sheath cells in leaves of C_3 plants. J Exp Bot 59: 1663–1673

Libik-Konieczny M, Konieczny R, Surówka E, Ślesak I, Michalec Ż, Rozpądek P, Miszalski Z (2012) Pathways of ROS homeostasis regulation in *Mesembryanthemum crystallinum L. calli* exhibiting differences in rhizogenesis. Plant Cell Tiss Org 110:123–131

Lundquist PK, Rosar C, Bräutigam A, Weber APM (2014) Plastid signals and the bundle sheath: mesophyll development in reticulate mutants. Mol Plant 7:14–29

Lunn JE, Furbank RT (1997) Localisation of sucrose-phosphate synthase and starch in leaves of C_4 plants. Planta 202:106–111

Lüttge U (2008) Clusia: Holy Grail and enigma. J Exp Bot 59:1503–1514

Lüttge U (2011) Photorespiration in phase III of Crassulacean acid metabolism: evolutionary and ecophysiological implications. Prog Bot 72:371–384

Manetas Y (2004) Probing corticular photosynthesis through *in vivo* chlorophyll fluorescence measurements: evidence that high internal CO_2 levels suppress electron flow and increase the risk of photoinhibition. Physiol Plant 120:509–517

Miller G, Schlauch K, Tam R, Cortes D, Torres MA, Shulaev V, Dangl JL, Mittler R (2009) The plant NADPH oxidase RBOHD mediates rapid systemic signaling in response to diverse stimuli. Sci Signal 2:ra45

Miszalski Z, Kornas A, Rozpądek P, Fischer-Schliebs E, Lüttge U (2013) Independent fluctuations of malate and citrate in the CAM species *Clusia hilariana* Schltdl. Under low light and high light in relation to photoprotection. J Plant Physiol 170:453–458

Miszalski Z, Niewiadomska E, Ślesak I, Lüttge U, Kluge M, Ratajczak R (2001) The effect of irradiation on carboxylating/decarboxylating enzymes and fumarase activities in *Mesembryanthemum crystallinum* L. exposed to salinity stress. Plant Biol 3:17–23

Miszalski Z, Skoczowski A, Silina E, Dymova O, Golovko T, Kornas A, Strzalka K (2016) Photosynthetic activity of vascular bundles in *Plantago media* leaves. J Plant Physiol 204: 36–43

Miszalski Z, Ślesak I, Niewiadomska E, Baczek-Kwinta R, Lüttge U, Ratajczak R (1998) Subcellular localization and stress responses of superoxide dismutase isoforms from leaves in the C_3-CAM intermediate halophyte *Mesembryanthemum crystallinum* L. Plant Cell Environ 21:169–179

Miyake H (2016) Starch accumulation in the bundle sheaths of C_3 plants: a possible pre-condition for C_4 photosynthesis. Plant Cell Physiol 57:890–896

Miyake H, Maeda E (1976) Development of bundle sheath chloroplasts in rice seedlings. Can J Bot 54:556–565

Monson R, Rawsthorne S (2000) CO_2 assimilation in C_3–C_4 intermediate plants. In: Leegood R, Sharkey T, von Caemmerer S (eds) Photosynthesis, vol 9. Springer, Dordrecht, The Netherlands, pp 533–550

Morison JIL, Gallouët E, Lawson T, Cornic G, Herbin R, Baker NR (2005) Lateral diffusion of CO_2 in leaves is not sufficient to support photosynthesis. Plant Physiol 139:254–266

Myburg AA, Sederoff RR (2001) Xylem structure and function. Nature Publishing Group. www.els.net

Niewiadomska E, Borland AM (2008) Crassulacean acid metabolism: a cause or consequence of oxidative stress in planta? In: Lüttge UE, Beyschlag W, Murata J (eds) Progress in botany, vol 69. Springer, New York, pp 247–266

Niewiadomska E, Bilger W, Gruca M, Mulisch M, Miszalski Z, Krupinska K (2011) CAM-related changes in chloroplastic metabolism of *Mesembryanthemum crystallinum* L. Planta 233:275–285

Niewiadomska E, Polzien L, Desel C, Rozpadek P, Miszalski Z, Krupinska K (2009) Spatial patterns of senescence and development-dependent distribution of reactive oxygen species in tobacco (*Nicotiana tabacum*) leaves. J Plant Physiol 166:1057–1068

Nosek M, Kornaś A, Kuźniak E, Miszalski Z (2015a) Plastoquinone redox state modifies plant response to pathogen. Plant Physiol Biochem 96:163–170

Nosek M, Rozpądek P, Kornaś A, Kuźniak E, Schmitt A, Miszalski Z (2015b) Veinal–mesophyll interaction under biotic stress. J Plant Physiol 185:52–56

Notaguchi M, Okamoto S (2015) Dynamics of long-distance signaling via plant vascular tissues. Front Plant Sci 6:161

Pastori GM, Mullineaux PM, Foyer CH (2000) Post-transcriptional regulation prevents accumulation of glutathione reductase protein and activity in the bundle sheath cells of maize. Plant Physiol 122:667–675

Paul MJ, Foyer CH (2001) Sink regulation of photosynthesis. J Exp Bot 52:1383–1400

Peoples MB, Beilharz VC, Waters SP, Simpson RJ, Dalling MJ (1980) Nitrogen redistribution during grain growth in wheat (*Triticum aestivum* L.) II. Chloroplast senescence and the degradation of ribulose-l,5-bisphosphate carboxylase. Planta 149:241–251

Pfannschmidt T, Bräutigem K, Wagner R, Dietzel L, Schröter Y, Steiner S, Nykytenko A (2009) Potential regulation of gene expression in photosynthetic cells by redox and energy state: approaches towards better understanding. Ann Bot (Lond) 103:599–607

Pfanz H, Aschan G, Langenfeld-Heyser R, Wittmann C, Loose M (2002) Ecology and ecophysiology of tree stems: corticular and wood photosynthesis. Naturwissenschaften 89:147–162

Pieruschka R, Schurr U, Jensen M, Wolff WF, Jahnke S (2006) Lateral diffusion of CO_2 from shaded to illuminated leaf parts affects photosynthesis inside homobaric leaves. New Phytol 169:779–788

Pilarski J, Tokarz K, Kocurek M (2008) Optical properties of the cork of stems and trunks of beech (*Fagus sylvatica* L.) Polish J Environ Stud 17:773–779

Rustérucci C, Espunya MC, Díaz M, Chabannes M, Martínez MC (2007) S-nitrosoglutathione reductase affords protection against pathogens in *Arabidopsis*, both locally and systemically. Plant Physiol 143:1282–1292

Rut G, Krupa J, Miszalski Z, Rzepka A, Ślesak I (2008) Crassulacean acid metabolism in the epiphytic fern *Platycerium bifurcatum*. Photosynthetica 46:156–160

Sage RF (2013) Photorespiratory compensation: a driver for biological diversity. Plant Biol 15: 624–638

Sage RF, Christin PA, Edwards EJ (2011) The C_4 plant lineages of planet earth. J Exp Bot 62: 3155–3169

Sage RF, Khoshravesh R, Sage TL (2014) From proto-Kranz to C_4 Kranz: building the bridge to C_4 photosynthesis. J Exp Bot 65:3341–3356

Sage TL, Sage RF (2009) The functional anatomy of rice leaves: implications for refixation of photorespiratory CO_2 and efforts to engineer C_4 photosynthesis into rice. Plant Cell Physiol 50: 756–772

Saxena I, Srikanth S, Chen Z (2016) Cross talk between H_2O_2 and interacting signal molecules under plant stress response. Front Plant Sci 7:570

Scheibe R, Backhausen JE, Emmerlich V, Holtgrefe S (2005) Strategies to maintain redox homeostasis during photosynthesis under changing conditions. J Exp Bot 56:1481–1489

Ślesak I, Libik M, Miszalski Z (2008a) The foliar concentration of hydrogen peroxide during salt-induced C3-CAM transition in *Mesembryanthemum crystallinum* L. Plant Sci 174:221–226

Ślesak I, Ślesak H, Libik M, Miszalski Z (2008b) Antioxidant response system in the short-term post-wounding effect in *Mesembryanthemum crystallinum* leaves. J Plant Physiol 165:12–137

Ślesak I, Libik M, Karpinska B, Karpinski S, Miszalski Z (2007) The role of hydrogen peroxide in regulation of plant metabolism and cellular signalling in response to environmental stresses. Acta Biochim Pol 54:39–50

Slewinski TL (2013) Using evolution as a guide to engineer Kranz-type C_4 photosynthesis. Front Plant Sci 4:212

Solhaug KA, Haugen J (1998) Seasonal variation of photoinhibition of photosynthesis in bark from *Populus tremula* L. Photosynthetica 35:411–417

Sørhagen K, Laxa M, Peterhänsel C, Reumann S (2013) The emerging role of photorespiration and non-photorespiratory peroxisomal metabolism in pathogen defence. Plant Biol 15:723–736

Stitt M (1986) Limitation of photosynthesis by carbon metabolism. I. Evidence for excess electron transport capacity in leaves carrying out photosynthesis in saturating light and CO_2. Plant Physiol 81:1115–1122

Sun Q, Yoda K, Suzuki M, Suzuki H (2003) Vascular tissue in the stem and roots of woody plants can conduct light. J Exp Bot 54:1627–1635

Sun W, Ubierna N, Ma J-Y, Walker BJ, Kramer DM, Cousins AB (2014) The coordination of C_4 photosynthesis and the CO_2-concentrating mechanism in maize and *Miscanthus* x *giganteus* in response to transient changes in light quality. Plant Physiol 164:1283–1292

Surówka E, Dziurka M, Kocurek M, Goraj S, Rapacz M, Miszalski Z (2016) Effects of exogenously applied hydrogen peroxide on antioxidant and osmoprotectant profiles and the C_3-CAM shift in the halophyte *Mesembryanthemum crystallinum* L. J Plant Physiol 200:102–110

Szechynska-Hebda M, Kruk J, Gorecka M, Karpinska B, Karpinski S (2010) Evidence for light wavelength-specific photoelectrophysiological signaling and memory of excess light episodes in *Arabidopsis*. Plant Cell 22:2201–2218

Tanou G, Filippou P, Belghazi M, Job D, Diamantidis G, Fotopoulos V, Molassiotis A (2012) Oxidative and nitrosative-based signaling and associated post-translational modifications orchestrate the acclimation of citrus plants to salinity stress. Plant J 72:585–599

Tobin AK, Thorpe JR, Hylton CM, Rawsthorne S (1991) Spatial and temporal influences on the cell-specific distribution of glycine decarboxylase in leaves of wheat (*Triticum aestivum* L.) and pea (*Pisum sativum* L.) Plant Physiol 91:1219–1225

Tsutsumi K, Kawasaki M, Taniguchi M, Miyake H (2008) Gene expression and accumulation of RubisCO in bundle sheath and mesophyll cells during leaf development and senescence in rice, a C_3 plant. Plant Prod Sci 11:336–343

Vidović M, Morina F, Prokić L, Milić-Komić S, Živanović B, Veljović Jovanović B (2016) Antioxidative response in variegated *Pelargonium zonale* leaves and generation of extracellular H_2O_2 in (peri)vascular tissue induced by sunlight and paraquat. J Plant Physiol 206: 25–39

Vlot AC, Klessig DF, Park S-W (2008) Systemic acquired resistance: the elusive signal(s). Curr Opin Plant Biol 11:436–442

Voss I, Sunil B, Scheibe R, Raghavendra AS (2013) Emerging concept for the role of photorespiration as an important part of abiotic stress response. Plant Biol 15:713–722

Voznesenskaya EV, Koteyeva NK, Edwards GE, Ocampo G (2010) Revealing diversity in structural and biochemical forms of C_4 photosynthesis and a C_3–C_4 intermediate in genus *Portulaca* L. (Portulacaceae). J Exp Bot 61:3647–3662

Wientjes E, van Amerongen H, Croce R (2013) LHCII is an antenna of both photosystems after long-term acclimation. Biochim Biophys Acta 1827:420–426

Williams ML, Farrar JF, Pollock CJ (1989) Cell specialization within the PBS of barley. Plant Cell Environ 12:909–918

Willis KJ, McElwain JC (2002) The evolution of plants. Oxford University Press, Oxford, UK

Wittmann C, Pfanz H, Loreto F, Centritto M, Pietrini F, Alessio G (2006) Stem CO_2 release under illumination: corticular photosynthesis, photorespiration or inhibition of mitochondrial respiration? Plant Cell Environ 29:1149–1158

Yamane K, Mitsuya S, Taniguchi M, Miyake H (2012) Salt-induced chloroplast protrusion is the process of exclusion of ribulose-1,5-bisphosphate carboxylase/oxygenase from chloroplasts into cytoplasm in leaves of rice. Plant Cell Environ 35:1663–1671

Yiotis C, Manetas Y (2010) Sinks for photosynthetic electron flow in green petioles and pedicels of *Zantedeschia aethiopica*: evidence for innately high photorespiration and cyclic electron flow rates. Planta 232:523–531

Yiotis C, Petropoulou Y, Manetas Y (2009) Evidence for light-independent and steeply decreasing PSII efficiency along twig depth in four tree species. Photosynthetica 47:223–231

Drought Stress and the Recovery from Xylem Embolism in Woody Plants

Andrea Nardini, Tadeja Savi, Patrizia Trifilò, and Maria A. Lo Gullo

Abstract Water is transported from roots to foliage through the xylem under negative pressure (=tension). Under this metastable status, water is prone to sudden phase change to water vapor (cavitation). In plants, air can also be aspirated into functioning xylem conduits through inter-conduit pit membranes, and the resulting embolism blocks water transport through the conduit and reduces plant hydraulic conductance and productivity. Xylem embolism and hydraulic failure are major factors contributing to tree mortality and forest decline under global-change-type droughts. However, some plants can tolerate even high embolism levels under drought, recovering hydraulic functionality upon partial or total rehydration via embolism repair and xylem refilling. Here, we review current evidence of embolism repair, highlighting possible physiological mechanisms and suggesting some functional and anatomical determinants making embolism reversal a feasible and successful drought resilience mechanism in some plants.

Contents

Communicated by Rainer Matyssek

A. Nardini (✉) and T. Savi
Dipartimento di Scienze della Vita, Università di Trieste, Via L. Giorgieri 10, 34127 Trieste, Italy
e-mail: nardini@units.it

P. Trifilò and M.A. Lo Gullo
Dipartimento di Scienze Chimiche, Biologiche, Farmaceutiche e Ambientali, Università di Messina, Salita F. Stagno D'Alcontres 31, 98166 Messina, Italy

Progress in Botany (2018) 79: 197–232, DOI 10.1007/124_2017_11,
Published online: 12 April 2017

1 Introduction

The frequency and the severity of anomalous drought and heat waves have apparently increased in several areas of the globe over recent decades (Easterling et al. 2000; Trenberth et al. 2014; Diffenbaugh et al. 2015), as a likely effect of ongoing climate warming. Considering that water availability is a major limiting factor for plant growth and survival (He et al. 2014), it is not surprising that such extreme events have already caused significant impacts on natural vegetation (Allen et al. 2015) as well as on crops (Lesk et al. 2016). In particular, forest decline processes associated with massive dieback and mortality of some tree species are being reported with increasing frequency (Allen et al. 2010; Carnicer et al. 2011; Ma et al. 2012; Cailleret et al. 2017), although some studies have proposed that drought might not be the only nor the major factor triggering such events (Sánchez-Salguero et al. 2012; Giuggiola et al. 2013; Steinkamp and Hickler 2015). Tree mortality has important cascade effects on ecosystems and related services (Anderegg et al. 2013a; Hanewinkel et al. 2013; Meisner et al. 2013), including reduction of the carbon sink's potential of forests (Ma et al. 2012; Reichstein et al. 2013). Hence, recent years have seen a steady increase in the number of studies investigating patterns, processes, and mechanisms leading to tree decline and mortality, with the aim of improving our capacity to predict and mitigate the impacts of severe droughts on forest ecosystems and woody crops. Yet, several key questions are still unresolved and ahead of us: why are some species more vulnerable to extreme drought than others? Which traits allow some species to resist severe drought and to recover upon stress relief? Is there significant genotypic variability within species in terms of resistance/resilience to extreme drought?

2 Drought Stress and Tree Mortality

Why does a tree die under drought? Although seemingly trivial, this question has no simple answer, as physiological mechanisms leading to tree decline and finally death under prolonged and/or intense water shortage are only partially understood.

Several mechanisms and processes have been proposed to play a role in the progressive drought-induced decline or sudden failure of tree's health. These can be summarized using three "key words" that have become more and more frequent in the literature dealing with impacts of drought on forests: "hydraulic failure," "carbon starvation," and "biotic attack."

2.1 *Xylem Embolism as a Primary Effect of Drought*

Terrestrial plants lose enormous amounts of water vapor to the atmosphere when stomata open to allow CO_2 diffusion into the photosynthetic tissue (Kramer 1983; Schulze et al. 2005). To keep their hydration status relatively stable and avoid lethal desiccation, plants must replace water lost with other water absorbed at the root level and transported to leaves via xylem conduits (Brodribb 2009). The forces necessary to uplift water are generated by water evaporation at the mesophyll level. Here, the adhesive and cohesive forces of water, coupled to the ultrastructure of cell walls, generate curved menisci where significant negative pressure (=tension) develops and is transmitted down to the root system via the continuous network of water-filled xylem conduits, forming a hydraulic continuum (Dixon 1914; Oertli 1971; Tyree 2003). Lumen size and wall adhesive properties of xylem conduits favor the maintenance of water in the liquid phase even under negative pressure. Still, water under tension is metastable and can undergo a sudden phase change to water vapor in a process known as cavitation (Crum 1979; Pickard 1981; Caupin and Herbert 2006). The cavitation threshold of water has been experimentally shown to average −25 MPa at relatively low temperatures in water-filled glass capillary tubes spinning in a centrifuge (Briggs 1950; Caupin and Herbert 2006). Such values are much more negative than pressures commonly recorded in xylem conduits, with lowest values reported being generally above −10 MPa when measured with a pressure chamber on detached leaves/twigs (Choat et al. 2012). Hence, "spontaneous (homogeneous)" water cavitation is unlikely to represent a common occurrence in plants. However, xylem conduits are not perfectly "sealed," as small apertures on their walls, known as pits, put them in contact with adjacent living cells or with other xylem conduits. The presence of pits is essential to assure continuity of the water column, as xylem conduits have finite lengths and no single conduit actually connects directly root tips to leaf minor veins (Jacobsen et al. 2012). Hence, water flow through the xylem involves the passage across several different conduits interconnected at the pits level which are essential structures for a functional xylem network (Choat et al. 2008). At the same time, inter-conduit pits represent the Achilles' heel of plant water transport systems (Rockwell et al. 2014). In fact, pits are potential air-entry points whenever a functioning water-filled conduit faces an air-filled compartment. The pit membrane is a porous structure, with pore diameters in the nanoscale (Choat et al. 2003). These pores' dimensions are small enough to trap an air-water interface (i.e., meniscus) at the given high surface tension of water against substantial pressure differences (Choat et al. 2008).

However, when xylem tension exceeds a critical threshold, air is unavoidably aspirated into the functioning conduit, a process which is known as air-seeding. This leads to breakage of the cohesive forces of water and to formation of embolism that occupies the xylem conduit lumen and prevents water transport through the embolized elements, thus reducing plant hydraulic conductance. Eventually, embolism can propagate to other adjacent conduits through other pits, leading to complete blockage of the long-distance water transport system and to plant "hydraulic failure." In conifers, Torus–Margo pit structures act as safety valves, as pressure differences between gas-filled and water-filled conduits deflect the membrane so that the central thickened torus plugs the pit aperture limiting the spread of the gas phase to adjacent conduits (Choat et al. 2008). Even in this case, however, embolism can propagate whenever the pressure difference is large enough to allow the gas phase to pass through the contact point between the torus and the pit aperture (Mayr et al. 2014a).

Another factor potentially leading to embolism accumulation and disruption of xylem water transport is represented by freeze-thaw cycles (Mayr and Sperry 2010). However, this chapter is focused on drought-induced embolism, and readers are referred to Mayr and Améglio (2016) for a recent overview on the effects of freeze stress on xylem conduits.

It is important to note that the terms "cavitation" and "embolism" are not synonyms. As noted above, cavitation is the spontaneous initiation of vapor bubbles in metastable water under tension, but it is unlikely to occur in xylem at physiologically relevant xylem pressure (Ψ_{xyl}) values. Thus, in accordance with Zwieniecki and Secchi (2015), in this chapter the term embolism will be used to refer to the process of formation of a gas phase into xylem conduits by the air-seeding mechanism.

Because embolism formation and propagation occur below species-specific Ψ_{xyl} thresholds, any environmental factor leading to a progressive drop of Ψ_{xyl} can potentially lead to partial or total hydraulic impairment. Over the short term, high evaporative demand associated with high air temperatures lowering relative humidity in the central hours of the day can lead to a transient drop of Ψ_{xyl}, eventually causing day-time formation of xylem embolism (McCully et al. 1998; Melcher et al. 2001; Zufferey et al. 2011; Trifilò et al. 2015). Over a prolonged drought, soil water content progressively declines, leading to a sustained decrease of Ψ_{xyl}. This pressure drop can be more or less pronounced as a function of plant structural/functional traits and responses, like rooting depth or stomatal closure under drought (Nardini et al. 2014), but it has the capacity for inducing massive embolism (Trifilò et al. 2003; Tombesi et al. 2014) and, in the worst scenario, plant desiccation and death (Davis et al. 2002; Barigah et al. 2013; Nardini et al. 2013; Anderegg et al. 2015).

Ψ_{xyl} values triggering air entry into conduits and further spread of embolism in the xylem network are known to be species-specific, with apparently narrow intra-specific variability (López et al. 2016), although only few species have been explicitly tested so far for genotypic variability of vulnerability to xylem embolism of different provenances. However, the well-known large inter-specific variability in xylem vulnerability suggests that different species can sustain different tension levels in their hydraulic system before facing the risk of hydraulic failure. The vulnerability of xylem to embolism formation and spread is generally quantified by measuring "vulnerability curves" (VCs), i.e., plots of plant (or organ) hydraulic conductance as a function of Ψ_{xyl} values reached in the same plant/organ. These curves are generally sigmoidal in shape, especially when measured on stem segments (Cochard et al. 2013), although different shapes are possible (Sperry et al. 2012). From VCs, several reference parameters can be derived, like P_{12}, P_{50}, and P_{88}, i.e., Ψ_{xyl} values inducing 12, 50, or 88% loss of hydraulic conductance. P_{12} is a useful proxy for the minimum tension leading to early embolism formation, while P_{88} indicates the tension values inducing enough embolism to pose plant hydraulic functioning and survival of several Angiosperms at serious risk (Urli et al. 2013). However, P_{50} is the generally reported reference value to synthetically describe the vulnerability to xylem embolism of different species/organs. This is because this value generally corresponds to the steepest region of the VC, so that even minor drops of Ψ_{xyl} below this threshold can imply development of nearly lethal embolism levels. Indeed, for several Gymnosperms P_{50} can already represent a lethal threshold (Brodribb et al. 2010).

Data on vulnerability to xylem embolism at the stem level and leaf level are now available for several hundred species around the globe (Choat et al. 2012; Nardini and Luglio 2014). According to a meta-analysis by Choat et al. (2012), P_{50} values range from -0.1 to -14 MPa. P_{50} values are correlated with mean annual precipitation of the species' habitat of origin (Gleason et al. 2016), with more embolism-resistant plants (lower P_{50}) generally thriving in drier areas. This however does not mean that species from different biomes are necessarily exposed to different risks toward hydraulic failure. In fact, when considering the difference between P_{50} and actual minimum seasonal Ψ_{xyl} (often indicated as Ψ_{min}) experienced by plants (the so-called "safety margin"), Choat et al. (2012) have shown that all woody plants from all major biomes converge toward a risky hydraulic behavior, with safety margins generally <1 MPa in Angiosperms and around 2 MPa for Gymnosperms. Clearly, this pattern suggests that embolism is not at all a rare occurrence in terrestrial plants (Nardini and Salleo 2000), and also that anomalous droughts might easily force several tree species to surpass critical tension thresholds leading to hydraulic failure (Savi et al. 2015; Anderegg et al. 2016).

Figure 1 reports an example of VC measured for *Ostrya carpinifolia* Scop., a widespread species in South-central and southern Europe that has suffered extensive crown desiccation in several areas following recent extreme drought events (Piškur et al. 2011; Nardini et al. 2016; Pollastrini et al. 2016). The curve reveals that *O. carpinifolia* suffers 50% loss of stem hydraulic conductance at Ψ_{xyl} of about -2.4 MPa, suggesting that this species is relatively vulnerable to drought-induced embolism. Moreover, the safety margin calculated as a function of Ψ_{min} values experienced by field growing individuals during a "normal" summer drought (Nardini et al. 2016) is negative, suggesting that the species routinely suffers embolism accumulation during summer and is at risk of hydraulic failure under anomalous drought condition, as in fact observed following extreme summer 2003 and summer 2012 in SE Europe (Nardini et al. 2016). It is also of interest to note

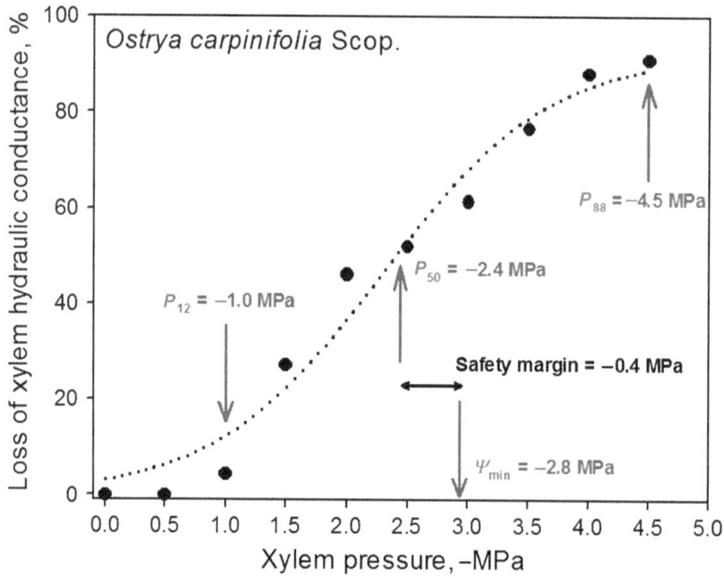

Fig. 1 Vulnerability curve of stems of *Ostrya carpinifolia* Scop. reporting percentage loss of xylem hydraulic conductance as a function of xylem pressure. Each point represents the mean of five samples, and the *dotted curve* represents the interpolation based on a 3-parameter sigmoidal. *Grey arrows* indicate xylem pressure values inducing 12% (P_{12}), 50% (P_{50}), or 88% (P_{88}) loss of hydraulic conductance, as well as minimum seasonal xylem pressure (Ψ_{min}) as reported by Nardini et al. (2016). The safety margin, calculated as ($\Psi_{min} - P_{50}$), is also indicated by a *black arrow*

that the curve predicts a 60% loss of stem hydraulic conductance at xylem pressures corresponding to Ψ_{min}, consistent with embolism levels recorded by Nardini et al. (2016) in field growing plants. The VC for *O. carpinfolia*, coupled to field measurements of the hydraulic status of the species under drought, confirms the powerful predictive value of this tool when assessing species-specific risk of drought-induced hydraulic failure and mortality.

2.2 Carbon Starvation

Plants adopt different water use strategies to cope with severe and/or prolonged drought stress. Some species rapidly respond to water shortage by closing stomata thus preserving water while limiting photosynthesis and carbon gain (isohydry), while others sustain gas exchange and maximize photosynthesis while tolerating large drops of Ψ_{xyl} (anisohydry). In reality, hydraulic strategies can fall anywhere in the continuum spectrum ranging from perfect isohydry to anisohydry (Klein 2014). Even species living in the same habitat can adopt quite contrasting water use strategies to cope with drought stress, also as a function of rooting depth, water storage, deciduous vs evergreen leaf habitus, and several other structural/functional traits (Nardini et al. 2003, 2016; Moreno-Gutiérrez et al. 2012). In any case, both isohydric and anisohydric species can apparently experience comparably high embolism levels under drought (Choat et al. 2012), and also suffer comparably severe stomatal closure at the peak of drought (Quero et al. 2011; Nardini et al. 2016). Hence, beside the risk of hydraulic failure, plants subjected to severe drought stress unavoidably face a reduction in photosynthetic carbon gain. Under such conditions, the maintenance of primary and secondary metabolism relies on consumption of stored nonstructural carbohydrates (NSC), potentially leading to exhaustion of NSC pools under prolonged water shortage. This condition is commonly referred to as "carbon starvation," and has been proposed as a possible mechanism leading to tree decline and death under extreme drought events (McDowell et al. 2008; Adams et al. 2009).

Research efforts aimed at verifying the occurrence and magnitude of carbon starvation in drought-stressed and declining plants, and particularly in trees, have produced contrasting results (Hoch 2015). In some cases, NSC content was found to be still high even in trees suffering extensive crown die-back (Gruber et al. 2012; Hartmann et al. 2013; Rowland et al. 2015), while other studies found significant drops of plant carbohydrate reserves under drought (Adams et al. 2013; Galvez et al. 2013; Zhao et al. 2013; Maguire and Kobe 2015). The discrepancies in the conclusions reached by different studies might arise for several different reasons. First of all, NSC in trees are represented by different pools, stored for different time scales in different organs (Hartmann and Trumbore 2016). Hence, different studies sampling different organs in different species might easily come to contrasting conclusions, also as a factor of time scales of observations. Moreover, trees are known to store NSC for quite long time intervals, so that under stress conditions the pools of NSC mobilized to sustain metabolism can be even decades old (Carbone et al. 2013). As a consequence, measurements of NSC content in relatively young organs like terminal branches might fail to detect a significant drop, even though the plant is indeed consuming long-term storage carbohydrates. Finally, the extraction and quantification of NSC from woody samples, if not based on standard protocols, can lead to substantial variation of data collected from independent laboratories, as recently shown by Quentin et al. (2015).

The fact that high NSC content can be detected even in trees subjected to long-term drought stress does not exclude the possibility that carbon starvation is indeed a mechanism contributing to tree decline. In fact, carbon limitation might occur even in case of physiological impairment of the machinery deputed to NSC mobilization, from enzymatic processes affecting carbohydrate metabolism (Ahmadi and Baker 2001) to long-distance transport of sugars via the phloem (Mencuccini et al. 2013). In a recent study, Sevanto (2014) has highlighted the importance of "phloem failure" as a possible component of the impact of drought on tree survival. Phloem impairment under drought can occur because of increased viscosity of phloem sap, or as a consequence of inability to maintain phloem turgor (Sevanto 2014). In both cases, carbon starvation might develop even under conditions of still relatively high NSC concentration in some plant sectors.

As a final note, it should be mentioned that both carbon starvation and phloem failure might also have important secondary effects on plant hydraulic functioning by affecting processes of xylem repair and refilling following embolism formation and drought relief (Salleo et al. 1996; Hölttä et al. 2006; McDowell 2011; Nardini et al. 2011a), which are thought to depend on availability of soluble sugars and phloem integrity. Under such a scenario, which is discussed in the following sections, carbon starvation and phloem failure might be viewed as cofactors further pushing trees toward hydraulic failure.

2.3 *Biotic Attacks*

Ongoing climate changes are affecting the distribution and abundance of forest insects known to act as agents of disturbance over trees (Weed et al. 2013), and biotic agents have been proposed as important drivers of observed tree and forest decline (Ayres and Lombardero 2000). It is beyond the scope of this chapter to revise current evidence about the relationships among climate change, insect outbreaks, and tree mortality. However, it is worth noting that some studies have causally linked anomalous insect attacks to the water status of attacked trees (Jactel et al. 2012; Gaylord et al. 2013), as an effect of likely depletion of carbon reserves and impairment of secondary metabolism and tree's defense capability.

3 Techniques to Quantify Xylem Vulnerability: Current Controversies

Xylem embolism is currently recognized as a major trigger of drought-induced tree dieback and forest decline. Even when nonlethal, xylem embolism poses serious threats to productivity of forest trees and crops, potentially affecting the carbon sink capacity of forests as well as food, wood, and fiber production for human needs. Hence, our capacity to accurately measure species vulnerability to xylem embolism is fundamental in order to make reliable predictions of the impact of future climate scenarios on global primary productivity, as well as to accurately measure the differential drought resistance of different genotypes in large-scale breeding programs.

The occurrence of embolism in xylem conduits, and its potential threats to water transport and plant productivity, was already well recognized when Dixon presented the extended version of the Cohesion-Tension Theory (CTT) for the ascent of sap (Dixon and Joly 1894; Dixon 1914; Tyree and Zimmermann 2002; Angeles et al. 2004). Hence, it is somehow surprising that over one century later, scientists have not found convergence and agreement on the most reliable experimental techniques to be used to verify the occurrence and quantify the impact of xylem embolism on plant hydraulics (Rockwell et al. 2014; Jansen et al. 2015).

Several experimental approaches have been proposed and tested so far to monitor xylem embolism and its impact on plant hydraulic conductance. These range from detection of acoustic waves generated by embolism formation events during progressive plant or organ dehydration (Milburn 1973; Tyree and Dixon 1983; Kikuta et al. 1997; Hölttä et al. 2005; Nolf et al. 2015), to direct quantification of air volumes extracted from branches exposed to progressively lowered Ψ_{xyl} (Pereira et al. 2016). However, the most commonly adopted techniques rely on measurements of the hydraulic conductance of stems, roots, and leaves exposed to progressively lower Ψ_{xyl} to induce embolism (Sperry et al. 1988; Nardini et al. 2001; Trifilò et al. 2016). Embolism is generally induced by progressively air-dehydrating whole plants or detached branches (Sperry 1986; Tyree et al. 1992), or via pressurization to force gas to pass through pit membranes (Cochard et al. 1992; Salleo et al. 1992), or finally by using centrifugal forces (Alder et al. 1997; Cochard 2002; Li et al. 2008). Both air pressurization and centrifugation can potentially induce embolism at Ψ_{xyl} values different from those causing hydraulic failure under natural dehydration (Ennajeh et al. 2011; Wang et al. 2014; Knipfer et al. 2015a), so that the reliability of these techniques to test plant vulnerability to embolism has been questioned and is currently debated (Torres-Ruiz et al. 2014; Hacke et al. 2015; Pratt et al. 2015). On the other hand, progressive air dehydration of large detached branches (or better whole plants) is agreed to be the best method to induce embolism and investigate its impact on xylem hydraulics (Cochard et al. 2013).

Embolism rates in air dehydrated plants and branches can be quantified with either direct or indirect techniques. Indirect techniques most commonly rely on measurements of water flow rates through excised stem or root segments (Sperry et al. 1988; Melcher et al. 2012) while perfusing samples with a low-concentration solution (Zimmermann 1978; Zwieniecki et al. 2001; Gascò et al. 2006, 2007; Nardini et al. 2007a) under relatively low pressure (<10 kPa) to avoid embolism displacement. This allows to measure "native" hydraulic conductivity (k_i) as $k_i = (F/P) \times L$, where F is the measured flow rate, P is the pressure applied, and L is sample length. Embolism is then removed by applying high pressure (in the range 0.1–0.2 MPa) "flushes," followed by measurements of flow rates under low pressure to obtain the "maximum" sample hydraulic conductivity (k_{max}). Finally, the embolism-induced percentage loss of hydraulic conductivity (PLC) of the sample is calculated as $PLC = [1 - (k_i/k_{max})] \times 100$. Plots of PLC or k_i vs Ψ_{xyl} (VCs) are then used to visualize the species-specific or organ-specific vulnerability to xylem embolism (Melcher et al. 2012), and extrapolate synthetic indexes that can describe numerically the pressure thresholds favoring hydraulic failure.

Several controversies around the validity of classical hydraulic measurements, and related data interpretation, have been raised in recent years (Rockwell et al. 2014). First of all, it has been questioned whether plotting PLC versus Ψ_{xyl} actually provides a reliable picture of species' vulnerability to xylem embolism. In fact, the high-pressure flush applied to samples during hydraulic measurements is expected to remove all embolism from the xylem network, including eventual gas bubbles present in old xylem conduits upon which the plant is not relying anymore for water supply to the foliage. In some cases, these old xylem conduits have become much more vulnerable to embolism as a consequence of "cavitation fatigue" phenomena, i.e., the progressive increase of vulnerability to embolism of conduits that have undergone cycles of embolism formation and repair (Stiller and Sperry 2002; Hillabrand et al. 2016). Under such a scenario, plotting PLC versus Ψ_{xyl} would lead to overestimate the vulnerability to xylem embolism. Hence, the opportunity has been suggested to represent vulnerability curves as plots of k_i versus Ψ_{xyl} (Schreiber et al. 2016). Secondly, and most importantly, the validity of hydraulic measurements has been questioned by a study (Wheeler et al. 2013) suggesting the occurrence of artifactual increases of PLC when stems are excised from plants, even if under water, while the xylem is under substantial tension. To avoid this possible artifact, it was suggested that plants/branches should be rehydrated before excising subsamples for hydraulic measurements, in order to "relax" xylem tension and avoid air aspiration into conduits upon cutting. Indeed, the procedure used by Wheeler et al. (2013) to demonstrate the "cutting" artifact was unusual, as samples were cut under water and directly connected to a hydraulic apparatus. Dating back to the original studies of Zimmermann (1978), routine laboratory procedures have consisted in recutting samples under water several times by trimming a few centimeters at both ends after the initial cut. This standard technique is apparently sufficient to release xylem tension within a short time frame, and it has been recently confirmed to prevent the "cutting" artifact (Beikircher and Mayr 2015; Venturas et al. 2015). Moreover, other studies have shown that in several cases the cutting artifact does not occur at all (Scoffoni and Sack 2015) or results in only minor impacts on estimated PLC values (Torres-Ruiz et al. 2015; Ogasa et al. 2016). Finally, the rehydration/relaxation procedure recommended by Wheeler et al. (2013) as a pretreatment to avoid the cutting artifact has later been shown to promote the recovery of embolized conduits in some species (Trifilò et al. 2014a; Venturas et al. 2015; Knipfer et al. 2016). In other words, an alternative conclusion that can be drawn on the basis of results reported by Wheeler et al. (2013) is that in some species hydraulic measurements can be biased by a "refilling" artifact when samples are left rehydrating for minutes/hours before connecting samples to the hydraulic apparatus (Knipfer et al. 2016). On this basis, while it is likely that measurements of embolism rates based on classical hydraulic methods and precautions (Zimmermann 1978) are not affected to a significant level by the "cutting" artifact, it is also important that future studies carefully check for the occurrence of cutting/refilling artifacts, especially when long-vessel species are investigated.

In any case, the recent controversy around the validity of hydraulic techniques and derived PLC data has stimulated new rounds of experiments aimed at direct visualization of the functional status of xylem conduits in vivo (Fig. 2), based on magnetic resonance imaging (MRI) (Holbrook et al. 2001; Zwieniecki et al. 2013) or X-ray microtomography (microCT) (Lee and Kim 2008; Suuronen et al. 2013; Choat et al. 2015). To date, only few studies have estimated plant vulnerability to xylem embolism on the basis of microCT observations (e.g., McElrone et al. 2012;

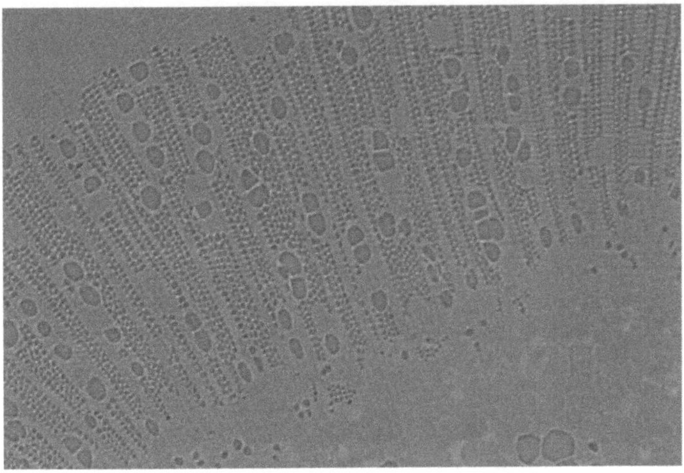

Fig. 2 Cross-section image, obtained by X-ray microtomography of a *Larus nobilis* L. stem. Embolized xylem conduits in the wood ring can be recognized by their *dark color*, compared to *light gray* of still water-filled and functional conduits

Bouche et al. 2016; Choat et al. 2016). Moreover, to the best of our knowledge only two studies have actually compared PLC values based on microCT observations to those obtained on the same plant material with classical hydraulic techniques (Nardini et al. 2017; Nolf et al. 2017). Although these studies have reported substantial agreement between the two techniques, thus offering support to the validity of destructive hydraulic methods, future tests of procedures and protocols on different species/organs should be mostly welcome. It should also be noted that imaging techniques do not allow direct measurements of hydraulic parameters. These can only be calculated on the basis of the geometry of conduits, a procedure that can result in important errors as inter-conduits pit resistances are ignored in this case. In any case, converging on accepted handling and measuring techniques when assessing embolism rates in plant samples, and the relative impact on plant- and organ-level water transport, is an important step to move forward our understanding of plant hydraulic functioning.

4 Is Embolism Forever? Experimental Evidence for Embolism Repair

> All living organisms have to a greater or less degree the ability to repair injuries
> (Minot 1885)

The capacity to repair injuries is considered among the essential properties of living organisms, as the quote above by Charles S. Minot highlights. Still today, basic biology textbooks stress this concept in the very first chapters. Embolism is a potentially severe injury to the water transport system of plants, and the risks associated with tension-driven xylem water transport were already clear to the early proponents of the Cohesion–Tension Theory (Dixon and Joly 1894). Of course, seasonal or continuous production of new xylem conduits by secondary growth allows some plants to substitute embolized conduits with newly generated water pathways connecting roots to foliage (Améglio et al. 2002; Spicer and Groover 2010). But in this chapter we will not address this "growth mechanism" to cope with negative effects of xylem embolism. Rather, we will focus on the more controversial issue of repair of embolized conduits, which has to be necessarily based on some mechanism allowing removal of the gas phase from the conduit lumen, and replenishment with water. A quote from Dixon (1914) will help to further introduce this crucial aspect of water transport in plants:

> It is also to be remembered that the periodic flooding of tracheae with water forced upwards by root-pressure will bring the bubbles into solution and will re-establish the conditions for tension throughout the water-tracts

The sentence above is very useful to highlight the important concept that positive pressures are necessary to push water in a gas-filled xylem conduit to restore the continuity of the water column from roots to leaves. More precisely, Ψ_{xyl} has to rise above a critical value to dissolve embolism. This critical value is dictated by Henry's law, so that embolism recovery will be possible when $\Psi_{xyl} > -2\tau/r$, where τ is the surface tension of water and r is the radius of the conduit containing the embolus (Yang and Tyree 1992). Hence, in small conduits spontaneous refilling is possible even at slightly negative Ψ_{xyl} (e.g., −5.8 kPa for a conduit 50 μm in diameter), although over quite long time intervals (Vesala et al. 2003). Readers are referred to the classical book by Tyree and Zimmermann (2002) for a more detailed analysis of expected physical constraints on bubble dissolution.

The notion that bulk positive pressures can build up in the xylem of some plant species by "root pressure" developing over seasonal or diurnal time scales is noncontroversial and commonly accepted (Priestley 1920). The phenomenon is apparently based on an osmotic mechanism triggered by solute accumulation in the root stele under conditions of low transpiration demand and high soil water content (Kramer 1983; Singh 2016), although alternative or additional mechanisms have been proposed to explain this process (Enns et al. 2000; Wegner 2014). In some cases, positive pressures up to 150 kPa have been recorded (Fisher et al. 1997), and experimental evidence for embolism removal based on root pressure has been provided by several studies, e.g., in grapevine (Sperry et al. 1987), bamboo (Yang et al. 2012), walnut (Améglio et al. 2002), and poplar (Leng et al. 2013). However, root pressures do not develop at all in several species. As an example, in a survey of 29 dycotiledons from ten different families in a tropical forest, only three exhibited positive xylem pressure at predawn, while Ψ_{xyl} in the remaining 26 species remained always negative (Ewers et al. 1997). Moreover, positive pressures commonly recorded in most species might explain at best the dissolution of embolisms up to a few meters of height (Ewers et al. 1997), making this process an unlikely mechanism to restore hydraulic continuity in the most distal branches of tall trees, due to gravity-induced tensions. Notably, it is in the distal aerial parts of plants that most negative Ψ_{xyl} develops during the day (Tyree et al. 1991), and peripheral organs are frequently more vulnerable to xylem embolism (Tyree et al. 1993; Salleo et al. 2000; Choat et al. 2005; Savi et al. 2016a). Finally, even in species known to develop root pressure, embolism reversal has been sometimes reported to be independent on the actual generation of such pressures (Knipfer et al. 2016).

Is embolism reversal possible even in the absence of root pressure, or whenever root pressures are not large enough to dissolve the gas phase in the most distal plant sectors? Examples of embolism repair in the absence of root pressure have been reported for some species that develop the so-called "stem pressure," including *Acer saccharum* (Sperry et al. 1988), *Acer pseudoplatanus* (Hacke and Sauter 1996), *Juglans regia* (Améglio et al. 2001, 2004), and *Betula pendula* (Hacke and Sauter 1996). In all these cases, positive xylem pressures are generated in the absence of root pressure. Still, the bulk of xylem at stem level is subjected to positive pressure, favoring embolism dissolution and restoring hydraulic continuity in conduits that have suffered winter embolism (Mayr and Améglio 2016). Although the mechanisms allowing build-up of bulk stem pressure have not been elucidated in detail, available experimental evidence converges toward an osmotic mechanism based on the release and accumulation of sugars, like sucrose, in conduits' lumen (Améglio et al. 2004). This would allow generation of osmotic pressures due to the presence of semipermeable barriers, likely located at the level of plasmalemma of conduit-associated parenchymatic cells, where over-expression of aquaporins would also favor water release into xylem conduits (Sakr et al. 2003; Laur and Hacke 2014).

All the above observations are consistent with the notion that embolism repair can be achieved by plants by generation of positive xylem pressure at the root or stem level. Over the last 25 years, however, an increasing number of reports have documented partial or even complete embolism reversal in plants experiencing negative Ψ_{xyl} in still functioning conduits, well below the critical value set by $-2\tau/r$. Evidence for embolism reversal under negative pressure was initially derived on hydraulic measurements of PLC performed on drought stressed and partially re-irrigated plants (Salleo and Lo Gullo 1989). Similar measurements have been repeated by independent laboratories, revealing that several species undergo cycles of embolism formation and repair with a diurnal rhythm (Zwieniecki and Holbrook 1998; Bucci et al. 2003; Domec et al. 2006; Taneda and Sperry 2008; Zufferey et al. 2011; Christman et al. 2012; Melcher and Zwieniecki 2013; Trifilò et al. 2015), on a seasonal scale (Jaquish and Ewers 2001; Mayr et al. 2014b; Trifilò et al. 2014b), or upon cycles of drought/rewatering (Salleo et al. 1996; Tyree et al. 1999; Hacke and Sperry 2003; Stiller et al. 2005; Nardini et al. 2008; Secchi and Zwieniecki 2010; Ogasa et al. 2013; Chitarra et al. 2014; Martorell et al. 2014; Ganthaler and Mayr 2015; Savi et al. 2016b), in all cases without apparent generation of root pressure or bulk stem pressure and under conditions of moderate tension in still functioning conduits. In most cases refilling under tension was observed when Ψ_{xyl} approached values in the range of −0.3 to −0.5 MPa (Bucci et al. 2003; Nardini et al. 2008; Secchi and Zwieniecki 2010; Christman et al. 2012; Melcher and Zwieniecki 2013; Chitarra et al. 2014), with only some reports documenting partial embolism reversal even at Ψ_{xyl} down to −1.0 MPa in some species (Salleo et al. 1996; Stiller et al. 2005; Zufferey et al. 2011; Trifilò et al. 2015). Interestingly, some reports also highlighted a number of species that were apparently unable to reverse xylem embolism (Ogasa et al. 2013; Trifilò et al. 2015), an important point that will be further discussed in the following sections.

The validity of all these reports has been questioned (Cochard and Delzon 2013; Sperry 2013) following the observations by Wheeler et al. (2013). Recent controversies around this methodological issue have been discussed in a previous section and will not be addressed again here. Rather, we will focus on the burden of proof around embolism recovery under negative pressures arising from direct observations of the functional status of xylem conduits.

Early attempts to visualize the functional status of xylem conduits under different physiological conditions have been based on Cryo-SEM observations (Canny 1997; Pate and Canny 1999). This technique is based on rapid freezing of roots, stems, or petioles still attached to the plant, by immersion in liquid nitrogen. Samples are then excised and cross sections are observed under a cryo-scanning electron microscope at a temperature of −90°C. Cryo-SEM observations of xylem conduits performed at different daytimes or seasons have revealed dynamic changes in their functional status, providing visual evidence for the process of xylem refilling (Canny 1997; McCully et al. 1998; Utsumi et al. 1998; Tyree et al. 1999; Melcher et al. 2001). Moreover, in several cases embolism repair was observed in plants where xylem pressures were substantially negative (Canny 1997; Tyree et al. 1999), apparently supporting observations based on indirect hydraulic techniques. Cryo-SEM observations also provided important insights into the putative process of water delivery into embolized conduits. In fact, images of refilling conduits revealed water droplets apparently being extruded through pits from neighboring conduit-associated parenchyma cells (Tyree et al. 1999). Also the validity of these observations is nowadays in doubt, because control experiments have revealed that the process of freezing, even if by immersion in liquid nitrogen, is not fast enough to avoid spurious embolism appearance, depending on the tension level (Cochard et al. 2000; Umebayashi et al. 2016). Although the conclusions derived from these control tests have been debated (Canny et al. 2001), in more recent studies cryo-SEM has been only occasionally used to monitor xylem embolism (Hukin et al. 2005; Johnson et al. 2009; Charra-Vaskou et al. 2012).

The most compelling evidence for embolism reversal under moderately negative Ψ_{xyl} has been obtained by techniques allowing in vivo imaging of the functional status of xylem conduits. On the basis of magnetic resonance imaging (MRI), Holbrook et al. (2001) reported refilling of embolized xylem conduits in intact grapevine plants at Ψ_{xyl} of about −0.2 MPa, without evidence of positive root pressure. Interestingly, this study highlighted that refilling was initiated only when artificial lighting was switched off, suggesting that low or null transpiration is a prerequisite for embolism repair. More recently, Zwieniecki et al. (2013) reported

MRI observations in intact *Acer rubrum* plants, documenting embolism refilling at Ψ_{xyl} of about −0.5 MPa.

MicroCT observations performed in intact plants of grapevine (Brodersen et al. 2010) confirmed occurrence of refilling of embolized conduits, and also showed that the process is initiated by water droplets apparently forming on the interior surface of xylem conduit walls, consistent with early cryo-SEM images of refilling conduits (Tyree et al. 1999) and with the suggested role of axial parenchyma in the process (Salleo et al. 2004, 2009). Most importantly, embolism reversal was observed to initiate in plants still experiencing negative Ψ_{xyl} ranging from −0.7 to −1.6 MPa, consistent with the range of values reported in several studies investigating embolism repair on the basis of indirect hydraulic techniques. Another important observation by Brodersen et al. (2010) was related to refilling times that were calculated to range between 2 and 17 h for conduits with a diameter of 20 and 150 μm, respectively. These values are in accordance with both theoretical considerations (Vesala et al. 2003) and reported changes of PLC on a diurnal time scale or following irrigation of drought stressed plants. In a follow-up study, Knipfer et al. (2015b) also reported in vivo reversal of embolism in two out of three *Vitis* species. In these cases, embolism reversal was initiated while xylem pressures in still functioning conduits were still around −1.0 MPa. Also in this case, refilling of embolized conduits was driven by radial water influx, as reported also in other independent studies based on microCT (Lee et al. 2013; Ryu et al. 2016). Just like in the case of studies based on hydraulic methods, these and other reports based on microCT also revealed a number of cases and species where embolism reversal under tension was not observed (Choat et al. 2015; Knipfer et al. 2015a, b).

5 Possible Mechanisms of Embolism Reversal and Related Experimental Evidence

5.1 *Embolism Refilling: An Osmotic Mechanism?*

Observations of embolism reversal under negative pressure are puzzling, as basic laws of thermodynamics suggest that this process should not occur at all. Hence, the most parsimonious hypothesis to explain such observations and reports is that local positive pressures can be generated in refilling conduits even while the bulk of still functioning xylem conduits experience moderate tension, as also suggested by observations of droplets entering conduits in the early phases of the repair process (Brodersen et al. 2010). Positive pressures in the xylem, like root pressure and stem pressure, are most likely generated by an osmotic mechanism. In fact, the current paradigm explaining embolism reversal postulates that solutes are extruded into refilling conduits, thus locally pushing the osmotic potential of xylem sap to values lower than the water potential of surrounding cells (Nardini et al. 2011a). The presence of semi-permeable barriers, probably represented by membranes of living cells surrounding xylem conduits (Salleo et al. 2004), would allow generation of an osmotic pressure, leading to replenishment of the embolized conduit with water.

Alternative explanations have been proposed in the past as well as in recent years. Canny (1998) proposed that generation of turgor pressure in living cells, by accumulation or production of solutes, would "squeeze" water into embolized conduits. However, Canny's hypothesis has been proved to be unlikely from a thermodynamic and mechanical point of view (Comstock 1999; Stiller and Sperry 1999). More recently, Rolland et al. (2015) have shown that pure capillary forces might be involved in the rapid refilling of embolized protoxylem conduits, but their experiments did not control for Ψ_{xyl} values during embolism formation and reversal. Hence, the relevance of this finding for the understanding of embolism reversal under tension remains unknown. Finally, Wegner (2014) has proposed a mechanism based on co-transporters of water and solutes (Zeuthen 2010) to explain "energetically uphill" water flow in refilling conduits. Albeit interesting and intriguing, these ideas still await experimental testing.

Experiments aimed at verifying an eventual rise in solute concentration in refilling conduits initially focused on inorganic ions, and showed an increase in xylem sap osmolality due to an increase in K^+ and Ca^{2+} content, concurrent with embolism repair (Tyree et al. 1999). However, the osmotic potential apparently assured by these ions was not negative enough to generate the necessary pressures required to refill embolized conduits. The discrepancy was likely due to the fact that these early experiments were not able to separately collect sap from functional and refilling conduits, so that any eventual rise of solute content to high values would have been masked by dilution with low-concentration sap originating from still water-filled conduits. Also, follow-up experiments highlighted a different role for the rise of potassium concentration in the xylem of drought stressed plants, related to the enhancement of residual hydraulic conductance and partial compensation of embolism-induced reduction of plant hydraulic conductance (Trifilò et al. 2008, 2011, 2014b; Cochard et al. 2010; Nardini et al. 2010; Oddo et al. 2011, 2014) via interaction with inter-conduit pit membranes (Zwieniecki et al. 2001; Gascò et al. 2006; Gortan et al. 2011; Nardini et al. 2007b, 2012; Jansen et al. 2011). This phenomenon has been often indicated in the literature as "ionic effect" (Nardini et al. 2011b).

The lack of strong support for a role of inorganic ions in the refilling process raised the hypothesis that organic solutes could be involved. Two independent studies (Bucci et al. 2003; Salleo et al. 2004) provided the first support to this hypothesis, by showing that embolism reversal is indeed accompanied by starch degradation in conduit-associated parenchyma cells. Hence, it was hypothesized that starch degradation leads to production of soluble sugars that are extruded into refilling conduits and contribute to the generation of osmotic pressures (Salleo et al. 2009). A convincing proof of the role of carbohydrates in embolism repair was provided by Secchi and Zwieniecki (2012), on the basis of a new experimental set-up designed to separately collect and analyze the xylem sap from functional conduits and nonfunctional conduits undergoing refilling in *Populus nigra*. These experiments revealed that both inorganic ions and sugars were present at high concentrations in refilling conduits, but not in functional ones. While the osmotic potential of the sap in refilling conduits (about −0.4 MPa) could not account for reversal of embolism in drought stressed plants at $\Psi_{xyl} = -1.5$ MPa, it was negative enough to explain the observed reversal of embolism in plants re-irrigated and recovering from drought, even if still experiencing moderately negative Ψ_{xyl}. This observation suggests that embolism repair under tension is possible, but successful refilling requires a reduction of the level of xylem tension and the release of an adequate concentration of solutes in refilling conduits. In a follow-up study (Secchi and Zwieniecki 2016), the same authors provided experimental evidence for a mechanism of sugar release and accumulation in refilling conduits based on xylem sap acidification under drought stress, a mechanism that would "prime" the xylem system for prompt restoration of functionality of xylem conduits upon partial rehydration and recovery from drought.

The physiological mechanisms described above, although still hypothetical, are consistent with molecular evidence related to gene expression regulation in drought stressed plants. In fact, an analysis of the transcriptome of xylem parenchyma cells of *Populus trichocarpa* upon embolism formation and recovery revealed significant up-regulation of genes encoding for carbohydrate metabolic pathways and ion transporters (Secchi et al. 2011). Consistent with these results, up-regulation of genes coding for enzymes involved in starch degradation and mobilization was recorded in grapevine petioles recovering from drought-induced embolism (Perrone et al. 2012; Chitarra et al. 2014). Interestingly, all the above studies also highlighted a significant up-regulation of genes coding for aquaporins (Chaumont and Tyerman 2014), and this finding is consistent with the requirements of high membrane water permeability to assure water delivery to refilling conduits and shorten recovery times (Secchi et al. 2016). Moreover, localization of aquaporins during recovery from embolism in needles of *Picea glauca* (Laur and Hacke 2014) and *Picea abies* (Mayr et al. 2014b) is consistent with their role in facilitating radial water movement towards the vascular tissue.

Some studies documenting embolism reversal under moderate xylem tension also reported that some chemical treatments can prevent xylem refilling. In particular, orthovanadate was reported to inhibit PLC recovery (Salleo et al. 2004; Trifilò et al. 2014a). Because orthovanadate is an inhibitor of the activity of proton pumps, its effects on the refilling process are consistent with the model proposed for this phenomenon, and in particular with the postulated involvement of xylem sap acidification to drive sucrose efflux into refilling conduits. It has also been suggested that decreased xylem sap pH would be responsible for the activation of an apoplastic invertase that would degrade sucrose into glucose and fructose. This would further increase solute concentration in conduits to be replenished with water under osmotically generated positive pressures (Secchi and Zwieniecki 2016), and the inhibitory effect of orthovanadate would be also consistent with this scenario.

5.2 The Importance of Phloem Functioning

Some of the studies documenting embolism repair under tension have reported that physical treatments affecting phloem functioning, including annular girdling or longitudinal incisions of the bark, cause inhibition of the refilling process (Bucci et al. 2003; Salleo et al. 2004; Christman et al. 2012; Trifilò et al. 2014a). Hence, it has been suggested that phloem provides solutes or water, or even both, to sustain the osmotic refilling mechanisms. Salleo et al. (2006) reported that the inhibition of refilling by girdling could be eliminated by supplying artificial radial pressure to drought stressed stems of *Laurus nobilis*, possibly suggesting that the main role of phloem would be to deliver water to refilling conduits. Indeed, previous theoretical calculations had suggested that water stored in xylem parenchyma cells and eventually released during embolism repair processes might not be sufficient to refill embolized conduits (Tyree et al. 1999), supporting the view of positive phloem pressure as an essential requisite to deliver water through radial parenchyma to embolized conduits (Hölttä et al. 2006). An important corollary of these observations is that phloem functioning and turgor pressure are important determinants of the capacity of a plant to recover hydraulic functionality during or after drought stress. The consequence of this mechanism and its relationship with carbon status of plants during severe droughts will be further discussed in Sect. 6.

5.3 Hydraulic Isolation of Refilling Conduits

Embolism reversal can apparently occur while still functioning conduits are experiencing moderately negative pressure. This poses a special and challenging problem to the proposed refilling mechanism. In fact, the water potential difference between refilling and water-filled conduits should cause water to be immediately drawn into functioning conduits as soon as partial reconnection is obtained between them, possibly leading to failure of the refilling process (Zwieniecki and Holbrook 2009). It has been proposed that refilling conduits are hydraulically isolated from neighboring and water-filled conduits thanks to the partially hydrophobic nature of inter-conduit pits, leading to nonzero contact angle between water and pit aperture (Holbrook and Zwieniecki 1999). This would lead to formation of a convex air-water interface at the entrance of the pit chamber and would prevent water flow to functioning conduits during refilling of embolized conduits. Experimental evidence for nonzero contact angle of water at the pit entrance has been provided by Zwieniecki and Holbrook (2000). On the basis of careful microscopy observations, it was concluded that the resulting air-water interface could stabilize a pressure difference between 0.07 and 0.3 MPa, a magnitude sufficient to account for transient hydraulic isolation of fluids under positive pressure in the refilling conduit from nearby compartments under tension. The partially hydrophobic behavior of pit apertures would be supported by immunolabelling experiments showing that lignin is apparently abundant in pit membranes (Herbette et al. 2015), and this finding would support a role for pit membranes as "valve-like" structures transiently isolating refilling conduits from nearby compartments under tension. In conifers, the valve-mechanism of pits would also provide hydraulic isolation of refilling conduits (Mayr et al. 2014a, b). In any case, such isolation mechanisms would require hydraulic reconnection between refilled and functioning conduits to be simultaneous at the level of all pit apertures on the xylem conduit wall surface, and how this could be achieved still deserves a convincing explanation. Indeed, Brodersen et al. (2010) reported microCT observations of grapevine vessels that apparently failed to refill with water, or that were progressively emptied of water despite evidence of co-occurring refilling in the same conduit. These observations are not consistent with the concept of hydraulic isolation of refilling conduits and further studies are required to solve this critical issue of the refilling mechanism.

5.4 Anatomical and Functional Traits Favoring Embolism Reversal

In a recent survey of studies reporting evidence for embolism repair (or lack of) in different species, Brodersen and McElrone (2013) highlighted that xylem refilling under tension has been most frequently reported for species displaying abundant parenchyma fraction in their wood, with special reference to axial parenchyma. Although this output might be affected by the fact that experimental evidence for lack of refilling is less frequently reported than evidence supporting this phenomenon, the occurrence of refilling in species with abundant xylem parenchyma would be consistent with the proposed role of conduit-associated cells in the process. Vascular plants display considerable variation in the amount of parenchyma in their wood. In a recent global analysis, Morris et al. (2016) have reported a 29-fold variation in parenchyma fraction across 1,439 species of seed plants. This survey revealed several interesting trends, including a negative correlation between parenchyma fraction and mean annual precipitation. This would suggest that anatomical traits hypothetically favoring recovery from xylem embolism are most common in species growing in habitats where risks of drought-induced hydraulic failure are possibly higher. Another recent study (Plavcová et al. 2016) also highlighted a positive correlation between xylem parenchyma fraction and concentration of nonstructural carbohydrates in the wood. These findings support the hypothesis that living cells are important for embolism repair as sources of organic solutes (sugars) to generate positive osmotic pressure in refilling conduits. This conclusion is indirectly supported by the negative relationships between xylem recovery capacity and wood density (the ratio between wood dry weight and volume) reported by Ogasa et al. (2013) among seven temperate trees as well as by Trifilò et al. (2015) among nine dry forest trees. In fact, wood density is inversely correlated with wood capacitance (the amount of water released by the wood per unit change in water potential) (Scholz et al. 2007; Meinzer et al. 2008; Trifilò et al. 2015). In turn, wood capacitance might be interpreted as an indirect estimate of the amount of living cells in the wood, so that the reported relationships between wood capacitance and refilling efficiency might reflect the crucial role of sufficient availability of soluble sugars to fuel the osmotically based repair mechanism.

The above hypotheses around the anatomical and functional determinants of plants' capacity to reverse xylem embolism are supported by a comparison of Angiosperms versus Gymnosperms. Embolism reversal has been more frequently reported for the former group than for the latter, although some conifers are known to perform significant xylem refilling even under partial tension (Mayr et al. 2014b; Klein et al. 2016). Johnson et al. (2012) reported that mean parenchyma area fraction in seven different Angiosperms averaged 25%, while in seven different Gymnosperms this value was only 8%. Such anatomical differences translated into a twofold to eightfold higher NSC content in Angiosperm branches compared to Gymnosperms. This might explain the lack of evidence for xylem refilling in several Gymnosperms and would also provide a possible interpretation for different safety margins recorded in the two groups. In fact, the difference between minimum seasonal Ψ_{xyl} values and critical tension inducing 50% (P_{50}) or 88% (P_{88}) loss of hydraulic conductance has been reported to be significantly larger in Gymnosperms than in Angiosperms (Choat et al. 2012; Johnson et al. 2012). This large safety margins would be fundamental for Gymnosperms in order to preserve the integrity of their water transport system, due to the apparently limited ability to reverse embolism over the short term. On the other hand, the large volume of wood occupied by parenchyma, and the resulting high NSC content, would provide Angiosperms with a mean to generate the forces necessary to reverse embolism, thus making feasible a more "risky" hydraulic behavior to maximize gas exchange and photosynthetic gain even when facing the possibility of embolism appearance.

6 The Possible Role of Embolism Repair in Plant Resistance/Resilience to Drought Stress: Links with Plant Carbon Status

Embolism accumulation and hydraulic failure have been reported as major causes of tree mortality and forest decline under anomalous drought and heat waves over recent decades. Yet, embolism accumulation is not always lethal. As an example, Nardini et al. (2013) reported that PLC of trees of different species suffering extensive die-back during an extreme summer drought was as high as 80–90%. At the same time, conspecific individuals still apparently healthy under the same conditions had only slightly lower PLC, around 70%. This means that from a hydraulic perspective trees can survive a single extreme drought episode, provided PLC does not surpass critical thresholds (Urli et al. 2013). The fate of trees surviving extreme drought is less clear. Repeated drought events might lead to accumulation of embolism, finally leading to lethal hydraulic failure. It has also been suggested that repeated drought events might increase the tree vulnerability to xylem embolism, thus inducing a progressive hydraulic deterioration (Hacke et al. 2001; Anderegg et al. 2013b; Savi et al. 2015; Hillabrand et al. 2016).

However, some tree species or individuals might also display significant resilience to drought stress, provided they can recover hydraulic functionality after stress relief, and hence restore photosynthetic carbon gain. An example will illustrate the importance of considering resilience mechanisms, and not only drought tolerance or resistance, in explaining and forecasting the impact of severe drought on different tree species. During summer 2012, a severe drought occurred in SE-Europe, and specifically in the Mediterranean basin (Sippel and Otto 2014). Crown desiccation and die-back were recorded in several native species (Nardini et al. 2013), and also impacted some alien species. In particular, several individuals of *Robina pseudoacacia* and *Ailanthus altissima* showed severe wilting signs. However, at the end of the drought period *R. pseudoacacia* did not recover and extensive branch die-back was observed over following years in this species. On the other hand, *A. altissima* fully recovered from the stress event, and only minor signs of die-back were observed in individuals of this species over following years (personal observations). These observations might suggest different levels of vulnerability to xylem embolism in the two species. Indeed, this is not the case, as a recent study investigating the responses of these species to a prolonged controlled drought (Savi et al. 2016b) revealed that both *R. pseudoacacia* and *A. altissima* suffered high embolism levels at comparable xylem tension values reached at the end of the experimental treatment. However, upon re-irrigation embolism levels remained very high in *R. pseudoacacia*, but returned to prestress values in *A. altissima*, again despite similar recovery of Ψ_{xyl} in the two species. Hence, the differential impact of the extreme summer 2012 drought on these two species was likely not due to different embolism resistance levels, but to different resilience mechanisms, with *A. altissima* being highly resilient to drought stress by efficient embolism repair, and *R. pseudoacacia* being unable to reverse embolism, in accordance with a previous study on the same species (Wang et al. 2014).

Savi et al. (2016b) also showed that the different magnitude of embolism recovery across four different species, including *R. pseudoacacia* and *A. altissima*, was correlated with the amount of soluble carbohydrates available at the stem level at the end of the drought period, consistent with the proposed role of sugars in embolism reversal processes. Some simple calculations and theoretical considerations based on the findings of Savi et al. (2016b) will help to further highlight this important point. At the end of the imposed drought stress, total stem NSC were approximately 10 and 50 mg g^{-1} in *R. pseudoacacia* and *A. altissima*, respectively. The PLC suffered by the two species was close to 80%, and both species experienced minimum xylem tension of about -1.5 MPa at the peak of drought. Taking into account that xylem conduits lumina occupy approximately 5% of stem volume in saplings used for experiments (unpublished data), and that wood density was about 0.45 g cm^{-3}, it is possible to calculate the amount of NSC necessary to generate an osmotic potential negative enough to counterbalance the residual negative Ψ_{xyl} following post-drought irrigation, thus making possible embolism repair. Both *R. pseudoacacia* and *A. altissima* recovered to a maximum Ψ_{xyl} of −0.5 MPa after irrigation. Let's assume a stem segment with a diameter of 8 mm and total length of 40 mm, for a total volume of 2 cm^3, representative of the size of samples in Savi et al. (2016b). In order to generate an osmotic potential of at least −0.6 MPa in the conduit volume corresponding to PLC of 80%, approximately 4 mg g^{-1} of glucose would be required. Even assuming that all NSC can be converted into glucose, experimental data and calculations suggest that *R. pseudoacacia* should invest about 50% of residual carbohydrate reserves to reverse embolism. Although it is unknown which percentage of carbohydrate reserves can be mobilized to fuel refilling, this value represents a very high share of residual NSC in *R. pseudoacacia*, and in fact PLC remained at high levels following irrigation in this species. In the case of *A. altissima*, the amount of glucose required to osmotically reverse xylem embolism would correspond to only about 7% of total residual NSC, and this would explain the prompt embolism recovery in this species. Clearly, the minimum amount of NSC theoretically required to energize complete embolism reversal will depend on cumulated PLC at the end of drought, and maximum Ψ_{xyl} following stress relief and plant rehydration. Figure 3 illustrates different scenarios for stem segments with the size reported above, and again assuming that 5% of stem volume is occupied by conduit lumina.

Of course, the scenario depicted in Fig. 3 is an over-simplification. First of all, it is unlikely that all residual NSC can be converted into simple sugars and released into refilling conduits. Moreover, metabolic impairment could prevent NSC usage

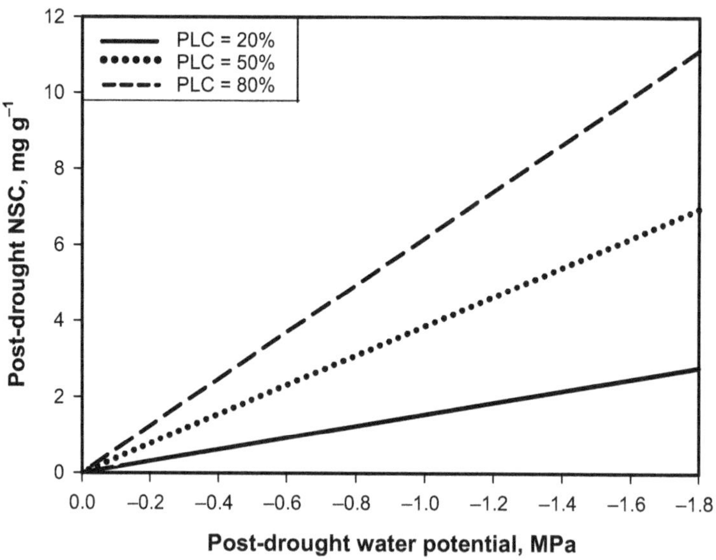

Fig. 3 Minimum residual concentration of stem nonstructural carbohydrates following drought exposure, theoretically required to generate sufficient osmotic forces to reverse a PLC of 20% (*solid line*), 50% (*dotted line*), or 80% (*dashed line*) at different rehydration levels, as indicated by different post-drought xylem water pressure values (see text for details)

even if sufficiently high NSC values are recorded at the end of drought. Finally, the necessary solutes might be provided by inorganic ions or even other NSC storage pools, like roots, rather than stems, provided that phloem transport is not impaired as well. Nonetheless, these calculations and Fig. 3 illustrate the importance of correlations between carbon and water metabolism in a scenario of drought resilience based on reversal of xylem embolism. In particular, Fig. 3 suggests that resilience mechanisms are more likely to occur in species with high stem NSC content, i.e., those having high wood parenchyma fraction (Plavcová et al. 2016), in accordance with conclusions by Brodersen and McElrone (2013).

Plants facing prolonged drought stress are exposed not only to the risk of embolism and hydraulic failure, but also suffer stomatal closure and reduced photosynthetic gain. In order to maintain basic metabolism, NSC reserves have to be used, progressively reducing the amount of carbohydrates potentially available for embolism repair processes. Short and intense drought events, leading to high embolism levels but not causing important NSC depletion, could be faced by plants by adopting a resilience mechanism based on embolism reversal that could theoretically operate even under conditions of substantial residual xylem tension (Fig. 3). This could possibly explain patterns of embolism formation during the day and repair during the night in plants exposed to diurnal dynamic drop of Ψ_{xyl}, as these plants would not face important carbohydrate limitation. However, prolonged drought stress leading to significant carbohydrate depletion or metabolic impairment limiting the ability to access and mobilize NSC reserves would result in inhibition of the refilling process when NSC content drops below the critical threshold (Fig. 3). Under this scenario, the likelihood of embolism reversal would finally depend on PLC levels, residual NSC, and post-drought Ψ_{xyl}. Hence, plant resistance and resilience to severe drought would result from the interplay of water and carbon metabolism, and this would possibly explain contrasting reports about the occurrence and magnitude of embolism reversal even in the same species in different seasons, environmental conditions, and laboratories. In fact, any factor affecting carbon gain and storage, like light, temperature, nutrients, exposure to previous biotic and abiotic stress factors, will likely impact the size and availability of NSC pools, thus affecting the embolism repair capability of the plant. We recommend that future studies addressing the topic of embolism repair be coupled to careful analysis of xylem anatomy (conduit volume and parenchyma fractions) and to reliable estimates of total NSC before, during, and after drought stress. This set of hypotheses and recommendations provides a theoretical and methodological framework for future experiments aimed at improving our capacity to predict not only species-specific vulnerability to drought stress, but also species- and individual-specific chances of resilience to drought events.

7 Conclusions

More than 20 years after the publication of the most detailed early report of embolism reversal in stems of Laurel experiencing still negative Ψ_{xyl} in functioning conduits (Salleo et al. 1996), this phenomenon is still awaiting the definition of a comprehensive mechanistic scenario to explain all the tricky processes and conditions required to repair this important injury to the water transport system of plants. Further evidence around the occurrence of embolism reversal has accumulated in recent years, and the role of nonstructural carbohydrate and plant carbon status is emerging as a critical issue, that we have only started to understand in its physiological and quantitative aspects. Recent criticisms and challenges to the current hypotheses about xylem refilling are stimulating, and correctly call for generation of new experimental evidence supporting or contrasting current paradigms of embolism recovery (Cochard et al. 2015; Nardini et al. 2017). Hopefully, new imaging techniques like microCT applied to model plants will help to solve some of the basic physiological questions around the mechanisms allowing some plants to pressurize water in individual xylem conduits while negative pressures are co-existing in neighboring ones. At the same time, enlarging the efforts to collect field measurements on the dynamics of embolism formation/repair in woody crops and forest trees from different species and biomes will likely allow to increase our understanding of the ecological correlates and adaptive values of embolism repair as a resilience strategy adopted by plants to heal the injuries of severe droughts.

References

Adams HD, Guardiola-Claramonte M, Barron-Gafford GA, Villegas JC, Breshears DD, Zou CB, Troch PA, Huxman TE (2009) Temperature sensitivity of drought-induced tree mortality portends increased regional die-off under global-change-type drought. Proc Natl Acad Sci U S A 106:7063–7066

Adams HD, Germino MJ, Breshears DD, Barron-Gafford BA, Guardiola-Claramonte M, Zou CB, Huxman TE (2013) Nonstructural leaf carbohydrate dynamics of *Pinus edulis* during drought-induced tree mortality reveal role for carbon metabolism in mortality mechanism. New Phytol 197:1142–1151

Ahmadi A, Baker DA (2001) The effect of water stress on the activities of key regulatory enzymes of the sucrose to starch pathway in wheat. Plant Growth Regul 35:81–91

Alder NN, Pockman WT, Sperry JS, Nuismer S (1997) Use of centrifugal force in the study of xylem cavitation. J Exp Bot 48:665–674

Allen CD, Macalady AK, Chenchouni H, Bachelet D, McDowell NG, Vennetier M, Kitzberger T, Rigling A, Breshears DD, Hogg EH et al (2010) A global overview of drought and heath-induced tree mortality reveals emerging climate change risks for forests. For Ecol Manage 259:660–684

Allen CD, Breshears DD, McDowell NG (2015) On underestimation of global vulnerability to tree mortality and forest die-off from hotter drought in Anthropocene. Ecosphere 6:129

Améglio T, Ewers FW, Cochard H, Martignac M, Vandame M, Bodet C, Cruiziat P (2001) Winter stem xylem pressure in walnut trees: effects of carbohydrates, cooling and freezing. Tree Physiol 21:387–394

Améglio T, Bodet C, Lacointe A, Cochard H (2002) Winter embolism, mechanisms of xylem hydraulic conductivity recovery and springtime growth patterns in walnut and peach trees. Tree Physiol 22:1211–1220

Améglio T, Decourteix M, Alves G, Valentin V, Sakr S, Julien JL, Petel G, Guilliot A, Lacointe A (2004) Temperature effects on xylem sap osmolarity in walnut trees: evidence for a vitalistic model of winter embolism repair. Tree Physiol 24:785–793

Anderegg WRL, Kane JM, Anderegg LDL (2013a) Consequences of widespread tree mortality triggered by drought and temperature stress. Nat Clim Chang 3:30–36

Anderegg WRL, Plavcová L, Anderegg LDL, Hacke UG, Berry JA, Field CB (2013b) Drought's legacy: multiyear hydraulic deterioration underlies widespread aspen forest die-off and portends increased future risk. Glob Chang Biol 19:1188–1196

Anderegg WRL, Flint A, Huang C, Flint L, Berry JA, Davis FW, Sperry JS, Field CB (2015) Tree mortality predicted from drought-induced vascular damage. Nat Geosci 8:367–371

Anderegg WRL, Klein T, Bartlett M, Sack L, Pellegrini AFA, Choat B, Jansen S (2016) Meta-analysis reveals that hydraulic traits explain cross-species patterns of drought-induced tree mortality across the globe. Proc Natl Acad Sci U S A 113:5024–5029

Angeles G, Bond B, Boyer JS, Brodribb T, Brooks JR, Burns MJ, Cavender-Bares J, Clearwater M, Cochard H, Comstock J et al (2004) The cohesion-tension theory. New Phytol 163:451–452

Ayres MP, Lombardero MJ (2000) Assessing the consequences of global change for forest disturbance from herbivores and pathogens. Sci Total Environ 262:263–286

Barigah TS, Charrier O, Douris M, Bonhomme M, Herbette S, Améglio T, Fichot R, Brignolas F, Cochard H (2013) Water stress-induced xylem hydraulic failure is a causal factor of tree mortality in beech and poplar. Ann Bot 112:1431–1437

Beikircher B, Mayr S (2015) Avoidance of harvesting and sampling artefacts in hydraulic analyses: a protocol tested on *Malus domestica*. Tree Physiol 36:797–803

Bouche PS, Delzon S, Choat B, Badel E, Brodribb TJ, Burlett R, Cochard H, Charra-Vaskou K, Lavigne B, Li S et al (2016) Are needles of *Pinus pinaster* more vulnerable to xylem embolism than branches? New insights from X-ray computed tomography. Plant Cell Environ 39:860–870

Briggs LJ (1950) Limiting negative pressure of water. J Appl Phys 21:721–722

Brodersen CR, McElrone AJ (2013) Maintenance of xylem network transport capacity: a review of embolism repair in vascular plants. Front Plant Sci 4:108

Brodersen CR, McElrone AJ, Choat B, Matthews MA, Shackel KA (2010) The dynamics of embolism repair in xylem: *in vivo* visualizations using high-resolution computed tomography. Plant Physiol 154:1088–1095

Brodribb TJ (2009) Xylem hydraulic physiology: the functional backbone of terrestrial plant productivity. Plant Sci 177:245–251

Brodribb TJ, Bowman DJMS, Nichols S, Delzon S, Burlett R (2010) Xylem function and growth rate interact to determine recovery rates after exposure to extreme water deficit. New Phytol 188:533–542

Bucci SJ, Scholz FG, Goldstein G, Meinzer FC, Sternberg SL (2003) Dynamic changes in hydraulic conductivity in petioles of two savanna tree species: factors and mechanisms contributing to the refilling of embolized vessels. Plant Cell Environ 26:1633–1645

Cailleret M, Jansen S, Robert EMR, Desoto L, Aakala T, Antos JA, Beikircher B, Bigler C, Bugmann H, Caccianiga M et al (2017) A synthesis of radial growth patterns preceding tree mortality. Glob Chang Biol 23:1675–1690

Canny MJ (1997) Vessel contents during transpiration – embolisms and refilling. Am J Bot 84:1223–1230

Canny MJ (1998) Applications of the compensating pressure theory of water transport. Am J Bot 85:897–909

Canny MJ, Huang CX, McCully ME (2001) The cohesion theory debate continues. Trends Plant Sci 6:454–455

Carbone MS, Czimczik CI, Keenan TF, Murakami PF, Pederson N, Schaberg PG, Xu X, Richardson AD (2013) Age, allocation and availability of nonstructural carbon in mature red maple trees. New Phytol 200:1145–1155

Carnicer J, Coll M, Ninyerola M, Pons X, Sánchez J, Peñuelas J (2011) Widespread crown condition decline, food web disruption, and amplified tree mortality with increased climate change-type drought. Proc Natl Acad Sci U S A 108:1474–1478

Caupin F, Herbert E (2006) Cavitation in water: a review. C R Phys 7:1000–1017

Charra-Vaskou K, Badel E, Burlett R, Cochard H, Delzon S, Mayr S (2012) Hydraulic efficiency and safety of vascular and non-vascular components in *Pinus pinaster* leaves. Tree Physiol 32:1161–1170

Chaumont F, Tyerman SD (2014) Aquaporins: highly regulated channels controlling plant water relations. Plant Physiol 164:1600–1618

Chitarra W, Balestrini R, Vitali M, Pagliarani C, Perrone I, Schubert A, Lovisolo C (2014) Gene expression in vessel-associated cells upon xylem embolism repair in *Vitis vinifera* L. petioles. Planta 239:887–899

Choat B, Ball M, Luly J, Holtum J (2003) Pit membrane porosity and water stress-induced cavitation in four co-existing dry rainforest tree species. Plant Physiol 131:41–48

Choat B, Lahr EC, Melcher PJ, Zwieniecki MA, Holbrook NM (2005) The spatial pattern of air seeding thresholds in mature sugar maple trees. Plant Cell Environ 28:1082–1089

Choat B, Cobb A, Jansen S (2008) Structure and function of bordered pits: new discoveries and impacts on whole plant hydraulic function. New Phytol 177:608–626

Choat B, Jansen S, Brodribb TJ, Cochard H, Delzon S, Bhaskar R, Bucci SJ, Feild TS, Gleason SM, Hacke UG et al (2012) Global convergence in the vulnerability of forests to drought. Nature 491:752–755

Choat B, Brodersen CR, McElrone AJ (2015) Synchrotron X-ray microtomography of xylem embolism in *Sequoia sempervirens* saplings during cycles of drought and recovery. New Phytol 205:1095–1105

Choat B, Badel E, Burlett R, Delzon S, Cochard H, Jansen S (2016) Noninvasive measurement of vulnerability to drought-induced embolism by X-ray microtomography. Plant Physiol 170:273–282

Christman MA, Sperry JS, Smith DD (2012) Rare pits, large vessels and extreme vulnerability to cavitation in a ring-porous tree species. New Phytol 193:713–720

Cochard H (2002) A technique for measuring xylem hydraulic conductance under high negative pressures. Plant Cell Environ 25:815–819

Cochard H, Delzon S (2013) Hydraulic failure and repair are not routine in trees. Ann For Sci 70:659–661

Cochard H, Cruiziat P, Tyree MT (1992) Use of positive pressures to establish vulnerability curves. Further support for the air-seedling hypothesis and implications for pressure-volume analysis. Plant Physiol 100:205–209

Cochard H, Bodet C, Améglio T, Cruiziat P (2000) Cryo-scanning electron microscopy observations of vessel contents during transpiration in walnut petioles. Facts or artifacts? Plant Physiol 124:1191–1202

Cochard H, Herbette S, Hernández E, Hölttä T, Mencuccini M (2010) The effects of sap ionic composition on xylem vulnerability to cavitation. J Exp Bot 61:275–285

Cochard H, Badel E, Herbette S, Delzon S, Choat B, Jansen S (2013) Methods for measuring plant vulnerability to cavitation: a critical review. J Exp Bot 64:4779–4791

Cochard H, Delzon S, Badel E (2015) X-ray microtomography (micro-CT): a reference technology for high-resolution quantification of xylem embolism in trees. Plant Cell Environ 38:201–206

Comstock JP (1999) Why Canny's theory doesn't hold water. Am J Bot 86:1077–1081

Crum LA (1979) Tensile strength of water. Nature 278:148–149

Davis SD, Ewers FW, Sperry JS, Portwood KA, Crocker MC, Adams GC (2002) Shoot dieback during prolonged drought in *Ceanothus* (Rhamnaceae) chaparral of California: a possible case of hydraulic failure. Am J Bot 89:820–828

Diffenbaugh NS, Swain DL, Touma D (2015) Anthropogenic warming has increased drought risk in California. Proc Natl Acad Sci U S A 112:3931–3936

Dixon HH (1914) Transpiration and the ascent of sap in plants. Macmillan, London

Dixon HH, Joly J (1894) On the ascent of sap. Philos Trans R Soc Lond 186:563–576

Domec JC, Scholz FG, Bucci SJ, Meinzer FC, Goldstein G, Villalobos-Vega R (2006) Diurnal and seasonal variation in root xylem embolism in neotropical savanna woody species: impact on stomatal control of plant water status. Plant Cell Environ 29:26–35

Easterling DR, Meehl GA, Parmesan C, Changnon SA, Karl TR, Mearns LO (2000) Climate extremes: observations, modeling, and impacts. Science 289:2068–2074

Ennajeh M, Simões F, Khemira H, Cochard H (2011) How reliable is the double-ended pressure sleeve technique for assessing xylem vulnerability to cavitation in woody angiosperms? Physiol Plant 142:205–210

Enns LC, Canny MJ, McCully ME (2000) An investigation of the role of solutes in the xylem sap and in the xylem parenchyma as the source of root pressure. Protoplasma 211:183–197

Ewers FW, Cochard H, Tyree MT (1997) A survey of root pressures in vines of a tropical lowland forest. Oecologia 110:191–196

Fisher JB, Angeles G, Ewers FW, Lopez Portillo J (1997) Survey of root pressures in tropical vines and woody species. Int J Plant Sci 158:44–50

Galvez DA, Landhäusser SM, Tyree MT (2013) Low root reserve accumulation during drought may lead to winter mortality in poplar seedlings. New Phytol 198:139–148

Ganthaler A, Mayr S (2015) Dwarf shrub hydraulics: two *Vaccinium* species (*Vaccinium myrtillus*, *Vaccinium vitis-idaea*) of the European Alps compared. Physiol Plant 155:424–434

Gascò A, Nardini A, Gortan E, Salleo S (2006) Ion-mediated increase in the hydraulic conductivity of Laurel stems: role of pits and consequences for the impact of cavitation on water transport. Plant Cell Environ 29:1946–1955

Gascò A, Salleo S, Gortan E, Nardini A (2007) Seasonal changes in the ion-mediated increase of xylem hydraulic conductivity in stems of three evergreens: any functional role? Physiol Plant 129:597–606

Gaylord ML, Kolb TE, Pockman WT, Plaut JA, Yepez EA, Macalady AK, Pangle RE, McDowell NG (2013) Drought predisposes piñon-juniper woodlands to insect attacks and mortality. New Phytol 198:567–578

Giuggiola A, Bugmann H, Zingg A, Dobbertin M, Rigling A (2013) Reduction of stand density increases drought resistance in xeric Scots pine forests. For Ecol Manage 310:827–835

Gleason SM, Westoby M, Jansen S, Choat B, Hacke UG, Pratt RB, Bhaskar R, Brodribb TJ, Bucci SJ, Cao KF et al (2016) Weak tradeoff between xylem safety and xylem-specific hydraulic efficiency across the world's woody plant species. New Phytol 209:123–136

Gortan E, Nardini A, Salleo S, Jansen S (2011) Pit membrane chemistry influences the magnitude of ion-mediated enhancement of xylem hydraulic conductivity in four Lauraceae. Tree Physiol 31:48–58

Gruber A, Pirkebner D, Florian C, Oberhuber W (2012) No evidence for depletion of carbohydrate pools in Scots pine (*Pinus sylvestris* L.) under drought stress. Plant Biol 14:142–148

Hacke UG, Sauter JJ (1996) Xylem dysfunction during winter and recovery of hydraulic conductivity in diffuse-porous and ring-porous trees. Oecologia 105:435–439

Hacke UG, Sperry JS (2003) Limits to xylem refilling under negative pressure in *Laurus nobilis* and *Acer negundo*. Plant Cell Environ 26:303–311

Hacke UG, Stiller V, Sperry JS, Pittermann J, McCulloh KA (2001) Cavitation fatigue. Embolism and refilling cycles can weaken the cavitation resistance of xylem. Plant Physiol 125:779–786

Hacke UG, Venturas MD, MacKinnon MD, Jacobsen AL, Sperry JS, Pratt RB (2015) The standard centrifuge method accurately measures vulnerability curves of long-vesselled olive stems. New Phytol 205:116–127

Hanewinkel M, Cullmann DA, Schelhaas MJ, Nabuurs GJ, Zimmermann NE (2013) Climate change may cause severe loss in the economic value of European forest land. Nat Clim Change 3:203–207

Hartmann H, Trumbore S (2016) Understanding the roles of nonstructural carbohydrates in forest trees – from what we can measure to what we want to know. New Phytol 211:386–403

Hartmann H, Ziegler W, Kolle O, Trumbore S (2013) Thirst beats hunger – declining hydration during drought prevents carbon starvation in Norway spruce saplings. New Phytol 200:340–349

He B, Cui X, Wang H, Chen A (2014) Drought: the most important physical stress of terrestrial ecosystems. Acta Ecol Sin 34:179–183

Herbette S, Bouchet B, Brunel N, Bonnin E, Cochard H, Guillon F (2015) Immunolabelling of intervessel pits for polysaccharides and lignin helps in understanding their hydraulic properties in *Populus tremula* x *alba*. Ann Bot 115:187–199

Hillabrand RM, Hacke UG, Lieffers VJ (2016) Drought-induced xylem pit membrane damage in aspen and balsam poplar. Plant Cell Environ 39:2210–2220

Hoch G (2015) Carbon reserves as indicators for carbon limitation in trees. Prog Bot 76:321–346

Holbrook NM, Zwieniecki MA (1999) Xylem refilling under tension. Do we need a miracle? Plant Physiol 120:7–10

Holbrook NM, Ahrens ET, Burns MJ, Zwieniecki MA (2001) *In vivo* observation of cavitation and embolism repair using magnetic resonance imaging. Plant Physiol 126:27–31

Hölttä T, Vesala T, Nikinmaa E, Perämäki M, Siivola E, Mencuccini M (2005) Field measurements of ultrasonic acoustic emissions and stem diameter variations. New insight into the relationship between xylem tensions and embolism. Tree Physiol 25:237–243

Hölttä T, Vesala T, Perämäki M, Nikinmaa E (2006) Refilling of embolised conduits as a consequence of 'Münch water' circulation. Funct Plant Biol 33:949–959

Hukin D, Cochard H, Dreyer E, Le Thiec D, Bogeat-Triboulot MB (2005) Cavitation vulnerability in roots and shoots: does *Populus euphratica* Oliv., a poplar from arid areas of Central Asia, differ from other poplar species? J Exp Bot 56:2003–2010

Jacobsen AL, Pratt RB, Tobin MF, Hacke UG, Ewers FW (2012) A global analysis of xylem vessel length in woody plants. Am J Bot 99:1583–1591

Jactel H, Petit J, Desprez-Loustau ML, Delzon S, Piou D, Battisti A, Koricheva J (2012) Drought effects on damage by forest insects and pathogens: a meta-analysis. Glob Chang Biol 18:267–276

Jansen S, Gortan E, Lens F, Lo Gullo MA, Salleo S, Scholz A, Stein A, Trifilò P, Nardini A (2011) Do quantitative vessel and pit characters account for ion-mediated changes in the hydraulic conductance of angiosperm xylem? New Phytol 189:218–228

Jansen S, Schuldt B, Choat B (2015) Current controversies and challenges in applying plant hydraulic techniques. New Phytol 205:961–964

Jaquish LL, Ewers FW (2001) Seasonal conductivity and embolism in the roots and stems of two clonal ring-porous trees, *Sassafras albidum* (Lauraceae) and *Rhus typhina* (Anacardiaceae). Am J Bot 88:206–212

Johnson DM, Meinzer FC, Woodruff DR, McCulloh KA (2009) Leaf xylem embolism, detected acoustically and by cryo-SEM, corresponds to decreases in leaf hydraulic conductance in four evergreen species. Plant Cell Environ 32:828–836

Johnson DM, McCulloh KA, Woodruff DR, Meinzer FC (2012) Hydraulic safety margins and embolism reversal in stems and leaves: why are conifers and angiosperms so different? Plant Sci 195:48–53

Kikuta SB, Lo Gullo MA, Nardini A, Richter H, Salleo S (1997) Ultrasound acoustic emissions from dehydrating leaves of deciduous and evergreen trees. Plant Cell Environ 20:1381–1390

Klein T (2014) The variability of stomatal sensitivity to leaf water potential across tree species indicates a continuum between isohydric and anisohydric behaviours. Funct Ecol 28:1313–1320

Klein T, Cohen S, Paudel I, Preisler Y, Rotenberg E, Yakir D (2016) Diurnal dynamics of water transport, storage and hydraulic conductivity in pine trees under seasonal drought. iForest 9:710–719

Knipfer T, Brodersen CR, Zedan A, Kluepfel DA, McElrone AJ (2015a) Patterns of drought-induced embolism formation and spread in living walnut saplings visualized using X-ray microtomography. Tree Physiol 35:744–755

Knipfer T, Eustis A, Brodersen CR, Walker AM, McElrone AJ (2015b) Grapevine species from varied native habitats exhibit differences in embolism formation/repair associated with leaf gas exchange and root pressure. Plant Cell Environ 38:1503–1513

Knipfer T, Cuneo IF, Brodersen CR, McElrone AJ (2016) *In situ* visualization of the dynamics in xylem embolism formation and removal in the absence of root pressure: a study on excised grapevine stems. Plant Physiol 171:1024–1036

Kramer PJ (1983) Water relations of plants. Academic Press, London

Laur J, Hacke UG (2014) Exploring *Picea glauca* aquaporins in the context of needle water uptake and xylem refilling. New Phytol 203:388–400

Lee SJ, Kim Y (2008) *In vivo* visualization of the water-refilling process in xylem vessels using X-ray micro-imaging. Ann Bot 101:595–602

Lee SJ, Hwang BG, Kim HK (2013) Hydraulic characteristics of water-refilling process in excised roots of *Arabidopsis*. Planta 238:307–315

Leng H, Lu M, Wan X (2013) Variation in embolism occurrence and repair along the stem in drought-stressed and re-watered seedlings of a poplar clone. Physiol Plant 147:329–339

Lesk C, Rowhani P, Ramankutty N (2016) Influence of extreme weather disasters on global crop production. Nature 529:84–87

Li Y, Sperry JS, Taneda H, Bush SE, Hacke UG (2008) Evaluation of centrifugal methods for measuring xylem cavitation in conifers, diffuse- and ring-porous angiosperms. New Phytol 177:558–568

López R, Cano FJ, Choat B, Cochard H, Gil L (2016) Plasticity in vulnerability to cavitation of *Pinus canariensis* occurs only at the driest end of an aridity gradient. Front Plant Sci 7:769

Ma Z, Peng C, Zhu Q, Chen H, Yu G, Li W, Zhou X, Wang W, Zhang W (2012) Regional drought-induced reduction in the biomass carbon sink of Canada's boreal forests. Proc Natl Acad Sci U S A 109:2423–2427

Maguire AJ, Kobe RK (2015) Drought and shade deplete nonstructural carbohydrate reserves in seedlings of five temperate tree species. Ecol Evol 5:5711–5721

Martorell S, Diaz-Espejo A, Medrano H, Ball MC, Choat B (2014) Rapid hydraulic recovery in *Eucalyptus pauciflora* after drought: linkages between stem hydraulics and leaf gas exchange. Plant Cell Environ 37:617–626

Mayr S, Améglio T (2016) Freezing stress in tree xylem. Prog Bot 77:381–414

Mayr S, Sperry JS (2010) Freeze-thaw-induced embolism in *Pinus contorta*: centrifuge experiments validate the "thaw-expansion hypothesis" but conflict with ultrasonic emission data. New Phytol 185:1016–1024

Mayr S, Kartusch B, Kikuta S (2014a) Evidence for air-seeding: watching the formation of embolism in conifer xylem. J Plant Hydraul 1:e0004

Mayr S, Schmid P, Laur J, Rosner S, Charra-Vaskou K, Dämon B, Hacke UG (2014b) Uptake of water via branches helps timberline conifers refill embolized xylem in late winter. Plant Physiol 164:1731–1740

McCully ME, Huang CX, Ling LE (1998) Daily embolism and refilling of xylem vessels in the roots of field-grown maize. New Phytol 138:327–342

McDowell NG (2011) Mechanisms linking drought, hydraulics, carbon metabolism, and vegetation mortality. Plant Physiol 155:1051–1059

McDowell NG, Pockman WT, Allen CD, Breshears DD, Cobb N, Kolb T, Plaut J, Sperry J, West A, Williams DG et al (2008) Mechanisms of plant survival and mortality during drought: why do some plants survive while others succumb to drought? New Phytol 178:719–739

McElrone AJ, Brodersen CR, Alsina MM, Drayton WM, Matthews MA, Shackel KA, Wada H, Zufferey V, Choat B (2012) Centrifuge technique consistently overestimates vulnerability to water stress-induced cavitation in grapevines as confirmed with high-resolution computed tomography. New Phytol 196:661–665

Meinzer FC, Campanello PI, Domec JC, Gatti MG, Goldstein G, Villalobos-Vega R, Woodruff DR (2008) Constraints on physiological function associated with branch architecture and wood density in tropical forest trees. Tree Physiol 28:1609–1617

Meisner A, De Deyna GB, de Boer W, van der Putten WH (2013) Soil biotic legacy effects of extreme weather events influence plant invasiveness. Proc Natl Acad Sci U S A 110:9835–9838

Melcher PJ, Zwieniecki MA (2013) Functional analysis of embolism induced by air injection in *Acer rubrum* and *Salix nigra*. Front Plant Sci 4:368

Melcher PJ, Goldstein G, Meinzer FC, Yount DE, Jones TJ, Holbrook NM, Huang CX (2001) Water relations of coastal and estuarine *Rhizophora mangle*: xylem pressure potential and dynamics of embolism formation and repair. Oecologia 126:182–192

Melcher PJ, Holbrook NM, Burns MJ, Zwieniecki MA, Cobb AR, Brodribb TJ, Choat B, Sack L (2012) Measurements of stem xylem hydraulic conductivity in the laboratory and field. Methods Ecol Evol 3:685–694

Mencuccini M, Hölttä T, Sevanto S, Nikinmaa E (2013) Concurrent measurements of change in the bark and xylem diameters of trees reveal a phloem-generated turgor signal. New Phytol 198:1143–1154

Milburn JA (1973) Cavitation studies on whole *Ricinus* plants by acoustic detection. Planta 112:333–342

Minot CS (1885) The formative force of organisms. Science 6:4–6

Moreno-Gutiérrez C, Dawson TE, Nicolás E, Querejeta JI (2012) Isotopes reveal contrasting water use strategies among coexisting plant species in a Mediterranean ecosystem. New Phytol 196:489–496

Morris H, Plavcová L, Cvecko P, Fichtler E, Gillingham MAF, Martínez-Cabrera HI, McGlinn DJ, Wheeler E, Zheng J, Ziemińska K, Jansen S (2016) A global analysis of parenchyma tissue fractions in secondary xylem of seed plants. New Phytol 209:1553–1565

Nardini A, Luglio J (2014) Leaf hydraulic capacity and drought vulnerability: possible trade-offs and correlations with climate across three major biomes. Funct Ecol 28:810–818

Nardini A, Salleo S (2000) Limitation of stomatal conductance by hydraulic traits: sensing or preventing xylem cavitation? Trees 15:14–24

Nardini A, Tyree MT, Salleo S (2001) Xylem cavitation in the leaf of *Prunus laurocerasus* and its impact on leaf hydraulics. Plant Physiol 125:1700–1709

Nardini A, Salleo S, Trifilò P, Lo Gullo MA (2003) Water relations and hydraulic characteristics of three woody species co-occurring in the same habitat. Ann For Sci 60:297–305

Nardini A, Gascò A, Trifilò P, Lo Gullo MA, Salleo S (2007a) Ion-mediated enhancement of xylem hydraulic conductivity is not always suppressed by the presence of Ca^{2+} in the sap. J Exp Bot 58:2609–2615

Nardini A, Gascò A, Cervone F, Salleo S (2007b) Reduced content of homogalacturonan does not alter the ion-mediated increase in xylem hydraulic conductivity in tobacco. Plant Physiol 143:1975–1981

Nardini A, Ramani M, Gortan E, Salleo S (2008) Vein recovery occurs under negative pressure in leaves of sunflower (*Helianthus annuus*). Physiol Plant 133:755–764

Nardini A, Grego F, Trifilò P, Salleo S (2010) Changes of xylem sap ionic content and stem hydraulics in response to irradiance in *Laurus nobilis*. Tree Physiol 30:628–635

Nardini A, Lo Gullo MA, Salleo S (2011a) Refilling embolized xylem conduits: is it a matter of phloem unloading? Plant Sci 180:604–611

Nardini A, Salleo S, Jansen S (2011b) More than just a vulnerable pipeline: xylem physiology in the light of ion-mediated regulation of plant water transport. J Exp Bot 62:4701–4718

Nardini A, Dimasi F, Klepsch M, Jansen S (2012) Ion-mediated enhancement of xylem hydraulic conductivity in four *Acer* species: relationships with ecological and anatomical features. Tree Physiol 32:1434–1441

Nardini A, Battistuzzo M, Savi T (2013) Shoot desiccation and hydraulic failure in temperate woody angiosperms during an extreme summer drought. New Phytol 200:322–329

Nardini A, Lo Gullo MA, Trifilò P, Salleo S (2014) The challenge of the Mediterranean climate to plant hydraulics: responses and adaptations. Environ Exp Bot 103:68–79

Nardini A, Casolo V, Dal Borgo A, Savi T, Stenni B, Bertoncin P, Zini L, McDowell NG (2016) Rooting depth, water relations and non-structural carbohydrate dynamics in three woody angiosperms differentially affected by an extreme summer drought. Plant Cell Environ 39:618–627

Nardini A, Savi T, Losso A, Petit G, Pacilè S, Tromba G, Mayr S, Trifilò P, Lo Gullo MA, Salleo S (2017) X-ray microtomography observations of xylem embolism in stems of *Laurus nobilis* are consistent with hydraulic measurements of percentage loss of conductance. New Phytol 213:1068–1075

Nolf M, Beikircher B, Rosner S, Nolf A, Mayr S (2015) Xylem cavitation resistance can be estimated based on time-dependent rate of acoustic emissions. New Phytol 208:625–632

Nolf M, Lopez R, Peters JMR, Flavel RJ, Koloadin LS, Young IM, Choat B (2017) Visualization of xylem embolism by X-ray microtomography: a direct test against hydraulic measurements. New Phtyol 214:890–898

Oddo E, Inzerillo S, La Bella F, Grisafi F, Salleo S, Nardini A (2011) Short-term effects of potassium fertilization on the hydraulic conductance of *Laurus nobilis* L. Tree Physiol 31:131–138

Oddo E, Inzerillo S, Grisafi F, Sajeva M, Salleo S, Nardini A (2014) Does short-term potassium fertilization improve recovery from drought stress in laurel? Tree Physiol 34:906–913

Oertli JJ (1971) The stability of water under tension in the xylem. Z Pflanzenphysiol 65:195–209

Ogasa MY, Miki NH, Murakami Y, Yoshikawa K (2013) Recovery performance in xylem hydraulic conductivity is correlated with cavitation resistance for temperate deciduous tree species. Tree Physiol 33:335–344

Ogasa MY, Utsumi Y, Miki NH, Yazaki K, Fukuda K (2016) Cutting stems before relaxing xylem tension induces artefacts in *Vitis coignetiae*, as evidenced by magnetic resonance imaging. Plant Cell Environ 39:329–337

Pate JS, Canny MJ (1999) Quantification of vessel embolisms by direct observation: a comparison of two methods. New Phytol 141:33–43

Pereira L, Bittencourt PRL, Oliveira RS, Junior MBM, Barros FV, Ribeiro RV, Mazzafera P (2016) Plant pneumatics: stem air flow is related to embolism – new perspectives on methods in plant hydraulics. New Phytol 211:357–370

Perrone I, Pagliarani C, Lovisolo C, Chitarra W, Roman F, Schubert A (2012) Recovery from water stress affects grape leaf petiole transcriptome. Planta 235:1383–1396

Pickard WF (1981) The ascent of sap in plants. Prog Biophys Mol Biol 37:181–229

Piškur B, Pavlic D, Slippers B, Ogris N, Maresi G, Wingfield MJ, Jurc D (2011) Diversity and pathogenicity of *Botryosphaeriaceae* on declining *Ostrya carpinifolia* in Slovenia and Italy following extreme weather conditions. Eur J For Res 130:235–149

Plavcová L, Hoch G, Morris H, Ghiasi S, Jansen S (2016) The amount of parenchyma and living fibers affects storage of nonstructural carbohydrates in young stems and roots of temperate trees. Am J Bot 103:603–612

Pollastrini M, Feducci M, Bonal D, Fotelli M, Gessler A, Grossiord C, Guyot V, Jactel H, Nguyen D, Radoglou K, Bussotti F (2016) Physiological significance of forest tree defoliation: results from a survey in a mixed forest in Tuscany (central Italy). For Ecol Manage 361:170–178

Pratt RB, MacKinnon ED, Venturas MD, Crous CJ, Jacobsen AL (2015) Root resistance to cavitation is accurately measured using a centrifuge technique. Tree Physiol 35:185–196

Priestley JH (1920) The mechanism of root pressure. New Phytol 19:189–200

Quentin AG, Pinkard EA, Ryan MG, Tissue DT, Baggett LS, Adams HD, Maillard P, Marchand J, Landhäusser SM, Lacointe A et al (2015) Non-structural carbohydrates in woody plants compared among laboratories. Tree Physiol 35:1146–1165

Quero JL, Sterck FJ, Martínez-Vilalta J, Villar R (2011) Water-use strategies of six co-existing Mediterranean woody species during a summer drought. Oecologia 166:45–57

Reichstein M, Bahn M, Ciais P, Frank D, Mahecha MD, Seneviratne SI, Zscheischler J, Beer C, Buchmann N, Frank DC et al (2013) Climate extremes and the carbon cycle. Nature 500:287–295

Rockwell FE, Wheeler JK, Holbrook NM (2014) Cavitation and its discontents: opportunities for resolving current controversies. Plant Physiol 164:1649–1660

Rolland V, Bergstrom DM, Lenné T, Bryant G, Chen H, Wolfe J, Holbrook NM, Stanton DE, Ball MC (2015) Easy come, easy go: capillary forces enable rapid refilling of embolized primary xylem vessels. Plant Physiol 168:1636–1647

Rowland L, da Costa ACL, Galbraith DR, Oliveira RS, Binks OJ, Oliveira AAR, Pullen AM, Doughty CE, Metcalfe DB, Vasconcelos SS et al (2015) Death from drought in tropical forests is triggered by hydraulics not carbon starvation. Nature 528:119–122

Ryu J, Hwang BG, Lee SJ (2016) *In vivo* dynamic analysis of water refilling in embolized xylem vessels of intact *Zea mays* leaves. Ann Bot 118:1033–1042

Sakr S, Alves G, Morillon R, Maurel K, Decourteix M, Guilliot A, Fleurat-Lessard P, Julien JL, Chrispeels MJ (2003) Plasma membrane aquaporins are involved in winter embolism recovery in walnut tree. Plant Physiol 133:630–641

Salleo S, Lo Gullo MA (1989) Xylem cavitation in nodes and internodes of *Vitis vinifera* L. plants subjected to water stress. Limits of restoration of water conduction in cavitated xylem conduits. In: Kreeb KH, Richter H, Hinckley TM (eds) Structural and functional responses to environmental stresses: water shortage. SPB Academic Publishing, The Hague, pp 33–42

Salleo S, Hinckley TM, Kikuta SB, Lo Gullo MA, Weilgony P, Yoon TM, Richter H (1992) A method for inducing xylem embolism *in situ*: experiments with a field-grown tree. Plant Cell Environ 15:491–497

Salleo S, Lo Gullo MA, De Paoli D, Zippo M (1996) Xylem recovery from cavitation-induced embolism in young plants of *Laurus nobilis*: a possible mechanism. New Phytol 132:47–56

Salleo S, Nardini A, Pitt F, Lo Gullo MA (2000) Xylem cavitation and hydraulic control of stomatal conductance in Laurel (*Laurus nobilis* L.) Plant Cell Environ 23:71–79

Salleo S, Lo Gullo MA, Trifilò P, Nardini A (2004) New evidence for a role of vessel-associated cells and phloem in the rapid xylem refilling of cavitated stems of *Laurus nobilis* L. Plant Cell Environ 27:1065–1076

Salleo S, Trifilò P, Lo Gullo MA (2006) Phloem as a possible major determinant of rapid cavitation reversal in stems of *Laurus nobilis* (laurel). Funct Plant Biol 33:1063–1074

Salleo S, Trifilò P, Esposito S, Nardini A, Lo Gullo MA (2009) Starch-to-sugar conversion in wood parenchyma of field-growing *Laurus nobilis* plants: a component of the signal pathway for embolism repair? Funct Plant Biol 36:815–825

Sánchez-Salguero R, Navarro-Cerrillo RM, Swetnam TW, Zavala MA (2012) Is drought the main decline factor at the rear edge of Europe? The case of southern Iberian pine plantations. For Ecol Manage 271:158–169

Savi T, Bertuzzi S, Branca S, Tretiach M, Nardini A (2015) Drought-induced xylem cavitation and hydraulic deterioration: risk factors for urban trees under climate change? New Phytol 205:1106–1116

Savi T, Marin M, Luglio J, Petruzzellis F, Mayr S, Nardini A (2016a) Leaf hydraulic vulnerability protects stem functionality under drought stress in *Salvia officinalis*. Funct Plant Biol 43:370–379

Savi T, Casolo V, Luglio J, Bertuzzi S, Trifilò P, Lo Gullo MA, Nardini A (2016b) Species-specific reversal of stem xylem embolism after a prolonged drought correlates to endpoint concentration of soluble sugars. Plant Physiol Biochem 106:198–207

Scholz FG, Bucci SJ, Goldstein G, Meinzer FC, Franco AC, Miralles-Wilhelm F (2007) Biophysical properties and functional significance of stem water storage tissues in Neotropical savanna trees. Plant Cell Environ 30:236–248

Schreiber SG, Hacke UG, Chamberland S, Lowe CW, Kamelchuk D, Bräutigam K, Campbell MM, Thomas BR (2016) Leaf size serves as a proxy for xylem vulnerability to cavitation in plantation trees. Plant Cell Environ 39:272–281

Schulze ED, Beck E, Müller-Hohenstein K (2005) Plant ecology. Springer, Berlin

Scoffoni C, Sack L (2015) Are leaves 'freewheelin'? Testing for a wheeler-type effect in leaf xylem hydraulic decline. Plant Cell Environ 38:534–543

Secchi F, Zwieniecki MA (2010) Patterns of PIP gene expression in *Populus trichocarpa* during recovery from xylem embolism suggest a major role for the PIP1 aquaporin subfamily as moderators of refilling process. Plant Cell Environ 33:1285–1297

Secchi F, Zwieniecki MA (2012) Analysis of xylem sap from functional (nonembolized) and nonfunctional (embolized) vessels of *Populus nigra*: chemistry of refilling. Plant Physiol 160:955–964

Secchi F, Zwieniecki MA (2016) Accumulation of sugars in the xylem apoplast observed under water stress conditions is controlled by xylem pH. Plant Cell Environ 39:2350–2360

Secchi F, Gilbert ME, Zwieniecki MA (2011) Transcriptome response to embolism formation in stems of *Populus trichocarpa* provides insight into signaling and the biology of refilling. Plant Physiol 157:1419–1429

Secchi F, Pagliarani C, Zwieniecki MA (2016) The functional role of xylem parenchyma cells and aquaporins during recovery from severe water stress. Plant Cell Environ. doi:10.1111/pce.12831

Sevanto S (2014) Phloem transport and drought. J Exp Bot 65:1751–1759

Singh S (2016) Root pressure: getting to the root of pressure. Prog Bot 77:105–150

Sippel S, Otto FEL (2014) Beyond climatological extremes – assessing how the odds of hydrometeorological extreme events in the South-East Europe change in a warming climate. Clim Change 125:381–398

Sperry JS (1986) Relationship of xylem embolism to xylem pressure potential, stomatal closure, and shoot morphology in the palm *Rhapis excelsa*. Plant Physiol 80:110–116

Sperry JS (2013) Cutting-edge research or cutting-edge artefact? An overdue control experiment complicates the xylem refilling story. Plant Cell Environ 36:1916–1918

Sperry JS, Holbrook NM, Zimmermann MH, Tyree MT (1987) Spring filling of xylem vessels in wild grapevine. Plant Physiol 83:414–417

Sperry JS, Donnelly JR, Tyree MT (1988) Seasonal occurrence of xylem embolism in sugar maple (*Acer saccharum*). Am J Bot 75:1212–1218

Sperry JS, Christman MA, Torres-Ruiz JM, Taneda H, Smith DD (2012) Vulnerability curves by centrifugation: is there an open vessel artefact, and are 'r' shaped curves necessarily invalid? Plant Cell Environ 35:601–610

Spicer R, Groover A (2010) Evolution of development of vascular cambia and secondary growth. New Phytol 186:577–592

Steinkamp J, Hickler T (2015) Is drought-induced forest dieback globally increasing? J Ecol 103:31–43

Stiller V, Sperry JS (1999) Canny's compensating pressure theory fails a test. Am J Bot 86:1082–1086

Stiller V, Sperry JS (2002) Cavitation fatigue and its reversal in sunflower (*Helianthus annuus* L.) J Exp Bot 53:1155–1161

Stiller V, Sperry JS, Lafitte R (2005) Embolized conduits of rice (*Oryza sativa*, Poaceae) refill despite negative xylem pressure. Am J Bot 92:1970–1974

Suuronen JP, Peura M, Fagerstedt K, Serimaa R (2013) Visualizing water-filled versus embolized status of xylem conduits by desktop x-ray microtomography. Plant Methods 9:11

Taneda H, Sperry JS (2008) A case-study of water transport in co-occurring ring- versus diffuse-porous trees: contrasts in water-status, conducting capacity, cavitation and vessel refilling. Tree Physiol 28:1641–1651

Tombesi S, Nardini A, Farinelli D, Palliotti A (2014) Relationships between stomatal behavior, xylem vulnerability to cavitation and leaf water relations in two cultivars of *Vitis vinifera*. Physiol Plant 152:453–464

Torres-Ruiz JM, Cochard H, Mayr S, Beikircher B, Diaz-Espejo A, Rodriguez-Dominguez CM, Badel E, Fernández JE (2014) Vulnerability to cavitation in *Olea europaea* current-year shoots: further evidence of an open-vessel artifact associated with centrifuge and air-injection techniques. Physiol Plant 152:465–474

Torres-Ruiz JM, Jansen S, Choat B, McElrone AJ, Cochard H, Brodribb TJ, Badel E, Burlett R, Bouche PS, Brodersen CR (2015) Direct X-ray microtomography observation confirms the induction of embolism upon xylem cutting under tension. Plant Physiol 167:40–43

Trenberth KE, Dai A, van der Schrier G, Jones PD, Barichivich J, Briffa KR, Sheffield J (2014) Global warming and changes in drought. Nat Clim Change 4:17–22

Trifilò P, Nardini A, Lo Gullo MA, Salleo S (2003) Vein cavitation and stomatal behaviour of sunflower (*Helianthus annuus*) leaves under water limitation. Physiol Plant 119:409–417

Trifilò P, Lo Gullo MA, Salleo S, Callea K, Nardini A (2008) Xylem embolism alleviated by ion-mediated increase in hydraulic conductivity of functional xylem: insights from field measurements. Tree Physiol 28:1505–1512

Trifilò P, Nardini A, Raimondo F, Lo Gullo MA, Salleo S (2011) Ion-mediated compensation for drought-induced loss of xylem hydraulic conductivity in field-growing plants of *Laurus nobilis*. Funct Plant Biol 38:606–613

Trifilò P, Raimondo F, Lo Gullo MA, Barbera PM, Salleo S, Nardini A (2014a) Relax and refill: xylem rehydration prior to hydraulic measurements favours embolism repair in stems and generates artificially low PLC values. Plant Cell Environ 37:2491–2499

Trifilò P, Barbera PM, Raimondo F, Nardini A, Lo Gullo MA (2014b) Coping with drought-induced xylem cavitation: coordination of embolism repair and ionic effects in three Mediterranean evergreens. Tree Physiol 34:109–122

Trifilò P, Nardini A, Lo Gullo MA, Barbera PM, Savi T, Raimondo F (2015) Diurnal changes in embolism rate in nine dry forest trees: relationships with species-specific xylem vulnerability, hydraulic strategy and wood traits. Tree Physiol 35:694–705

Trifilò P, Raimondo F, Savi T, Lo Gullo MA, Nardini A (2016) The contribution of vascular and extra-vascular water pathways to drought-induced decline of leaf hydraulic conductance. J Exp Bot 67:5029–5039

Tyree MT (2003) The ascent of water. Nature 423:923

Tyree MT, Dixon MA (1983) Cavitation events in *Thuja occidentalis* L. Ultrasonic acoustic emissions from the sapwood can be measured. Plant Physiol 72:1094–1099

Tyree MT, Zimmermann MH (2002) Xylem structure and the ascent of sap. Springer, Berlin

Tyree MT, Snyderman DA, Wilmot TR, Machado JL (1991) Water relations and hydraulic architecture of a tropical tree (*Schefflera morototoni*): data, models, and a comparison with two temperate species (*Acer saccharum* and *Thuja occidentalis*). Plant Physiol 96:1105–1113

Tyree MT, Alexander J, Machado JL (1992) Loss of hydraulic conductivity due to water stress in intact juveniles of *Quercus rubra* and *Populus deltoides*. Tree Physiol 10:411–415

Tyree MT, Cochard H, Cruiziat P, Sinclair B, Ameglio T (1993) Drought-induced leaf shedding in walnut: evidence for vulnerability segmentation. Plant Cell Environ 7:879–882

Tyree MT, Salleo S, Nardini A, Lo Gullo MA, Mosca R (1999) Refilling of embolized vessels in young stems of Laurel. Do we need a new paradigm? Plant Physiol 120:11–21

Umebayashi T, Ogasa MY, Miki NH, Utsumi Y, Haishi T, Fukuda K (2016) Freezing xylem conduits with liquid nitrogen creates artefactual embolisms in water-stressed broadleaf trees. Trees 30:305–316

Urli M, Porté AJ, Cochard H, Guengant Y, Burlett R, Delzon S (2013) Xylem embolism threshold for catastrophic hydraulic failure in angiosperm trees. Tree Physiol 33:672–683

Utsumi Y, Sano Y, Fujikawa S, Funada R, Ohtani J (1998) Visualization of cavitated vessels in winter and refilled vessels in spring in diffuse-porous trees by cryo-scanning electron microscopy. Plant Physiol 117:1463–1471

Venturas MD, Mackinonn ED, Jacobsen AL, Pratt RB (2015) Excising stem samples underwater at native tension does not induce xylem cavitation. Plant Cell Environ 38:1060–1068

Vesala T, Hölttä T, Perämäki M, Nikinmaa E (2003) Refilling of a hydraulically isolated embolized xylem vessel: model calculations. Ann Bot 91:419–428

Wang R, Zhang L, Zhang S, Cai J, Tyree MT (2014) Water relations of *Robinia pseudoacacia* L.: do vessels cavitate and refill diurnally or are R-shaped curves invalid in *Robinia*? Plant Cell Environ 37:2667–2678

Weed AS, Ayres MP, Hicke JA (2013) Consequences of climate change for biotic disturbances in North American forests. Ecol Monogr 83:441–470

Wegner LH (2014) Root pressure and beyond: energetically uphill water transport into xylem vessels? J Exp Bot 65:381–393

Wheeler JK, Huggett BA, Tofte AN, Rockwell FE, Holbrook NM (2013) Cutting xylem under tension or supersaturated with gas can generate PLC and the appearance of rapid recovery from embolism. Plant Cell Environ 36:1938–1949

Yang S, Tyree MT (1992) A theoretical model of hydraulic conductivity recovery from embolism with comparison to experimental data on *Acer saccharum*. Plant Cell Environ 15:633–643

Yang SJ, Zhang YJ, Sun M, Goldstein G, Cao KF (2012) Recovery of diurnal depression of leaf hydraulic conductance in a subtropical woody bamboo species: embolism refilling by nocturnal root pressure. Tree Physiol 32:414–422

Zeuthen T (2010) Water-transporting proteins. J Membr Biol 234:57–73

Zhao J, Hartmann H, Trumbore S, Ziegler W, Zhang Y (2013) High temperature causes negative whole-plant carbon balance under mild drought. New Phytol 200:330–339

Zimmermann MH (1978) Hydraulic architecture of some diffuse-porous trees. Can J Bot 56:2286–2295

Zufferey V, Cochard H, Améglio T, Spring JL, Viret O (2011) Diurnal cycles of embolism formation and repair in petioles of grapevine (*Vitis vinifera* cv. Chasselas). J Exp Bot 62:3885–3894

Zwieniecki MA, Holbrook NM (1998) Diurnal variation in xylem hydraulic conductivity in white ash (*Fraxinus americana* L.), red maple (*Acer rubrum* L.) and red spruce (*Picea rubens* Sarg.) Plant Cell Environ 21:1173–1180

Zwieniecki MA, Holbrook NM (2000) Bordered pit structure and vessel wall surface properties. Implications for embolism repair. Plant Physiol 123:1015–1020

Zwieniecki MA, Holbrook NM (2009) Confronting Maxwell's demon: biophysics of xylem embolism repair. Trends Plant Sci 14:530–534

Zwieniecki MA, Secchi F (2015) Threats to xylem hydraulic function of trees under 'new climate normal' conditions. Plant Cell Environ 38:1713–1724

Zwieniecki MA, Melcher PJ, Holbrook NM (2001) Hydrogel control of xylem hydraulic resistance in plants. Science 291:1059–1062

Zwieniecki MA, Melcher PJ, Ahrens ET (2013) Analysis of spatial and temporal dynamics of xylem refilling in *Acer rubrum* L. using magnetic resonance imaging. Front Plant Sci 4:265

A Resource Centric View of Climate and Mast Seeding in Trees

Robert B. Allen, Peter Millard, and Sarah J. Richardson

Abstract Seed production by trees has consequences for regeneration, nutrient cycling, trophic interactions, invasive species impacts and various diseases influencing human health. Effective management of these consequences requires a mechanistic understanding of tree seeding patterns. It has long been known that inter-annual variability and synchrony in seed production relate to among year differences in climate. A central question to resolve is whether such seeding patterns relate to climate-driven resource dynamics. To answer this question, we first assess the role of resources (emphasising mineral nutrients and carbohydrates) during the reproductive cycle. We conclude that trees not only utilise remobilised resources stored at the beginning of a reproductive cycle but also resources taken up and recycled subsequently. Secondly, we consider resource dynamics in relation to climate-driven processes. A sequence of climate-resource relationships potentially influences the maximum seed production set at initiation as well as the level to which this maximum is achieved throughout the rest of the reproductive cycle. Thirdly, climate-resource relationships are used to appraise the assumptions of resource-based models commonly employed to enhance our understanding of tree seeding. We conclude that these models need to accommodate a wider set of processes including resource uptake and interrelationships among multiple resources throughout the reproductive cycle. Finally, we consider how climate-driven resource dynamics may have facilitated selection for mast seeding. We propose that climate-driven resource dynamics in an antecedent population could have synchronised seeding and directly advantaged individuals. These advantages were

Communicated by Rainer Matyssek

R.B. Allen (✉)
Private Researcher, Lincoln, New Zealand
e-mail: roballen1706@gmail.com

P. Millard and S.J. Richardson
Landcare Research, Lincoln, New Zealand
e-mail: millardp@landcareresearch.co.nz; richardsons@landcareresearch.co.nz

Progress in Botany (2018) 79: 233–268, DOI 10.1007/124_2017_8,
Published online: 30 March 2017

subsequently amplified through the economies of scale fitness benefits seen today in contemporary tree populations.

Contents

1 Introduction

Tree seeding has long attracted the interest of foresters, botanists and ecologists. Writings as far back as the tenth century considered the significance of tree seeding to animals as a food source and qualitative records of annual seeding events have been systematically recorded for centuries in Europe (Maurer 1964; Hilton and Packham 2003). Today, quantitative seeding records exist for the last half century at some locations. This interest is sustained because tree seeding has consequences for regeneration, nutrient cycling, trophic interactions, invasive species impacts and various diseases influencing human health (Janzen 1971; Wardle 1984; Jones et al. 1998; Zackrisson et al. 1999; Ostfeld and Keesing 2000; Tersago et al. 2009). The future of these consequences is uncertain because seed production is responsive to environmental changes such as climate warming (e.g. Tersago et al. 2009; Allen et al. 2014; Kroiss and HilleRisLambers 2015). Effective management requires a mechanistic understanding of patterns in tree seeding.

That tree populations produce seed crops with distinctive inter-annual patterns is of particular interest (e.g. Silvertown 1980; Owens 1995; Pearse et al. 2016). Patterns include temporal trends in seed crop size (Allen et al. 2014; Buechling et al. 2016), a degree of spatial synchronicity (e.g. Burrows and Allen 1991; LaMontagne and Boutin 2007; Fearer et al. 2008) and inter-annual periodicity (e.g. Sork et al. 1993; Allen et al. 2012; cf. Greene and Johnson 2004), as well as negative temporal autocorrelation (e.g. Koenig et al. 2003; Crone et al. 2011) and high variability (Fig. 1a, b; Herrera et al. 1998; Greene and Johnson 2004; Richardson et al. 2005; Crone et al. 2011). In the extreme, tree seeding that is either large or often nil implies a bimodality in the frequency distribution of seed crop size (but see Allen et al. 2012).

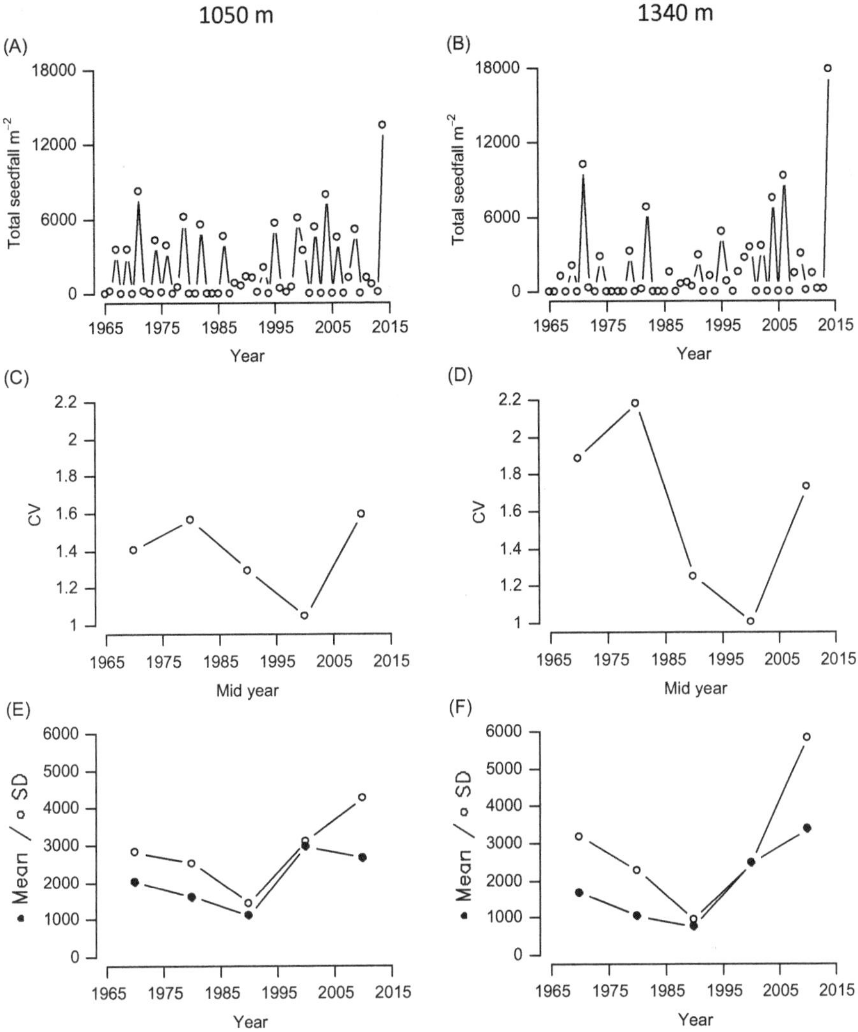

Fig. 1 Seeding patterns in a *Nothofagus* population at each of 1,050 and 1,340 m elevation over a 50-year period (1965–2014; see Allen et al. 2014). (**a**, **b**) Total annual seed production (raw data; seedfall m^{-2}) in each year at 1,050 and 1,340 m elevation respectively; (**c**, **d**) coefficient of variation among years in 10-yearly windows and plotted as mid-point of each window (Mid year); (**e**, **f**) mean seed production and standard deviation among years in 10-yearly windows

High variability in population-level annual seed production (hereafter seed production) among years, coupled with spatial synchronicity, has often, but not exclusively, been used to define mast seeding in tree populations (Silvertown 1980; Kelly 1994). This variability is strongly related to individual-level variability and synchrony among years (Koenig et al. 2003). The coefficient of variation in seed production among years has gained some traction as a metric of mast seeding (Herrera et al. 1998;

Kelly and Sork 2002; Koenig et al. 2003; cf. Crone et al. 2011). However, this metric fluctuates temporally in *Nothofagus* tree populations and, for example, varies by >100% among 10-year time intervals (Fig. 1c, d). The coefficient of variation itself is a product of variability (standard deviation) divided by the mean seed production and so the fluctuations result from a complex interplay between varying mean seed production and the standard deviation (Fig. 1e, f). The tendency towards mast seeding also varies among species and populations (Kelly and Sork 2002). Greater seed production variability can be found at higher elevations within a species (e.g. Kelly et al. 2001) and in those species occurring at lower latitudes in Northern Hemisphere forests (Koenig and Knops 2000). Seed production synchronisation is found among individuals within a population and among populations (e.g. Ims 1990; Burrows and Allen 1991; Koenig et al. 2003; Fearer et al. 2008) although there can be a marked distance-decay in synchronicity at regional (e.g. Fearer et al. 2008) and local (e.g. Allen and Platt 1990) scales.

Inter-annual variation in seed production, and the flowering that precedes it, has often been related to inter-annual variation in climate (e.g. Eis 1973; Law et al. 2000; Schauber et al. 2002; Cook et al. 2012; Roland et al. 2014). These relationships have been reported for various seasonal climate variables (hereafter climate variables) the specifics of which vary with environment (Richardson et al. 2005; Fearer et al. 2008; Smaill et al. 2011; Canham et al. 2014), species (Fearer et al. 2008; Canham et al. 2014; Pérez-Ramos et al. 2015) and stage in the reproductive cycle (Piovesan and Adams 2001; Roland et al. 2014; Allen et al. 2014). Decadal-level trends in seed production have also been related to temporal trends in climate variables (Allen et al. 2014; Buechling et al. 2016).

The mechanisms by which climate variables relate to tree seeding patterns are not well understood and several non-mutually exclusive options exist (Moreira et al. 2015). Evolutionary mechanisms all involve some economies of scale fitness benefits from climatic synchronisation of flowering and seeding events. These have received the most attention in the literature (Crone and Rapp 2014; Pearse et al. 2016). Under such selective mechanisms, any climatic cue would suffice and this could explain why so many climate variables are involved (Koenig and Knops 2014; Crone and Rapp 2014). We expect that a climatic cue would operate before floral initiation if trees do not sustain high levels of reproductive buds. Unambiguously determining fitness benefits that accrue from climatic synchronisation is challenging if climate cues themselves relate to resource supply (Crone and Rapp 2014). While climate imposes strong constraints on resource dynamics little attention has focussed on how climate variables influence the specific resources required during the reproductive cycle in trees. It is also increasingly apparent that such resources relate to environmental constraints other than climate and, as such, decouple seeding patterns from climate (Smaill et al. 2011; Canham et al. 2014; Miyazaki et al. 2014).

In this review we focus on how climate-driven resource dynamics determine inter-annual patterns in tree seeding (e.g. Smaill et al. 2011; Crone and Rapp 2014). Our attention is focussed on that literature pertaining to uncultivated trees. We first review the role of critical resources, particularly mineral nutrients and carbohydrates, during the reproductive cycle. A greater understanding of resource supply and use is required to explain tree seeding patterns (e.g. Newbery et al. 2006; Crone and Rapp 2014; Miyazaki et al. 2014; Allen et al. 2014). We evaluate whether trees store sufficient resources at the beginning of the reproductive cycle by assessing whether trees also rely on uptake and recycling. Secondly, we assess how temporal variability in climate variables might drive internal and external processes that determine the supply of critical resources (e.g. Smaill et al. 2011; Koenig et al. 2015). We assess whether a sequence of climate-resource relationships first influence the potential maximum seed production set at initiation and then the level to which this maximum is achieved throughout the rest of the reproductive cycle. To understand how resource dynamics control inter-annual variability in seed production theoretical ecologists have developed resource budget models (e.g. Isagi et al. 1997; Satake and Iwasa 2000). Thirdly, we review the resource orientated assumptions in these models and suggest improvements to their mechanistic basis including the need to account for multiple resources. Finally, we consider how climate-driven resource dynamics in an antecedent population may have facilitated the selection for mast seeding in some tree species. Overall we bridge a critical gap between the relatively common climate-orientated studies and the relatively few resource-based studies as a means of improving our understanding of when and where climate-driven resource dynamics may be a primary force driving patterns in tree seeding.

2 Role of Resources During the Reproductive Cycle

Trees require a range of resources to reproduce, including a direct allocation of resources to seeds and related structures (e.g. flowers, cones), as well as indirectly through an allocation to processes supporting and enhanced demand for resources (e.g. carbohydrate use by mycorrhiza). We often do not know which mineral nutrients will be most limiting to a particular population, but emphasise N and P as widely occurring candidates (e.g. Newbery et al. 2006; Smaill et al. 2011; Ichie and Nakagawa 2013; Han et al. 2014). Although seed counts are widely used as proxies for resources (e.g. negative temporal autocorrelations; Koenig et al. 2003) the mineral nutrient content of reproductive structures is a more appropriate measure of resources (Fernández-Martínez et al. 2017). This is because individual seed mass, and their nutrient concentrations, can vary markedly within species (e.g. Ledgard and Cath 1983; Moles and Westoby 2003; Lamont and Groom 2013). We also emphasise carbohydrates as a derived resource (produced from water, solar radiation and CO_2) because they are commonly thought to control seeding by trees (e.g. Isagi et al. 1997; Richardson et al. 2005; Alla et al. 2012; Hoch et al. 2013). Non-structural carbohydrates are the currency of interest. We assess an expectation that the reproductive cycle utilises multiple resources with strong interrelationships.

It is well established that tree growth in any given year uses carbohydrates, N and P taken up in previous years, due to the internal cycling of resources (e.g. for a review of carbohydrates and N, Millard and Grelet 2010; for P, Proe and Millard 1995). The ability of trees to internally cycle resources is important for several reasons. First, it allows for growth when external resources are constrained, as often seen in temperate regions where spring growth can commence while resource availability is still constrained by cool air and soil temperatures (Millard and Grelet 2010), so that in a range of evergreen and deciduous temperate species, bud-burst and the growth of new leaves can be supplied predominantly from remobilisation of stored N (Neilsen et al. 1997), typically for 20–30 days before root uptake of any soil N (Millard and Proe 1993; Tagliavini et al. 1997; Millard et al. 2001, 2006; El Zein et al. 2011). Secondly, resource storage allows trees to recover from events such as defoliation by insects or browsing ungulates (Millard et al. 2001). Thirdly, stored resources are important for supporting reproduction. For example, N from storage is the primary source for the spring flowering of a range of fruit trees (Sanchez et al. 1991; Muñoz et al. 1993; Tagliavini et al. 1997) and there is increasing evidence that stored N contributes to masting events (e.g. Sala et al. 2012; Ichie and Nakagawa 2013; Miyazaki et al. 2014).

Thus in any year trees can rely upon both external and internal resources for growth and development (Fig. 2). Uptake of external N and P (as well as some other mineral nutrients) comes from the mineralisation of soil organic matter and decomposition of the litter layer, often via mycorrhizal fungi (Chapman et al. 2006), biological N_2 fixation and, in some forests, atmospheric N deposition. Carbohydrates come from photosynthesis. Internal resources come from storage, through the processes of remobilisation and recycling (Fig. 2). When trees have an abundance of a particular resource there can also be sequestration, representing a metabolic 'dead-end' which does not allow the resource to be reused. Examples of sequestration include starch recovered in the roots (e.g. Bhupinderpal-Singh et al. 2003) or trunk (Regler et al. 2010) below girdling that has killed trees, or arginine accumulation in the foliage of trees experiencing very high atmospheric N deposition due to air pollution (Näsholm 1994) or P limitation (Tarvainen et al. 2016). Detrital material (e.g. flowers) represents a resource loss by the tree (Fig. 2).

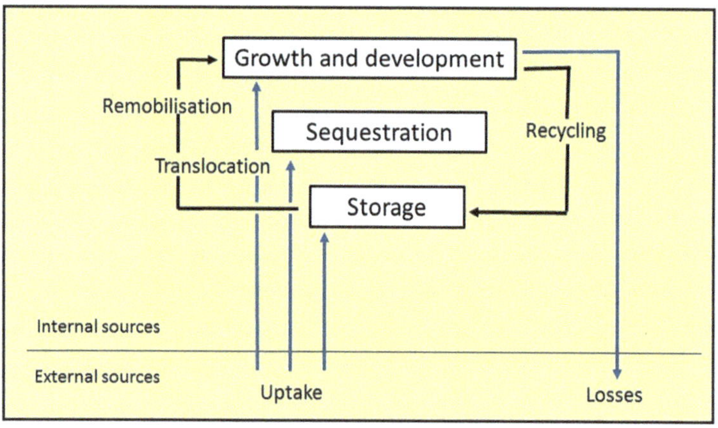

Fig. 2 Schematic of resource dynamics by a tree (modified from Millard and Grelet (2010)). Exchange of resources between external sources and internal pools in the tree is shown as *blue lines*: for uptake directly into storage, sequestration or use in growth and reproduction or other metabolism; and for losses via senescence and abscission of leaves and roots. Exchange of resources among internal pools is shown in *black lines* and represented by remobilisation or recycling

For a resource to be considered stored, it is necessary to demonstrate that it can be remobilised and reused for further growth, maintenance (e.g. respiration) or defence. Trees accumulate large amounts of carbon within most tree tissues, mainly as non-structural carbohydrates and lipids (e.g. Hoch et al. 2003; Würth et al. 2005). During periods of rapid growth a proportion of these carbohydrates disappear, so in temperate trees, concentrations are often greatest in late summer or autumn and smallest in late spring or early summer after new canopy growth has occurred (Hoch et al. 2003). Non-structural carbohydrates are never fully depleted (Körner 2003), which leads to an important distinction between storage and accumulation. Accumulation occurs if a resource is taken up in excess of the current or future needs (i.e. luxury consumption) and so does not imply reuse, rather sequestration (Fig. 2). In contrast, to assess resource storage it is necessary to determine the subsequent remobilisation and reuse, while recognising that storage can occur when the tree is resource limited.

In contrast to C, trees store N in specific organs, dependent upon phenology (Millard and Grelet 2010). Deciduous trees store N during winter in the bark or wood of the roots or trunk, often as specific bark storage proteins, while coniferous evergreens store N in their needles (Millard and Grelet 2010). Storage does not preclude the resource also being used actively at the same time. A good example of this is how many trees store N in the photosynthetic protein, Rubisco (Millard et al. 2007). During the autumn senesce of deciduous leaves, the N from Rubisco is recycled back into storage and can make a major contribution to canopy growth the next year (Fig. 2; Millard and Thomson 1989). In evergreen species, Rubisco is used as storage in winter and turnover of the protein (independently of any leaf senescence) contributes N to the development of buds and new leaf growth the following spring (reviewed by Millard et al. 2007).

Given that storage is dynamic, it is difficult to infer the amount of resource storage from a static measurement of concentration. This is especially the case for carbohydrates. Trees accumulate large pools of non-structural carbohydrates in stems and branches, enough to replace the entire leaf canopy four times in temperate species (Hoch et al. 2003). Even severe defoliation by browsing animals over several years does not deplete carbohydrate levels, (Palacio et al. 2008), suggesting that tree growth is not C limited (Millard et al. 2007). Much of the non-structural carbohydrates found in trees are likely sequestered (Millard and Grelet 2010), or at any rate cannot be fully utilised as they serve several purposes (Martínez-Vilalta et al. 2016). Thus the proportion of non-structural carbohydrates that are stored remains obscure. Moreover, our knowledge remains incomplete because most of the studies quantifying storage and remobilisation of resources have used isotopic tracers (e.g. ^{13}C, ^{15}N ^{32}P) only in young seedlings or saplings. Thus for N, for example, there have been relatively few studies on large, field grown trees (e.g. Millard 1994; Millard et al. 2006; El Zein et al. 2011).

The focus here is on the dynamics of labile resources required throughout the tree reproductive cycle (Miyazaki 2013). We delineate four stages in the reproductive cycle of trees that include, sequentially, the periods leading up to initiation (including the process of induction), anthesis, and maturation, as well as the period during, and immediately after seedfall (Table 1). Initiation is a critical stage, setting the maximum potential reproduction and a baseline for flower and seed production (Owens 1995). This is because subsequent stages limit the extent to which the maximum is reached through floral abortion, ovule abortion, embryo abortion and the failure of seeds to mature (Owens 1995). Most tree species have a period of dormancy between floral initiation and anthesis, although other types of flowering periodicities occur in tropical tree species (Owens 1995). The duration of each stage can vary among and within tree species (Table 1; Foster and Gifford 1974). The majority of temperate tree species, for example, pollinate (included in anthesis; Table 1) and fertilise (included in maturation; Table 1) reproductive structures in the same season, but, in *Pinus*, this can take over 1 year. Because tree populations of mast seeding species sometimes produce seeds in consecutive years (Fig. 1a, b) it may be that resource dynamics reflect a complex set of requirements among cohorts. Variation within species is exemplified by cork oak (*Quercus suber* L.) where individuals can produce annual acorns that mature in the same year as

Table 1 Broad stages in the reproductive cycle of trees

Stage	Description	Timing and duration
Initiation	Up to, and including, the irreversible morphological conversion of vegetative buds (i.e. initiation) into differentiated reproductive buds	Conditions up to at least 1 year before floral induction and subsequent initiation can influence the differentiation of reproductive buds (e.g. Owens 1995; Piovesan and Adams 2001)
Anthesis	Development of the reproductive buds (i.e. primordia) into flowers and cones that are fully opened and functional, including pollination	Most temperate, and many tropical tree species, have a period of dormancy after initiation (often lasting 6 months), although the period from flowering to pollination is usually short (e.g. Owens 1995)
Maturation	Includes the period from pollination to fertilisation, the complete development of embryos, seeds, cones and fruits, as well as their maturation	Most tree species fertilise reproductive structures in the same season as pollination, but, in *Pinus*, this can take over 1 year (Owens 1995). The duration of maturation can also vary markedly within species and with environment (e.g. Pons and Pausas 2012)
Seedfall	Period of mature seed and fruit dispersal	Timing and duration of seedfall varies markedly among species. Seeds of temperate tree species can fall in the autumn (e.g. Wardle 1984), spring (e.g. Hannerz et al. 2002) or both (e.g. Burrows and Allen 1991) following maturation. Seeds can also be stored in the canopy for consecutive years (e.g. Yates et al. 1994)

Four sequential stages are defined which span the period leading up to the differentiation of reproductive buds through to the period immediately after propagule dispersal. This cycle involves the production of tree seeds and fruits, as well as a range of associated structures, and their dispersal. Variations in the timing and duration of each stage are also indicated

anthesis or biennial acorns that mature in autumn of the following year, even when found within the same populations (e.g. Pons and Pausas 2012). In addition, the duration of the reproductive cycle can vary with environment (Foster and Gifford 1974; Pons and Pausas 2012). For such reasons it is more meaningful to examine resource dynamics at stages in the reproductive cycle rather than absolute time.

2.1 Initiation

Because floral induction and subsequent initiation are not physically obvious in trees they are often overlooked in studies of tree seeding patterns. A range of environmental constraints (e.g. temperature, moisture, light) and internal dynamics (e.g. resources, hormones, genes) relate to floral induction and the ability of trees to initiate the irreversible conversion of vegetative buds into a differentiated reproductive state (e.g. Sedgley and Griffin 1989; Meilan 1997). As a consequence, it is unlikely that trees possess one simple regulatory system for this conversion and many mechanisms need to be explored, particularly with regard to how nutritional resources affect induction and initiation (Owens 1995; Danielson and Frommer 2013).

Resource supply, through uptake and remobilisation from storage, controls the abundance of vegetative buds, their phenology, and facilitates floral initiation (Fig. 3; Miyazaki et al. 2014). The application of mineral nutrients, particularly of N, has long been known to increase tree flower and seed production (e.g. Steinbrenner et al. 1960; Sperens 1997; Williams et al. 2003; Davis et al. 2004; Montesinos et al. 2012) and is, in the first instance, likely a consequence of heightened induction and initiation. Certainly there are increased requirements for nucleic acids and proteins, as well as carbohydrates, for metabolic activities such as cell division (Sedgley and Griffin 1989). Uncertainty about the timing of induction and initiation, as well as the opportunities for alternative pathways for resource use (e.g. canopy growth), make it challenging to demonstrate the explicit links to resource supply (Sedgley and Griffin 1989; Miyazaki et al. 2014). Recently, Miyazaki et al. (2014) used flowering gene expression as markers to indicate the timing of floral initiation in *Fagus crenata.* N fertilisation (improved uptake) facilitated higher expression of flowering genes, increased N concentration in current year shoots, but not the concentrations of sugars or starch, and boosted the proportion of vegetative buds that became reproductive buds. Certainly, the proportion of vegetative buds that later switch to become reproductive in crop trees is dependent upon nutrient supply, particularly of N (e.g. Sedgley and Griffin 1989). Increased N resources may also enhance reproductive output through stimulating the formation of growth hormones (Wahl et al. 2013). It is possible that resources other than N might take a similar functional role leading up to initiation in other species (Wahl et al. 2013; Danielson and Frommer 2013). The relationships between resource supply and floral initiation support a view that sufficient resources have a mechanistic role in regulating the level of flowering (Wahl et al. 2013; Miyazaki et al. 2014).

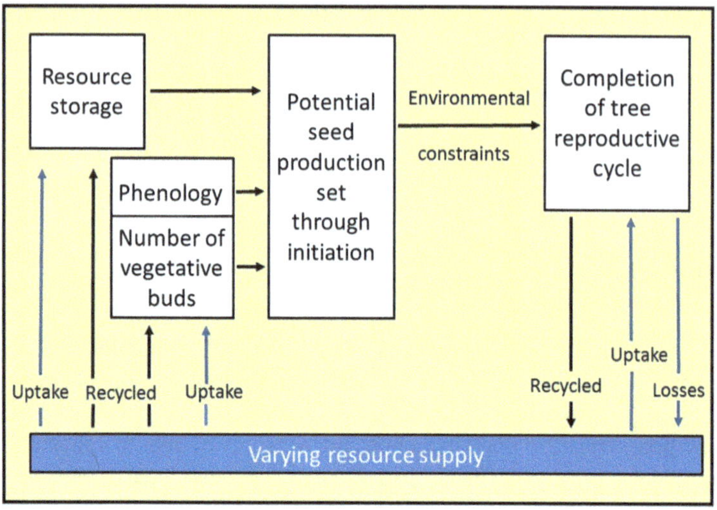

Fig. 3 Depiction of the resource dynamics before and after the potential seed production is set through initiation by a tree. It represents that period in the life of a tree where resource supply through uptake (*blue lines*) or remobilised from storage (*black lines*) determine the number of vegetative buds (including the proportion of which turn into reproductive buds) and the amount of resources allocated to storage. These resources set the potential seed production which is seldom realised because of environmental constraints throughout the reproductive cycle. Resource losses (*blue lines*) and recycling (*black lines*) recharge resource supply. The total time period involved varies and depends upon, for example, species and environmental constraints

2.2 *Anthesis*

Reproductive primordia developing into mature floral structures, including their pollination, create a significant demand for resources. Some researchers have asserted the importance of N for floral development in mesic temperate forests (Han et al. 2008; Müller-Haubold et al. 2015) and P in tropical rain forests on infertile soils (Ichie and Nakagawa 2013). Greater N content, and dry matter, in developing flower buds, rather than vegetative buds, is thought to relate to the resource requirements for developing flower primordia in *Fagus* (Han et al. 2008). N and P can accumulate in high amounts in developing strobili of gymnosperms, along with various micronutrients (Fe, B and Cu), which are later translocated into developing seeds (Dickmann and Kozlowski 1969). Theoretical and empirical studies have concluded that female reproductive tissue requires more carbohydrates than male tissue, although photosynthesis by female reproductive tissue can contribute to their own needs. As male structures usually do not photosynthesise their carbohydrate demands are met entirely by vegetative tissue (Bazzaz et al. 1979; Linder and Troeng 1981; McDowell et al. 2000). Hoch et al. (2013) used ^{13}C labelled trees of *Carpinus*, *Fagus* and *Quercus* species to show that developing flowers carried a significant label from C remobilised from previous growing season storage. There can be a decline in the starch reserves in reproductive branchlets at the time of anthesis (Miyazaki et al. 2002; Hoch et al. 2013). This, combined with the use of ‘old’ C at anthesis, is likely due to high carbohydrate usage at this stage in the reproductive cycle.

Resource dynamics at anthesis can be used to portray the relationships which occur among different resources. Trees allocate carbohydrates to roots or mycorrhiza to increase the uptake of mineral nutrients, particularly of N and P, that can then be stored and remobilised when needed (Fig. 3; Marschner and Dell 1994). There is also increasing evidence of ericoid and ectomycorrhizal uptake of organic forms of these nutrients on poor soils (e.g. Read and Perez-Moreno 2003; Högberg et al. 2010). Masting tree species are commonly ectomycorrhizal and this could be an important factor in acquiring nutrients (Newbery 2005; Newbery et al. 2006). Carbohydrate stimulation of mycorrhizal (and root) uptake may be why trees at anthesis can have high mineral nutrient concentrations in reproductive branches, and sometimes their leaves, which are later used in the maturation stage (e.g. Cipollini and Stiles 1991; Sala et al. 2012; Ichie and Nakagawa 2013).

Many flowers can abort before reaching maturity (e.g. Stephenson 1981; Wardle 1984; Sork and Bramble 1993). In *Nothofagus* the ratio of flower to seed production mass can be 0.65 in a heavy seeding year (Davis et al. 2004). However, there are years where flower mass is much greater than seed mass (ratio of 1.69), indicating that many flowers must have been aborted (Davis et al. 2004). As current uptake, and stored resources, support floral development they can potentially determine the level to which flowering is inhibited below the potential set by reproductive buds. Because flowers can represent only a small part of total resource investment in reproduction their abortion could be a beneficial bet-hedging strategy to maximise reproductive output in the longer term in spite of unpredictable events (Goubitz et al. 2002; Montesinos et al. 2012). Abortion in *Pinus halepensis* appears mediated by both pollen quantity and resource availability although the relative importance of these two is flexible, varying between trees and years (Goubitz et al. 2002). It would be useful to know how pollen quality varies with resources. If flowers compete for limited resources, it is also reasonable to expect that early flowers on an individual would subsequently have greater seed production (Stephenson 1981). A pattern of selective development may minimise resource losses due to abscission of floral structures. This may be why N fertiliser addition to crop trees can increase the proportion of viable flowers (Stephenson 1981).

2.3 Maturation

Development of embryos, seeds, cones and fruits following fertilisation requires a period of maturation (Owens 1995). Maturation entails developmental, physiological and biochemical changes that commonly produce protein storage products, but in some seeds carbohydrates or lipids, and lead to structures enriched in mineral nutrients relative to other plant tissues (Reekie and Bazzaz 1987; Owens 1995; Sala et al. 2012). Species in the Proteaceae and, not surprisingly, the Leguminaceae have some of the highest average protein concentrations (>30%) in seeds of the 114 families studied by Barclay and Earle (1974), whereas seeds of Myrtaceae and Sapotaceae species are among the lowest (<10%). In the highly P-impoverished soils of south-western Australia up to 48% of total aboveground P can be found in the seeds of Proteaceae species (Groom and Lamont 2010). This P content is higher, and largely stored in phytin, on the most impoverished soils. Mineral nutrient enrichment in seeds is essential because these nutrients, and energy, can be decisive for seedling establishment (Leishman et al. 2000). Early seedling development In *Quercus ilex* is more dependent on N remobilised from acorn storage than from uptake (Villar-Salvador et al. 2010). Compared to carbohydrates, mineral nutrient allocation to developing seeds, particularly N and P, are critical resources because they are usually limiting in the environment (Reekie and Bazzaz 1987) and reproductive structures can manufacture part of the photosynthetic cost of their production. Nutrient concentrations found in fully developed seeds also vary markedly among populations within species (Barclay and Earle 1974).

The nutritional status of a tree has consequences for the number of developing reproductive structures, as well as the mass and viability of the individual seeds produced (Chandler 1938; Karlsson and Örlander 2002; Smaill et al. 2011; Pérez-Ramos et al. 2014). Seed maturation can utilise N and P from uptake or remobilisation from various storage tissues, or both (Fig. 4; Han et al. 2008; Sala et al. 2012). Greater N supply increases seeding directly by more allocation to reproductive tissue development

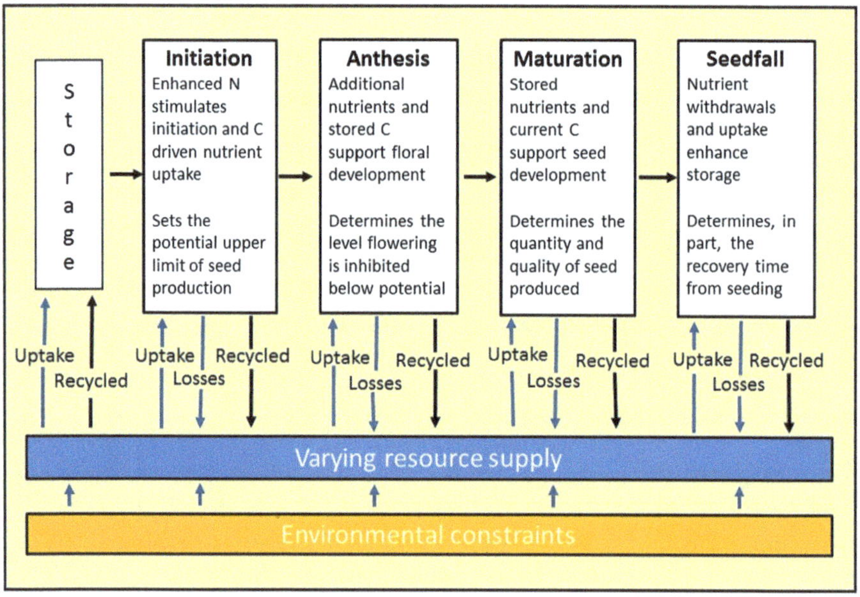

Fig. 4 Resource dynamics representation of a mesic temperate forest tree reproductive cycle. External and internal sources of resource supply for storage and growth are controlled by environmental constraints at each stage in the cycle. At each stage in the cycle, supply is recharged by uptake and recycling of resources. The time period involved in the reproductive cycle of a tree varies and depends upon, for example, species and environmental constraints

(e.g. Chandler 1938; Davis et al. 2004; Han et al. 2008, 2014) or by enhancing photosynthesis and the supply of carbohydrates for resource capture growth (e.g. Waring and Schlesinger 1985). Maturing seeds contain exclusively current growing season uptake of photo-assimilates in *Carpinus*, *Fagus* and *Quercus* species, rather than the stored reserves as found at anthesis (Hoch et al. 2013). Similar conclusions were reached by Ichie et al. (2013) using radiocarbon data and Han et al. (2016) by interpreting natural abundance ^{13}C signatures in tree tissues on a seasonal basis. Hoch et al.'s (2013) isotope study showed that late in seed maturation all of the starch reserves in the branchlet xylem was also from current years assimilates. Clearly carbohydrate uptake during maturation is important. Isotope studies could also help resolve the relative importance of uptake and remobilisation of mineral nutrients.

In *Pinus albicaulis*, seed maturation appeared to first reduce N and P concentrations in the needles and sapwood branches bearing cones, but eventually in all tissues (Sala et al. 2012). Whereas the source of soluble sugars, starch and N did not appear to come from current-year, acorn-bearing shoots in two *Quercus* species (Alla et al. 2012). There was instead higher absolute N remobilisation from foliage, relative to branch sapwood, but roughly similar P extraction from both sources. Sala et al. (2012) suggested P extraction and allocation to seeds were detected later during seed maturation than it was for N as a strategy to maximise P use. Developing seeds, and winter buds (for leaves and flowers), appear to use the same N source and, because the seeds require a large proportion of the available N, the source for further differentiating floral primordia is restricted (Han et al. 2008).

Reproductive allocation theory assumes that the availability of resources is fixed and that trade-offs exist between the costs of reproduction and the vegetative growth of individuals. This is because a preferential allocation to a given function reduces the resources available for other competing functions. However, resource trade-offs between reproduction and storage within trees may be more consistent than between reproduction and growth (Stearns 1989; Ehrlén and van Groenendael 2001; Sala et al. 2012), so that although a negative trade-off between reproduction and main stem diameter growth has been used to infer resource-switching in trees it is only found in some instances (e.g. Yasumura et al. 2006; Monks and Kelly 2006; Sala et al. 2012). In *Quercus*, the stem diameters of acorn-bearing branches can be greater than that of non-bearing branches and diameter was positively related to seed set and size (Sánchez-Humanes et al. 2011; Alla et al. 2012). This may be because robust vegetative structures on reproductive shoots can have a major influence on the proportion of flowers that lead to maturing seeds (Watson 1986; Alla et al. 2012).

2.4 Seedfall

The demands of large seed crops can reduce tissue nutrient concentrations and storage at the time of seedfall (e.g. Sala et al. 2012; Ichie and Nakagawa 2013; Han et al. 2014; Müller-Haubold et al. 2015) although substantial stores can remain (Han et al. 2014). The presence of a first order autoregressive correlation structure in inter-annual seeding has often been used to suggest internal resource depletion (Owens 1969; Kon et al. 2005; Crone et al. 2011; Sala et al. 2012; Allen et al. 2014). This may be because trees take time to recover from seeding events (e.g. Allen and Platt 1990; Han et al. 2008; Crone et al. 2011). Trees can respond to resource depletion by recycling nutrients from senescing reproductive structures before they are discarded (Dickmann and Kozlowski 1969) and by increasing uptake of soil nutrients made available through decomposition, mineralisation and leachates from the high quality reproductive litter (Pregitzer and Burton 1991; Zackrisson et al. 1999; Han et al. 2008). The level of external uptake is likely influenced by the length of time involved for a reproductive event (Table 1). As expected the periodic fertilisation of trees can reduce the time interval between seeding events and diminish any reduction in diameter growth with heavy seeding (Han et al. 2008).

3 Influence of Climatic Constraints on Resource Availability

Resources that limit growth and development in trees vary markedly in response to environmental constraints (Kozlowski et al. 1991). Climatic constraints regulate external processes that drive resource supply to trees through uptake and internal metabolic processes that drive recycling, storage and remobilisation (Fig. 2). Variability in these climatic constraints can drive inter-annual pulses in resource availability (Yang et al. 2008) and directional change at longer timescales (e.g. Waring and Schlesinger 1985; Wardle 2002). The processes by which climatic constraints drive resources can be indirect and have relationships with a wider set of environmental constraints (Orwin et al. 2015). There is an emerging view that disentangling these relationships will help clarify how temporal and spatial variability in climate variables drive resources critical to reproduction in trees (e.g. Smaill et al. 2011; Crone et al. 2011; Cook et al. 2012; Roland et al. 2014; Koenig and Knops 2014; Allen et al. 2014).

Climatic constraints may also restrict seed production through damaging events. High precipitation at anthesis appears to interfere with pollen dispersal and wind pollination (e.g. Allen and Platt 1990; Pearse et al. 2016). This impact may be extensive because of spatial synchrony in seasonal climate patterns whereas unusual frosts may damage reproductive structures locally and result in some spatial asynchrony in seed production (Ågren 1988; Chang-Yang et al. 2016). But as such restrictions are not resource related they will not be further considered here.

3.1 *Inter-Annual Variability in Climatic Constraints*

Inter-annual variation in climate variables has often been related to inter-annual variation in seed production (Cook et al. 2012; Koenig and Knops 2014; Crone and Rapp 2014; Allen et al. 2014). Such relationships between seed production and a specific climate variable are usually not particularly strong (coefficients of determination often <0.40; e.g. Selås et al. 2002; Richardson et al. 2005; Kelly et al. 2013). Where a sequence of different climate variables from throughout the reproductive cycle has been included in one analysis they can cumulatively explain much of the inter-annual variation in seed production (coefficients of determination can be >0.80; Allen and Platt 1990; Selås et al. 2002; Smaill et al. 2011; Buechling et al. 2016). One expectation is that the sequence of seasonal climate variables will be different in different environments but somewhat similar across taxa in similar environments. Koenig et al. (2016) showed a similarity in climatic variables for *Quercus* taxa while controlling for phylogeny and species traits. Below we review the role of inter-annual variability in climate variables for resource dynamics and seeding. We carefully match the timing of climate variables with explicit stages in the reproductive cycle. We emphasise mesic temperate forests because they are dominated by relatively well-studied angiosperm (e.g. *Fagus*, *Nothofagus*, *Quercus*) and gymnosperm (e.g. *Pseudotsuga*, *Abies*) tree species. Their seeding patterns are related broadly to a similar sequence of climate variables (e.g. Eis 1973; Piovesan and Adams 2001; Clement et al. 2010; Allen et al. 2014; Buechling et al. 2016) although the relative importance of specific climate variables may vary along environmental gradients (e.g. Allen and Platt 1990; Fearer et al. 2008; Masaki et al. 2008).

3.1.1 Initiation

In mesic temperate forests, relatively wet, cool growing seasons leading up to induction have been related to subsequent high seed production (e.g. Lowry 1966; van Vredenburch and la Bastide 1969; Eis 1973; Piovesan and Adams 2001; Richardson et al. 2005; Fearer et al. 2008; Buechling et al. 2016). Richardson et al. (2005) showed greater modelled net C availability under wet, cool growing conditions in a New Zealand *Nothofagus* forest. Relatively high growing season precipitation can also lead to more litter mass loss (e.g. Upadhyay et al. 1989), enhanced nutrient mineralisation and uptake (e.g. Paul et al. 2003; Schaeffer et al. 2013). However, this enhancement has limits because the relationship between soil moisture and decomposition is parabolic (Waring and Schlesinger 1985; Wardle 2002). Smaill et al. (2011) showed that greater summer precipitation leading up to initiation in the *Nothofagus* forest studied by Richardson et al. (2005) elevated foliar N in the summer months and related this to increased subsequent seed production. The addition of N fertiliser markedly reduced the influence of summer precipitation on subsequent seeding in this forest (Davis et al. 2004; Smaill et al. 2011). Increased mineral nutrients under wet, cool conditions could enhance photosynthesis and the supply of carbohydrates for storage and growth (Waring and Schlesinger 1985) or be allocated to stimulate root and mycorrhizal activity, that would further increase nutrient uptake (e.g. Smith and Read 2008; Högberg et al. 2010). Fine root growth is stimulated by enhanced soil N availability in low N forest soils (Hutchings et al. 2003; Platt et al. 2004). Increased fine root and mycorrhizal activity could thus further increase storage that can be remobilised in subsequent reproductive cycle stages. The positive effect of wet, cool growing seasons on N uptake (Smaill et al. 2011; Allen et al. 2014), net carbon availability (Richardson et al. 2005) and subsequent seed production appeared most pronounced at high elevations in the New Zealand *Nothofagus* forest.

A link between soil moisture and nutrient supply, leading up to initiation, has also been suggested for some tropical rainforests. Newbery et al. (2006) found that a relatively 'wet' dry season increased the probability of heavy seeding in an N-fixing, ectomycorrhizal central African species (*Microberlinia bisulcata* A. Chev.) found on nutrient poor, well-drained soils. These authors suggest that roots and ectomycorrhiza have greater access to carbohydrates at initiation and use these to take up more nutrients than in a 'dry' dry season. A 'wet' dry season appears to allow this N-fixing species to adjust the P balance within the tree for subsequent stages in the reproductive cycle (Newbery et al. 2006). Masting by such large leguminous, ectomycorrhizal tropical trees occurs widely in central Africa and South America forests on low P soils (e.g. Henkel et al. 2005; Newbery 2005). Nutrient limitation is not confined to poor soils and even in a relatively fertile lowland tropical forest in Panama a nutrient addition experiment showed that P increased the mass of reproductive structures produced in 8 of 11 years (Wright et al. 2011).

3.1.2 Anthesis

Higher than average summer temperatures during the early phase of floral primordia development, which can be more than a year before seed maturation, often relates to subsequent high seed production in mesic temperate forests (e.g. van Vredenburch and la Bastide 1969; Eis 1973; Piovesan and Adams 2001; Richardson et al. 2005; Fearer et al. 2008). An assumption that temperature relates positively to carbohydrate production (e.g. Kelly and Sork 2002; Carevic et al. 2010) may not always be so in trees because net C gain depends, in part, upon the species-specific slopes of photosynthetic and respiration response curves to temperature (Waring and Schlesinger 1985; Richardson et al. 2005). Richardson et al. (2005) modelled that net tree C availability (without altering N availability) in fact declined with increasing summer temperatures in the New Zealand *Nothofagus* forest. This would support a dependence on stored C reserves at anthesis (Hoch et al. 2013). The relationship between microbially mediated mineralisation can be monotonic and positive with temperature over a wide range of values (Waring and Schlesinger 1985) so that in the short term there can be a positive response of soil microbial respiration and soil organic matter decomposition to temperature (Wardle 2002). We are unaware of any studies specifically showing that higher summer temperatures increase soil nutrient availability, uptake and subsequent seed production in forests. However, in an alpine grassland, Tanentzap et al. (2012) showed that soil N mineralisation and flowering increased with late spring temperatures, but suggested that plants still require other favourable growing conditions to utilise the increased N.

There is no reason to expect that the same climate variables drive resources in the same way in markedly different environments. In tropical rainforests, Newbery et al. (2006) had a relatively 'dry' dry season at flowering (and maturation) being positively related to seed production. Irrigation in such forests, throughout dry seasons, failed to alter reproduction by canopy trees (Wright and Cornejo 1990). It may be that dry climate is associated with a lack of clouds and higher irradiance. Increased solar radiation in 'dry' dry seasons appears to enhance photosynthesis (e.g. van Schaik et al. 1993; Wright and van Schaik 1994). Canham et al. (2014) characterised spatial and temporal variation in seed production by temperate tree populations on adjacent landforms that varied markedly in surface age and thus soil fertility (Coomes et al. 2005). While very similar climates were expected across adjacent landforms there was asynchrony in the two species observed across landforms (Canham et al. 2014). Fernández-Martínez et al. (2017) recently showed that foliar Zn and P concentrations, measured at one point in time, as well as modelled gross primary production, were better geographic predictors of seed production averaged across years than precipitation and temperature variables for common masting tree species in European forests. These authors argued that the positive relationship between foliar Zn (and P) and average seed production occurred because Zn is important for metabolic processes, including protein synthesis. Moreover, gross primary production and foliar Zn were negatively related to variability in seed production among years (Fernández-Martínez et al. 2017). An insightful extension to such a geographic study would be to also simultaneously measure year-to-year variation in resource supply and climate and relate these to inter-annual variability in seed production (e.g. Smaill et al. 2011).

3.1.3 Maturation

Higher than average summer temperatures during maturation have also often been related to increased seed production (e.g. Lowry 1966; Sork et al. 1993; Richardson et al. 2005; Fearer et al. 2008; Buechling et al. 2016). This stage in the reproductive cycle uses stored minerals and currently produced carbohydrates (Hoch et al. 2013). Although foliage photosynthetic rates can respond to growing season climate (e.g. Iio et al. 2008), the whole-tree model of Richardson et al. (2005) suggested for *Nothofagus* that high temperatures alone did not heighten current carbohydrate production. However, this could be counteracted if some of the stored nutrients were allocated to foliage and enhanced photosynthetic rates. This conflicts, however, with the decline in foliage nutrients found on reproductive branches (Miyazaki et al. 2002; Sala et al. 2012). Temperature may instead influence production through driving mineralisation of soil nutrients (Wardle 2002). This effect may weaken in the longer term if there is thermal adaptation of microbes (Bradford et al. 2008). Maintaining a temperature-induced increase in soil nutrient mineralisation at maturation, and the early part of anthesis in the previous year, could also be difficult because of depletion in the soil (Bradford et al. 2008). However, there may also be considerable nutrient return to the soil through redundant flowers (e.g. Davis et al. 2004). Because Smaill et al. (2011) did not detect greater *Nothofagus* stand-level foliage N in warm summers, even though warm summers at maturation were related to seeding events, it is possible that any enhancement of nutrient uptake is allocated to the developing seeds.

There is no reason to expect that climate variables are fixed, in the same environment, for functionally different species. Moisture constraints in a dry season can impact on forest productivity in a low precipitation environment (Waring et al. 2008). In a meta-analysis of summer dry Mediterranean oak species, seed production was most strongly correlated with spring and summer water availability (Pérez-Ramos et al. 2015). Spring water stress reduces growth and storage with potentially negative consequences for the remobilisation of resources for reproduction (Carevic et al. 2010; Pérez-Ramos et al. 2015). It is not surprising then that *Quercus* species found on dry sites develop thick stems on acorn-bearing shoots that may provide higher total water conductivity and greater structural support (Sánchez-Humanes et al. 2011; Alla et al. 2012). Pérez-Ramos et al. (2015) showed that acorn production was positively related to moisture availability at the time of maturation in the evergreen *Quercus suber*, but at an earlier stage in the reproductive cycle for the co-occurring summer deciduous *Q. canariensis*. This led to asynchrony in seed production by these two oak species. *Q. suber* appears to have a different functional strategy than *Q. canariensis*, including an extended maturation stage (biennial) in particularly dry Mediterranean forests (Pons and Pausas 2012).

3.1.4 Seedfall

Seedfall signals the end of a sequence of resource investments in developing reproductive structures. At best, only a few of the viable seeds produced ultimately survive and develop into seedlings, and most seeds die and decompose (can be >90%; Nielsen 1977). Seeds form a high quality litter from which mineralised nutrients are released at a rate which again depends upon climate and other factors (Wardle 2002). Little attention has focussed on how specific climate variables influence decomposition and resource uptake following seeding events (Sala et al. 2012). This will have consequences for initiation of subsequent seeding events.

3.2 *Trends in Climatic Constraints*

A simple expectation is that decadal-level trends in climatic variables will induce changes in resource availability and lead to decadal-level trends tree seeding. Kelly et al. (2013) showed relationships between relative differences in a seasonal temperature variable across years and flower or seed production in plants with various growth forms including trees, but inferred reproduction will not respond to a changing climate as there was no need to invoke a link of climate to resources. Monks et al.

(2016) countered this view, using long-term data from long-lived grasses and a combination of empirical relationships and resource-based modelling, to suggest that resources are critical, reflect climate and that a changing climate will impact on reproduction. Certainly decadal-level trends have emerged in flower and seed production and these do relate to trends in climate variables (e.g. Wright and Calderón 2006; Allen et al. 2014; Buechling et al. 2016). A 45-year trend of increasing seed production in *Nothofagus* populations was particularly strong at higher elevations (Allen et al. 2014). Greater phenological sensitivity of plants to climate variation at higher elevations has also been shown in Scottish mountains (Chapman 2013). In the *Nothofagus* populations there were also significant changes in four climate variables over the same time interval. Regression coefficients of relationships between climate variables and seed production were consistent with a trend in climate variables leading to increased seed production (Allen et al. 2014).

Determining what causes temporal trends in tree demographic processes, such as seed production, is challenging and controversial (e.g. Lewis et al. 2009; Lines et al. 2010). A suite of environmental constraints can co-vary with climatic trends, with both influencing resource supply. Just how environmental constraints, including climate, affect resource supply and seeding depends upon a complex set of interrelationships, many of which are poorly understood. The rise in atmospheric CO_2 partial pressures has the potential to increase CO_2 uptake and carbohydrate storage, and hence seed production. Often trees exposed to high CO_2 in Free-Air CO_2 Enrichment (FACE) experiments and Open Top Chamber experiments have been found to produce more seeds than at ambient CO_2 (e.g. Stiling et al. 2004). Average seed production was increased short term for *Pinus taeda* trees growing in a high-CO_2 environment, compared with ambient conditions, but not average seed mass, viability or nutrient content (e.g. LaDeau and Clark 2001; Way et al. 2010). However, maintaining this long term assumes that mineral nutrients are not limiting, which is widely not the case (e.g. Körner 2006; Huang et al. 2007; Millard et al. 2007; McCarthy et al. 2010; Palacio et al. 2014; Hoch 2015). FACE experiments also show that trees sometimes alter their photosynthetic response to elevated atmospheric CO_2, depending upon their N status, and acclimate to elevated atmospheric CO_2 by down-regulating their photosynthetic capacity (Ainsworth and Long 2005). Any positive effects of rising atmospheric CO_2 on photoassimilates will also be moderated where concurrent warming, and reduced moisture availability, stimulate respiration (Clark et al. 2003).

Oxidised forms of gases in the atmosphere, such as NO_2 and SO_2, also vary through time, particularly in those parts of the world with intense volcanic or industrial activity. These gases readily dissolve in water and can influence, for example, ecosystem level N uptake, change demand for other nutrients (e.g. P) and influence carbohydrate allocation (e.g. root growth; e.g. Jonard et al. 2009). Atmospheric N deposition is considered one factor leading to the greater frequency and magnitude of seeding events in *Fagus* forests in southern Sweden (Övergaard et al. 2007). However, any positive effects on seed quantity, and quality, may be negated by N-mediated changes in biotic interactions (e.g. seed predation; Bogdziewicz et al. 2017). Experimental additions of such nutrients can enhance the number, size and total mass of seeds produced annually in most years (Davis et al. 2004; Callahan et al. 2008; Smaill et al. 2011; Wright et al.

2011; Bogdziewicz et al. 2017) and reduce masting (population-level coefficient of variation) by causing a smaller relative increase in the standard deviation of seed mass production than mean seed production (Table 2; Smaill et al. 2011). There is a tendency for such fertiliser additions to have minimal effect in what are ultimately low seed production years when compared with unfertilised controls (e.g. Smaill et al. 2011; Wright et al. 2011). This means fertiliser addition (boosted uptake) did not enhance one or more of initiation, anthesis and maturation. It may be that seed production at one of these stages better reflects remobilisation from storage rather than uptake in low seed production years (Fig. 2).

Compositional and structural trends in forests following disturbance also have strong impacts on resource supply. Canopy disturbance can increase radiation, moisture and nutrient availability on the forest floor. As a consequence, partial tree removal experiments have increased foliar nutrient concentrations (N, P, K) and cone production by residual trees in pine stands on infertile soils (e.g. Karlsson 2000; Karlsson and Örlander 2002). As even-aged stands develop following canopy disturbance nutrients are sequestered in aggrading biomass and this can lead to a gradual decline in soil nutrient availability (Ryan et al. 1997; Clinton et al. 2002). The experimental addition of N fertiliser increased average flower, seed and fine root production to a greater degree in large biomass, older even-aged stands than younger stands (Davis et al. 2004; Smaill et al. 2011). This may be because the large biomass in older stands allowed greater N storage mass which can be remobilised to meet the nutritional demands of reproduction (Yasumura et al. 2006). Increasing basal area dominance, and the changing levels of solar irradiance due to drier and warmer conditions created by increasing El Niño events, have been related to increased flower but not seed production by tree species on Barro Colorado Island, Panama (Wright and Calderón 2006). Below-ground disturbance may also alter seed production. *Acacia* shrubs in the south-western USA, with basal ant nests, have higher soil water, N and P availability than those without ant nests and drives tree to tree variation in resource availability as well as synchrony and average seed production (Wagner 1997).

Table 2 Seed production metrics for replicated *Nothofagus* tree populations with and without N fertiliser addition (see Davis et al. 2004; Smaill et al. 2011)

Metric	Mean for unfertilised plots	Mean for fertilised plots	$F_{2,13}$	P
CV	1.66 (0.06)	1.48 (0.06)	2.72	0.045
SD	2.95 (0.17)	5.33 (0.17)	31.76	0.026
Mean	1.78 (0.14)	3.61 (0.14)	46.51	0.004

The coefficient of variation (CV), standard deviation (SD) and mean seed production are given across 8 years based upon the mean of replicate plot values. Linear models testing for an effect of fertiliser addition on seed production metrics over 8 years ($N = 16$, with eight fertilised and eight unfertilised plots) have associated F and P values given. Pre-fertilisation seed production (from 1999) was included as a covariate in all models (Smaill et al. 2011)

4 Progressing Resource Budget Models

What then are the implications of climate-driven resource dynamics for resource budget models? These models focus on how resource gain, subsequent costs of reproduction and internal resource depletion act during the reproductive cycle of an individual tree (e.g. Isagi et al. 1997; Satake and Iwasa 2000; Satake and Bjørnstad 2008). A tree accumulates a fixed amount of resources in each year in these models, which is added to that stored at the beginning of the year, and unless the summed level of resources exceeds a threshold level required for floral induction and initiation the tree does not produce any flowers and stores the acquired resource (Abe et al. 2016). This is thought to drive the negative temporal autocorrelation sometimes found among years in seeding but not climate. When the level of resource storage exceeds a threshold trees instigate flowering. These stored resources are sequentially invested in anthesis and maturation with an ongoing depletion of storage. The models generate the temporal resource dynamics by which a tree might produce variable seeding (Isagi et al. 1997; Satake and Iwasa 2000). They suggest that at locations where the cost of reproduction is low, relative to annual resource gain, trees may produce seeds in every year, and if the cost of reproduction is high, relative to annual resource gain, seeds are produced at super-annual intervals (e.g. Crone and Rapp 2014). Seed crops are thought to be synchronised at a population-level to improve pollination success although few studies have directly tested for density-dependent pollen limitation among years (Crone and Rapp 2014). Increasingly sophisticated models have been used to portray an interplay between internal resource dynamics, pollination processes and climate (Satake and Iwasa 2002; Venner et al. 2016; Abe et al. 2016).

Resource budget models represent a significant advance in our conceptual understanding of mast seeding although they currently oversimplify resource dynamics (Sala et al. 2012; Crone and Rapp 2014; Abe et al. 2016). These models usually assume non-structural carbohydrates are the resource of interest and that these provide a source of energy (e.g. Isagi et al. 1997; Satake and Iwasa 2000; Miyazaki 2013; Venner et al. 2016). Empirical studies focussed on non-structural carbohydrates have given mixed results with evidence for tissue storage at only some stages in the reproductive cycle (Miyazaki et al. 2002; Hoch et al. 2003; Ichie et al. 2005, 2013). This is not surprising as mineral nutrient resources are thought to be more critical to reproduction than carbohydrates (Ashman 1994; Hemborg and Karlsson 1998; Wheelwright and Logan 2004). Trees have the ability to remobilise significant amounts of mineral nutrients stored in their relatively large branch, stem and root biomass when compared to other perennial plants (Fig. 2; Millard and Grelet 2010). Ichie and Nakagawa (2013) calculated for a tropical, masting tree, using a nutrient budget approach, that stored P accounted for 68% of total P requirement for reproduction, while for N the figure was only 20%.

Abe et al. (2016) recently developed a resource budget model capable of identifying the roles of specific resource-types. This form of the model showed that N dynamics in reproductive structures between anthesis and maturation much better explained masting patterns in *Fagus* than those based upon carbohydrates at these stages. Such evidence suggests that resource budget models be interpreted more generally as the dynamics of a range of resources that limit reproduction at various stages in the reproductive cycle (Sala et al. 2012; Abe et al. 2016). It may not be about whether mineral nutrients or carbohydrates are the currency of interest, but when in the reproductive cycle each resource has a role, as well as what relationships occur among resources (Crone and Rapp 2014). It is worth noting that much of the empirical evidence for a resource dynamics role comes from measurements of specific tissues yet resource budget models operate at the individual tree level.

If resource budget models are to provide a more general framework for understanding tree seeding they need to accommodate a wider range of internal and external processes that determine resource supply variation among years. We propose a model for how these processes might determine the reproductive cycle in a mesic temperate tree (Fig. 4). Initially the critical factors are the level of resource storage resulting from uptake and remobilisation and the number of vegetative buds formed that can later be converted to reproductive buds (Fig. 3). Subsequently, it is temporal variability in uptake, particularly a pulse of N, which might set the level of induction and initiation (Fig. 4; Smaill et al. 2011; Miyazaki et al. 2014). However the explicit relationships between inter-annual variability in resource supply and variability in induction and initiation are largely unexplored (Crone and Rapp 2014). In resource budget models floral initiation is sometimes set by a temperature threshold that is effectively binary, rather than the variability in a specific resource with a mechanistic influence on initiation. If subsequent climatic conditions are suitable, combined with the enhanced N supply, increased photo-assimilates are generated and allocated to structures that enhance uptake of externally sourced mineral nutrients (Fig. 4). Resource budget models do not yet incorporate such interrelationships driven by both internal and external processes. Both remobilisation and uptake of mineral nutrients, as well as stored carbohydrates, are used for floral development leading up to anthesis (Fig. 4). There can then be appreciable nutrient losses through higher than average floral litter fall (Fig. 4; e.g. Davis et al. 2004). The seed maturation stage is one in which stored mineral nutrients, along with currently produced carbohydrates, are used in the developing seeds (Fig. 4). This stage is particularly demanding for both remobilised and current uptake of nutrients (e.g. Abe et al. 2016). Limited supply can further reduce seed production below the potential set at initiation. Over-allocation to reproduction, when resource uptake is low, due to environmental constraints, might impact on tree survival (Owens 1995). At maturation there is a reliance on current uptake of photo-assimilates (Fig. 4) suggesting that this stage coincides with depleted carbohydrate reserves. However, it is also noted that trees can upregulate photosynthetic gene expression to provide additional carbohydrates in response to a reproductive sink (Paul and Foyer 2001). To some level this decouples carbohydrate supply from environmental constraints. Finally, an individual tree can withdraw resources from some reproductive structures (e.g. cones) once their function is completed and also capture nutrients from litter fall, to minimise depletion and reduce recovery time (Fig. 4). The depletion of resource storage during the reproductive cycle appears common, but not universal (Crone and Rapp 2014).

Although we expect a similar set of processes to operate during the reproductive cycle in other types of forest, the nature of the environmental constraints could be very different. The impact of environmental constraints will also vary among species with different traits. The significance of being an N-fixer, ectomycorrhizal or cluster rooted with high P uptake under nutrient poor conditions, given the roles of N and P in seeding, makes such taxa deserving of more attention. This is particularly so if comparisons are made among tree taxa having different nutrient uptake strategies along soil fertility gradients (Lambers et al. 2008). There remains much to understand about the internal recycling of resources, as well as their uptake and loss, if we are to better understand the role of resources during the reproductive cycle in perennial plants (e.g. Millard and Grelet 2010; Smaill et al. 2011; Miyazaki et al. 2014).

5 Climate Cues, Resources and Selection

Much of the evidence for mast seeding increasing plant fitness comes from studies focussed on the abundance of flowers and seeds, as well as their implications for seed dispersal (Piovesan and Adams 2005; Crone and Rapp 2014; Pearse et al. 2016). Of the economies of scale explanations for mast seeding the most widely tested include increased pollination efficiency because flowers are less pollen limited in high-flowering years (e.g. Norton and Kelly 1988), seed predator satiation (e.g. Janzen 1971) and enhanced seed dispersal in high seeding years (Crone and Rapp 2014; Pearse et al. 2016). Each of these explanations requires the essential elements of masting – synchronicity and variability. It is however challenging for fitness benefits to become embedded as a heritable trait in mature offspring when the average fitness contribution of an individual to the next generation is subject to many subsequent developmental processes, filters and stochastic events (e.g. Westoby et al. 2002; Myers and Harms 2009). For example, very few seeds lead to established seedlings and very few established seedlings achieve reproductive status (Peet and Christensen 1987). Under some conditions, such as competition from existing vegetation, high seedling survival is attributable to larger initial seedling size (Westoby et al. 2002). Given that many studies have shown that initial seedling size is positively related to seed size, both across and within species, it may not always be the abundance of seeds that is important to fitness but instead seed size (Westoby et al. 2002).

It is challenging to isolate whether an adaptive trait that varies among populations, or species, is a consequence of ecological sorting in contrasting environments, in space and time, or adaptation by a heritable change in a species population (Ackerly 2003). It is tempting to give natural selection a primary role because of the view that evolution provides the ‘ultimate’ cause of phenotypic variation in contrast to ‘proximate’ developmental and ecological causes (Mayr 1982; Ackerly 2003).

5.1 Climate as the Primary Determinant

Easily measured and readily available climate variables are often used to examine climate-seeding relationships in tree populations. Usually, in the absence of more difficult to measure resource dynamics, interpretations gravitate towards climatic explanations for inter-annual variability in seeding. Some have argued that climate acts only as a cue directly signalling initiation of the reproductive cycle, without a need to invoke any consideration of resources (Kelly et al. 2013; Koenig and Knops 2014). The merits of climate, particularly temperature, as a widespread cue for mast seeding have been argued (Kelly et al. 2013). Under this explanation, individual trees, in a contemporary population, react similar to the climatic cue and benefit from the associated economies of scale. However, this line of research is bedevilled with correlations being interpreted as causation. For example, one line of evidence for pollination efficiency is of increasing seed quality (assumed to reflect pollination) with a climatically related increase in seed crop size (assumed to reflect total number of flowers; e.g. Norton and Kelly 1988; Allen and Platt 1990). Yet an unmeasured co-varying resource supply driven by climate could also impact on seed quality. It is becoming increasingly untenable to assume climatic variation can occur without indirectly affecting variation in mineral nutrient and carbohydrate resources.

A more nuanced view of the climate cue perspective is that the primary role of climate is modified when also associated with variability in resource supply. So that when a cue is correlated with resources, these resources are in turn allocated to enhance variation in seed production among years (Crone and Rapp 2014) and offer the advantage of minimising storage costs (e.g. Kelly and Sork 2002). However, the evidence is often unclear because it does not attempt to separate the role of climate from resources. For example, it has been assumed that cooler than average temperatures would relate to decreased resources for masting dipterocarp trees (Kelly and Sork 2002), yet, as we have previously discussed, assuming how climate variables relate to carbohydrate resources in trees is tenuous (Waring and Schlesinger 1985) and cool temperatures can increase soil-available nutrients for uptake by a tree (e.g. Upadhyay et al. 1989; Paul et al. 2003; Smaill et al. 2011; Schaeffer et al. 2013).

5.2 Resources as the Ancestral Determinant

It is proving difficult to justify an evolutionary basis for climate-seeding relationships if climate cues themselves do not provide a direct advantage, at an individual-level, in an antecedent population for a selected trait (Waller 1993; Moreira et al. 2015; Koenig et al. 2015). The emergence of mast seeding from an antecedent population is favoured for those tree species where a level of flowering and seeding synchrony already conferred advantages to individuals. For example, the selection of masting in a dioecious species, such as the strongly masting podocarp *Dacrydium cupressinum* (Norton and Kelly 1988) or *Juniperus thurifera* (Montesinos et al.

2012), required synchronisation among individual male and female trees within an antecedent population. Furthermore, trees primarily outcross as a consequence of mechanisms that enforce allogamy such as self-incompatibility to overcome inbreeding depression (Petit and Hampe 2006) and some strongly masting species appear self-sterile (e.g. *Nothofagus*; Wardle 1984). The level of outcrossing directly affects progeny vigour and population demography and there are benefits to maintaining outcrossing rates through the synchronisation of flowering phenology among individuals (Petit and Hampe 2006; Wright et al. 2013). The benefits do depend upon the level to which short-term reproductive assurance through selfing compensates for any negative consequences of reduced outcrossing (Igic and Busch 2013).

The weight of evidence suggests that the origin of mast seeding could have been a consequence of an ecological relationship that existed in an antecedent population which was subsequently amplified through selection to embed the trait in contemporary populations. We postulate that inter-annual variation in seed production in an antecedent population was synchronised by inter-annual variation in resources critical to reproduction and reflected climate-driven processes. Given that the spatial scale of climate variation is often large we would expect a resource-driven level of synchronisation among individuals within a population. Climate in some years does constrain the availability of carbohydrates and mineral nutrients essential to seed production in trees (e.g. Paul et al. 2003; Smaill et al. 2011; Schaeffer et al. 2013). In contemporary populations, we know that an increase in the types of resources constrained by climate can increase the level of seed production (e.g. Le Tacon and Oswald 1977; Fahey et al. 1998; Karlsson and Örlander 2002; Davis et al. 2004). In addition, a trade-off between current and previous annual seed production being moderated by seasonal climate suggests a resource-based climate control on recovery from seeding events (Moreira et al. 2015).

A resource-driven level of synchrony among individuals would provide an advantage to certain species, for example those that outcross or are dioecious. Those individuals in an antecedent population that produced the most flowers and seeds, in response to a climate-driven resource pulse, would couple and have progeny that are well resourced and numerically dominant. A resource pulse would be effective at floral initiation by setting the potential seed production or at anthesis and maturation by minimising the level to which actual seed production falls below the potential. For a reproductive trait like flowering, a level of assortive mating between individuals is inevitable (Forrest and Miller-Rushing 2010). Successful offspring from synchronised individuals could progressively enhance the linkage between climate and mast seeding. We might expect masting to develop in species found in environments with a scarcity of critical resources and subject to resource pulsing. This, for example, might partly explain why mast seeding is so developed in New Zealand (Webb and Kelly 1993) where widespread infertile soils have been used to explain other traits in the tree flora (e.g. evergreeness; McGlone et al. 2004). More widely, many mast seeding tree species are ectomycorrhzal (e.g. *Nothofagus* and *Pinus*) and Newbery et al. (2006) argued that uptake of resource pulses occur in response to climatically-related activity of roots and ectomycorrhiza. It does remain challenging to determine the level to which selection enhances mast seeding beyond that explained by a complex view of resource-climate dynamics.

6 Conclusions and Directions for Research

Climate-driven resource dynamics may well have a central and diverse role in determining the distinctive inter-annual patterns of tree seed production (Crone and Rapp 2014). This role has sometimes been questioned in favour of a direct climatic link to fitness benefits, although the evidence for both is limited and incomplete. While challenging, clarifying the role of resources in tree seed production patterns is a more tractable problem than resolving the influence of natural selection in long-lived organisms like trees. Recently, considerable progress has been made in clarifying the functional role of resource supply, including uptake, in processes leading up to initiation (e.g. Miyazaki et al. 2014). Such a role in initiation needs to be more widely substantiated but it is important to also further understand the role of resource dynamics post-initiation. That different climate variables relate to various stages in the reproductive cycle, that climate variables relate to external resource availability, that resource supply involves uptake during this cycle and that resource-driven abortion at several stages lowers reproductive output all point to uptake complementing remobilisation post-initiation. Confirming how tree seeding patterns are, in part, determined by a sequence of climate-driven resources dynamics (see Fig. 4 for a view on mesic temperate forests) requires:

1. Evaluating whether the proportion of reproductive buds that ultimately produce flowers (anthesis) or seeds (maturation) is influenced by resource supply. This entails having a means of assessing the abundance of reproductive buds that have been initiated as well as a way of altering resource supply post-initiation (e.g. fertiliser addition);
2. Establishing the level to which resource supply to reproductive tissues reflects uptake, storage and remobilisation, or both, at all stages in the reproductive cycle (potentially using isotopes; Fig. 4). When uptake has a role we also then need to understand what regulates uptake. Trees have the capacity to up- and down-regulate carbohydrates through photosynthesis and control mineral nutrient uptake through root and mycorrhizal growth and activity;
3. Determining how climate variables determine the supply of multiple resources at various stages in the reproductive cycle. We need to move away from the obfuscation caused by the common use of climate as an assumed resource-related proxy (Crone and Rapp 2014). Smaill et al. (2011) showed that inter-annual variability in rainfall-driven N supply appears to occur within tree populations dispersed over a 3.6 km^2 area and it would be insightful to know more fully at what scales such relationships exist.

A suitable design to progress these ideas would involve the experimental manipulation of resource dynamics at various stages in the reproductive cycle with complementary measurements of inter-annual variation in climate, resources as well reproductive buds, flowers and seeds. Such experiments would help clarify whether trees accumulate resources in each year, add these to storage, and that only when storage exceeds a threshold does floral initiation occur and set the level of seed production (Abe et al. 2016). This may help explain any periodicity in seeding (Allen et al. 2012).

The same climate-driven resources dynamics (in terms of specific climate variables and resources) are not expected to determine seed production patterns in regions with markedly different climates. Some have considered that the involvement of different seasonal climate variables means they have only been chosen as a cue (Koenig and Knops 2014; Crone and Rapp 2014). In this review we did not find an inordinate number of climatic variables involved and we instead suggest there is the potential for each to have a role in resource dynamics. In addition, we do not necessarily expect that species with functionally different traits will have the same climate-resource relationships at a location, even when subjected to the same climate regime. Given the functional role of N in initiation, for example, we might expect an N-fixing tree species to exhibit very different resource dynamics than a tree species without this trait. As a consequence, there is much merit in undertaking the type of experiment outlined above for functionally different species in the same environment and contrasting these relationships for the same species in markedly different environments. The ability to contrast the results of such experiments will have strong implications for the advancement of resource budget models and integrating the role of proximate factors with fitness benefits to explain seed production patterns in trees.

Acknowledgements This work was funded by Landcare Research, the former Foundation for Research, Science and Technology's Ecosystem Resilience Outcome Based Investment grant (contract number C09X0502) and one of the authors. Input by Matt McGlone and Bill Lee improved the presentation of ideas in this chapter.

References

Abe T, Tachiki Y, Kon H, Nagasaka A, Onodera K, Minamino K, Han Q, Satake A (2016) Parameterisation and validation of a resource budget model for masting using spatiotemporal flowering data of individual trees. Ecol Lett 19:1129–1139

Ackerly DD (2003) Community assembly, niche conservatism, and adaptive evolution in changing environments. Int J Plant Sci 164(S3):S165–S184

Ågren J (1988) Between-year variation in flowering and fruit set in frost-prone and frost-sheltered populations of dioecious *Rubus chamaemorus*. Oecologia 76:175–183

Ainsworth EA, Long SP (2005) What have we learned from 15 years of free-air CO2 enrichment (FACE)? A meta-analytic review of the responses of photosynthesis, canopy properties and plant production to rising CO2. New Phytol 165:351–372

Alla AQ, Camarero JJ, Maestro-Martínez M, Montserrat-Martí G (2012) Acorn production is linked to secondary growth but not to declining carbohydrate concentrations in current-year shoots of two oak species. Trees 26:841–850

Allen RB, Platt KH (1990) Annual seedfall variation in *Nothofagus solandri* (Fagaceae), Canterbury, New Zealand. Oikos 57:199–206

Allen RB, Mason NWH, Richardson SJ, Platt KH (2012) Synchronicity, periodicity, and bimodality in inter-annual tree seed production along an elevation gradient. Oikos 121:367–376
Allen RB, Hurst JM, Portier J, Richardson SJ (2014) Elevation-dependent responses of tree seed production to climate change over 45 years. Ecol Evol 4:3525–3537
Ashman TL (1994) A dynamic perspective on the physiological cost of reproduction in plants. Am Nat 144:300–316
Barclay AS, Earle FR (1974) Chemical analyses of seeds III. Oil and protein content of 1253 species. Econ Bot 28:178–236
Bazzaz FA, Carlson RW, Harper JL (1979) Contribution to reproductive effort by photosynthesis of flowers and fruits. Nature 279:554–555
Bhupinderpal-Singh NA, Lofvenius MO, Högberg MN, Mellander PE, Högberg P (2003) Tree root and soil heterotrophic respiration as revealed by girdling of boreal scots pine forest: extending observations beyond the first year. Plant Cell Environ 26:1287–1296
Bogdziewicz M, Crone EE, Steele MA, Zwolak R (2017) Effects of nitrogen deposition on reproduction in a masting tree: benefits of higher seed production are trumped by negative biotic interactions. J Ecol 105:310–320
Bradford MA, Davies CA, Frey SD, Maddox TR, Melillo JM, Mohan JE, Reynolds JF, Treseder KK, Wallenstein MD (2008) Thermal adaptation of soil microbial respiration to elevated temperature. Ecol Lett 11:1316–1327
Buechling A, Martin PH, Canham CD, Shepperd WD, Battaglia MA (2016) Climate drivers of seed production in *Picea engelmannii* and response to warming temperatures in the southern Rocky Mountains. J Ecol 104:1051–1062
Burrows LE, Allen RB (1991) Silver beech (Nothofagus menziesii (Hook. f.) Oerst.) seedfall patterns in the Takitimu range, South Island, New Zealand. New Zeal J Bot 29:361–365
Callahan HS, Del Fierro K, Patterson AE, Zafar H (2008) Impacts of elevated nitrogen inputs on oak reproductive and seed ecology. Glob Chang Biol 14:285–293
Canham CD, Ruscoe WA, Wright EA, Wilson DJ (2014) Spatial and temporal variation in tree seed production and dispersal in a New Zealand temperate rainforest. Ecosphere 5(4):49
Carevic FS, Fernández M, Alejano R, Vázquez-Piqué J, Tapias R, Corral E, Domingo J (2010) Plant water relations and edaphoclimatic conditions affecting acorn production in a holm oak (*Quercus ilex* L. ssp. ballota) open woodland. Agrofor Syst 78:299–308
Chandler RF (1938) The influence of nitrogenous fertilizer applications upon seed production of certain deciduous forest trees. J For 36:761–766
Chang-Yang CH, Sun I, Tsai CH, Lu CL, Hsieh CF (2016) ENSO and frost codetermine decade-long temporal variation in flower and seed production in a subtropical rain forest. J Ecol 104: 44–54
Chapman DS (2013) Greater phenological sensitivity to temperature on higher Scottish mountains: new insights from remote sensing. Glob Chang Biol 19:3463–3471
Chapman SK, Langley JA, Hart SC, Koch GW (2006) Plants actively control nitrogen cycling: uncorking the microbial bottleneck. New Phytol 169:27–34
Cipollini ML, Stiles EW (1991) Costs of reproduction in *Nyssa sylvatica*: sexual dimorphism in reproductive frequency and nutrient flux. Oecologia 86:585–593
Clark DA, Piper SC, Keeling CD, Clark DB (2003) Tropical rain forest tree growth and atmospheric carbon dynamics linked to interannual temperature variation during 1984–2000. Proc Natl Acad Sci U S A 100:5852–5857
Clement J, Maes P, de Strihou CVY, van der Groen G, Barrios JM, Verstraeten WW, van Ranst M (2010) Beechnuts and outbreaks of nephropathia epidemica (NE): of mast, mice and men. Nephrol Dial Transplant 25:1740–1746
Clinton PW, Allen RB, Davis MR (2002) Nitrogen storage and availability during stand development in a New Zealand *Nothofagus* forest. Can J For Res 32:344–352
Cook BI, Wolkovich EM, Parmesan C (2012) Divergent responses to spring and winter warming drive community level flowering trends. Proc Natl Acad Sci U S A 109:9000–9005

Coomes DA, Allen RB, Bentley WA, Burrows LE, Canham CD, Fagan L, Forsyth DM, Gaxiola-Alcantar A, Parfitt RL, Ruscoe WA, Wardle DA, Wilson DJ, Wright EF (2005) The hare, the tortoise, and the crocodile: the ecology of angiosperm dominance, conifer persistence and fern filtering. J Ecol 93:918–935
Crone EE, Rapp JM (2014) Resource depletion, pollen coupling, and the ecology of mast seeding. Ann N Y Acad Sci 1322:21–34
Crone EE, McIntire EJ, Brodie J (2011) What defines mast seeding? Spatio-temporal patterns of cone production by whitebark pine. J Ecol 99:438–444
Danielson JÅH, Frommer WB (2013) Jack of all trades: master of flowering. Science 339:659–660
Davis MR, Allen RB, Clinton PW (2004) The influence of N addition on nutrient content, leaf carbon isotope ratio, and productivity in a *Nothofagus* forest during stand development. Can J For Res 34:2037–2048
Dickmann DI, Kozlowski TT (1969) Seasonal changes in the macro-and micro-nutrient composition of ovulate strobili and seeds of *Pinus resinosa*. Can J Bot 47:1547–1554
Ehrlén J, van Groenendael J (2001) Storage and the delayed costs of reproduction in the understorey perennial *Lathyrus vernus*. J Ecol 89:237–246
Eis S (1973) Cone production of Douglas-fir and Grand fir and its climatic requirements. Can J For Res 3:61–70
El Zein R, Bréda N, Gérant D, Zeller B, Millard P (2011) Nitrogen sources for current-year shoot growth in 50-year-old sessile oak trees: an in situ N labeling approach. Tree Physiol 31:1390–1400
Fahey TJ, Battles JJ, Wilson GF (1998) Responses of early successional northern hardwood forests to changes in nutrient availability. Ecol Monogr 68:183–212
Fearer TM, Norman GW, Pack JC, Bittner S, Healy WM (2008) Influence of physiographic and climatic factors on spatial patterns of acorn production in Maryland and Virginia, USA. J Biogeogr 35:2012–2025
Fernández-Martínez M, Vicca S, Janssens IA, Espelta JM, Peñuelas J (2017) The role of nutrients, productivity and climate in determining tree fruit production in European forests. New Phytol 213:669–679
Forrest J, Miller-Rushing AJ (2010) Toward a synthetic understanding of the role of phenology in ecology and evolution. Philos Trans R Soc Lond B Biol Sci 365:3101–3112
Foster AS, Gifford EM (1974) Comparative morphology of vascular plants, 2nd edn. WH Freeman and Company, San Francisco, CA
Goubitz S, Werger MJ, Shmida A, Ne'eman G (2002) Cone abortion in *Pinus halepensis*: the role of pollen quantity, tree size and cone location. Oikos 97:125–133
Greene DF, Johnson EA (2004) Modelling the temporal variation in the seed production of North American trees. Can J For Res 34:65–75
Groom PK, Lamont BB (2010) Phosphorus accumulation in Proteaceae seeds: a synthesis. Plant and Soil 334:61–72
Han Q, Kabeya D, Iio A, Kakubari Y (2008) Masting in *Fagus crenata* and its influence on the nitrogen content and dry mass of winter buds. Tree Physiol 28:1269–1276
Han Q, Kabeya D, Iio A, Inagaki Y, Kakubari Y (2014) Nitrogen storage dynamics are affected by masting events in *Fagus crenata*. Oecologia 174:679–687
Han Q, Kagawa A, Kabeya D, Inagaki Y (2016) Reproduction-related variation in carbon allocation to woody tissues in *Fagus crenata* using a natural ^{13}C approach. Tree Physiol 36:1343–1352
Hannerz M, Almqvist C, Hornfeldt R (2002) Timing of seed dispersal in *Pinus sylvestris* stands in central Sweden. Silva Fenn 36:757–765
Hemborg ÅM, Karlsson PS (1998) Somatic costs of reproduction in eight subarctic plant species. Oikos 82:149–157
Henkel TW, Mayor JR, Woolley LP (2005) Mast fruiting and seedling survival of the ectomycorrhizal, monodominant *Dicymbe corymbosa* (Caesalpiniaceae) in Guyana. New Phytol 167:543–556
Herrera CM, Jordano P, Guitián J, Traveset A (1998) Annual variability in seed production by woody plants and the masting concept: reassessment of principles and relationship to pollination and seed dispersal. Am Nat 152:576–594

Hilton GM, Packham JR (2003) Variation in the masting of common beech (*Fagus sylvatica* L.) in northern Europe over two centuries (1800–2001). Forestry 76:319–328

Hoch G (2015) Carbon reserves as indicators for carbon limitation in trees. In: Progress in botany. Springer International Publishing, Switzerland, pp 321–346

Hoch G, Richter A, Körner C (2003) Non-structural carbon compounds in temperate forest trees. Plant Cell Environ 26:1067–1081

Hoch G, Siegwolf RT, Keel SG, Körner C, Han Q (2013) Fruit production in three masting tree species does not rely on stored carbon reserves. Oecologia 171:653–662

Högberg MN, Briones MJI, Keel SG, Metcalfe DB, Campbell C, Midwood AJ, Thornton B, Hurry V, Linder S, Näsholm T, Högberg P (2010) Quantification of effects of season and nitrogen supply on tree below-ground carbon transfer to ectomycorrhizal fungi and other soil organisms in a boreal pine forest. New Phytol 187:485–493

Huang JG, Bergeron Y, Denneler B, Berninger F, Tardif J (2007) Response of forest trees to increased atmospheric CO_2. Crit Rev Plant Sci 26:265–283

Hutchings MJ, John EA, Wijesinghe DK (2003) Toward understanding the consequences of soil heterogeneity for plant populations and communities. Ecology 84:2322–2334

Ichie T, Nakagawa M (2013) Dynamics of mineral nutrient storage for mast reproduction in the tropical emergent tree *Dryobalanops aromatica*. Ecol Res 28:151–158

Ichie T, Kenzo T, Kitahashi Y, Koike T, Nakashizuka T (2005) How does *Dryobalanops aromatica* supply carbohydrate resources for reproduction in a masting year? Trees 19:704–711

Ichie T, Igarashi S, Yoshida S, Kenzo T, Masaki T, Tayasu I (2013) Are stored carbohydrates necessary for seed production in temperate deciduous trees? J Ecol 101:525–531

Igic B, Busch JW (2013) Is self-fertilization an evolutionary dead end? New Phytol 198:386–397

Iio A, Yokoyama A, Takano M, Nakamura T, Fukasawa H, Nose Y, Kakubari Y (2008) Interannual variation in leaf photosynthetic capacity during summer in relation to nitrogen, leaf mass per area and climate within a *Fagus crenata* crown on Naeba Mountain, Japan. Tree Physiol 28:1421–1429

Ims RA (1990) The ecology and evolution of reproductive synchrony. Trends Ecol Evol 5:135–140

Isagi Y, Sugimura K, Sumida A, Ito H (1997) How does masting happen and synchronize? J Theor Biol 187:231–239

Janzen DH (1971) Seed predation by animals. Annu Rev Ecol Evol Syst 2:465–492

Jonard M, André F, Dambrine E, Ponette Q, Ulrich E (2009) Temporal trends in the foliar nutritional status of the French, Walloon and Luxembourg broad-leaved plots of forest monitoring. Ann For Sci 66:1–10

Jones CG, Ostfeld RS, Richard MP, Schauber EM, Wolff JO (1998) Chain reactions linking acorns to gypsy moth outbreaks and Lyme disease risk. Science 279:1023–1026

Karlsson C (2000) Seed production of *Pinus sylvestris* after release cutting. Can J For Res 30: 982–989

Karlsson C, Örlander G (2002) Mineral nutrients in needles of *Pinus sylvestris* seed trees after release cutting and their correlations with cone production and seed weight. For Ecol Manage 166:183–191

Kelly D (1994) The evolutionary ecology of mast seeding. Trends Ecol Evol 9:465–470

Kelly D, Sork VL (2002) Mast seeding in perennial plants: why, how, where? Annu Rev Ecol Evol Syst 33:427–447

Kelly D, Hart DE, Allen RB (2001) Evaluating the wind-pollination benefits of mast seeding. Ecology 82:117–126

Kelly D, Geldenhuis JA, Holland EP, Plank MJ, Brockie RE, Cowan PE, Harper GA, Lee WG, Maitland MJ, Mark AF, Mills JA, Wilson PR, Byrom AE (2013) Of mast and mean: differential-temperature cue makes mast seeding insensitive to climate change. Ecol Lett 16:90–98

Koenig WD, Knops JMH (2000) Patterns of annual seed production by northern hemisphere trees: a global perspective. Am Nat 155:59–69

Koenig WD, Knops JM (2014) Environmental correlates of acorn production by four species of Minnesota oaks. Popul Ecol 56:63–71

Koenig WD, Kelly D, Sork VL, Duncan RP, Elkinton JS, Peltonen MS, Westfall RD (2003) Dissecting components of population-level variation in seed production and the evolution of masting behavior. Oikos 102:581–591
Koenig WD, Knops JM, Carmen WJ, Pearse IS (2015) What drives masting? The phenological synchrony hypothesis. Ecology 96:184–192
Koenig WD, Alejano R, Carbonero MD, Fernández-Rebollo P, Knops JM, Marañón T, Padilla-Díaz CM, Pearse IS, Pérez-Ramos IM, Vázquez-Piqué J, Pesendorfer MB (2016) Is the relationship between mast-seeding and weather in oaks related to their life-history or phylogeny? Ecology 97:2603–2615
Kon H, Noda T, Terazawa K, Koyama H, Yasaka M (2005) Proximate factors causing mast seeding in *Fagus crenata*: the effects of resource level and weather cues. Can J Bot 83:1402–1409
Körner C (2003) Carbon limitation in trees. J Ecol 91:4–17
Körner C (2006) Plant CO_2 responses: an issue of definition, time and resource supply. New Phytol 172:393–411
Kozlowski TT, Kramer PJ, Pallardy SG (1991) The physiological ecology of woody plants. Academic Press, New York, NY
Kroiss SJ, HilleRisLambers J (2015) Recruitment limitation of long-lived conifers: implications for climate change responses. Ecology 96:1286–1297
LaDeau S, Clark JS (2001) Rising CO_2 levels and the fecundity of forest trees. Science 292:95–99
Lambers H, Raven JA, Shaver GR, Smith SE (2008) Plant nutrient-acquisition strategies change with soil age. Trends Ecol Evol 23:95–103
Lamont BB, Groom PK (2013) Seeds as a source of carbon, nitrogen, and phosphorus for seedling establishment in temperate regions: a synthesis. Am J Plant Sci 4 No. 5A. Article ID:32249, 11 pp
Lamontagne JM, Boutin S (2007) Local-scale synchrony and variability in mast seed production patterns of *Picea glauca*. J Ecol 95:991–1000
Law B, Mackowski C, Schoer L, Tweedie T (2000) Flowering phenology of myrtaceous trees and their relation to climatic, environmental and disturbance variables in northern New South Wales. Austral Ecol 25:160–178
Le Tacon F, Oswald H (1977) Influence de la fertilisation minérale sur la fructification du Hêtre (*Fagus sylvatica*). Ann For Sci 34:89–109
Ledgard NJ, Cath PW (1983) Seed of New Zealand *Nothofagus* species: studies of seed weight, viability, shape and the effect of varying stratification periods. New Zeal J For 28:150–162
Leishman MR, Wright IJ, Moles AT, Westoby M (2000) The evolutionary ecology of seed size. In: Seeds, the ecology of regeneration in plant communities. CAB International, Wallingford, pp 31–57
Lewis SL, Lloyd J, Sitch S, Mitchard ETA, Laurence WF (2009) Changing ecology of tropical forests: evidence and drivers. Annu Rev Ecol Evol Syst 40:529–549
Linder S, Troeng E (1981) The seasonal course of respiration and photosynthesis in strobili of Scots pine. For Sci 27:267–276
Lines ER, Coomes DA, Purves DW (2010) Influences of forest structure, climate and species composition on tree mortality across the eastern US. PLoS One 5(10):e13212
Lowry WP (1966) Apparent meteorological requirements for abundant cone crops in Douglas fir. For Sci 2:328–332
Marschner H, Dell B (1994) Nutrient uptake in mycorrhizal symbiosis. Plant and Soil 159:89–102
Martínez-Vilalta J, Sala A, Asenio D, Galliano L, Hoch G, Palacio S, Piper FI, Lloret F (2016) Dynamics of non-structural carbohydrates in terrestrial plants: a global synthesis. Ecol Monogr 86:495–516
Masaki T, Oka T, Osumi K, Suzuki W (2008) Geographical variation in climatic cues for mast seeding of *Fagus crenata*. Popul Ecol 50:357–366
Maurer E (1964) Buchen- und Eichensamenjahre in Unter-franken wahrend der letzten 100 Jahre. Allgemeine Forst Zeitschrift 31:469–470
Mayr E (1982) The growth of biological thought. Harvard University Press, Cambridge/Massachusetts

McCarthy HR, Oren R, Johnsen KH, Gallet-Budynek A, Pritchard SG, Cook CW, LaDeau SL, Jackson RB, Finzi AC (2010) Re-assessment of plant carbon dynamics at the Duke free-air CO_2 enrichment site: interactions of atmospheric [CO_2] with nitrogen and water availability over stand development. New Phytol 185:514–528
McDowell SC, McDowell NG, Marshall JD, Hultine K (2000) Carbon and nitrogen allocation to male and female reproduction in Rocky Mountain Douglas-fir (*Pseudotsuga menziesii var. glauca*, Pinaceae). Am J Bot 87:539–546
McGlone M, Dungan R, Hall GMJ, Allen RB (2004) Winter leaf loss in the New Zealand woody flora. New Zeal J Bot 42:1–19
Meilan R (1997) Floral induction in woody angiosperms. New For 14:179–202
Millard P (1994) Measurement of the remobilization of nitrogen for spring leaf growth of trees under field conditions. Tree Physiol 14:1049–1054
Millard P, Grelet G-A (2010) Nitrogen storage and remobilization by trees: ecophysiological relevance in a changing world. Tree Physiol 30:1083–1095
Millard P, Proe MF (1993) Nitrogen uptake, partitioning and internal cycling in *Picea sitchensis* (Bong.) Carr. as influenced by nitrogen supply. New Phytol 125:113–119
Millard P, Thomson CM (1989) The effect of autumn senescence of leaves on the internal cycling of nitrogen for the spring growth of apple trees. J Exp Bot 40:1285–1289
Millard P, Hester A, Wendler R, Baillie G (2001) Interspecific defoliation responses of trees depend on sites of winter nitrogen storage. Funct Ecol 15:535–543
Millard P, Wendler R, Grassi G, Grelet G-A, Tagliavini M (2006) Translocation of nitrogen in the xylem of field-grown cherry and poplar trees during remobilization. Tree Physiol 26:527–536
Millard P, Sommerkorn M, Grelet G-A (2007) Environmental change and carbon limitation in trees: a biochemical, ecophysiological and ecosystem appraisal. New Phytol 175:11–28
Miyazaki Y (2013) Dynamics of internal carbon resources during masting behavior in trees. Ecol Res 28:143–150
Miyazaki Y, Hiura T, Kato E, Funada R (2002) Allocation of resources to reproduction in *Styrax obassia* in a masting year. Ann Bot 89:767–772
Miyazaki Y, Maruyama Y, Chiba Y, Kobayashi MJ, Joseph B, Shimizu KK, Mochida K, Hiura T, Kon H, Satake A (2014) Nitrogen as a key regulator of flowering in *Fagus crenata*: understanding the physiological mechanism of masting by gene expression analysis. Ecol Lett 17:1299–1309
Moles A, Westoby M (2003) Latitude, seed predation and seed mass. J Biogeogr 30:105–128
Monks A, Kelly D (2006) Testing the resource-matching hypothesis in the mast seeding tree *Nothofagus truncata* (Fagaceae). Austral Ecol 31:366–375
Monks A, Monks JM, Tanentzap AJ (2016) Resource limitation underlying multiple masting models makes mast seeding sensitive to future climate change. New Phytol 210:419–430
Montesinos D, García-Fayos P, Verdú M (2012) Masting uncoupling: mast seeding does not follow all mast flowering episodes in a dioecious juniper tree. Oikos 121:1725–1736
Moreira X, Abdala-Roberts L, Linhart YB, Mooney KA (2015) Effects of climate on reproductive investment in a masting species: assessment of climatic predictors and underlying mechanisms. J Ecol 103:1317–1324
Müller-Haubold H, Hertel D, Leuschner C (2015) Climatic drivers of mast fruiting in European beech and resulting C and N allocation shifts. Ecosystems 18:1083–1100
Muñoz N, Guerri J, Legaz F, Primo-Millo E (1993) Seasonal uptake of N-nitrate and distribution of absorbed nitrogen in peach trees. Plant and Soil 150:263–269
Myers JA, Harms KE (2009) Seed arrival, ecological filters, and plant species richness: a meta-analysis. Ecol Lett 12:1250–1260
Näsholm T (1994) Removal of nitrogen during needle senescence in Scots pine (*Pinus sylvestris* L.) Oecologia 99:290–296
Neilsen D, Millard P, Neilsen GH, Hogue EJ (1997) Sources of N for leaf growth in a high-density apple (*Malus domestica*) orchard irrigated with ammonium nitrate solution. Tree Physiol 17:733–739
Newbery DM (2005) Ectomycorrhizas and mast fruiting in trees: linked by climate-driven tree resources? New Phytol 167:324–326

Newbery DM, Chuyong GB, Zimmermann L (2006) Mast fruiting of large ectomycorrhizal African rain forest trees: importance of dry season intensity, and the resource-limitation hypothesis. New Phytol 170:561–579

Nielsen BO (1977) Beech seeds as an ecosystem component. Oikos 29:268–274

Norton DA, Kelly D (1988) Mast seeding over 33 years by *Dacrydium cupressinum* Lamb.(rimu) (Podocarpaceae) in New Zealand: the importance of economies of scale. Funct Ecol 2:399–408

Orwin KH, Stevenson BA, Smaill SJ, Kirschbaum MU, Dickie IA, Clothier BE, Garrett LG, Weerden TJ, Beare MH, Curtin D, Klein CA (2015) Effects of climate change on the delivery of soil-mediated ecosystem services within the primary sector in temperate ecosystems: a review and New Zealand case study. Glob Chang Biol 21:2844–2860

Ostfeld RS, Keesing F (2000) Pulsed resources and community dynamics of consumers in terrestrial ecosystems. Trends Ecol Evol 15:232–237

Övergaard R, Gemmel P, Karlsson M (2007) Effects of weather conditions on mast years in beech (*Fagus sylvatica* L.) in Sweden. Forestry 80:555–565

Owens JN (1969) The relative importance of initiation and early development on cone production in Douglas fir. Can J Bot 47:1039–1049

Owens JN (1995) Constraints to seed production: temperate and tropical forest trees. Tree Physiol 15:477–484

Palacio S, Hester AJ, Maestro M, Millard P (2008) Browsed *Betula pubescens* trees are not carbon-limited. Funct Ecol 22:808–815

Palacio S, Hoch G, Sala A, Körner C, Millard P (2014) Does carbon storage limit tree growth? New Phytol 201:1096–1100

Paul MJ, Foyer CH (2001) Sink regulation of photosynthesis. J Exp Bot 52:1383–1400

Paul KI, Polglase PJ, O'Connell AM, Carlyle JC, Smethurst PJ, Khanna PK (2003) Defining the relation between soil water content and net nitrogen mineralization. Eur J Soil Sci 54:39–47

Pearse IS, Koenig WD, Kelly D (2016) Mechanisms of mast seeding: resources, weather, cues, and selection. New Phytol 212:546–562

Peet RK, Christensen NL (1987) Competition and tree death. Bioscience 37:586–595

Pérez-Ramos IM, Aponte C, García LV, Padilla-Díaz CM, Marañón T (2014) Why is seed production so variable among individuals? A ten-year study with oaks reveals the importance of soil environment. PLoS One 9(12):e115371

Pérez-Ramos IM, Padilla-Díaz CM, Koenig WD, Maranon T (2015) Environmental drivers of mast-seeding in Mediterranean oak species: does leaf habit matter? J Ecol 103:691–700

Petit RJ, Hampe A (2006) Some evolutionary consequences of being a tree. Annu Rev Ecol Evol Syst 37:187–214

Piovesan G, Adams JM (2001) Masting behaviour in beech: linking reproduction and climatic variation. Can J Bot 79:1039–1047

Piovesan G, Adams JM (2005) The evolutionary ecology of masting: does the environmental prediction hypothesis also have a role in mesic temperate forests? Ecol Res 20:739–743

Platt KH, Allen RB, Coomes DA, Wiser SK (2004) Mountain beech seedling responses to removal of below-ground competition and fertiliser addition. New Zeal J Ecol 28:289–293

Pons J, Pausas JG (2012) The coexistence of acorns with different maturation patterns explains acorn production variability in cork oak. Oecologia 169:723–731

Pregitzer K, Burton A (1991) Sugar maple seed production and nitrogen in litterfall. Can J For Res 21:1148–1153

Proe MF, Millard P (1995) Effect of P supply upon seasonal growth and internal cycling of P in Sitka spruce (*Picea sitchensis* (Bong.) Carr.) seedlings. Plant and Soil 168–169:313–317

Read DJ, Perez-Moreno J (2003) Mycorrhizas and nutrient cycling in ecosystems – a journey towards relevance? New Phytol 157:475–492

Reekie EG, Bazzaz FA (1987) Reproductive effort in plants. 2. Does carbon reflect the allocation of other resources? Am Nat 129:897–906

Regler N, Streb S, Zeeman SC, Frey B (2010) Seasonal changes in starch and sugar content of poplar (*Populus deltoides* x nigra cv. Dorskamp) and the impact of stem girdling on carbohydrate allocation to roots. Tree Physiol 30:979–987

Richardson SJ, Allen RB, Whitehead D, Carswell FE, Ruscoe WA, Platt KH (2005) Climate and net carbon availability both determine seed production in a temperate tree species. Ecology 86: 972–981

Roland CA, Schmidt JH, Johnstone JF (2014) Climate sensitivity of reproduction in a mast-seeding boreal conifer across its distributional range from lowland to treeline forests. Oecologia 174:665–677

Ryan MG, Binkley D, Fownes JH (1997) Age-related decline in forest productivity: pattern and process. Adv Ecol Res 27:213–262

Sala A, Hopping K, McIntire EJB, Delzon S, Crone EE (2012) Masting in whitebark pine (*Pinus albicaulis*) depletes stored nutrients. New Phytol 196:189–199

Sanchez EE, Righetti TL, Sugar D, Lombard PB (1991) Recycling of nitrogen in field-grown 'Comice' pears. J Hortic Sci 66:479–486

Sánchez-Humanes B, Sork VL, Espelta JM (2011) Trade-offs between vegetative growth and acorn production in *Quercus lobata* during a mast year: the relevance of crop size and hierarchical level within the canopy. Oecologia 166:101–110

Satake A, Bjørnstad ON (2008) A resource budget model to explain intraspecific variation in mast reproductive dynamics. Ecol Res 23:3–10

Satake A, Iwasa Y (2000) Pollen coupling of forest trees: forming synchronized and periodic out of chaos. J Theor Biol 203:63–84

Satake A, Iwasa Y (2002) The synchronized and intermittent reproduction of forest trees is mediated by the Moran effect, only in association with pollen coupling. J Ecol 90:830–838

Schaeffer SM, Sharp E, Schimel JP, Welker JM (2013) Soil-plant N processes in a high arctic ecosystem, NW Greenland are altered by long-term experimental warming and higher rainfall. Glob Chang Biol 19:3529–3539

Schauber EM, Kelly D, Turchin P, Simon C, Lee WG, Allen RB, Payton IJ, Wilson PR, Cowan PE, Brockie RE (2002) Masting of eighteen New Zealand plant species: the role of temperature as a synchronizing cue. Ecology 83:1214–1225

Sedgley M, Griffin AR (1989) Sexual reproduction of tree crops. Academic Press, London

Selås V, Piovesan G, Adams JM, Bernabei M (2002) Climatic factors controlling reproduction and growth of Norway spruce in southern Norway. Can J For Res 32:217–225

Silvertown J (1980) The evolutionary ecology of mast seeding in trees. Biol J Linn Soc 14:235–250

Smaill SJ, Clinton PW, Allen RB, Davis MR (2011) Climate cues and resources interact to determine seed production by a masting species. J Ecol 99:870–877

Smith SE, Read DJ (2008) Mycorrhizal symbiosis. Academic Press, Cambridge

Sork VL, Bramble JE (1993) Prediction of acorn crops in three species of North American oaks: *Quercus alba*, *Q. rubra* and *Q. velutina*. Ann Sci For 50:128s–136s

Sork VL, Bramble J, Sexton O (1993) Ecology of mast-fruiting in three species of North American deciduous oaks. Ecology 74:528–541

Sperens U (1997) Long-term variation in, and effects of fertiliser addition on, flower, fruit and seed production in the tree *Sorbus aucuparia* (Rosaceae). Ecography 20:521–534

Stearns SC (1989) Trade-offs in life-history evolution. Funct Ecol 3:259–268

Steinbrenner EC, Duffield JW, Campbell RK (1960) Increased cone production of young Douglas-fir following nitrogen and phosphorus fertilization. J For 58:105–110

Stephenson AG (1981) Flower and fruit abortion: proximate causes and ultimate functions. Annu Rev Ecol Syst 12:253–279

Stiling P, Moon D, Hymus G, Drake B (2004) Differential effects of elevated CO_2 on acorn density, weight, germination, and predation among three oak species in a scrub oak forest. Glob Chang Biol 10:228–232

Tagliavini M, Quartieri M, Millard P (1997) Remobilised nitrogen and root uptake of nitrate for spring leaf growth, flowers and developing fruits of pear (*Pyrus communis* L.) trees. Plant and Soil 195:137–142

Tanentzap AJ, Lee WG, Coomes DA (2012) Soil nutrient supply modulates temperature-induction cues in mast-seeding grasses. Ecology 93:462–469

Tarvainen L, Lutz M, Rantfors M, Näsholm T, Wallin G (2016) Increased needle nitrogen contents did not improve shoot photosynthetic performance of mature nitrogen-poor Scots pine trees. Front Plant Sci 7:1051

Tersago K, Verhagen R, Servais A, Heyman P, Ducoffre G, Leirs H (2009) Hantavirus disease (*Nephropathia epidemica*) in Belgium: effects of tree seed production and climate. Epidemiol Infect 137:250–256

Upadhyay VP, Singh JS, Meentemeyer V (1989) Dynamics and weight loss of leaf litter in central Himalayan forests: abiotic versus litter quality influences. J Ecol 77:147–161

van Schaik CP, Terborgh JW, Wright SJ (1993) The phenology of tropical forests: adaptive significance and consequences for primary consumers. Annu Rev Ecol Syst 24:353–377

Van Vredenburch CLH, la Bastide JGA (1969) The influence of meteorological factors on the cone crop of Douglas fir in the Netherlands. Silvae Genet 18:182–186

Venner S, Siberchicot A, Pélisson PF, Schermer E, Bel-Venner MC, Nicolas M, Débias F, Miele V, Sauzet S, Boulanger V, Delzon S (2016) Fruiting strategies of perennial plants: a resource budget model to couple mast seeding to pollination efficiency and resource allocation strategies. Am Nat 188:66–75

Villar-Salvador P, Heredia N, Millard P (2010) Remobilization of acorn nitrogen for seedling growth in holm oak (*Quercus ilex*), cultivated with contrasting nutrient availability. Tree Physiol 30:257–263

Wagner D (1997) The influence of ant nests on Acacia seed production, herbivory and soil nutrients. J Ecol 85:83–93

Wahl V, Ponnu J, Schlereth A, Arrivault S, Langenecker T, Franke A, Feil R, Lunn JE, Stitt M, Schmid M (2013) Regulation of flowering by trehalose-6-phosphate signaling in *Arabidopsis thaliana*. Science 339:704–707

Waller DM (1993) How does mast-fruiting get started? Trends Ecol Evol 8:122–123

Wardle JA (1984) The New Zealand beeches. New Zealand Forest Service, Christchurch

Wardle DA (2002) Communities and ecosystems: linking the aboveground and belowground components. In: Monographs in population biology, vol 34. Princeton University Press, Princeton

Waring RH, Schlesinger WH (1985) Forest ecosystems: concepts and management. Academic Press, Orlando

Waring R, Nordmeyer A, Whitehead D, Hunt J, Newton M, Thomas C, Irvine J (2008) Why is the productivity of Douglas-fir higher in New Zealand than in its native range in the Pacific Northwest, USA? For Ecol Manage 255:4040–4046

Watson MA (1986) Integrated physiological units in plants. Trends Ecol Evol 12:119–123

Way DA, Ladeau SL, McCarthy HR, Clark JS, Oren RAM, Finzi AC, Jackson RB (2010) Greater seed production in elevated CO_2 is not accompanied by reduced seed quality in *Pinus taeda* L. Glob Chang Biol 16:1046–1056

Webb CJ, Kelly D (1993) The reproductive biology of the New Zealand flora. Trends Ecol Evol 8:442–447

Westoby M, Falster DS, Moles AT, Vesk PA, Wright IJ (2002) Plant ecological strategies: some leading dimensions of variation between species. Annu Rev Ecol Syst 33:125–159

Wheelwright NT, Logan BA (2004) Previous-year reproduction reduces photosynthetic capacity and slows lifetime growth in females of a neotropical tree. Proc Natl Acad Sci U S A 101:8051–8055

Williams DR, Potts BM, Smethurst PJ (2003) Promotion of flowering in *Eucalyptus nitens* by paclobutrazol was enhanced by nitrogen fertilizer. Can J For Res 33:74–81

Wright SJ, Calderón O (2006) Seasonal, El Niño and longer term changes on flower and seed production in a moist tropical forest. Ecol Lett 9:35–44

Wright SJ, Cornejo FH (1990) Seasonal drought and leaf fall in a tropical forest. Ecology 71: 1165–1175

Wright SJ, van Schaik CP (1994) Light and the phenology of tropical trees. Am Nat 143:192–199

Wright SJ, Yavitt JB, Wurzburger N, Turner BL, Tanner EV, Sayer EJ, Santiago LS, Kaspari M, Hedin LO, Harms KE, Garcia MN (2011) Potassium, phosphorus, or nitrogen limit root allocation, tree growth, or litter production in a lowland tropical forest. Ecology 92:1616–1625

Wright SI, Kalisz S, Slotte T (2013) Evolutionary consequences of self-fertilization in plants. Proc R Soc Lond B Biol Sci 280(1760):20130133

Würth MKR, Peláez-Riedl S, Wright SJ, Körner C (2005) Non-structural carbohydrate pools in a tropical forest. Oecologia 143:11–24

Yang LH, Bastow JL, Spence KO, Wright AN (2008) What can we learn from resource pulses? Ecology 89:621–634

Yasumura Y, Hikosaka K, Hirose T (2006) Resource allocation to vegetative and reproductive growth in relation to mast seeding in *Fagus crenata*. For Ecol Manage 229:228–233

Yates CJ, Hobbs RJ, Bell RW (1994) Factors limiting the recruitment of *Eucalyptus salmonophloia* in remnant woodlands. I. Pattern of flowering, seed production and seed fall. Aust J Bot 42: 531–542

Zackrisson O, Nilsson MC, Jäderlund A, Wardle DA (1999) Nutritional effects of seed fall during mast years in boreal forest. Oikos 84:17–26

Terrestrial Microalgae: Novel Concepts for Biotechnology and Applications

Michael Lakatos and Dorina Strieth

Abstract In the emerging field of algal biotechnology, optimization of algal production, engineering challenges, and scale-up of photobioreactors are urgent demands. Emphasis is placed on reducing cultivation limitations regarding, e.g., mass transfer, thermostability, photoinhibition, as well as expenses for energy and resource investments. Until now, almost all cultivation techniques are processed under submerged conditions with aquatic microalgae. The biotechnological implementation of terrestrial microalgae, however, exhibits several physiological and technological advantages for an efficient production in biofilm photobioreactors. Their outstanding performance and considerable advantages for biotechnology may reduce several of the current limitations and provide new principles in bioengineering. Can they outcompete the capacity of commercial algal strain due to their thermostability, light utilization, or desiccation tolerance? How terrestrial microalgae could highly contribute to technological and economic improvement of microalgal biotechnology is discussed reviewing their species diversity, physiology, valuable products, bioengineering processes, biofilm photobioreactors, and some visions of potential developments. Moreover, the overview may allow choosing interesting organisms for further studies.

Communicated by Ulrich Lüttge

M. Lakatos (✉)
Applied Pharmacy, University of Applied Sciences Kaiserslautern, Carl-Schurz-Str. 10 – 16, Pirmasens 66953, Germany
e-mail: michael.lakatos@hs-kl.de

D. Strieth
University of Kaiserslautern, Gottlieb-Daimler-Str. 49, Kaiserslautern 67663, Germany
e-mail: strieth@mv.uni-kl.de

Progress in Botany (2018) 79: 269–312, DOI 10.1007/124_2017_10,
Published online: 28 May 2017

Contents

1 Introduction

Biotechnology, using microalgae, offers innovative solutions to meet the growing demand for bioactive ingredients, food and feedstock with low resource consumption (e.g., Chisti 2007; Spolaore et al. 2006; Mata et al. 2010; Brennan and Owende 2010; Borowitzka 2013, 2016). To date, however, limitations in physiology and bioreactor design have prevented an economically feasible transfer of numerous microalgae-based production processes into an industrial scale. The most widely used systems for production of microalgae biomass are open ponds (Brennan and Owende 2010). Closed systems like tubular (Broneske et al. 1998) and plate (Münkel et al. 2013; Posten 2009) photobioreactors are also used in pre-industrial pilot projects. The latter have a lower risk of contamination and the light source can be sunlight or artificial light (Brennan and Owende 2010). All those systems have in common that the cultures are submerged in liquid medium. Due to the submerged cultivation techniques mainly aquatic microalgae have been used. Up to now submerged cultivations lead to several limitations, e.g.: (1) limited gas exchange caused by diffusion resistance, (2) growth reducing O_2:CO_2 ratio (Kliphuis et al. 2011), (3) limiting light transmission (Kliphuis et al. 2010), and (4) high expenses for mixing and separation. Recent results demonstrate that some of these limitations and expenses can be avoided by the use of novel photobioreactor designs using more air-exposed production processes and microalgae immobilized aş biofilms (Kuhne et al. 2014a; Gross et al. 2015; Berner et al. 2015; Podola et al. 2017). The immobilized cultivation technique meets the requirements of a more efficient production of microalgae. Moreover, these novel photobioreactor designs are close to satisfy the cultivation requirements of terrestrial microalgae. Therefore, the next generation technology will be the introduction of terrestrial microalgae for bioengineering. The utilization of terrestrial microalgae will initiate the access to new valuable products and may allow a paradigm change in algal biotechnology.

Terrestrial microalgae, composed by prokaryotic cyanobacteria and eukaryotic algae, are characterized by a great variety of valuable products, which expand that of aquatic microalgae. Additionally, they offer physiological features such as desiccation tolerance and stable growth over a broad range of temperatures, light intensities, and high concentrated gases (e.g., CO_2, H_2, H_2S, N_2). This physiological performance permits stable production processes under a high variety of artificial and natural conditions. Furthermore, their physiological activation can be induced by small amounts of liquid water or even high humidity which could conserve water resources for biotechnological applications. Currently, the exploitation of their valuables and the production of larger quantities of their metabolites are not realized because the industrial cultivation technology is not available yet. Some new photobioreactor developments arrange microalgae as biofilms on different substrates. This cultivation technique solves the demands for more efficient production processes and satisfies the prerequisites of exploiting not only aquatic but also terrestrial microalgae. Reviews for the current application of aquatic microalgae are attributing, e.g., bioreactor design and cost-effectiveness (Chisti 2013; Muffler et al. 2014; Olivieri et al. 2014; Gross et al. 2015), harvesting (Vandamme et al. 2013; Barros et al. 2015), biofuels (Rawat et al. 2013; Singh et al. 2014), sewage treatment (Kesaano and Sims 2014; Maity et al. 2014), and biorefinery (Gerardo et al. 2015; Trivedi et al. 2015). With the focus on terrestrial microalgae, a current overview of advantages of the species diversity, physiology, valuable products, bioengineering processes, biofilm photobioreactors, and some visions of potential developments will be presented here.

2 Biodiversity of Terrestrial Microalgae

Biodiversity forms the basis for a large pool of potential products and productive organisms offered by terrestrial microalgae. But what comprises terrestrial microalgae and how divers are they?

Terrestrial microalgae, also called aeroterrestrial, aerophil, or aerial, survive on land and are composed by prokaryotic photosynthesizing bacteria (cyanobacteria) and eukaryotic algae (review: Büdel 2011a, b; Hoffmann 1989). Terrestrial cyanobacteria belong to the order Chroococcales (coccoid unicells or colonial coccoids; e.g., *Chroococcus, Cyanothece, Gloeocapsa, Myxosacrina*), Oscillatoriales (filamentous; e.g., *Microcoleus, Coleofasciculus, Trichocoleus*), Pseudanabaenales (filamentous; e.g., *Schizothrix*), and Nostocales (filamentous with N_2-fixing heterocysts; e.g., *Calothrix, Dichothrix, Hyphomorpha, Nostoc, Nostochopis, Petalonema, Scytonema, Stigonema, Tolypothrix*). They colonize rocks (epilithic and endolithic), soils (e.g., biological soil crusts), fresh and marine waters, and some are lichenized – living in symbioses with fungi (mainly ascomycetes) as lichens (Büdel 2011a). A few genera are also epiphytic and grow on tree bark or on leaves of bryophytes (Adams 2002; Solheim and Zielke 2002; Lindo et al. 2013).

Within the eukaryotic algae terrestrial genera can be found among the green algae (Chlorophyta) in the classes Trebouxiophyceae, Chlorophyceae, and Ulvophyceae (Leliaert et al. 2012). Members of the Trebouxiophyceae are mainly coccoid photobionts of lichens, but also occur commonly in terrestrial habitats (e.g., *Trebouxia, Asterochloris, Apatococcus, Chlorella, Diplosphaera, Stichococcus*; Lewis and Lewis 2005; Darienko et al. 2010; Friedl and Rybalka 2012; Karsten et al. 2007). Within the Chlorophyceae terrestrial forms occur unicellular, multicellular, or filamentous in rocks (endolithic; e.g., *Coccobotrys*), on rocks (epilithic; e.g., *Desmococcus, Haematococcus*), several green-algal species in or on soils (endo- and epidaphic; e.g., *Fritschiella, Macrochloris*) and on plants (epiphytic; e.g., *Coccomyxa, Neochloris*) (Büdel 2011b). Unicellular to filamentous and mainly epilithic and epiphytic genera of the Ulvophyceae are, for example, *Trentepohlia, Cephaleuros, Desmochloris*, and *Phycopeltis* and are also commonly associated with lichen-forming fungi (Nelsen et al. 2011). Also among the Charophyta – a division of freshwater green algae belonging to the Streptophyta – terrestrial forms are found in the Chlorokybophyceae (*Chlorokybus* growing subaerially; Lewis and McCourt 2004), the Klebsormidiophyceae (e.g., *Klebsormidium*; Karsten et al. 2010; Karsten and Holzinger 2012; Holzinger et al. 2014), the Zygnematophyceae (e.g., *Zygogonium ericetorum*, Lewis and McCourt 2004; Holzinger et al. 2010), and the Coleochaetophyceae (Cook and Graham 2016). Some terrestrial species are also known from Xanthophyta (yellow–green algae, e.g., *Botrydiopsis, Bumilleria, Bumilleriopsis, Chloridella, Ellipsoidion, Heterococcus, Heterothrix, Pleurochloris, Tribonema, Vaucheria*) and Eustigmatophyta (e.g., Gerrath et al. 2000; Mikhailyuk 2008). Quite often, terrestrial biofilms are also composed by Bacillariophyta (e.g., *Achnanthes, Cyclotella, Cymbella, Fragilaria, Eunotia, Hantzschia, Luticola, Navicula, Pinnularia, Stauroneis*, and *Surirella*). These terrestrial diatoms are sometimes cosmopolitan (e.g. *Achnanthes coarctata, Hantzschia amphioxys, Pinnularia borealis, Mayamaea atomus*) and are widely distributed from the Antarctic (Souffreau et al. 2013) to the tropics. In the latter, diatoms even inhabit lichens thalli growing between the filaments of the mycobiont (e.g., *Diadesmis, Luticola, Melosira, Nitzschia, Orthoseira, Stauroneis*; Lakatos et al. 2004).

The enormous diversity of terrestrial microalgae implies that many species exist which could be interesting for biotechnological approaches but are not screened yet (Lang et al. 2011). Right now, only around 20 microalgae species are commercially used. Most of them are aquatic microalgae and are cultivated under submerged conditions. Some prominent examples of green algae are *Dunaliella salina* (ß-carotene production), *Dunaliella tertiolecta* (e.g., ATCC 30929, lipid production), *Haematococcus pluvialis* (astaxanthin production), *Chlorella vulgaris* (food and food additives), and the cyanobacteria *Arthrospira platensis* (formerly *Spirulina*; food additives). The green algae *Chlamydomonas reinhardtii* (wild type and different mutant strains) and the cyanobacteria *Synechocystis* PCC 6803 and *Synechococcus* PCC 7002 are often used as model organisms for scientific research. These and other microalgae used for biotechnological applications and research are of aquatic origin and desiccation sensitive (e.g., Raanan et al. 2016). Besides the terrestrial microalgae extremophilic algae also exhibit high potential contributions in biotechnology (for a review, see Varshney et al. 2015).

The characterization of many different terrestrial microalgae species is important for a better exploitation of the potentials of their diversity. In some cases, further process-oriented and application-oriented processing of interesting cultures is required for biotechnological production. This involves close cooperation between basic and application-related research as well as engineering sciences, taking into account the requirements of the potential market.

3 Physiological Advantages

The previous focus on aquatic microalgae for biotechnological approaches might be due to the fact that they are highly abundant in freshwater and marine habitats displaying the main driver of aquatic primary production. Moreover, they are – at first glance – easy to cultivate in liquid cultures. The outstanding role of terrestrial microalgae, e.g., in nutrient-limited ecosystems as soil stabilizer and nitrogen supplier, was recognized since three decades (Medina 1993; Büdel et al. 1997; Belnap and Lange 2001). Together with bryophytes, terrestrial microalgae – living free, in symbioses (lichens), or in association with bryophytes – may account for 7% of net primary production by terrestrial vegetation. Moreover, the ability of terrestrial cyanobacteria to fix nitrogen (N_2) from the atmosphere might be responsible for half of the biological nitrogen fixation on land (Elbert et al. 2012). These surprising functions together with their physiological ability to survive extreme environmental conditions bring them into the focus of potential biotechnological applications. Two features are particularly remarkable for terrestrial microalgae to grow on land and for the biotechnological utilization – their broad physiological performance as well as their desiccation tolerance. They occupy terrestrial habitats from the cold polar to the hot desert, and are adapted to the extreme fluctuations in temperature, light, and water availability.

3.1 Temperature: Thermostability as Considerable Approach

Temperature is an important factor determining growth and mortality of organisms. The control and maintenance of optimal temperature conditions by cooling and heating are considerable investments for the efficient production of microalgae in photobioreactors. Most commercially used microalgae survive only within a narrow range of temperatures. If they exceed their species-specific temperature limit, they are going to die. Therefore, the majority of outdoor productivity in temperate habitats is restricted to calm weather conditions in spring and autumn. In summer, productivity is associated with highly expensive cooling investments while productivity most often is stopped during winter – a loss of production that one would actually want to avoid. Do terrestrial microalgae exhibit a higher thermostability to outperform commercial microalgae?

Regarding their optimal growth temperature ($Temp_{opt}$), organisms are classified as psychrophiles ($Temp_{opt}$ below 15 °C), mesophiles (intermediate $Temp_{opt}$ 15–50 °C), thermophiles ($Temp_{opt}$ above 50 °C), and hyperthermophiles ($Temp_{opt}$ exceeding 80 °C). The upper temperature limit for eukaryotic algae is 62 °C (Rothschild and Mancinelli 2001). The cyanobacteria *Synechococcus* cf. *lividus* exists above this temperature surviving 45–73 °C (Miller and Castenholz 2000) as microbial mat in alkaline hot springs. This outstanding thermostability is caused by four amino acid substitutes stabilizing the important enzyme RuBisCo (Miller et al. 2013; Pedersen and Miller 2016). Further aquatic thermophile microalgae are, for example, the cyanobacteria *Synechococcus elongatus* (temperature limit: 60 °C; Miyairi 1995), *Phormidium boryanum*, *Thermosynechococcus elongatus*, *T. vulcanus*, and *Oscillatoria terebriformis* ($Temp_{limit}$: 52–58 °C; Varshney et al. 2015), as well as the red algae *Cyanidium caldarium*, *Cyanidioschyzon merolae*, and *Galdieria sulphuraria* ($Temp_{limit}$ around: 56 °C; Doemel and Brock 1971). Most microalgae used in biotechnology are mesophiles from upper temperature limits of around 45–50 °C (*Desmodesmus* sp. F51, *Chlorella kessleri*, *Chlorella sorokiniana*, *Scenedesmus obliquus*, *Chroococcidiopsis thermalis*) down to optimal growth temperatures of 15 °C (e.g., *Chlamydomonas raudensis*).

It is interesting that most terrestrial microalgae, even when isolated from hot habitats, exhibit $Temp_{opt}$ between 25 and 35 °C such as, e.g., *Nostoc flagelliforme* (Gao 1998), *Trichocoleus* (formerly *Microcoleus*) *sociatus*, *Nostoc commune* (Sand-Jensen 2014), *N. muscorum*, *Coleofasciculus* (formerly *Microcoleus*) *chtonoplastes* (Table 1). The strategy of terrestrial microalgae to survive surface temperatures which often exceed the optimal growth temperature (e.g., Rascher et al. 2003) is to downregulate the cell physiology by desiccation which protects the organism from respiration loss and photodamage (review, e.g., Lüttge 2011). For example, temperature treatments from −15 to 45 °C in a dehydrated state for 24 h did not affect the dark respiration and net photosynthesis of *Nostoc flagelliforme* after rewetting. Pretreatment at 95 °C lowered photosynthesis to only 8% while respiration was 47% compared to the control even at 107 °C – pretreatment at 117 °C was lethal for the terrestrial cyanobacteria. In contrast, when *Nostoc flagelliforme* was pretreated at a hydrated state, temperatures of 45 °C lead to an inactivation of photosynthesis while respiration reached zero at lethal temperatures of already 65 °C (Gao 1998). The broader physiological range of growth temperatures also seems to be a physiological feature of terrestrial microalgae. *Nostoc commune,* for example, tolerates temperature ranges from 6 to 43 °C (Møller et al. 2014).

Psychrophilic microalgae are considered as interesting candidates for outdoor bioreactors in cold habitats and seasons (Varshney et al. 2015). In polar and alpine habitats terrestrial snow and ice microalgae inhabit frozen water at the ice–air interface and are considered here as terrestrial organisms in contrast to marine cryobiontic algae which inhabit the ice–water interface and are therefore aquatic.

Table 1 Cultivation parameter upon species, lifestyle, cultivation system, and photosynthetic performance of terrestrial and aquatic microalgae

Species	Life style	Cultivation system	Method	Photosynthesis	$Temp_{opt}$ (°C)	PFD_{comp} (µmol m^{-2} s^{-1})	PFD_{sat} (µmol m^{-2} s^{-1})	Growth (µ day^{-1})	References
Apatococcus lobatus (Chodat) Petersen	TG	In situ	CO_2	32 (µl CO_2 g^{-1} h^{-1})	20				(Bertsch 1966)
Desmococcus sp. Brand/ *Apatococcus* sp. Brand	TG	In situ	ETR	51 ETR			1,445		(Lüttge and Büdel 2010)
Klebsormidium BIOTA 14613.5e	TG	S	O_2	140 (µmol O_2 mg^{-1} Chl h^{-1}) 16 ETR	30	4	20 (I_k)		(Karsten et al. 2016)
Klebsormidium BIOTA 14614.18.24	TG	S	O_2	140 (µmol O_2 mg^{-1} Chl h^{-1})	30	6	14 (I_k)		(Karsten et al. 2016)
Klebsormidium BIOTA 14614.7	TG	S	O_2	79 (µmol O_2 mg^{-1} Chl h^{-1}) 19 ETR	35	15	19 (I_k)		(Karsten et al. 2016)
Klebsormidium crenulatum (Kütz.) Lokhorst	TG	S	O_2 & ETR	38 (µmol O_2 mg^{-1} Chl h^{-1}) 14 ETR	35	6	22 (I_k)[a] 58 (I_k ETR)	0.15–0.37	(Karsten et al. 2010)
Klebsormidium dissectum (Gay) Ettl & Gärtner	TG	S	O_2 & ETR	115 (µmol O_2 mg^{-1} Chl h^{-1}) 24.1 ETR	35	1.8	18[a]	0.2–0.52	(Karsten and Holzinger 2012)
Klebsormidium SAG 384–1	TG	S	O_2	150 (µmol O_2 mg^{-1} Chl h^{-1}) 11 ETR	15	10	11 (I_k)		(Karsten et al. 2016)
Trentepohlia umbrina (Kützing) Bornet	TG	In situ	ETR	59 ETR			1,910		(Lüttge and Büdel 2010)
Trentepohlia odorata (F.H.Wiggers) Wittrock	TG	In situ	O_2	50.6 (µmol O_2 mg^{-1} Chl h^{-1})	25	56	170		(Ong et al. 1992)
Trichocoleus sociatus (West & G.S. West) Anagnostidis	TC	E	CO_2	2,124 (nmol CO_2 cm^{-2} h^{-1})	30	90	800	0.984	(Kuhne et al. 2014a)

(continued)

Table 1 (continued)

Species	Life style	Cultivation system	Method	Photosynthesis	$Temp_{opt}$ (°C)	PFD_{comp} (μmol m^{-2} s^{-1})	PFD_{sat} (μmol m^{-2} s^{-1})	Growth (μ day^{-1})	References
Chroococcidiopsis sp. Geitler	TC	E	CO_2	720 (nmol CO_2 cm^{-2} h^{-1})				0.35–0.63[b]	(Büdel et al. 2009)
Coleofasciculus chthonoplastes (Thuret ex Gomont) Siegesmund, Johansen & Friedl	TC	E	CO_2			60	450		Unpublished
Microcoleus vaginatus Gomont ex Gomont	TC	S	O_2	160–270 (μmol O_2 mg^{-1} Chl h^{-1})	25		600–800		(Ohad et al. 2010)
Nostoc commune Vaucher ex Bornet & Flahault	TC		O_2	353–376 (nmol O_2 cm^{-2} h^{-1}) 69 (μmol O_2 mg^{-1} Chl h^{-1})	25	9.5[c]	n.n.		(Sand-Jensen 2014)
Nostoc flagelliforme[d] (Berk. et Curt.) Berk. & Curt. ex Harvey	TC	In situ	CO_2	14.1[e] (mg CO_2 g^{-1} day^{-1})	25–30	40–90 225[e]	1,000–1,200		(Dingji et al. 1992; Mei and Cheng 1990)
Nostoc muscorum C. Agardh ex Bornet & Flahault	TC	E	CO_2		24–30	90	300	1.15	Unpublished
Scytonema ocellatum Lyngbye ex Bornet & Flahault	TC	E	CO_2			230	480		Unpublished
Scytonema stuposum Bornet ex Bornet & Flahault	TC	E	CO_2			50	430		Unpublished

Stigonema ocellatum (Dillwyn) Thur.	TC	In situ	CO_2	97 (μmol O_2 mg^{-1} Chl h^{-1})	32		1,300		(Büdel 1999)
Stigonema panniforme Bornet & Flahault	TC	In situ	CO_2	108 (μmol CO_2 mg^{-1} Chl h^{-1})	25		1,800		(Büdel 1999)
Chlorella saccharophila (Krüger) Migula	STG	S	O_2	81 (μmol O_2 mg^{-1} Chl h^{-1})	30		130	0.48 (10 °C)	(Vona et al. 2004)
Zygogonium ericetorum Kützing (green strain)	STG	In situ	ETR	10.9 ETR			90		(Aigner et al. 2013)
Zygogonium ericetorum Kützing (purple strain)	STG	In situ	ETR	32.8 ETR			300[a]		(Aigner et al. 2013)
Chlamydomonas reinhardtii Dangeard	STG	E	G		24			1.17	Unpublished
Nostoc punctiforme Hariot	STC	S	O_2	552–654 (nmol O_2 cm^{-2} h^{-1}) 60 (μmol O_2 mg^{-1} Chl h^{-1})			610		(Raun et al. 2009)
Koliella antarctica SAG 2030	AG	S	O_2	105 (μmol O_2 mg^{-1} Chl h^{-1})	25		130	>0.3 (15 °C)	(Vona et al. 2004)
Chlorella sorokiniana Shihira & Krauss, strain 211/8 K	AG	S	O_2	123 (μmol O_2 mg^{-1} Chl h^{-1})	35		250	3.18	(Vona et al. 2004; Andreoli et al. 1998)
Arthrospira platensis Gomont	AC	E	CO_2		25-30[f]	80	850		Unpublished
Arthrospira platensis Gomont	AC	S	O_2	614–645 (μmol O_2 mg^{-1} Chl h^{-1})	25–30		150–200	0.46–0.58	(Vonshak 1997)
Anabaena variabilis Kützing ex Bornet & Flahault	AC	S	G		35			1.1	(Robarts and Zohary 1987)
Nostoc zetterstedtii Areschoug ex Bornet & Flahault	AC	S	O_2	206–409 (nmol O_2 cm^{-2} h^{-1}) 20 (μmol O_2 mg^{-1} Chl h^{-1})		19.3[c]	200		(Sand-Jensen et al. 2009)

(continued)

Table 1 (continued)

Species	Life style	Cultivation system	Method	Photosynthesis	$Temp_{opt}$ (°C)	PFD_{comp} (μmol m^{-2} s^{-1})	PFD_{sat} (μmol m^{-2} s^{-1})	Growth (μ day^{-1})	References
Spirulina maxima (Setchell & Gardner) Geitler	AC	S	G		30–35		460–550[g]	0.26–0.45	(Vonshak 1997)

Microalgae are distinguished as cyanobacteria (C) and green algae (G) and their lifestyle is classified as terrestrial (T), subterrestrial or subaquatic (ST), and aquatic (A). Cultivations before measurements indicate if microalgae were cultivated under submerged (S) or emerse, air-exposed, conditions (E). Specimens directly measured on air-exposed samples (collected) in the field are indicated (In situ). Photosynthetic performances are characterized as methodology of measurement [gas exchange of CO_2 and O_2 as well as chlorophyll *a* fluorescence of electron transfer rate (ETR) and growth studies (G)], maximal photosynthesis, optimal growth temperature ($Temp_{opt}$), light compensation point (PFD_{comp}), light saturation point (PFD_{sat}), and specific growth rate μ per day (μ day^{-1}). Reference values of maximal photosynthesis are related to method as well as dry mass, chlorophyll content (Chl), ETR or area of measured samples

(I_k) Two models exist to calculate the light saturation point by photosynthesis-irradiance (PI) curves. One method calculates the irradiance at 95% of light-saturated photosynthesis overestimating the light intensity. The second method is based on the mathematical photosynthesis model of Webb et al. (1974), which allowed the calculation of the *initial* value of light-saturated photosynthesis to determine the initial light-saturation intensity (I_k) by dividing the light-saturated photosynthesis by the linear slope (α) at limiting photon fluence rates, which underestimates the saturating light intensity

[a]No photoinhibition could be detected in the study

[b]Conversion doubling time = $\ln 2/\mu$

[c]Calculated at low irradiance (after Sand-Jensen 2014)

[d]*Nostoc flagelliforme* Harvey ex Molinari-Novoa, Calvo-Pérez & Guiry nom. nov. (see Calvo-Pérez E.A., Molinari-Novoa E.A. & Guiry M.D. (2016). Validation of *Nostoc flagelliforme* (Nostocaceae, Cyanobacteria). *Notulae Algarum*, 2, 1–2)

[e](Cui 1985)

[f](Oliveira et al. 1999)

[g]Conversion factor Lux to PPF (μmol m^{-2} s^{-1}) of 0.0185

These terrestrial cryophiles grow in the surface layers of snow and ice occupying the spaces between ice crystals. Terrestrial snow and ice microalgae are adapted to repeated freezing and thawing cycles with temperatures that may increase to 1.3 °C during daytime (Hoham 1975) because of light absorption by the pigments. Thus, they should be suited to grow at temperatures near zero degrees Celsius (Leya 2013). From 354 species recorded from snow and ice (Kol and Flint 1968), 46% belong to the Chlorophyta, most of them being unicellular flagellates of the family Chlamydomonadaceae, 26% to the Cyanophyta, and 19% are Bacillariophyta. Many of these terrestrial microalgae produce red carotenoids. Regarding this coloring Kol and Flint (1968) distinguish five types of colored snow and ice communities (red, green, yellow, blue, purple–brown) each characterized by dominant algal species. Particularly, the red snow community could be interesting for biotechnology regarding the carotenoid production. This community is composed of, e.g., *Chlamydomonas antarcticus*, *C. bolyaianus*, *C. nivalis*, *C. pichincha*, *C. sanguineus*, *Smithsonimonas abbotii*, *Sphaerellopsis rubra*, *Scotiella nivalis*, *S. nivalis* var. *californica*, *S. nivalis* var. *nipponica*, *S. tatrae*, *Trochiscia rubra*, and *T. americana*. Today the cosmopolitan subterrestrial/subaquatic snow alga *Haematococcus pluvialis* is commercially cultivated for astaxanthin and other carotenoid production. Differences in growth temperature ranges, morphology, and molecular phylogeny suggest the division of this species (Allewaert et al. 2015). *Haematococcus pluvialis*, on the one hand, grows well from 8 to 28 °C. At temperatures above 28 °C growth is strongly inhibited and between 33 and 35 °C the cultures die (Borowitzka et al. 1991; Wan et al. 2014). On the other hand, an increase in temperature stimulates the formation of astaxanthin-containing palmella cells with highest astaxanthin contents at 28 °C. Here, the narrow window of suitable temperature between highest growth rate as well as astaxanthin production and mortality is close together indicating precise thermoregulation of the commercial production process.

With regard to temperature, terrestrial microalgae perform physiological thermostability to grow over broad temperature ranges and at temperatures around zero as well as near 45 °C. This offers process stability particularly at fluctuating environmental conditions and access to a pool of species suitable for the operation in cold and hot environments. Their ability to survive dehydration processes may be employed for downregulation processes at strong irradiances instead of intensive cooling by thermoregulation.

3.2 *Irradiance: Adaptation Ability for Efficient Light Utilization*

Light conditions are changing during the day and season and are influencing biomass productivity particularly in outdoor bioreactors. Moreover, submerged microalgal-cultivations require intensive mixing to promote essential CO_2:O_2

ratio and nutrient supply (mass transfer between cells and culture medium). In this turbulent suspension, the photosynthesizing cell is frequently exposed to high light condition at the surface and low light condition at the inner zone due to the strong light gradient within the microalgae suspension (Jassby and Platt 1976). Aquatic microalgae are mostly adapted to low and intermediate light intensities and frequently changing light can reduce photolimitation at high light intensities and at light-time/dark-time fluctuations of around 0.1–10 Hz (Kok 1953; Grobbelaar 2009). Unfortunately, despite mixing photolimitation occurs and mixing also prevents adaptation to light, both effects reduce productivity (Grobbelaar 2010, 2013). Can these limitations be reduced by the application of terrestrial microalgae for a more efficient light utilization?

Often terrestrial microalgae are adapted to low light requirements with low light compensation points (when photosynthesis compensates respiration) and low light saturation points (when the increase of light intensities does not result in increased photosynthesis). But microalgae generally adapt to the light environment they are exposed to (Table 1). To receive an overview of the performance, the photosynthetic capacities of microalgae for around 30 strains of terrestrial and some studied subterrestrial as well as aquatic species are summarized (Table 1). The studies were conducted with various designs and methodologies and thus the comparison should be seen as a preliminary impression to be improved in further studies. Basically, the photosynthetic capacity is similar within the groups. The specific growth rates (μ day^{-1}) of aquatic cyanobacteria and algae are between 0.3 and 1.1 μ day^{-1} and represent the typical value of biotechnological strains between 0.1 and 1 μ day $^{-1}$ (e.g., Nascimento et al. 2013; Ho et al. 2011). Only a few studies report extraordinary specific growth rates exceeding 3 μ day^{-1} by *Chlorella sorokiniana* (Vona et al. 2004) and 5 μ day^{-1} by *Chlorella* sp. (Sakai et al. 1995). The growth rates of aquatic cyanobacteria are between 0.01 and 0.05 (*Oscillatoria* sp., Phlips and Zeman 1990), 0.22–0.45 (*Microcoleus chthonoplastes*, Karsten et al. 1996), 0.35–0.48 (*Chlorogloeopsis* sp., Garcia-Pichel et al. 1992), and 1.0–1.1 (*Oscillatoria* sp., Karsten et al. 1996). However, the growth rates of *Arthrospira platensis* with approximately 0.8–1.7 (Tomaselli et al. 1997) are rarely reached. Specific growth rates of terrestrial microalgae listed here are located in similar ranges between 0.37 and 1.15 μ day^{-1} (Table 1). Thus, these organisms are in the range of the growth rates of mammalian and insect cells, which generally achieve growth rates of about 0.12 and 0.72 μ day^{-1}, respectively (e.g., Jaluria et al. 2007; Ikonomou et al. 2003). Moreover, the growth of terrestrial microalgae is clearly above those of moss cells of *Physcomitrella patens* (0.36 μ day^{-1}), which is currently validated as promising expression systems for heterologous proteins (Perner-Nochta and Posten 2007). This emphasizes that terrestrial microalgae can be of interest as alternative production systems for high-quality products.

Also the photosynthetic capacity of aquatic versus terrestrial microalgae performs in similar ranges for light compensation (19–80 vs. 2–230 μmol m^{-2} s^{-1}) and light saturation (130–850 vs. 11–1,910 μmol m^{-2} s^{-1}). However, one interesting difference is obvious, the reaction of terrestrial microalgae under high light intensities often lacks photosynthetic reduction by photoinhibition. Cyanobacteria at exposed habitats from the deserts or the tropics exhibit light saturation around 1,000–1,200 μmol photons m^{-2} s^{-1} such as the biological soil crust of *Nostoc flagelliforme* (Dingji et al. 1992) or the epilithic community on an inselberg in French-Guiana composed mainly by *Gloeocapsa sanguinea*, *Scytonema myochrous*, and *Stigonema ocellatum* (Rascher et al. 2003). Even at higher irradiance these communities did not show photoinhibition (Table 1). This performance is not restricted to cyanobacteria only, because also terrestrial green algae collected in arid and alpine habitats displace no photoinhibition at light intensities exceeding 400 μmol m^{-2} s^{-1} (Karsten et al. 2010, 2014, 2016; Karsten and Holzinger 2012; Aigner et al. 2013). The absence of photolimitation at high light intensities might be a clear benefit of terrestrial microalgae for the biotechnological production at high or fluctuating irradiance and particularly if the organisms are immobilized and exposed as biofilms.

That many terrestrial microalgae seem to be adapted to low light conditions (low light compensation points) are the consequences of water availability. The activation of photosynthesis depends on the water content of the organisms. Cyanobacteria require liquid water from dew fall, rain, or fog. In contrast, some green algae can be reactivated by high humidity only (e.g., Bertsch 1966; Lüttge and Büdel 2010). In the morning when light intensities are low, water is often available and with rising sun and temperatures the water of the organisms evaporates until dehydration. During midday, when the irradiances are highest, the terrestrial microalgae are mostly desiccated. Thus, terrestrial microalgae undergo frequent drying and rehydration cycles. Dew fall during the night and high humidity in the morning activate photosynthesis and the rising light intensities can be used for photosynthesis until the terrestrial microalgae are dehydrated at high irradiance.

In the context of cultivations in outdoor photobioreactors or at high light intensities, terrestrial microalgae are well adapted to high irradiance with high light saturation values and low light compensation points (Table 1). Particularly the ability to overcome photolimitation at high light distinguishes this group and implies a better light utilization. Moreover, induced desiccation could also protect them from photoinhibition at high light intensities, which might be of importance particularly for immobilized biofilms.

Besides light quantity also the light quality influences the light adaptation process. Cyanobacteria and red algae possess a slightly different light harvesting complex. The harvesting complex of the photosystem is not constructed by carotenoids, chlorophyll *a* and chlorophyll *b* with an action spectrum of photosynthesis with maximum in the blue and red light regions. In order to enhance photosynthetic capacity, cyanobacteria and red algae employ large protein complexes called phycobilisomes for light harvesting addressing also the green wavelengths (Tandeau de Marsac 2003). Phycobilisomes are made up of phycobiliproteins (PBP) which are apo-proteins with covalently attached open-chain tetrapyrroles (phycobilins). PBPs in cyanobacteria are made up out of heterodimers, which consist of α- and β-subunits. The αβ-heterodimers build ring-shaped trimers (heterohexamers) or hexamers (heterododecamers) which are the building blocks for the phycobilisome rods which are connected by colorless linker proteins

(Grossman et al. 1994). Four major types of PBPs can be found: allophycocyanin (blue–green), phycocyanin (blue), phycoerythrocyanin (violet), and phycoerythrin (red) which mainly differ in the number and nature of the bound phycobilin chromophore resulting in their different coloration. The composition of these pigments is influenced by the light quality – more red light induces the increase of phycocyanin while green light raises the production of phycoerythrin. This acclimation process is called "complementary chromatic adaptation." Three affecting features are known to induce complementary chromatic adaptation – alteration due to the light quality (Overkamp and Frankenberg-Dinkel 2014), high temperatures (Pittera et al. 2017), and – according to our findings – drought stress. Regarding the biotechnological application at present, a blue extract of the aquatic cyanobacterium *Arthrospira platensis* (Spirulina-extract) containing the natural pigment phycocyanin is used to color food products. However, its widespread use is limited by its lack of acid and thermal stability below pH 4 and over 60 °C, respectively. Therefore, major obstacles for the coloration of beverages, confectionery and many other foods exist, resulting in the primary use of synthetic blue-dyes in food industry. Here, the utilization of terrestrial cyanobacteria could profit from stabilizing ingredients such as the stabilizing proteins HspA, WspA, and prolin (e.g., Nakamoto et al. 2000; Nakamoto and Honma 2006), which are induced by drought stress during desiccation. Moreover, the combination of a more suitable medium and the induction of drought stress by air-exposed cultivation via a novel emerse biofilm photobioreactor (see below) lead to an increased PBP production by ~33% compared to submerged conditions (preliminary results; Contreras-Porcia et al. 2011).

The ability of the PBP to exploit the green spectrum delivers one more application. Photobioreactors are often exposed to contaminations by other photosynthetic organisms. Providing solely green light as a source, cyanobacteria and red algae will survive while contamination by, e.g., green algae will be eliminated.

Concerning light conditions, the utilization of terrestrial microalgae in biofilm photobioreactors offers several production scenarios suitable for high light exposure. Their ability of photosynthetic adaptation and efficient light utilization (e.g., photosynthetic capacities, lack of photolimitation, chromatic adaptation) reduces limitations of conventional strains and photobioreactor-designs. Different light qualities could intensify the production of specific pigments and the green irradiance spectrum can be exploited by several terrestrial cyanobacteria. In addition, the purposeful induction of desiccation could prevent photoinhibition and photodamage by excessive light.

3.3 Desiccation Tolerance: Performance for Novel Biotechnological Processes

Desiccation tolerance is an outstanding performance of terrestrial microalgae (Lüttge et al. 2011). Until now, desiccation tolerance has not been in the focus of algae biotechnology, because biotechnological research and applications mainly operated with suspension-based cultivation systems since the 1940s. Cell suspensions are associated with low biomass density and large culture medium volume of around 0.05% (open ponds) to 0.6% dry matters (Davis et al. 2011). Several limitations arise in open raceway ponds and closed tubular or flat-panel photobioreactors due to the low surface-to-volume ratio and the large liquid volume (Fig. 1). Besides limited gas exchange and light transmission mentioned above, strong mechanical impacts by mixing are affecting the cell integrity and restrict the cultivation to robust microalgae strains, often possessing rigid cell walls. Moreover, the large liquid volume of the suspension is associated with high expenses for mixing and for separation of the algal biomass of low density (Christenson and Sims 2011a, b; Milledge and Heaven 2013; Barros et al. 2015). Mixing and dewatering account for around 20–40% of the biomass production cost (Davis et al. 2011; Norsker et al. 2011). Therefore, the reduction of water volume would implement several advantages. Could the desiccation tolerance of terrestrial microalgae initiate production processes that reduce volume related limitations?

Desiccation tolerance is well documented for lichens and bryophytes (for a review, see, e.g., Green et al. 2011; Lakatos 2011) while the mechanisms of cyanobacteria and green algae are only partly understood (for a review, see, e.g., Lüttge 2011; Holzinger and Karsten 2013). However, the tolerance to survive dehydration is species-specific and depends on the condition during desiccation and rehydration. During de- and rehydration processes the duration of desiccation, temperature, and light as well as the time period and surrounding conditions during dehydration influence the success of recovery (e.g., Lüttge et al. 1995; Gasulla et al. 2009; Garcia-Pichel et al. 2013).

Only a few studies compare the desiccation tolerance of aquatic and terrestrial microalgae under standardized conditions. Gray et al. (2007) examined desiccation tolerance of eleven aquatic and terrestrial green algae addressing the photosynthetic recovery following different duration of dehydration (1 day, 1 week, 4 weeks) and light treatments. Within cyanobacteria, genomic (Murik et al. 2017) and physiological comparisons (Raanan et al. 2016) of nine desiccation-sensitive and –tolerant strains used for biotechnological applications and research were studied. They showed different performances of the cyanobacteria *Synechococcus elongatus* PCC 7942, *Microcystis aeruginosa* NIES-843 (both are unicellular) as well as *Anabaena variabilis* ATCC 29413, *Cylindrospermopsis raciborskii*, and *Nostoc* PCC 7120 (all are filamentous) which are sensitive to dehydration. The latter is considered to be intermediate desiccation tolerant while *Leptolyngbya ohadii, Leptolyngbya* sp. NIES 2104, *Microcoleus vaginatus* PCC 9802, and

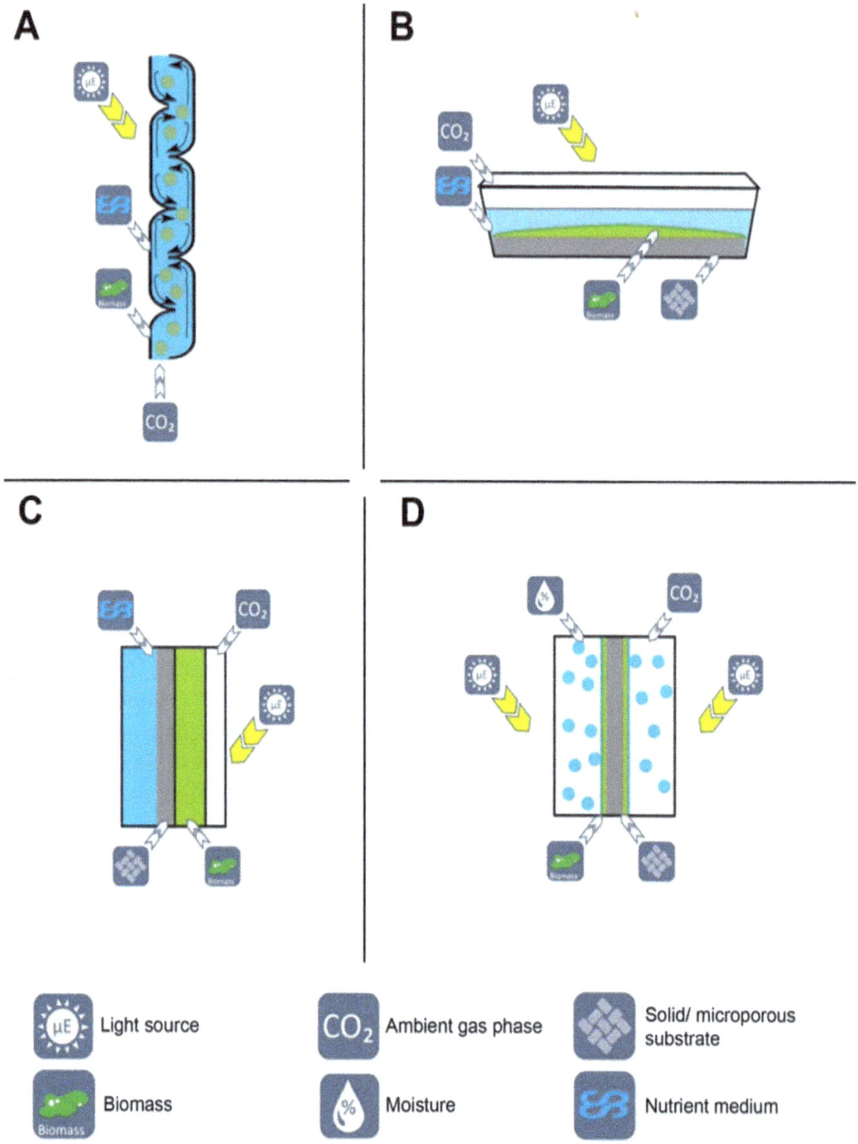

Fig. 1 Schematic drawing of different photobioreactor types for cultivation of microalgae under submerged and surface-associated conditions (copyright by Strieth D. & Schmidt T.). (**a**) Typical closed suspension plate photobioreactor [e.g., FPA-reactor developed by IGB Stuttgart, Germany (Münkel et al. 2013)]; (**b**) subterrestrial photobioreactor [e.g., attached microalgal growth system (Johnson and Wen 2010)]; (**c**) porous substrate biofilm bioreactor (e.g., PSBR (Podola et al. 2017); (**d**) aerosol-based photobioreactor [e.g., emerse photobioreactor ePBR (Kuhne et al. 2014a)]

Table 2 Desiccation experiments upon species, lifestyle, treatments, and recovery assessments of terrestrial and aquatic microalgae

Species	Life style	Pretreatment	Desiccation period	Treatment during desiccation	Recovery time	% of control	Full recovery after 24 h or indicated reaction	DDI	References
Zygogonium ericetorum Kützing (green strain)	STG	In situ Whatman filter	150 min	20–25 μE 22 °C	1 Day	5	53% after 8 days	0.95	(Aigner et al. 2013)
Hantzschia amphioxys (Ehrenberg) Grunow	TD	18 °C PAR: 25 12 light: 12 dark	10 min	–	Direct	0–5	Mortality (cell count)	0.95	(Souffreau et al. 2010)
Chlamydopodium sp. MLO301CT	AG	Submerse Glass beads & Petri dishes	24 h	Darkness 20 °C 25% RH	1 h	6	7% after 24 h	0.93	(Gray et al. 2007)
Pinnularia borealis Ehrenberg	TD	18 °C PAR: 25 12 light: 12 dark	10 min	–	Direct	0–10	Mortality (cell count)	0.90	(Souffreau et al. 2010)
Pinnularia borealis Ehrenberg	TD	18 °C PAR: 25 12 light: 12 dark	10 min	+ 30 °C	Direct	0–10	Mortality (cell count)	0.90	(Souffreau et al. 2010)
Zygogonium ericetorum Kützing (purple strain)	STG	In situ Whatman filter	150 min	20–25 μE 22 °C	1 Day	14	Died after 6 days	0.86	(Aigner et al. 2013)
Chlorogonium elongatum UTEX 11	AG	Submerse Glass beads & Petri dishes	24 h	Darkness 20 °C 25% RH	1 h	29	17% after 24 h	0.83	(Gray et al. 2007)
Monoraphidium braunii SAG 2006	AG	Submerse Glass beads & Petri dishes	24 h	Darkness 20 °C 25% RH	1 h	16	19% after 24 h	0.81	(Gray et al. 2007)

(continued)

Table 2 (continued)

Species	Life style	Pretreatment	Desiccation period	Treatment during desiccation	Recovery time	% of control	Full recovery after 24 h or indicated reaction	DDI	References
Klebsormidium BIOTA 14614.7	TG	Submerse Whatman filter	400 min	40 µE 22 °C	750 min	20	Still 20% after 2,000 min	0.80	(Karsten and Holzinger 2012)
Scenedesmus platydiscus UTEX 2457	AG	Submerse Glass beads & Petri dishes	24 h	Darkness 20 °C 25% RH	1 h	19	29% after 24 h	0.71	(Gray et al. 2007)
Chlorella sp. UTEX 318c	AG	Submerse Glass beads & Petri dishes	24 h	Darkness 20 °C 25% RH	1 h	33	48% after 24 h	0.52	(Gray et al. 2007)
Klebsormidium dissectum (Gay) Ettl & Gärtner	TG	Submerse Whatman filter	3 h	20–25 µE 22 °C	10 min 300 min	33 50	1 Week	0.50	(Karsten and Holzinger 2012)
Chlorella sp. BC4-VF9	TG	Submerse Glass beads & Petri dishes	24 h	Darkness 20 °C 25% RH	1 h	82	83% after 24 h	0.45	(Gray et al. 2007)
Myrmecia sp. BC8–8	TG	Submerse Glass beads & Petri dishes	24 h	Darkness 20 °C 25% RH	1 h	55	70% after 24 h	0.30	(Gray et al. 2007)
Navicula radiosa Kützing	TD	18 °C PAR: 25 12 light: 12 dark	10 min	–	Direct	0–74	Mortality (cell count)	0.26	(Souffreau et al. 2010)
Chlorosarcinopsis sp. SEV2-VF1	TG	Submerse Glass beads & Petri dishes	24 h	Darkness 20 °C 25% RH	1 h	81	78% after 24 h	0.22	(Gray et al. 2007)

Bracteacoccus sp. CNP2-VF25	TG	Submerse Glass beads & Petri dishes	24 h	Darkness 20 °C 25% RH	1 h	71	80% after 24 h	0.20	(Gray et al. 2007)
Navicula radiosa Kützing	TD	18 °C PAR: 25 12 light: 12 dark	10 min	+ 30 °C	Direct	0–84	Mortality (cell count)	0.16	(Souffreau et al. 2010)
Bracteacoccus sp. BC2–1	TG	Submerse Glass beads & Petri dishes	24 h	Darkness 20 °C 25% RH	1 h	86	87% after 24 h	0.13	(Gray et al. 2007)
Scenedesmus rotundus SEV3-VF49	TG	Submerse Glass beads & Petri dishes	24 h	Darkness 20 °C 25% RH	1 h	93	95% after 24 h	0.05	(Gray et al. 2007)
Klebsormidium BIOTA 14613.5e	TG	Submerse Whatman filter	350 min	40 µE 22 °C	200 min	50	25 h	0.05	(Karsten and Holzinger 2012)
Nostoc flagelliforme[a] (Berk. et Curt.) Berkeley & M.A. Curtis ex Harvey	TC	24 month dryness; 20 °C	270 min 20 °C 40% RH	10 µE (500 lux) +27°	7 h	83	23 h	0.00	(Scherer et al. 1984)
Nostoc commune Vaucher ex Bornet & Flahault	TC	5 Days 50% RH, 18–20 °C	580 min 20 °C 40% RH	10 µE (500 lux) +27 °C	7 h	83	23 h	0.00	(Scherer et al. 1984)
Nostoc sp. Vaucher ex Bornet & Flahault	TC	5 Days 50% RH, 18–20 °C	1,020 min 20 °C 40% RH	10 µE (500 lux) +27°	7 h	50	23 h	0.00	(Scherer et al. 1984)
Klebsormidium SAG 384-1	TG	Submerse Whatman filter	300 min	40 µE 22 °C	300 min	50	21 h	0.00	(Karsten and Holzinger 2012)

(continued)

Table 2 (continued)

Species	Life style	Pretreatment	Desiccation period	Treatment during desiccation	Recovery time	% of control	Full recovery after 24 h or indicated reaction	DDI	References
Klebsormidium BIOTA 14614.18.24	TG	Submerse Whatman filter	380 min	40 μE 22 °C	100 min	50	16.6 h	0.00	(Karsten and Holzinger 2012)
Klebsormidium crenulatum (Kütz.) Lokhorst	TG	Submerse	3 h	20–25 μE 22 °C	10 min 30 min 60 min	43 48 59	2 h	0.00	(Karsten et al. 2010)
Gloeocapsa sanguinea (ag.) Kütz. em. Jaag	TC	In situ (tropics)	30 min 1,800 μE 30–40 °C	1,800 μE Surface: 30–40 °C	7 min	50	30 min	0.00	(Rascher et al. 2003)
Stigonema ocellatum (Dillwyn) Thur.	TC	In situ (tropics)	45 min 1,800 μE 30–40 °C	1,800 μE Surface: 30–40 °C	4 min	50	30 min	0.00	(Rascher et al. 2003)
Scytonema myochrous (Dillwyn) Born. em Jaag	TC	In situ (tropics)	30 min 1,800 μE 30–40 °C	1,800 μE Surface: 30–40 °C	6 min	50	20 min	0.00	(Rascher et al. 2003)
Trentepohlia umbrina (Kützing) Bornet	TG	In situ (temperate)	Days 3 11 20 53	Dark Room temp	0.5 10 20 50 min	50	75 140 100 150 min	0.00	(Lüttge and Büdel 2010)

D *Desmococcus* sp. Brand/*Apatococcus* sp. Brand	TG	In situ (temperate)	Days 0 10 19 60 80	Dark Room temp	0.1 2 5 10 60 min	50	60 70 70 150 200 min	0.00	(Lüttge and Büdel 2010)

Studied species are displayed in descending order regarding the desiccation damage index (DDI). Microalgae are distinguished as cyanobacteria (C), green algae (G), and diatoms (D) and their lifestyle is classified as terrestrial (T), subterrestrial or subaquatic (ST), and aquatic (A). The conditions before (pretreatments) and during desiccation as well as the duration of desiccation process are indicated (light intensity [μmol photon m^{-2} s^{-1}] is indicated by the former expression μE). To assess the recovery, different perspectives of regeneration such as percentages of the control after specific recovery time and full recovery after 24 h (or the reaction after indicated time) as well as the referring DDI are extracted from the references

Chroococcidiopsis thermalis 7203 tolerate long dehydration periods (Raanan et al. 2016).

Some studies addressed the specific reactivation of photosynthesis after dehydration. The overview of photorecovery for diatoms, green algae, and cyanobacteria (Table 2) comprises the pretreatment, period of rehydration, and recovery after specific treatment. The experiments were conducted with various designs and methodologies. For comparison of desiccation tolerances among terrestrial microalgae a "desiccation damage index" (DDI; modified after Gray et al. 2007; Pardow and Lakatos 2013) was extracted from desiccation tolerance studies as the relative percentage of prior or maximal photosynthetic capacity (or percentage of alive cells, when mortality was studied) reached within 24 h after desiccation at −187 to −94 MPa (30–50% RH, around 20 °C). In general, terrestrial cyanobacteria recovered faster than green algae and diatoms. In one study of diatoms, all aquatic diatoms died due to rapid desiccation (10 min) and only some terrestrial species could survive this extreme dehydration procedure (Souffreau et al. 2010). Also aquatic green algae show high desiccation related damage with DDI of 0.95–0.5 (Table 2). Some (sub)terrestrial green algae isolated from soil or streamlet achieved suboptimal recovery or died (*Zygogonium ericetorum* Aigner et al. 2013; *Klebsormidium* sp. Karsten and Holzinger 2012). However, most terrestrial green algae and all cyanobacteria display good to optimal photorecovery with less or no desiccation damage (DDI 0.5 to zero; Table 2). Halftime recovery is achieved within seconds (e.g., *Trentepohlia umbrina*, *Desmococcus-Apatococcus* dominated epiphytic crust; Lüttge and Büdel 2010), minutes (e.g., epilithic cyanobacteria mats; Rascher et al. 2003) or 1 h (e.g., terrestrial green algae, Gray et al. 2007). Almost all of the studied terrestrial microalgae attain halftime recovery within 7 h and most are fully recovered within 24 h (Table 2). Even after 24 months in dryness, full recovery was reached within 23 h by *Nostoc flagelliforme* (Scherer et al. 1984). Considering these results and that terrestrial habitats are subject to frequent fluctuations in the level of humidity and water availability, it is still surprising that some terrestrial microalgae survive even longer periods of dehydration up to 87 years (e.g., *Nostoc commune*, *Chlorococcum humicola*, and *Stichococcus bacillaris;* Bristol 1919; Lipman 1941; Trainor and Gladych 1995). *Nostoc commune* could be recovered after 107 years yet using the same herbarium collection (Cameron 1962).

These performances of desiccation tolerance are outstanding and could be integrated in the bio-engineering processes to solve the current limitation of suspension-based cultivation. The most radical solution to reduce water volume would be the introduction of desiccation tolerant terrestrial microalgae as immobilized biofilm (avoiding mixing). In biofilm photobioreactors desiccation tolerant microalgae naturally adhere to surfaces, they need less water supply and they could desiccate to low cell water content for harvesting at high biomass to volume ratio (reducing dewatering). Moreover, desiccation induces the biosynthesis of different secondary metabolites which could be the target of high-value products. Therefore, the desiccation tolerance of terrestrial microalgae could initiate production processes that reduce volume related limitations. Admittedly, the design of photobioreactors has to be changed from a submerse cultivation towards an emerse – air-exposed – cultivation and from unicellular suspensions to immobilized biofilms. Furthermore, new products could be brought into focus, particularly those which are induced by desiccation.

4 Valuable Products of Terrestrial Microalgae

Nine million tons of microalgae per year are produced industrially with a turnover of six billion Euro per year worldwide. Microalgae generate a large number of valuable substances and this production by biotechnological processes has an enormous commercial potential due to increasing demand for pharmaceuticals, foodstuffs, and renewable raw materials (Mata et al. 2010; Wijffels and Barbosa 2010). The biomass can be used likewise as bioenergy, dietary supplement, and animal feed (Abed et al. 2009; Chisti 2007, 2013). In particular, microalgae produce high valuable products including carotenoids, long-chain polyunsaturated fatty acids, and phycobilins (for review, see, e.g., Borowitzka 2013; Draaisma et al. 2013; Wijffels et al. 2013; Leu and Boussiba 2014). Moreover, microalgae are a diverse source of pharmaceutical and biotechnological substances such as polysaccharides (exopolysaccharides), lipids, amino acids (cyanophycin), vitamins (pantothenates, B_{12}), sterols, enzymes, pharmaceuticals (cyto-, hepato-, neuro- and endotoxins), bioactive substances (Singh et al. 2005; Varshney et al. 2015; Vijayakumar and Menakha 2015), and other fine chemicals. The biosynthesis can be influenced, for example, by different wavelengths, temperatures, light intensities, and pH-values (Johnson et al. 2014). Most stress conditions inducing the production of high-value compounds are reviewed (Markou and Nerantzis 2013). The question arises, if terrestrial microalgae produce specific high-value compounds under typical stress conditions such as desiccation or high light?

Terrestrial microalgae are protected against rapid dehydration and associated anabiosis by an enveloping slime matrix (e.g., Shaw et al. 2003; Helm et al. 2000) consisting of hygroscopic extracellular polysaccharides (EPS). By releasing this EPS matrix, particularly terrestrial cyanobacteria gain mobility and the remaining EPS significantly contributes to soil particle stabilization, especially in sandy deserts (e.g., Belnap and Lange 2001). The EPS are also amplified under nitrogen and sodium chloride limitation as well as increased stressors such as red and blue light (Han et al. 2014) as well as temperature (e.g., Moreno et al. 1998; Otero and Vincenzini 2003). Another key factor seems to be the availability of CO_2 in unbalanced diazotrophic C/N ratios (Otero and Vincenzini 2004). Generally, terrestrial cyanobacteria are composed of a high proportion of 40–60% carbohydrates, which are predominantly in the form of polysaccharides (Takenaka et al. 1998). The EPS make up at least 10–20% of dry matter (Huang et al. 1998) and can be induced to 50% of DM by drought stress in emerse photobioreactors (Lakatos and Strieth unpublished). The EPS is defined as heteropolysaccharide composed by high amounts of glucose, galactose, and mannose, as well as smaller contents of fructose, ribose, xylose, arabinose, rhamnose, fucose, and glucuronic acid (e.g., Han et al. 2014). Often cyanobacterial EPS possess higher intrinsic viscosity and, for example, the EPS of *Nostoc flagelliforme* displayed pseudoplastic behavior in aqueous solution. Moreover, further characteristics, such as high intrinsic viscosity, good emulsification activity, and excellent flocculation capability, emphasize the EPS of terrestrial cyanobacteria as promising substances for numerous industrial applications (Han et al. 2014). Terrestrial cyanobacteria have attracted increasing pharmaceutical interest in recent years, because the antiviral substance Nostoflan was discovered within the EPS of *Nostoc flagelliforme* (Kanekiyo et al. 2005, 2007, 2008). But also other cyanobacteria exhibit bioactive substances (e.g., cryptophycins) and antiviral substances in *Nostoc ellipsosporum* (Cyanovirin-N, Boyd et al. 1997; Bewley et al. 1998) and *Scytonema* (Scytovirin, Bokesch et al. 2003) with bioactive action against HIV. In many species, EPS also contain the UV-protection pigment and anti-inflammatory drug scytonemin. Scytonemin is induced by desiccation, UV radiation, and nitrogen deficiency as reported e.g. for *Nostoc punctiforme* (Potts 1999; Fleming and Castenholz 2007, 2008; Pereira et al. 2009).

On the cellular level of cyanobacteria the desiccation process induces several physiological reactions such as synthesis of trehalose, sucrose, cyanophycin, polyhydroxyalkanoate, genes utilizing polyphosphate, ROS scavengers, sugar uptake, phycobilisome degradation, chaperones, glycogen breakdown, pentose phosphate pathway, proteases, and sigma factor regulation of cyanobacteria (Rajeev et al. 2013). In terrestrial green algae the biochemical protection strategy is the production of osmotically active carbohydrates such as polyols. These are particularly generated by green algae from the Trebouxiophyceae (e.g., Gustavs et al. 2010) while in Klebsormidiophyceae these compounds are missing in reasonable concentrations (Kaplan et al. 2012; Karsten and Holzinger 2014). As a protective strategy for cellular stabilization and against water loss, the terrestrial green algae *Apatococcus*, *Chloroidium*, *Coccomyxa*, *Prasiola*, *Stichococcus*, and *Trentepohlia* produce and accumulate high concentrations of polyols (Holzinger and Karsten 2013). In the cell, polyols function as organic osmolytes and compatible solutes. Moreover, they act as antioxidants, stabilizer of proteins, and as energy supply, e.g., to maintain repair processes (Yancey 2005; Karsten et al. 2007; and the references therein). Synthesized polyols are glycerol, erythritol, ribitol, arabitol, mannitol, sorbitol, and volemitol (Feige and Kremer 1980; Gustavs et al. 2010, 2011). Some adaptation processes of terrestrial cyanobacteria concerning light quality by complementary chromatic adaptation via phycobiline production and rearrangement are mentioned above. The subsection regarding temperature effects reveals how the production of the carotenoid astaxanthin is induced by the temperature. The carotenoid production of terrestrial microalgae is also influenced by light intensities. Tropical epilithic cyanobacteria displaced high contents of canthaxanthin at high light conditions while at low light ß-carotene dominated (Lakatos et al. 2001). Enhanced carotenoid production could also be induced in the biological soil crust *Leptolyngbya* sp. by high temperatures in an emerse photobioreactor (Kuhne et al. 2013, 2014b). Also the photobioreactor design seems to influence carotenoid production. Comparative analysis of growth and carotenoid accumulation of the epiphytic green algae *Trentepohlia arborum* showed higher production of carotenoids under aerial and subaerial conditions (4.88–67.7 mg m^{-2} s^{-1}, respectively) compared to submerse conditions (Chen et al. 2015). These results partly exceed carotenoid productivity of *Dunaliella salina* (13.5–90 mg m^{-2} s^{-1}) under submerse cultivation (Kleinegris et al. 2011; García-González et al. 2003).

The use of terrestrial microalgae for the production of valuable substances involves numerous different facets, in particular, the unique feature of synthesizing specific dehydration protective and photoprotective substances under the stress condition of dehydration, air-exposition, and high light. These recently introduced induction processes to stimulate the production of high-value substances are leading to interesting applications and could potentially develop new markets.

5 Bioengineering and Biofilm-Photobioreactors

Cultivation of terrestrial phototrophic microorganisms requires a different set-up in comparison to standard fermentation technologies because their growth is limited under submerged conditions (Kumar et al. 2010; Malcata 2011). Moreover, they show a better adaptability to extreme biotic and abiotic conditions due to their ecological impact in nutrient poor arid habitats (Dojani et al. 2007; Belnap and Lange 2001) and withstand high temperature fluctuations (see Table 1) and low water availability as mentioned above (see Table 2; Rascher et al. 2003, Sand-Jensen and Jespersen 2012). Currently, however, the most widely used standard fermentation systems for production of suspended microalgal biomass are open ponds (Brennan and Owende 2010). Because of the high contamination risk in open ponds three closed types of photobioreactors for a submerged cultivation of microalgae that are also used in industry were developed: tubular photobioreactors (Broneske et al. 1998), plate photobioreactors (or flat-panel) (Posten et al. 2011, Münkel et al. 2013), and foil photobioreactors (Cordes 2005). Most tubular and plate photobioreactors are still pilot-installations. The few commercial producers of microalgae products using closed systems mainly employed tubular photobioreactors. The company ALGOMED, for example, in cooperation with IGV GmbH (Potsdam, Germany) developed a tubular photobioreactor for the production of *Chlorella* powder. The scale-up of this reactor is located in Klötze, Germany (Broneske et al. 1998). A typical plate photobioreactor represents the hanging garden concept developed by the company ecoduna (Bruck an der Leitha, Austria) and is mainly used for biomass production. The reactor consists of communicating tubes that are gassed from below for a good intermixing of the algal suspension (Huber 2012). The company NOVAgreen together with an energy provider (RWE) built a foil photobioreactor set-up in Bergheim-Niederaussem (Germany) for microalgal production and CO_2-sequestration from dried flue gas (Cordes 2005). In comparison to open ponds the closed photobioreactors have a lower risk of contamination and a higher production of cell density (see Table 4). Moreover, the areal biomass productivity is higher (Jorquera et al. 2010) leading to higher CO_2 consumption rates (Chisti 2007). However, the energy input to produce 1 kg of biomass is 40% lower in open ponds (Chisti 2007, Jorquera et al. 2010). All those systems have in common that the cultures are submerged in liquid medium.

Due to the submerged cultivation techniques mainly aquatic microalgae have been used involving some disadvantages. As early as in the 1991 Gudin et al. described cell fragility as the key problem of microalgal mass production in closed photobioreactors (Gudin and Chaumont 1991). The surface-to-volume ratio is low in closed tubular photobioreactors. To receive high mass transfer rates high Reynolds numbers of medium turbulence are necessary. A weakly turbulent flow leads to an increasing availability of CO_2 and light (Bezerra et al. 2011). The result is the enhanced biomass productivity. However, with further increase of Reynolds number the number of cell fragility is increasing too (Gudin and Chaumont 1991; Camacho et al. 2000). On the one hand, this is because of hydrodynamic stress in bulk turbulent flow, mechanical agitation, and bubble break-up (Camacho et al. 1999) and, on the other hand the organisms are not able to adapt fast enough to the different light zones in the bioreactor – a crucial factor reducing productivity (Jassby and Platt 1976). Another disadvantage is that the intensive mixing of the cell suspension occurs energy intensive and is consequently an expensive process (Norsker et al. 2011).

The attempts to cultivate terrestrial microalgae submerged in an open pond or closed photobioreactors were less successful (e.g., Gao and Ye 2003). Many terrestrial microalgae are difficult to cultivate in liquid medium such as the green algae *Trentepohlia* or *Apatococcus*. Additionally, terrestrial microalgae are living in their natural form as immobilized biofilm often embedded in an extracellular polysaccharide-matrix (Tago and Aida 1977; Pereira et al. 2009) and are well adapted to xeric habitats (Lakatos et al. 2001). Therefore, some algae exhibit higher specific growth rates under air-exposed (emerse) conditions than submerged in suspension (Table 3; Gustavs et al. 2009; Kuhne et al. 2014a; Kuhne 2015).

The new generation of photobioreactors try to close the gap between the theoretical biological potential of phototrophic microorganisms and the biomass productivity. In the last years a new generation of photobioreactors were developed (see Tables 4 and 5) that enable an immobilized surface-associated growth of terrestrial microalgae (see Fig. 1). In subterrestrial cultivations the biomass is attached to solid substrates and imitates the natural habitat of the biofilm. The phototrophic microorganisms can form their matrix consisting of extracellular polymeric substances that is protecting the cell (Pereira et al. 2009; Tago and Aida 1977). Porous substrate photobioreactors, also known as twin-layer adhesion systems (Shi et al. 2007; Naumann et al. 2013) or attached cultivation systems (Liu et al. 2013) solve several volume-related problems (see Fig. 1c). The biofilm is immobilized by self-adhesion on a micro-porous surface that is separating the biofilm from the medium which is on the surface opposing side. Through this arrangement shear stress and cell fragility through inter-mixing the medium can be minimized. The gas phase is located on the other side which leads to an optimized supply of the biofilm with CO_2. The main changes from a submerge

Table 3 Comparison of maximal growth rates μ_{max} (day^{-1}) of different species in submerged and surface-associated cultivations (References a: Gustavs et al. 2009; b: Kuhne 2015; c: Kuhne et al. 2014a)

Microalgae	Submerged	Surface-associated
Apatococcus lobatus (Chodat) Boye Petersen	0.27a	0.24a
Chlamydomonas reinhardtii Dangeard	0.29b	0.34b
Chlorella (salina) gloriosa Molinari & Calvo-Pérez	1.09a	0.75a
Chlorella trebouxioides[a] Puncochárová	0.78a	0.62a
Chlorella vulgaris Beyerinck (Beijerinck)	1.12a	0.73a
Coccomyxa sp. Schmidle	0.48a	0.50a
Nostoc muscorum[b] Agardh ex Bornet & Flahault	0.29b	0.50b
Stichococcus sp. Nägeli	1.05a	0.71a
Trichocoleus sociatus (West & G.S. West) Anagnostidis	0.036b	0.46c

[a]This name is currently regarded as a taxonomic synonym of *Chloroidium ellipsoideum* (Gerneck) Darienko, Gustavs, Mudimu, Menendez, Schumann, Karsten, Friedl & Proschold
[b]This name is currently regarded as a taxonomic synonym of *Desmonostoc muscorum* (C. Agardh ex Bornet & Flahault) Hrouzek & Ventura

cultivation system to a porous substrate photobioreactor is the relocation of medium to the opposite side of the biofilm, light source, and gas supply. This arrangement allows a minimization of water and energy consumption (see Table 4) that is resulting in a minimization of costs. Another way of imitating the natural habitat of phototrophic biofilms is to mimic the tides. In these reactor types the biofilm is attached to a surface that is periodically supplied with medium. Accordingly, various types of rotating disc reactors were developed such as the rotating bioreactor including the Spool harvester (Christenson and Sims 2011a, b; Christenson and Sims 2012), the photorotating biological contactor PRBC (Orandi et al. 2012), or the rotating biological contactor-based photobioreactor (Blanken et al. 2014). All these rotating bioreactors have in common that the rotating cylindrical surface or the discs are representing the surfaces that are placed in a chamber which is half filled with medium. Through the rotation the microalgae are frequently moistened followed by exposure to the air. There are only two types of real surface-associated photobioreactor systems that mimic the rain or the conditions in the desert. Here the biofilm is attached to a surface and is getting moistened through spray nozzles or nebulized medium (aerosol). The company bioprocess algae developed the Grower Harvester™ – also a rotating photobioreactor – that imitates the rain by using spray nozzles for moistening the microorganisms with waste water (Haley et al. 2014). The construction is located in Omaha in pilot scale. The second one developed by Lakatos and colleagues is a stationary emerse photobioreactor (ePBR; see Fig. 1d) where only nebulized medium (aerosol) is pumped into the bioreactor (e.g., Kuhne et al. 2014a; Muffler et al. 2014). The air-exposed (emerse) microorganisms are attached to glass rods that are fixed in the cover and are inserted into a glass cylinder. The light source is located above the cover. Consequently the glass rods are simultaneously used as optical fiber and cultivation surface (Kuhne et al. 2014a). Kuhne et al. determined the growth rates per day of the terrestrial

Table 4 Comparison of open ponds, closed tubular systems, and biofilm reactors

	Raceway ponds	References	Closed tubular system	References	Biofilm reactor	References
Contamination risk	High		Low		Low	
Total system volume required to produce one kg of dry matter (L kg^{-1})	2,000–2,850	(Jorquera et al. 2010; Davis et al. 2011)	170–500	(Davis et al. 2011)	36–609	(Nowack et al. 2005; Johnson and Wen 2010; Ozkan et al. 2012; Naumann et al. 2013; Christenson and Sims 2012; Liu et al. 2013; Blanken et al. 2014, Boelee et al. 2014)
Investment costs	Low		Middle		Middle	
Harvest costs	High		High		Low	
Harvest of biomass	Dewatering process	(Jorquera et al. 2010)	Dewatering process	(Broneske et al. 1998; Cordes 2005; Huber 2012)	Scraping	(Johnson and Wen 2010; Christenson and Sims 2012; Blanken et al. 2014)
Cell density (g L^{-1})	0.35–0.5	(Jorquera et al. 2010; Davis et al. 2011)	1.02–6.01	(Jorquera et al. 2010; Davis et al. 2011)	36–300	(Nowack et al. 2005; Johnson and Wen 2010; Ozkan et al. 2012; Naumann et al. 2013; Christenson and Sims 2012; Liu et al. 2013, Blanken et al. 2014; Boelee et al. 2014)
Areal biomass productivity (g m^{-2} day^{-1})	11	(Jorquera et al. 2010)	25	(Jorquera et al. 2010)	20.1	(Blanken et al. 2014)
Energy input to produce 1 kg biomass (MJ kg^{-1})	9.18	(Chisti 2007; Jorquera et al. 2010)	386	(Chisti 2007; Jorquera et al. 2010)	4.71	(Ozkan et al. 2012)
CO_2 consumption[a] (g m^{-2} day^{-1})	20.7		47		37.8	

[a]Calculated from the biomass productivity according to the following equation: CO_2 fixation rate (P_{CO_2}) = 1.88 × biomass productivity (g m^{-2} day^{-1}), which is derived from the typical molecular formula of microalgal biomass, CO 0.48 H 1.83 N 0.11 P 0.01 (Chisti 2007)

Table 5 Comparison of different biofilm bioreactors (*PBR* photobioreactor, *FAME* fatty acid methyl ester, *COD* chemical oxygen demand, *EPS* extracellular polysaccharides, *NA* not applicable)

PBR	Substratum movement	Medium supply	Scale	Surface	Microorganism	Productivity (g m^{-2} day^{-1})	CO_2 consumption rate[a] (g m^{-2} day^{-1})	Harvest	Product	References
Attached algal culture system	Rocking	Liquid	Lab	Polystyrene foam	*Chlorella sp.*	2.57	4.83	Scraping	Biomass, FAME, removal COD	(Johnson and Wen 2010)
Spool harvester	Rotating	Liquid	Lab Pilot	Cotton rope	Polyculture	2.5	4.7	Scraping	Biomass, FAME, removal COD	(Christenson and Sims 2012)
PRBC (photorotating biological contactor)	Rotating	Liquid	Lab	Polyvinyl chloride	*Klebsormidium sp.*	0.45	0.85	Scraping	Biomass	(Orandi et al. 2012)
Grower Harvester™	Rotating Hanging	Spray nozzles	Lab Pilot	Various	*Chlorella pyrenoidosa, Phaeodactylum tricornutum, Chlamydomonas reinhardtii, Chlorella vulgaris. . .*	NA	NA	Washing	Biomass, removal COD	US 8809037 B2
Rotating biological contactor based photobioreactor	Rotating	Liquid	Lab	Stainless steel and polycarbonate	*Chlorella sorokiniana*	20.1	37.8	Scraping	Biomass, removal COD	(Blanken et al. 2014)
Concrete slab algal bio-film photobioreactor	None	Liquid	Lab	Concrete	*Botryococcus braunii*	0.71	1.33	Scraping	Biomass, FAME	(Ozkan et al. 2012)
BPBR (biofilm photobioreactor)	None	Liquid	Lab	Glass beads	*Rhodopseudomonas palustris*	NA	NA	NA	Biomass, hydrogen, removal COD	(Tian et al. 2010)
ePBR (emerse photobioreactor	None	Aerosol	Lab	Borosilicate glass	*Trichocoleus sociatus*	3.14	5.9	Scraping	Biomass, EPS	(Kuhne et al. 2014a)
ssPBR (suspended-solid photobioreactor)	None	Liquid	Lab	Cotton, linen, mohair	*Scenedesmus so.*	NA	NA	Catching solid carriers	Biomass	(Zhuang et al. 2014)
PSBR (porous substrate biofilm photobioreactor)	None	Liquid	Lab Pilot	Glass fiber filter	*Anabaena variabilis*	1.2	2.26	Mechanical detachment	Biomass, hydrogen, removal COD	(Podola et al. 2017)
Twin-layer photobioreactor	None	Liquid	Lab Pilot	Nylon filter sheets	*Halochlorella rubescens*	6.3	11.8	Scraping	Biomass, removal COD	(Shi et al. 2007)
Twin-layer: tube-type photobioreactor	None	Liquid	Lab	Paper filter	*Isochrysis sp. TISO, Phaeodactylum tricornutum, Tetraselmis suecica*	0.6 1.5 1.8	1.13 2.82 3.38	Scraping	Biomass	(Naumann et al. 2013)

(continued)

Table 5 (continued)

PBR	Substratum movement	Medium supply	Scale	Surface	Microorganism	Productivity (g m^{-2} day^{-1})	CO_2 consumption rate[a] (g m^{-2} day^{-1})	Harvest	Product	References
Twin-layer: solid-state photo-bioreactor	None	Liquid	Pilot	Unprinted newspaper	*Isochrysis sp. TISO, Phaeodactylum tricornutum, Tetraselmis suecica*	0.42 0.88 0.58	0.79 1.65 1.09	Scraping	Biomass	(Naumann et al. 2013)
Flow-cell concept	None	Liquid	Lab	Glass	*Nitzschia palea, Scenedesmus obliquus*	2.8	4.68	Scraping	Biomass, FAME	(Schnurr et al. 2013)
Parallel plate air lift reactor	None	Liquid	Lab	Cellulose acetate	Polyculture	2.08	3.91	Scraping	Biomass, FAME	(Genin et al. 2014)
Phototrophic biofilm pilot-scale reactor	None	Liquid	Lab Pilot	Polyfelt sheet	Polyculture	4.5	8.76	Scraping	Biomass, removal COD	(Boelee et al. 2014)
Roof-installed parallel plate microalgae biofilm reactor	None	Liquid	Lab	Polycarbonate sheet	*Scenedesmus obliquus*	2.5	4.7	Backwashing	Biomass, removal COD	(Zamalloa et al. 2013)
Algal turf scrubber	None	Liquid	Pilot	Polyethylene mesh	Polyculture	5.0	9.4	Scraping	Biomass, removal COD	(Mulbry and Wilkie 2001)
WWTP (wastewater treatment plant)	None	Liquid	Lab	Polypropylene	*Scenedesmus sp.*	NA	NA	NA	Biomass, removal COD	(He and Xue 2010)
Algal–bacterial biofilm bioreactor	None	Liquid	Lab	Polypropylene	Polyculture	NA	NA	NA	Biomass, removal COD	(Posadas et al. 2013)
BMPBR (algal biofilm membrane photobioreactor)	None	Liquid	Lab	Flexible fiber bundles	*Chlorella vulgaris*	NA	NA	Mechanical separator, solvent extraction	Biomass, removal COD	(Gao et al. 2015)

[a]Calculated from the biomass productivity according to the following equation: CO_2 fixation rate (Pco_2) = 1.88 × biomass productivity (g m^{-2} day^{-1}), which is derived from the typical molecular formula of microalgal biomass, CO 0.48 H 1.83 N 0.11 P 0.01 (Chisti 2007)

cyanobacterium *Trichocoleus sociatus* in shaking flasks (Kuhne 2015) and in the ePBR (Kuhne et al. 2014a). The growth rates per day were 24 times higher in the ePBR. So, terrestrial cyanobacteria show an enhanced growth under emerse conditions. Another advantage is that EPS production can be stimulated by the application of less quantities of aerosol (more drought stress) and is antiproportional associated with biomass production. High quantities of aerosol favor growth of biomass while the influence of drought stress stimulates the production of EPS to the amount up to 50% of DM (unpublished) due to low quantities of aerosol (Kuhne et al. 2014a).

In Table 4 the advantages and disadvantages of open ponds, closed systems, and biofilm reactors are summarized and classified into cultivation conditions: permanent submerse, submerse/emerse-cycle, and mainly emerse. In comparison to submerge photobioreactors the harvest of cells is much easier in biofilm photobioreactors. Less costly dewatering processes like centrifugation or flocculation (Christenson and Sims 2011a, b) are needed because the biofilm has a solid content between 10 and 20% (Johnson and Wen 2010; Gross et al. 2013) and can easily be scraped off the substrate (Christenson and Sims 2011a, b; Christenson and Sims 2012; Ozkan et al. 2012). The energy consumption for dewatering in downstream processing can be up to 82 MJ kg^{-1} biomass (Chisti 2007; Jorquera et al. 2010; Ozkan et al. 2012). In conclusion to produce 1 kg biomass an energy input of 385 MJ kg^{-1} is necessary in a tubular system (Chisti 2007; Jorquera et al. 2010) and only 4.7 MJ kg^{-1} in a biofilm reactor (Ozkan et al. 2012). In addition, in biofilm photobioreactors the cell density is up to 90 times higher in comparison to submerge technologies like open ponds or closed systems (Jorquera et al. 2010; Ozkan et al. 2012; Podola et al. 2017).

In principle the photosynthetic performance is composed of the photochemical reaction complex which mainly depends on the light availability and a whole slew of temperature dependent enzymatic reactions. Phototrophic organisms transform light energy into chemical energy that is stored in high energy compounds (Iverson 2006). As a consequence an optimal light distribution excites the biological activity of photosynthetic microorganisms (Zhang et al. 2014) and leads to a high biomass production. A uniform light distribution depends on the light path and biomass density or biofilm thickness (Zhang et al. 2014). Through a regularly scraping of the upper part of the phototrophic biofilm the light intensities can be regulated and an inefficient light entry can be prevented (Guzzon et al. 2008; Blanken et al. 2014). In contrast, high cell densities mainly lead to mutual shading between different cells under submerge conditions. Especially light scattering by cells and absorption by intracellular pigments influence the transmitted light (Wang et al. 2012). While the light intensities in the biofilm reactor can be controlled over the biofilm thickness under submerge conditions a frequent harvest of biomass would result in a diluted algal density with a large amount of water, which in turn will cause higher harvest costs.

Beneath the main influence factors light and temperature the availability of CO_2 is very important for the photosynthesis. To guarantee a sufficient mass transfer of CO_2 in submerge cultivation systems high Reynolds numbers are necessary that in turn cause cell damage. It could be shown that a separation between the gas phase and the medium supply leads to higher growth rates.

In conclusion, the efforts of new generation of biofilm photobioreactors (see Table 5) to reduce harvesting cost, water consumption, and energy input (see Table 4) lead stepwise to the orientation on the natural habitat of phototrophic biofilms. Also the light distribution can easily be controlled through a regularly scrape of the upper part of the phototrophic biofilm whereby the thickness of biofilm and an optimal light input can be controlled.

6 Visions of Systemic Approaches: Possible Steps into the Future

The advantages of biofilm photobioreactors in combination with the use of desiccation tolerant organisms open new perspectives and visions to face the future challenges. In recent years different disciplines from bioengineering, agro science, and environmental protection to ecology have come together for systemic approaches to create projects facing future bioeconomic challenges. For a projected population of 9.6 billion people in 2050 (Lelieveld et al. 2015), for example, an additional required agricultural area approximately the size of Brazil is predicted. This enormous agricultural area can be reached by only one tenth worldwide, even with the utilization of potential land. At the same time, as urbanization increases, 6.4 billion people – 66% of the world's population – will live in cities (Lelieveld et al. 2015), and the urban area will rise from 1 to 6% worldwide. The increased competition for acreages, the increasing use of renewable raw materials, and the scarcity of fossil fuels are forcing a rethinking of previous agricultural concepts. Urgently demanded are aspects of recirculation and intelligent networking of material flows (energy, resources, waste, processes, services), as well as the flexibility to produce next to food also proof-based and demand-oriented nutritional supplements. For this purpose, the development of new agricultural areas, for example, on sealed and anthropogenic transformed areas, is necessary through decentralized, regionally, and smart gridded micro-agro systems with the potential of multifunctional applications – a systemic innovation network. Could emerse photobioreactor designs in the combination with terrestrial microalgae substantially contribute to such micro-agro system for the future?

6.1 *Future Technology*

In urban agglomerations, horizontal cultivation areas are almost completely covered by anthropogenic services, while vertical areas, for example façades, are available and are not in competition with existing agricultural areas. On these vertical surfaces, conventional agriculture with typical agricultural crops can only be carried out with considerable effort (vertical farming). As an alternative, the combination of microalgae as "micro-agro crops" and photobioreactors (PBR) can be used as vertical agricultural systems. So far, as mentioned above, bioprocessing methods that are mostly used with aquatic microalgae under submersed conditions are inefficient in terms of resources and energy input. New opportunities are considered here by aerosol-based production processes with terrestrial microalgae. Compared to the conventional submerged producing bioreactors, the use of aerosols in emerse PBR can save approximately 90% of water and achieves an energy saving of 20–40% by eliminating mixing and drying processes. In comparison to submerse reactors, an emerse photobioreactor has a weight which is about 50–60% lighter and allows subsequent wall mounting even on existing buildings.

6.2 *Production Organism*

The aerosol-based method combined with immobilized terrestrial microalgae has high process stability, an optimal gas exchange simplifies the charging and absorption of emissions, as well as lower resource consumption due to the fixation of CO_2

Fig. 2 Conceptual frame of biotechnology with terrestrial microalgae. Integration of material flow management for recycling and loading with regional end-of-the-pipe material flows, for example, waste heat (temperature), sewerage (moistures), emissions (CO_2) increasing production rates and reducing expenses of photobioreactors (copyright by T. Schmidt)

and N_2 as well as the recycling of nutrient-containing process and waste water. Also, the bioavailability and composition of the ingredients, the high proportions of proteins (special amino acids), and unsaturated fatty acids as well as the tenfold higher production of polysaccharides (exopolysaccharides induced by desiccation stress) justify these “micro-agro crops” as excellent food producers (see Fig. 2). For this reason, they have been used, for example, in Asian cuisine for many centuries (*Nostoc flagelliforme*). Additional advantages are the coupling production of pharmaceutically active and valuable substances, a high and hitherto scarcely used species diversity and a high accessibility for the transformation with heterologous DNA for the targeted de novo synthesis of specific substances or the manipulated elimination of certain target products (this biotransformation facilitates downstream processes). As alternative to terrestrial microalgae, plant cell lines and other desiccation tolerant organisms (for example, fungi, bryophytes, lichens) can also be cultivated. Thus, food and/or higher value-added products can directly be produced and recycled at the location of the consumption. This significantly reduces transport distances and prevents the associated CO_2 emissions.

6.3 Networking of Value-Added Chains

The versatility of biomass production, the independence of seasonality, and the short-term harvesting periods enable the creation of an individual and functional composition of the food ingredients for the consumer. A food generation (food design, food 4.0) is based on demand, health aspects, and taste preferences. The integration of, for example, cyber-physical systems and cloud technologies helps to monitor, produce, process, and deliver in a controlled manner. Regarding the material flow management, the micro-agro systems are multifunctional and flexible because they can be used for agricultural production not only with different producers (microalgae and other desiccation tolerant organisms or cells), but also for recycling and loading with regional end-of-the-pipe material flows (e.g., waste heat, sewerage, emissions) to increase the production rate and to reduce expenses. Fine tuning of optimal growth conditions or objectives of the production can be set depending on whether they are exposed to tower buildings, industrial plants, or wind turbines. Accordingly, the investment of fertilizers and pesticides or thermoregulations could be largely reduced. In addition, natural (e.g., solar energy) as well as material and energy flows can be systemically optimized and automated (intelligent self-regulation by cyber-physical systems) via smart networking. Material flows are not lost, but are recycled to increase resource efficiency. Considering the regional networking, smart networking and material flow management also leads to an efficient regulation of regional-production-networks, on the one hand, by linking micro-agro systems with each other and/or with building technology, and, on the other hand, by linking with commerce, industry, energy provider, sewage treatment plants, and consumers (humans). Also the economy could benefit from this systemic approach. In such multifunctional, decentralized, and unconventional production-networks, new services (business models) would arise, allocated to fields of the production (e.g., harvesting, processing, food design, distribution), the process (e.g., technology development, implementation, process engineering, biotechnology, environmental technology, maintenance, suppliers) as well as the networking (e.g., IT, material flow management, environmental protection manager, building technology).

7 Conclusion

The implementation of terrestrial microalgae for biotechnology and new applications sounds at a first glance surprising. At closer view their outstanding performance and considerable advantages for biotechnology may reduce several current limitations regarding mass transfer, thermostability, photoinhibition as well as expenses for energy (e.g., mixing and dewatering) and resources (e.g., water and nutrients). The application of terrestrial microalgae and exploitation of their advantages comprise the shift from a suspension-based cultivation to an emerse, air-exposed, cultivation with immobilized biofilms. Due to the increased gas phase in emerse photobioreactor, the integration of external material flows are facilitated allowing the inflow and recycling of end-of-the-pipe material (e.g., waste heat, sewerage, CO_2-emissions) to increase the production rate. Additionally, the investment for fertilizers or thermoregulations could be reduced. Moreover, the desiccation processes induce the production of high-value substances opening a potential way to new markets.

The concept of an urban and smart micro-agro-network for evidence-based agricultural production through recycling of material flow, innovative emerse photobioreactor technology, and transformation of sealed surfaces into new agricultural areas might be just a vision far into the future. But novel concepts for biotechnology and new applications are provided by the implementation of terrestrial microalgae and further efforts should emphasize the biological and engineering challenges in this emerging field. The presented considerations and potentials of terrestrial microalgae may contribute to improve the technological and economic progress of microalgal biotechnology.

References

Abed RMM, Dobretsov S, Sudesh K (2009) Applications of cyanobacteria in biotechnology. J Appl Microbiol 106:1–12

Adams DG (2002) Cyanobacteria in symbiosis with hornworts and liverworts. In: Rai AN, Bergman B, Rasmussen U (eds) Cyanobacteria in symbiosis. Springer, Dordrecht, pp 117–135

Aigner S, Remias D, Karsten U et al (2013) Unusual phenolic compounds contribute to ecophysiological performance in the purple-colored green alga Zygogonium ericetorum (Zygnematophyceae, Streptophyta) from a high-alpine habitat. J Phycol 49:648–660

Allewaert CC, Vanormelingen P, Pröschold T et al (2015) Species diversity in European Haematococcus pluvialis (Chlorophyceae, Volvocales). Phycologia 54:583–598

Andreoli C, Lokhorst GM, Mani AM, Scarabel L, Moro I, La Rocca N, Tognetto L (1998) *Koliella antarctica sp. nov.* (*Klebsormidiales*) a new marine green microalga from the Ross Sea (Antarctica). Algol Stud/Arch Hydrobiol 90:1–8

Barros AI, Gonçalves AL, Simões M et al (2015) Harvesting techniques applied to microalgae: a review. Renew Sustain Energy Rev 41:1489–1500

Belnap J, Lange O (2001) Biological soil crusts: structure, function, and management. Springer, Heidelberg

Berner F, Heimann K, Sheehan M (2015) Microalgal biofilms for biomass production. J Appl Phycol 27:1793–1804

Bertsch A (1966) CO_2-Gaswechsel und Wasserhaushalt der Aerophilen Grünalge *Apatococcus lobatus*. Planta 70:46–72

Bewley CA, Gustafson KR, Boyd MR et al (1998) Solution structure of cyanovirin-N, a potent HIV-inactivating protein. Nat Struct Biol 5:571–578

Bezerra RP, Montoya EYO, Sato S et al (2011) Effects of light intensity and dilution rate on the semicontinuous cultivation of *Arthrospira* (*Spirulina*) *platensis*. A kinetic Monod-type approach. Bioresour Technol 102:3215–3219

Blanken W, Janssen M, Cuaresma M et al (2014) Biofilm growth of *Chlorella sorokiniana* in a rotating biological contactor based photobioreactor. Biotechnol Bioeng 111:2436–2445

Boelee N, Janssen M, Temmink H et al (2014) The effect of harvesting on biomass production and nutrient removal in phototrophic biofilm reactors for effluent polishing. J Appl Phycol 26: 1439–1452

Bokesch HR, O'keefe BR, Mckee TC et al (2003) A potent novel anti-HIV protein from the cultured cyanobacterium Scytonema varium. Biochemistry 42:2578–2584

Borowitzka MA (2013) High-value products from microalgae – their development and commercialisation. J Appl Phycol 25:743–756

Borowitzka MA (2016) Algal physiology and large-scale outdoor cultures of microalgae. In: Borowitzka MA, Beardall J, Raven JA (eds) The physiology of microalgae. Springer, Cham, pp 601–652

Borowitzka MA, Huisman JM, Osborn A (1991) Culture of the astaxanthin-producing green alga *Haematococcus pluvialis* 1. Effects of nutrients on growth and cell type. J Appl Phycol 3: 295–304

Boyd MR, Gustafson KR, Mcmahon JB et al (1997) Discovery of cyanovirin-N, a novel human immunodeficiency virus-inactivating protein that binds viral surface envelope glycoprotein gp120: potential applications to microbicide development. Antimicrob Agents Chemother 41: 1521–1530

Brennan L, Owende P (2010) Biofuels from microalgae – a review of technologies for production, processing, and extractions of biofuels and co-products. Renew Sustain Energy Rev 14: 557–577

Bristol BM (1919) On the retention of vitality byy algae from old stored soils. New Phytol 18: 92–107

Broneske J, Pulz O, Franke H et al (1998) Installation for carrying out photochemical and photocatalytic reactions and photoinduced processes. Patents EP0968273 B1

Büdel B (1999) Ecology and diversity of rock-inhabiting cyanobacteria in tropical regions. Eur J Phycol 34:361–370

Büdel B (2011a) Cyanobacteria: habitats and species. In: Lüttge U, Beck E, Bartels D (eds) Plant desiccation tolerance. Springer, Heidelberg, pp 11–21

Büdel B (2011b) Eukaryotic algae. In: Lüttge U, Beck E, Bartels D (eds) Plant desiccation tolerance. Springer, Heidelberg, pp 45–63

Büdel B, Becker U, Porembski S et al (1997) Cyanobacteria and cyanobacterial lichens from inselbergs of the ivory coast, Africa. Bot Acta 110:458–465

Büdel B, Schulz B, Reichenberger H et al (2009) Cryptoendolithic cyanobacteria from calcite marble rock ridges, Taylor Valley, Antarctica. Algol Stud 129:61–69
Calvo-Pérez EA, Molinari-Novoa EA, Guiry MD (2016) Validation of *Nostoc flagelliforme* (Nostocaceae, Cyanobacteria). Notulae Algarum 2:1–2
Camacho FGA, Gómez AC, Fernández FGA et al (1999) Use of concentric-tube airlift photobioreactors for microalgal outdoor mass cultures. Enzyme Microb Technol 24:164–172
Camacho FGA, Gómez AC, Sobczuk TM et al (2000) Effects of mechanical and hydrodynamic stress in agitated, sparged cultures of *Porphyridium cruentum*. Process Biochem 35:1045–1050
Cameron RE (1962) Species of Nostoc Vaucher occurring in the Sonoran Desert in Arizona. Trans Am Microsc Soc 81:379–384
Chen L, Zhang L, Zhang W et al (2015) Comparative analysis of growth and carotenoid accumulation of *Trentepohlia arborum* in aerial, subaerial, and aquatic cultivation. J Appl Phycol 27:1079–1087
Chisti Y (2007) Biodiesel from microalgae. Biotechnol Adv 25:294–306
Chisti Y (2013) Constraints to commercialization of algal fuels. J Biotechnol 167:201–214
Christenson L, Sims R (2011a) Production and harvesting of microalgae for wastewater treatment, biofuels, and bioproducts. Biotechnol Adv 29:686–702
Christenson L, Sims R (2011b) Rotating bioreactor and spool harvester apparatus for biomass production. Patent US20110217764 A1
Christenson LB, Sims RC (2012) Rotating algal biofilm reactor and spool harvester for wastewater treatment with biofuels by-products. Biotechnol Bioeng 109:1674–1684
Contreras-Porcia L, Thomas D, Flores V et al (2011) Tolerance to oxidative stress induced by desiccation in *Porphyra columbina* (Bangiales, Rhodophyta). J Exp Bot 62:1815–1829
Cook ME, Graham LE (2016) Chlorokybophyceae, Klebsormidiophyceae, Coleochaetophyceae. In: Archibald JM, Simpson AGB, Slamovits CH, Margulis L, Melkonian M, Chapman DJ, Corliss JO (eds) Handbook of the protists. Springer, Cham, pp 1–20
Cordes R (2005) Vorrichtung zur Zucht und Massenproduktion von Algen, Patent DE 102004007564 A1
Cui Z (1985) The distribution of *Nostoc flagelliforme* in Siziwangqi county, Inner Mongolia. Acta Phytoecologica et Geobotanica Sinica 9:223–230
Darienko T, Gustavs L, Mudimu O et al (2010) *Chloroidium*, a common terrestrial coccoid green alga previously assigned to *Chlorella* (Trebouxiophyceae, Chlorophyta). Eur J Phycol 45: 79–95
Davis R, Aden A, Pienkos PT (2011) Techno-economic analysis of autotrophic microalgae for fuel production. Appl Energy 88:3524–3531
Dingji S, Yuanyuan Q, Zepu Z (1992) Studies on photosynthesis, respiration and morphology of *Nostoc flagelliforme*. Acta Bot Sin 34:507–514
Doemel WN, Brock TD (1971) The physiological ecology of *Cyanidium caldarium*. Microbiology 67:17–32
Dojani S, Lakatos M, Rascher U et al (2007) Nitrogen input by cyanobacterial biofilms of an inselberg into a tropical rainforest in French Guiana. Flora 202:521–529
Draaisma RB, Wijffels RH, Slegers PM et al (2013) Food commodities from microalgae. Curr Opin Biotechnol 24:169–177
Elbert W, Weber B, Burrows S et al (2012) Contribution of cryptogamic covers to the global cycles of carbon and nitrogen. Nat Geosci 5:459–462
Feige GB, Kremer BP (1980) Unusual carbohydrate pattern in *Trentepohlia* species. Phytochemistry 19:1844–1845
Fleming ED, Castenholz RW (2007) Effects of periodic desiccation on the synthesis of the UV-screening compound, scytonemin, in cyanobacteria. Environ Microbiol 9:1448–1455
Fleming ED, Castenholz RW (2008) Effects of nitrogen source on the synthesis of the UV-screening compound, scytonemin, in the cyanobacterium Nostoc punctiforme PCC 73102. FEMS Microbiol Ecol 63:301–308
Friedl T, Rybalka N (2012) Systematics of the green algae: a brief introduction to the current status. In: Lüttge U, Beyschlag W, Büdel B, Francis D (eds) Prog bot. Springer, Berlin, Heidelberg, pp 259–280

Gao K (1998) Chinese studies on the edible blue-green alga, Nostoc flagelliforme: a review. J Appl Phycol 10:37–49

Gao K, Ye C (2003) Culture of the terrestrial cyanobacterium, *Nostoc flagelliforme* (Cyanophycae), under aquatic conditions. J Phycol 39:617–623

Gao F, Yang Z-H, Li C et al (2015) A novel algal biofilm membrane photobioreactor for attached microalgae growth and nutrients removal from secondary effluent. Bioresour Technol 179: 8–12

García-González M, Moreno J, Canavate J et al (2003) Conditions for open-air outdoor culture of *Dunaliella salina* in southern Spain. J Appl Phycol 15:177–184

Garcia-Pichel F, Sherry ND, Castenholz RW (1992) Evidence for an ultraviolet sunscreen role of the extracellular pigment scytonemin in the terrestrial cyanobacterium *Chlorogloeopsis* sp. Photochem Photobiol 56:17–23

Garcia-Pichel F, Loza V, Marusenko Y et al (2013) Temperature drives the continental-scale distribution of key microbes in topsoil communities. Science 340:1574–1577

Gasulla F, De Nova PG, Esteban-Carrasco A et al (2009) Dehydration rate and time of desiccation affect recovery of the lichenic algae *Trebouxia erici*: alternative and classical protective mechanisms. Planta 231:195–208

Genin SN, Stewart Aitchison J, Grant Allen D (2014) Design of algal film photobioreactors: material surface energy effects on algal film productivity, colonization and lipid content. Bioresour Technol 155:136–143

Gerardo ML, Van Den Hende S, Vervaeren H et al (2015) Harvesting of microalgae within a biorefinery approach: a review of the developments and case studies from pilot-plants. Algal Res 11:248–262

Gerrath JF, Gerrath JA, Matthes U et al (2000) Endolithic algae and cyanobacteria from cliffs of the Niagara Escarpment, Ontario, Canada. Can J Bot 78:807–815

Gray DW, Lewis LA, Cardon ZG (2007) Photosynthetic recovery following desiccation of desert green algae (*Chlorophyta*) and their aquatic relatives. Plant Cell Environ 30:1240–1255

Green T, Sancho LG, Pintado A (2011) Ecophysiology of desiccation/rehydration cycles in Mosses and Lichens. In: Lüttge U, Beck E, Bartels D (eds) Plant desiccation tolerance. Springer, Heidelberg, pp 89–120

Grobbelaar JU (2009) Upper limits of photosynthetic productivity and problems of scaling. J Appl Phycol 21:519–522

Grobbelaar JU (2010) Microalgal biomass production: challenges and realities. Photosynth Res 106:135–144

Grobbelaar JU (2013) Mass production of microalgae at optimal photosynthetic rates. In: Dubinsky Z (ed) Photosynthesis. InTech, Rijeka, pp 357–371

Gross R, Buehler K, Schmid A (2013) Engineered catalytic biofilms for continuous large scale production of n-octanol and (S)-styrene oxide. Biotechnol Bioeng 110:424–436

Gross M, Jarboe D, Wen Z (2015) Biofilm-based algal cultivation systems. Appl Microbiol Biotechnol 99:5781–5789

Grossman AR, Schaefer MR, Chiang GG et al (1994) The response of cyanobacteria to environmental conditions: light and nutrients. Kluwer Academic Publishers, Dordrecht

Gudin C, Chaumont D (1991) Cell fragility – the key problem of microalgae mass production in closed photobioreactors. Bioresour Technol 38:145–151

Gustavs L, Schumann R, Eggert A et al (2009) In vivo growth fluorometry: accuracy and limits of microalgal growth rate measurements in ecophysiological investigations. Aquat Microb Ecol 55:95–104

Gustavs L, Eggert A, Michalik D et al (2010) Physiological and biochemical responses of green microalgae from different habitats to osmotic and matric stress. Protoplasma 243:3–14

Gustavs L, Görs M, Karsten U (2011) Polyol patterns in biofilm-forming aeroterrestrial green algae (Trebouxiophyceae, Chlorophyta)1. J Phycol 47:533–537

Guzzon A, Bohn A, Diociaiuti M et al (2008) Cultured phototrophic biofilms for phosphorus removal in wastewater treatment. Water Res 42:4357–4367

Haley JW, Ahrens TD, Kitchner SR (2014) Systems, apparatuses and methods for treating wastewater. Patent US8809037 B2

Han P-P, Sun Y, Jia S-R et al (2014) Effects of light wavelengths on extracellular and capsular polysaccharide production by Nostoc flagelliforme. Carbohydr Polym 105:145–151

He S, Xue G (2010) Algal-based immobilization process to treat the effluent from a secondary wastewater treatment plant (WWTP). J Hazard Mater 178:895–899

Helm RF, Huang Z, Edwards D et al (2000) Structural characterization of the released polysaccharide of desiccation-tolerant *Nostoc commune* DRH-1. J Bacteriol 182:974–982

Ho S-H, Chen C-Y, Lee D-J et al (2011) Perspectives on microalgal CO_2-emission mitigation systems – a review. Biotechnol Adv 29:189–198

Hoffmann L (1989) Algae of terrestrial habitats. Bot Rev 55:77–105

Hoham RW (1975) Optimum temperatures and temperature ranges for Growth of snow algae. Arct Alp Res 7:13–24

Holzinger A, Karsten U (2013) Desiccation stress and tolerance in green algae: consequences for ultrastructure, physiological and molecular mechanisms. Front Plant Sci 4:1–18

Holzinger A, Tschaikner A, Remias D (2010) Cytoarchitecture of the desiccation-tolerant green alga *Zygogonium ericetorum*. Protoplasma 243:15–24

Holzinger A, Kaplan F, Blaas K et al (2014) Transcriptomics of desiccation tolerance in the streptophyte green alga *Klebsormidium* reveal a land plant-like defense reaction. PLoS One 9: e110630

Huang Z, Liu Y, Paulsen BS, Klaveness D (1998) Studies on polysaccharides from three edible species of *Nostoc (Cyanobacteria)* with different colony morphologies: comparison of monosaccharide compositions and viscosities of polysaccharides from field colonies and suspension cultures. J Phycol 34:962–968

Huber EW (2012) Hanging garden planter apparatus with integrated drainage system. Patent US20120247010 A1

Ikonomou L, Schneider Y-J, Agathos S (2003) Insect cell culture for industrial production of recombinant proteins. Appl Microbiol Biotechnol 62:1–20

Iverson TM (2006) Evolution and unique bioenergetic mechanisms in oxygenic photosynthesis. Curr Opin Chem Biol 10:91–100

Jaluria P, Betenbaugh M, Konstantopoulos K et al (2007) Enhancement of cell proliferation in various mammalian cell lines by gene insertion of a cyclin-dependent kinase homolog. BMC Biotechnol 7:71

Jassby A, Platt T (1976) Mathematical formulation of the relationship between photosynthesis and light for phytoplankton. Limnol Oceanogr 21:540–547

Johnson M, Wen Z (2010) Development of an attached microalgal growth system for biofuel production. Appl Microbiol Biotechnol 85:525–534

Johnson EM, Kumar K, Das D (2014) Physicochemical parameters optimization, and purification of phycobiliproteins from the isolated *Nostoc* sp. Bioresour Technol 166:541–547

Jorquera O, Kiperstok A, Sales EA et al (2010) Comparative energy life-cycle analyses of microalgal biomass production in open ponds and photobioreactors. Bioresour Technol 101: 1406–1413

Kanekiyo K, Lee J-B, Hayashi K et al (2005) Isolation of an antiviral polysaccharide, Nostoflan, from a terrestrial cyanobacterium, *Nostoc flagelliforme*. J Nat Prod 68:1037–1041

Kanekiyo K, Hayashi K, Takenaka H et al (2007) Anti-herpes simplex virus target of an acidic polysaccharide, Nostoflan, from the edible blue-green alga *Nostoc flagelliforme*. Biol Pharm Bull 30:1573–1575

Kanekiyo K, Hayashi K, Lee JB et al (2008) Structure and antiviral activity of an acidic polysaccharide from an edible blue-green alga, *Nostoc flagelliforme*. Yakugaku Zasshi 128:25–731

Kaplan F, Lewis LA, Wastian J et al (2012) Plasmolysis effects and osmotic potential of two phylogenetically distinct alpine strains of *Klebsormidium* (Streptophyta). Protoplasma 249: 789–804

Karsten U, Holzinger A (2012) Light, temperature, and desiccation effects on photosynthetic activity, and drought-induced ultrastructural changes in the green alga *Klebsormidium dissectum* (Streptophyta) from a high alpine soil crust. Microb Ecol 63:51–63

Karsten U, Klimant I, Holst G (1996) A new in vivo fluorimetric technique to measure growth of adhering phototrophic microorganisms. Appl Environ Microbiol 62:237–243

Karsten U, Lembcke S, Schumann R (2007) The effects of ultraviolet radiation on photosynthetic performance, growth and sunscreen compounds in aeroterrestrial biofilm algae isolated from building facades. Planta 225:991–1000

Karsten U, Lütz C, Holzinger A (2010) Ecophysiological performance of the aeroterrestrial green alga *Klebsormidium crenulatum* (Charophyceae, Streptophyta) isolated from analpine soil crust with emphasis on desiccation stress. J Phycol 46:1187–1197

Karsten U, Holzinger A (2014) Green algae in alpine biological soil crust communities: acclimation strategies against ultraviolet radiation and dehydration. Biodivers Conserv 23:1845–1858

Karsten U, Herburger K, Holzinger A (2014) Dehydration, temperature, and light tolerance in members of the aeroterrestrial green algal genus *Interfilum* (Streptophyta) from biogeographically different temperate soils. J Phycol 50:804–816

Karsten U, Herburger K, Holzinger A (2016) Living in biological soil crust communities of African deserts – physiological traits of green algal *Klebsormidium* species (Streptophyta) to cope with desiccation, light and temperature gradients. J Plant Physiol 194:2–12

Kesaano M, Sims RC (2014) Algal biofilm based technology for wastewater treatment. Algal Res 5:231–240

Kleinegris DMM, Janssen M, Brandenburg WA et al (2011) Continuous production of carotenoids from *Dunaliella salina*. Enzyme Microb Technol 48:253–259

Kliphuis AMJ, De Winter L, Vejrazka C et al (2010) Photosynthetic efficiency of *Chlorella sorokiniana* in a turbulently mixed short light-path photobioreactor. Biotechnol Prog 26: 687–696

Kliphuis AMJ, Martens DE, Janssen M et al (2011) Effect of O_2:CO_2 ratio on the primary metabolism of *Chlamydomonas reinhardtii*. Biotechnol Bioeng 108:2390–2402

Kok B (1953) Experiments on photosynthesis by *Chlorella* in flashing light. In: Burlew JS (eds) Algal culture: from laboratory to pilot plant. Vol. 600 Algal culture: from laboratory to pilot plant. Carnegie Institution of Washington Publ., Washington, pp 63–75

Kol E, Flint EA (1968) Algae in green ice from the Balleny Islands, Antarctica. N Z J Bot 6:249–261

Kuhne S (2015) In: Fermentation von phototrophen Organismen zur Produktion von biotechnologischen Wertstoffen. PhD thesis, University of Kaiserslautern, Cuvillier, Göttingen

Kuhne S, Lakatos M, Foltz S et al (2013) Characterization of terrestrial cyanobacteria to increase process efficiency in low energy consuming production processes. Sustainable Chem Processes 1:1–6

Kuhne S, Strieth D, Lakatos M et al (2014a) A new photobioreactor concept enabling the production of desiccation induced biotechnological products using terrestrial cyanobacteria. J Biotechnol 192:28–33

Kuhne S, Strieth D, Weber A et al (2014b) Screening of two terrestrial cyanobacteria for biotechnological production processes in shaking flasks, bubble columns, and stirred tank reactors. J Appl Phycol 26:1639–1648

Kumar A, Ergas S, Yuan X et al (2010) Enhanced CO_2 fixation and biofuel production via microalgae: recent developments and future directions. Trends Biotechnol 28:371–380

Lakatos M (2011) Lichens and bryophytes: habitats and species. In: Lüttge U, Beck E, Bartels D (eds) Plant desiccation tolerance. Springer, Heidelberg, pp 65–87

Lakatos M, Bilger W, Büdel B (2001) Carotenoid composition of terrestrial Cyanobacteria: response to natural light conditions in habitats in Venezuela. Eur J Phycol 36:367–375

Lakatos M, Lange-Bertalot H, Büdel B (2004) Diatoms living inside the thallus of the green algal lichen *Coenogonium linkii* in neotropical lowland rain forests. J Phycol 40:70–73

Lang I, Hodac L, Friedl T et al (2011) Fatty acid profiles and their distribution patterns in microalgae: a comprehensive analysis of more than 2000 strains from the SAG culture collection. BMC Plant Biol 11:1–16

Leliaert F, Smith DR, Moreau H et al (2012) Phylogeny and molecular Evolution of the green algae. Crit Rev Plant Sci 31:1–46

Lelieveld J, Evans JS, Fnais M et al (2015) The contribution of outdoor air pollution sources to premature mortality on a global scale. Nature 525:367–371

Leu S, Boussiba S (2014) Advances in the production of high-value products by microalgae. Ind Biotechnol 10:169–183

Lewis LA, Lewis PO (2005) Unearthing the molecular phylodiversity of desert soil green algae (Chlorophyta). Syst Biol 54:936–947
Lewis LA, Mccourt RM (2004) Green algae and the origin of land plants. Am J Bot 91:1535–1556
Leya T (2013) Snow algae: adaptation strategies to survive on snow and ice. In: Seckbach J, Oren A, Stan-Lotter H (eds) Polyextremophiles. Springer, Dordrecht, pp 401–423
Lindo Z, Nilsson M-C, Gundale MJ (2013) Bryophyte-cyanobacteria associations as regulators of the northern latitude carbon balance in response to global change. Glob Chang Biol 19: 2022–2035
Lipman CB (1941) The successful revival of *Nostoc commune* from a herbarium specimen eighty-seven years old. Bull Torrey Bot Club 68:664–666
Liu T, Wang J, Hu Q et al (2013) Attached cultivation technology of microalgae for efficient biomass feedstock production. Bioresour Technol 127:216–222
Lüttge U (2011) Cyanobacteria: multiple stresses, desiccation-tolerant photosynthesis and dinitrogen fixation. In: Lüttge U, Beck E, Bartels D (eds) Plant desiccation tolerance. Springer, Berlin, Heidelberg, pp 23–43
Lüttge U, Büdel B (2010) Resurrection kinetics of photosynthesis in desiccation-tolerant terrestrial green algae (Chlorophyta) on tree bark. Plant Biol 12:437–444
Lüttge U, Büdel B, Ball E et al (1995) Photosynthesis of terrestrial cyanobacteria under light and desiccation stress as expressed by chlorophyll fluorescence and gas exchange. J Exp Bot 46:309–319
Lüttge U, Beck E, Bartels D (2011) Plant desiccation tolerance. Springer, Heidelberg
Maity JP, Bundschuh J, Chen C-Y et al (2014) Microalgae for third generation biofuel production, mitigation of greenhouse gas emissions and wastewater treatment: present and future perspectives – a mini review. Energy 78:104–113
Malcata FX (2011) Microalgae and biofuels: a promising partnership? Trends Biotechnol 29: 542–549
Markou G, Nerantzis E (2013) Microalgae for high-value compounds and biofuels production: a review with focus on cultivation under stress conditions. Biotechnol Adv 31:1532–1542
Mata TM, Martins AA, Caetano NS (2010) Microalgae for biodiesel production and other applications: a review. Renew Sustain Energy Rev 14:217–232
Medina E (1993) Mineral nutrition: tropical savannas. Prog Bot 54:237–253
Mei J, Cheng Z (1990) Effects of temperature on physiological activities of *Nostoc flagelliforme* Born. et Flah. J Western-north Normal Univ 1:75–85
Mikhailyuk T (2008) Terrestrial lithophilic algae in a granite canyon of the Teteriv River (Ukraine). Biologia 63:824–830
Milledge JJ, Heaven S (2013) A review of the harvesting of micro-algae for biofuel production. Rev Environ Sci Biotechnol 12:165–178
Miller SR, Castenholz RW (2000) Evolution of Thermotolerance in Hot Spring Cyanobacteria of the Genus *Synechococcus*. Appl Environ Microbiol 66:4222–4229
Miller SR, Mcguirl MA, Carvey D (2013) The evolution of RuBisCO stability at the thermal limit of photoautotrophy. Mol Biol Evol 30:752–760
Miyairi S (1995) CO_2 assimilation in a thermophilic cyanobacterium. Energ Conver Manage 36:763–766
Møller CL, Vangsøe MT, Sand-Jensen K (2014) Comparative growth and metabolism of gelatinous colonies of three cyanobacteria, *Nostoc commune*, *Nostoc pruniforme* and *Nostoc zetterstedtii*, at different temperatures. Freshw Biol 59:2183–2193
Moreno J, Vargas A, Olivares H, Rivas J, Guerrero MMJ, Vargas MA, Olivares H, Rivas J, Guerrero MG (1998) Exopolysaccharide production by the *cyanobacterium Anabaena* sp. ATCC 33047 in batch and continuous culture. J Biotechnol 60:175–182
Muffler K, Lakatos M, Schlegel C et al (2014) Application of Biofilm Bioreactors in White Biotechnology. Adv Biochem Eng Biotechnol 146:123–161
Mulbry WW, Wilkie AC (2001) Growth of benthic freshwater algae on dairy manures. J Appl Phycol 13:301–306

Münkel R, Schmid-Staiger U, Werner A et al (2013) Optimization of outdoor cultivation in flat panel airlift reactors for lipid production by *Chlorella vulgaris*. Biotechnol Bioeng 110: 2882–2893

Murik O, Oren N, Shotland Y et al (2017) What distinguishes cyanobacteria able to revive after desiccation from those that cannot: the genome aspect. Environ Microbiol 19:535–550

Nakamoto H, Honma D (2006) Interaction of a small heat shock protein with light-harvesting cyanobacterial phycocyanins under stress conditions. FEBS Lett 580:3029–3034

Nakamoto H, Suzuki N, Roy SK (2000) Constitutive expression of a small heat-shock protein confers cellular thermotolerance and thermal protection to the photosynthetic apparatus in cyanobacteria. FEBS Lett 483:169–174

Nascimento IA, Marques SSI, Cabanelas ITD et al (2013) Screening microalgae strains for biodiesel production: lipid productivity and estimation of fuel quality based on fatty acids profiles as selective criteria. Bioenergy Res 6:1–13

Naumann T, Çebi Z, Podola B et al (2013) Growing microalgae as aquaculture feeds on twin-layers: a novel solid-state photobioreactor. J Appl Phycol 25:1413–1420

Nelsen MP, Plata ER, Andrew CJ et al (2011) Phylogenetic diversity of Trentepohlialean algae associated with lichen-forming fungi. J Phycol 47:282–290

Norsker N-H, Barbosa MJ, Vermuë MH et al (2011) Microalgal production – a close look at the economics. Biotechnol Adv 29:24–27

Nowack ECM, Podola B, Melkonian M (2005) The 96-well twin-layer system: a novel approach in the cultivation of microalgae. Protist 156:239–251

Ohad I, Raanan H, Keren N, Tchernov D, Kaplan A (2010) Light-induced changes within photosystem II protects *Microcoleus sp.* in biological desert sand crusts against excess light. PLoS One 5:e11000

Oliveira MD, Monteiro M, Robbs P et al (1999) Growth and chemical composition of *Spirulina maxima* and *Spirulina platensis* biomass at different temperatures. Aquac Int 7:261–275

Olivieri G, Salatino P, Marzocchella A (2014) Advances in photobioreactors for intensive microalgal production: configurations, operating strategies and applications. J Chem Technol Biotechnol 89:178–195

Ong B-L, Lim M, Wee Y-C (1992) Effects of desiccation and illumination on photosynthesis and pigmentation of an edaphic population of *Trentepohlia odorata (Chlorophyta)*. J Phycol 28: 768–772

Orandi S, Lewis D, Moheimani N (2012) Biofilm establishment and heavy metal removal capacity of an indigenous mining algal-microbial consortium in a photo-rotating biological contactor. J Ind Microbiol Biotechnol 39:1321–1331

Otero A, Vincenzini M (2003) Extracellular polysaccharide synthesis by *Nostoc* strains as affected by N source and light intensity. J Biotechnol 102:143–152

Otero A, Vincenzini M (2004) NOSTOC (CYANOPHYCEAE) goes nude: extracellular polysaccharides serve as a sink for reducing power under unbalanced C/N METABOLISM1. J Phycol 40:74–81

Overkamp KE, Frankenberg-Dinkel N (2014) Phycobiliproteins-biosynthesis, assembly and applications. In: Ferreira G, Kadish K, Smith K, Guilard R (eds) The porphyrin handbook. World Scientific, New Jersey, London, pp 187–226

Ozkan A, Kinney K, Katz L et al (2012) Reduction of water and energy requirement of algae cultivation using an algae biofilm photobioreactor. Bioresour Technol 114:542–548

Pardow A, Lakatos M (2013) Desiccation tolerance and global change: implications for tropical bryophytes in lowland forests. Biotropica 45:27–36

Pedersen D, Miller SR (2016) Photosynthetic temperature adaptation during niche diversification of the thermophilic cyanobacterium *Synechococcus* A/B clade. ISME J 11:1053–1057

Pereira S, Zille A, Micheletti E et al (2009) Complexity of cyanobacterial exopolysaccharides: composition, structures, inducing factors and putative genes involved in their biosynthesis and assembly. FEMS Microbiol Rev 33:917–941

Perner-Nochta I, Posten C (2007) Simulations of light intensity variation in photobioreactors. J Biotechnol 131:276–285

Phlips E, Zeman C (1990) Photosynthesis, growth and nitrogen fixation by epiphytic forms of filamentous cyanobacteria from pelagic Sargassum. Bull Mar Sci 47:613–621

Pittera J, Partensky F, Six C (2017) Adaptive thermostability of light-harvesting complexes in marine picocyanobacteria. ISME J 11:112–124

Podola B, Li T, Melkonian M (2017) Porous substrate bioreactors: a paradigm shift in Microalgal biotechnology? Trends Biotechnol 35:121–132

Posadas E, García-Encina P-A, Soltau A et al (2013) Carbon and nutrient removal from centrates and domestic wastewater using algal–bacterial biofilm bioreactors. Bioresour Technol 139: 50–58

Posten C (2009) Design principles of photo-bioreactors for cultivation of microalgae. Eng Life Sci 9:165–177

Posten C, Jacobi A, Steinweg C, Lehr F, Rosello R (2011) Photobioreactor. Patent US20110318804 A1

Potts M (1999) Mechanisms of desiccation tolerance in cyanobacteria. Eur J Phycol 34:319–328

Raanan H, Oren N, Treves H et al (2016) Towards clarifying what distinguishes cyanobacteria able to resurrect after desiccation from those that cannot: the photosynthetic aspect. Biochim Biophys Acta 1857:715–722

Rajeev L, Da Rocha UN, Klitgord N et al (2013) Dynamic cyanobacterial response to hydration and dehydration in a desert biological soil crust. ISME J 7:2178–2191

Rascher U, Lakatos M, Büdel B et al (2003) Photosynthetic field capacity of cyanobacteria of a tropical inselberg of the Guiana Highlands. Eur J Phycol 38:247–256

Raun A, Borum J, Sand-Jensen K (2009) Active accumulation of internal DIC pools reduces transport limitation in large colonies of *Nostoc pruniforme*. Aquat Biol 5:23–29

Rawat I, Ranjith Kumar R, Mutanda T et al (2013) Biodiesel from microalgae: a critical evaluation from laboratory to large scale production. Appl Energy 103:444–467

Robarts RD, Zohary T (1987) Temperature effects on photosynthetic capacity, respiration, and growth rates of bloom-forming cyanobacteria. N Z J Mar Freshw Res 21:391–399

Rothschild LJ, Mancinelli RL (2001) Life in extreme environments. Nature 409:1092–1101

Sakai N, Sakamoto Y, Kishimoto N et al (1995) Chlorella strains from hot springs tolerant to high temperature and high CO2. Energ Conver Manage 36:693–696

Sand-Jensen K (2014) Ecophysiology of gelatinous *Nostoc* colonies: unprecedented slow growth and survival in resource-poor and harsh environments. Ann Bot 114:17–33

Sand-Jensen K, Jespersen TS (2012) Tolerance of the widespread cyanobacterium *Nostoc commune* to extreme temperature variations (– 269 to 105°C), pH and salt stress. Oecologia 169: 331–339

Sand-Jensen K, Raun AL, Borum J (2009) Metabolism and resources of spherical colonies of *Nostoc zetterstedtii*. Limnol Oceanogr 54:1282–1291

Scherer S, Ernst A, Chen T-W et al (1984) Rewetting of drought-resistant blue-green algae: time course of water uptake and reappearance of respiration, photosynthesis, and nitrogen fixation. Oecologia 62:418–423

Schnurr PJ, Espie GS, Allen DG (2013) Algae biofilm growth and the potential to stimulate lipid accumulation through nutrient starvation. Bioresour Technol 136:337–344

Shaw E, Hill DR, Brittain N et al (2003) Unusual water flux in the extracellular polysaccharide of the cyanobacterium *Nostoc commune*. Appl Environ Microbiol 69:5679–5684

Shi J, Podola B, Melkonian M (2007) Removal of nitrogen and phosphorus from wastewater using microalgae immobilized on twin layers: an experimental study. J Appl Phycol 19:417–423

Singh S, Kate BN, Banerjee UC (2005) Bioactive compounds from cyanobacteria and microalgae: an overview. Crit Rev Biotechnol 25:73–95

Singh B, Guldhe A, Rawat I et al (2014) Towards a sustainable approach for development of biodiesel from plant and microalgae. Renew Sustain Energy Rev 29:216–245

Solheim B, Zielke M (2002) Associations between cyanobacteria and mosses. In: Rai AN, Bergman B, Rasmussen U (eds) Cyanobacteria in symbiosis. Springer, Dordrecht, pp 137–152

Souffreau C, Vanormelingen P, Verleyen E et al (2010) Tolerance of benthic diatoms from temperate aquatic and terrestrial habitats to experimental desiccation and temperature stress. Phycologia 49:309–324

Souffreau C, Vanormelingen P, Van De Vijver B et al (2013) Molecular evidence for distinct Antarctic lineages in the cosmopolitan terrestrial diatoms *Pinnularia borealis* and *Hantzschia amphioxys*. Protist 164:101–115

Spolaore P, Joannis-Cassan C, Duran E et al (2006) Commercial applications of microalgae. J Biosci Bioeng 101:87–96

Tago Y, Aida K (1977) Exocellular mucopolysaccharide closely related to bacterial floc formation. Appl Environ Microbiol 34:308–314

Takenaka H, Yamaguchi Y, Sakaki S et al (1998) Safety evaluation of *Nostoc flagelliforme* (nostocales, cyanophyceae) as a potential food. Food Chem Toxicol 36:1073–1077

Tandeau de Marsac N (2003) Phycobiliproteins and phycobilisomes: the early observations. Photosynth Res 76:193–202

Tian X, Liao Q, Zhu X et al (2010) Characteristics of a biofilm photobioreactor as applied to photo-hydrogen production. Bioresour Technol 101:977–983

Tomaselli L, Boldrini G, Margheri MC (1997) Physiological behaviour of *Arthrospira* (*Spirulina*) *maxima* during acclimation to changes in irradiance. J Appl Phycol 9:37–43

Trainor FR, Gladych R (1995) Survival of algae in a desiccated soil: a 35-year study. Phycologia 34:191–192

Trivedi J, Aila M, Bangwal DP et al (2015) Algae based biorefinery – how to make sense? Renew Sustain Energy Rev 47:295–307

Vandamme D, Foubert I, Muylaert K (2013) Flocculation as a low-cost method for harvesting microalgae for bulk biomass production. Trends Biotechnol 31:233–239

Varshney P, Mikulic P, Vonshak A et al (2015) Extremophilic micro-algae and their potential contribution in biotechnology. Bioresour Technol 184:363–372

Vijayakumar S, Menakha M (2015) Pharmaceutical applications of cyanobacteria – a review. JACME 5:15–23

Vona V, Di Martino RV, Lobosco O et al (2004) Temperature responses of growth, photosynthesis, respiration and NADH: nitrate reductase in cryophilic and mesophilic algae. New Phytol 163:325–331

Vonshak A (1997) *Spirulina*: growth, physiology and biochemistry. In: Vonshak A (ed) *Spirulina* platensis arthrospira: physiology, cell-biology and biotechnology. CRC Press, London, pp 43–65

Wan M, Zhang J, Hou D et al (2014) The effect of temperature on cell growth and astaxanthin accumulation of *Haematococcus pluvialis* during a light–dark cyclic cultivation. Bioresour Technol 167:276–283

Wang B, Lan CQ, Horsman M (2012) Closed photobioreactors for production of microalgal biomasses. Biotechnol Adv 30:904–912

Webb WL, Newton M, Starr D (1974) Carbon dioxide exchange of *Alnus rubra*: a mathematical model. Oecologia 17:281–291

Wijffels RH, Barbosa MJ (2010) An outlook on microalgal biofuels. Science 329:796–799

Wijffels RH, Kruse O, Hellingwerf KJ (2013) Potential of industrial biotechnology with cyanobacteria and eukaryotic microalgae. Curr Opin Biotechnol 24:405–413

Yancey PH (2005) Organic osmolytes as compatible, metabolic and counteracting cytoprotectants in high osmolarity and other stresses. J Exp Biol 208:2819–2830

Zamalloa C, Boon N, Verstraete W (2013) Decentralized two-stage sewage treatment by chemical–biological flocculation combined with microalgae biofilm for nutrient immobilization in a roof installed parallel plate reactor. Bioresour Technol 130:152–160

Zhang C, Zhang H, Zhang Z et al (2014) Effects of mass transfer and light intensity on substrate biological degradation by immobilized photosynthetic bacteria within an annular fiber-illuminating biofilm reactor. J Photochem Photobiol B Biol 131:113–119

Zhuang L-L, Hu H-Y, Wu Y-H et al (2014) A novel suspended-solid phase photobioreactor to improve biomass production and separation of microalgae. Bioresour Technol 153:399–402

Background Insect Herbivory: Impacts, Patterns and Methodology

Mikhail V. Kozlov and Elena L. Zvereva

Abstract Plants provide humans with oxygen, food, fibre and fuel, but their effectiveness in performing these roles is affected by herbivores. Historically, studies on insect herbivory have primarily addressed pest outbreaks, which have indisputable ecological and economic consequences. By contrast, less attention has been paid to background ('normal') insect herbivory (BIH), which inflicts minor damage but acts on plants continuously. In this review, we introduce BIH as a phenomenon of great ecological and evolutionary importance, summarize the current knowledge regarding the levels and patterns of BIH and the effects of BIH on individual plants, plant communities and ecosystem-level processes, and discuss the methodology of studies addressing BIH. In the long term, global terrestrial net primary production (NPP) is more strongly affected by BIH than by the outbreaks of eruptive insect species. Plant responses to BIH differ from their responses to severe damage, and abiotic drivers of global change may have different effects on background versus outbreak herbivory. Minor changes in BIH caused by human activities may have profound but imperfectly understood consequences for the structure and functions of terrestrial ecosystems. This justifies the urgent need to move the focus away from rare bouts of severe plant damage by insects to the ubiquitous phenomenon of BIH in ecosystem-level studies and away from episodic major damage to chronic minor damage in studies of plant–herbivore interactions.

Contents

Communicated by Rainer Matyssek

M.V. Kozlov (✉) and E.L. Zvereva
Section of Ecology, Department of Biology, University of Turku, 20014 Turku, Finland
e-mail: mikoz@utu.fi; elezve@utu.fi

Progress in Botany (2018) 79: 313–356, DOI 10.1007/124_2017_4,
Published online: 29 March 2017

1 Insect Herbivory in Terrestrial Ecosystems

The demand for plant-derived natural resources and for the ecosystem services that are supplied by plant communities is growing exponentially as the world's human population increases. Plants provide humans with oxygen, food, fibre and fuel; at the same time, plants are also an indispensable resource for the wealth of animals and microorganisms.

Herbivores are ubiquitous in terrestrial ecosystems: it is nearly impossible to find a mature plant that bears no marks of animal feeding. At the global scale, the larger part of plant biomass losses are due to insects – 'the little things that run the World' (Wilson 1987). As a result, on average, natural plant communities suffer greater effects from invertebrate than from vertebrate herbivores (Bigger and Marvier 1998), although some ecosystems are strongly shaped by mammalian grazers

(Pachzelt et al. 2015). Insect herbivory is one of the major driving forces in plant evolution (Crawley 1983; Linhart 1991); it affects the productivity of ecosystems, modifies nutrient cycling and maintains the diversity of plant communities (Lamarre et al. 2012; Bagchi et al. 2014; Metcalfe et al. 2014). Consequently, herbivory is one of the most intensively studied biotic interactions (Tylianakis et al. 2008; Morales-Castilla et al. 2015).

Much is known about insect herbivory: the voluminous literature reports the results of studies of different insect–plant systems, and the behaviour of some of these systems has been explored in great detail. However, quite naturally, researchers tend to select their study objects from among the eruptive species that reach high densities and impose severe damage on their host plants, particularly because outbreaks of these insects may be economically important and attract considerable attention from policy-makers and the general public. However, only a few plant-feeding insects exhibit eruptive population dynamics (Faeth 1987). For example, only 15 insect species (of the several thousand feeding on trees) were identified as the agents of biotic disturbance in North American forests (Ayres and Lombardero 2000). Moreover, peaks in density are relatively rare, are usually (with a few notable exceptions) local and are typically 1–3 years in duration. By contrast, the majority of insect herbivores remain permanently at low densities and collectively impose minor but chronic damage on plants.

The background insect herbivory (BIH hereafter; also called endemic, nominal or ambient herbivory) is defined as the damage imposed by a community of plant-feeding insects, which includes both species that never exhibit eruptive population dynamics and eruptive species whose populations are at 'normal' low densities (Kozlov et al. 2015b). In contrast to outbreak-level damage, background damage occurs continuously on plants across the globe. Direct losses of foliar biomass due to BIH are relatively minor (1–15% annually for different regions and plant communities: Coley and Aide 1991; Kozlov 2008; Turcotte et al. 2014; Kozlov et al. 2015b, d), and the consequences of these losses are often considered 'negligible', especially when compared to the severe defoliation associated with pest outbreaks. However, all insect outbreaks reported from boreal forests of the European part of Russia from 1953 to 1998 jointly caused an annual loss of 0.02% of the biomass of forest trees (Selikhovkin 2009). Thus, even an annual loss of 1% of the foliar biomass due to BIH at the regional or global scale would greatly exceed the annual loss caused by local outbreaks of forest pests. The challenge, therefore, is to explore the collective ecosystem role of the 'minor' plant-feeding players, which are currently 'out of sight, out of mind'.

The goal of this paper is to introduce BIH as a phenomenon of substantial ecological and evolutionary importance. We summarize the current knowledge regarding the levels of BIH, the impacts of BIH on individual plants, plant communities and ecosystems, and the plant responses to BIH. We also review the variations in BIH among plant functional types and geographical regions, identify research gaps and discuss methodological problems associated with studying BIH in terrestrial ecosystems.

2 Background and Outbreak Insect Herbivory

The levels of insect herbivory vary greatly across plant species, ecosystems and years, from extremely low damage to complete defoliation (Fig. 1). Intuitively, experienced researchers can easily distinguish between BIH and outbreak damage, but providing formal definitions is a more difficult task.

Labandeira (2012) defined an outbreak as at least a two- to threefold increase in the level of plant damage relative to the situation considered 'typical' for a given plant species and region. Similarly, in epidemiology, an outbreak is recognized as the occurrence of more cases of a disease than is normally expected within a specific place or group of people over a given time interval (Reingold 1998), although numerical thresholds differ among the studies. By contrast, Selikhovkin (2009) defined an outbreak of forest insects as a 100-fold or greater increase in their density relative to their non-outbreak density, with the implicit assumption that, between outbreaks, an eruptive species consumes less than 1% of the foliar biomass of its host plant. Berryman (1987) also stressed that outbreaks generally occur over relatively short time intervals.

Another important difference between background and outbreak herbivory concerns the diversity of damage. BIH is inflicted by a variety of species, usually representing different feeding guilds and belonging to different phenological groups. Thus, the background plant herbivory results from a combination of different patterns of plant damage, and it is therefore dispersed over growth season and within a plant. By contrast, outbreak herbivory is inflicted by one eruptive species, whose individuals simultaneously impose similar damage on a plant.

The list of eruptive species is obviously biased towards insects that compete with humans for food or fibre resources. This means that the outbreak damage has a low probability of being detected and reported for a plant species of no economic importance. Furthermore, many definitions presume that the outbreak damage should be

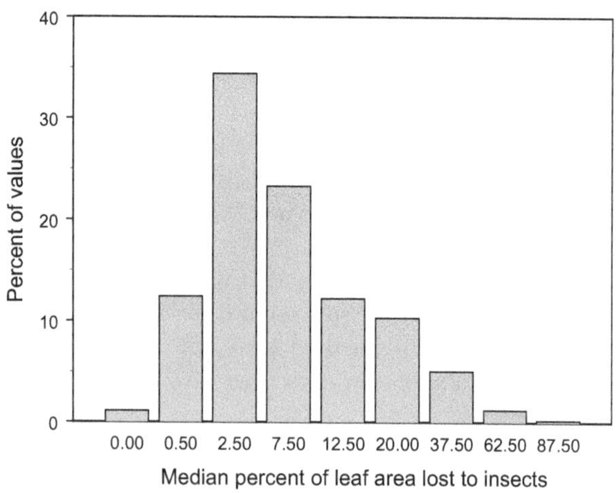

Fig. 1 Distribution of published values of losses of leaf area to defoliating insects ($N = 1{,}890$). Based on data from Zhang et al. (2016)

considerable; consequently, a 100-fold increase in losses of foliage from 0.01 to 1% will hardly be classified by applied entomologists as an outbreak.

Thus, an outbreak herbivory can be recognized only by comparison to the 'normal' situation (i.e. with BIH). In other words, only the analysis of long-term data allows a reliable distinction between background and outbreak herbivory. However, only 5% of published data on losses of woody plant foliage to insects (summarized by Kozlov et al. (2015b)) have been based on observations that lasted 4–9 years, whereas the majority (73%) of data were collected during a single year. Thus, an among-year comparison cannot be used to decide whether the most common (single-year) data represent background or outbreak herbivory. Therefore, we suggest a rule of thumb: any data reporting a loss of more than 20% of the plant foliage should be carefully examined for their typicality for the given plant species and environment, while losses exceeding 40% of plant foliage should without a doubt be attributed to outbreak herbivory.

3 Levels of Background Insect Herbivory

3.1 Aboveground Herbivory

3.1.1 Tissue Feeders

The first data on the background losses of woody plant foliage to defoliating insects in natural ecosystems, based on accurate measurements of leaf damage, were published in the mid-twentieth century, but their rapid accumulation started from the late 1990s (reviewed by Kozlov and Zvereva (2015)), and only very recently the empirical knowledge on BIH reached the level at which large-scale generalizations became possible. Several reviews estimate that the global losses of plant foliage to defoliating insects are from 4.7 to 8.8% (Turcotte et al. 2014; Kozlov et al. 2015b; Zhang et al. 2016).

The larger part of the background leaf damage (90%) in woody plants was due to externally feeding defoliators. Leaf miners contributed 8% to BIH; the remaining 2% of damaged leaf area was covered by galls (Kozlov et al. 2015b). The data on contribution of these feeding guilds of invertebrate herbivores to foliar damage of herbaceous plants are lacking.

The quantitative data on the background losses of stem biomass and reproductive organs to insects remain scarce and have not, to our knowledge, been summarized at the global scale.

3.1.2 Sap-Feeders

The background losses of plant biomass to sap-feeding insects, which remove assimilates from phloem and xylem or from individual plant cells, to our knowledge, are quantified only for four species of forest trees in the Northern Europe (Kozlov et al. 2015a). It was found that Norway spruce, *Picea abies*, and Scots pine, *Pinus sylvestris*, annually lose to sap-feeders 0.05–0.5% of their foliar biomass, while birches, *Betula pendula* and *Betula pubescens*, lose 1.0–3.5%. These data show that the background losses of plant biomass to sap-feeders are of the same order of magnitude as the background losses of these tree species to defoliators (Kozlov et al. 2015a). Still ecosystem-level studies generally ignore plant losses to sap-feeders and quantify insect herbivory based on the amount of foliage eaten away by defoliators (Hodkinson and Hughes 1982; Cyr and Pace 1993; Wolf et al. 2008; but see Andrew and Hughes 2005a). This approach, which is largely explained by the impossibility to measure plant losses to sap-feeding insects from their feeding marks, obviously underestimates the overall losses of NPP to insects.

3.2 Belowground Herbivory

3.2.1 Tissue Feeders

Although belowground NPP has been estimated to account for 40–70% of total terrestrial NPP (Vogt et al. 1996), the belowground herbivory in natural ecosystems has received relatively minor attention. The only available estimate (13%) for the average global losses of root biomass to both vertebrate and invertebrate herbivores was published long ago (Cyr and Pace 1993). The ecological field studies included in recent meta-analysis (Zvereva and Kozlov 2012) reported, on average, 25% losses of root biomass due to insect herbivory, i.e. much larger than the average loss of foliar biomass to defoliators. However, these studies were most likely conducted when densities of root-feeding insects were exceptionally high, and thus the obtained estimate of root losses exceeded the level of BIH. On the other hand, from the published data on the biomass of soil-dwelling insect herbivores and their food conversion efficiency, we estimated that insects in boreal forests consume 0.2–51% of belowground NPP, with a median value of 10% (M.V.K. and E.L.Z., unpublished data). Thus, in line with conclusion by Cyr and Pace (1993), our estimate suggests that BIH removes similar percentage of root biomass and foliar biomass from natural ecosystems.

3.2.2 Sap-Feeders

We are not aware of any information on the background levels of direct losses of plant biomass to Hemiptera, such as cicadas, aphids and scale insects, feeding on plant roots.

4 Impacts of Background Insect Herbivory on Plants

4.1 Growth and Reproduction of Individual Plants

4.1.1 Aboveground Herbivory

A long-held belief is that the annual loss of up to 10–20% of tree foliage has no ecological or economic importance (Price 1975; Crawley 1983). However, steadily accumulating data now reveal that minor damage can activate plant defences, trigger premature leaf abscission and affect plant growth and reproduction (Marquis and Whelan 1994; Björkman et al. 2008; Kozlov et al. 2012; Zvereva et al. 2012; Kozlov and Zvereva 2014; Zvereva and Kozlov 2014). Still only a small proportion of studies addressing the impacts of herbivory on plants considered low levels of plant damage by insects (Zvereva et al. 2012). Therefore the conclusions of meta-analyses, exploring impacts of aboveground herbivory on plants (Bigger and Marvier 1998; Hawkes and Sullivan 2001; Nykänen and Koricheva 2004; Zvereva et al. 2010a; Zvereva and Kozlov 2012), are greatly influenced by studies addressing the outbreak levels of plant damage, and the applicability of these conclusions to minor damage resulting from BIH remains unknown.

The studies exploring impacts of BIH on plants by either decreasing or increasing its levels remain scarce. Woody plants treated with insecticide, which demonstrated decrease in BIH from 8–13% to 3–6% of leaf area, produced 2.5–4.5 times more seeds (Crawley 1985) or increased their aboveground biomass by 10–19% (Balciunas and Burrows 1993; Marquis and Whelan 1994). Removal of sap-feeding insects alone resulted, on average, in 15% increase in aboveground growth and reproduction of woody plants (Zvereva et al. 2012). The consequences of herbivore removal from herbaceous plants experiencing BIH varied from non-measurable effects (Root 1996) to strong (1.5- to tenfold: Takahashi and Huntly 2010; Lehndal and Agren 2015) increase in aboveground growth and seed production.

The consumption of 3.3% of leaf area from cuttings of grey willow, *Salix cinerea*, by adult leaf beetles, *Phratora vulgatissima*, resulted in a 20% reduction in shoot growth and biomass production by the end of the growth season relative to undamaged controls (Björkman et al. 2008). Similar effects were demonstrated in experiments with simulated BIH: the removal of 10% of a plant's leaf area from the understory tropical wet forest shrub, *Piper arieianum*, resulted in decrease of branch growth and seed production by 25–40% relative to the control (Marquis 1992). The annual removal of 2, 4, 8 and 16% of the leaf area from saplings of mountain birch, *B. pubescens* subsp. *czerepanovii*, during a 7-year period (Fig. 2) resulted in a pronounced reduction of plant growth. Leaf size decreased first (already after 1 year of the 16% treatment), leading to the reduction of the total leaf area (Fig. 3). This effect was followed by a considerable decrease in the length of shoots in all treatments, while leaf number was maintained for a longer time (Zvereva et al. 2012).

Thus, relatively low levels of aboveground BIH can impose substantial detrimental effects on plants, and in perennial plants these effects can accumulate,

Fig. 2 Simulation of background insect herbivory (BIH) by punching holes in leaves of mountain birch, *Betula pubescens* subsp. *czerepanovii* (photo: V. Zverev)

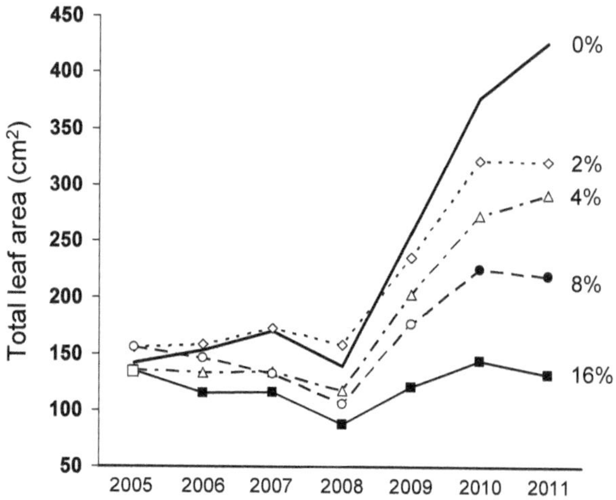

Fig. 3 Changes in the total leaf area of mountain birch, *B. pubescens* subsp. *czerepanovii*, saplings subjected to minor simulated herbivory (removal of 2, 4, 8 and 16% of leaf area annually) for 7 years. Means are each based on measurements of 16–21 saplings. *Solid symbols* indicate that the value differs significantly from undamaged (0%) plants (Dunnett one-tailed test, experiment-wide $P < 0.10$). Reproduced from Zvereva et al. (2012)

resulting in an increase of adverse effects over years. As a result, the overall effects of minor but chronic herbivory on birch growth appeared substantially larger than the effects of acute defoliation. For example, a single removal of 50% of the foliage from birch seedlings or saplings caused a 10–20% reduction in aboveground growth, whereas the removal of 8% of the foliage during 7 consecutive years (resulting in approximately the same total amount of removed foliage) caused a 45% growth reduction (Table 1). This comparison suggests that in the long term, BIH imposes stronger effects on the growth of woody plants than the short-term outbreaks of defoliators (Zvereva et al. 2012).

Table 1 Impacts of acute and chronic simulated herbivory on growth of birches (*Betula* spp.)

Betula species	Simulated level of herbivory	Period of damage (years)	Foliage removed (%)[a]	Measure of growth	Growth reduction (%)	Reference
B. pubescens	Outbreak	1	50	Biomass	20	Markkola et al. (2004)
B. pendula	Outbreak	1	50	Height	10	Anttonen et al. (2002)
B. pubescens	Background	7	56	Height	45	Zvereva et al. (2012)
B. pubescens	Outbreak	1	100	Biomass	44	Markkola et al. (2004)
B. pendula	Outbreak	1	100	Height	54	Anttonen et al. (2002)
B. pubescens	Background	7	112	Height	78	Zvereva et al. (2012)

[a]For chronic defoliation, this percentage was calculated by multiplying annual removal of foliage (8 and 16%) by the number of years (7) when the damage was applied

4.1.2 Belowground Herbivory

Some studies suggest that root damage may have more severe effects on plant performance than a similar loss of foliar biomass (Reichman and Smith 1991). However, meta-analyses demonstrated that the aboveground growth responses of woody plants to root herbivory acting alone (Hedge's $d = -0.98$: Zvereva and Kozlov 2012) were similar to the effects caused by defoliators ($d = -1.00$: Nykänen and Koricheva 2004) and sap-feeders ($d = -1.06$: Zvereva et al. 2010a). The ecological studies which were conducted in field conditions by either removal of root-feeding insects with insecticide or infestation by these insects (summarized by Zvereva and Kozlov (2012)) demonstrated that the significant loss of root biomass (25%; Hedge's $d = -0.38$; $CI_{95} = -0.89 \ldots -0.01$) resulted in non-significant decrease in aboveground growth and reproduction of plants (7%; Hedge's $d = -0.07$; $CI_{95} = -0.25 \ldots 0.10$).

Thus, the levels of root herbivory, that are above the current estimates of BIH (Sect. 3.2.1), did not cause significant reduction in aboveground plant growth. However, the adverse effects of belowground BIH may well emerge when the damage is applied during several years, like in the case of chronic damage of mountain birch (Sect. 4.1.1). Long-term experiments with exclusion of root herbivores from different types of natural ecosystems and with infestation of pot-growing plants with root-feeding insects are badly needed to clarify the ecological importance of belowground BIH.

4.2 Mechanisms Behind the Impacts of Background Insect Herbivory on Plants

4.2.1 Photosynthesis

Effects of Leaf-Feeders

Compensatory photosynthesis, defined as an increase in photosynthetic rates in partially defoliated plants relative to intact ones (Nowak and Caldwell 1984), is one of the most general and best-documented plant responses to simulated and natural defoliation (Welter 1989; Trumble et al. 1993; Nykänen and Koricheva 2004). However, defoliation caused an increase in photosynthesis only in 62% of studies reviewed by Welter (1989); the rest of studies demonstrated either no effect or decrease in photosynthesis. This variation is explained by a variety of habitat-, plant-, and damage-specific factors (Maschinski and Whitham 1989; Trumble et al. 1993), including the level of herbivory. For example, clipping a half of leaves from mountain birch (simulating outbreak levels of damage) caused an increase in photosynthetic rates in the remaining foliage (Hoogesteger and Karlsson 1992), whereas no changes in birch photosynthesis were detected at low (background) levels of damage (Lempa et al. 2004; Zvereva et al. 2012).

The local damage of a leaf may reduce the photosynthetic capacity in the surrounding (undamaged) tissues, and these indirect effects may be considerably greater than the direct effects of the removal of leaf area (Welter 1989; Zangerl et al. 2002; Nabity et al. 2009). For example, the removal of only 6% of the area of an individual wild parsnip, *Pastinaca sativa*, leaf by caterpillars reduced its photosynthetic capacity by 26% (Zangerl et al. 2002).

While compensatory photosynthesis is usually explained by increased sink demands of damaged tissues (Trumble et al. 1993), the suppression of photosynthetic capacity by minor damage is underpinned by several mechanisms, ranging from disruptions in fluid or nutrient transport to self-inflicted reductions in metabolic processes (Nabity et al. 2009), or may be related with defence signalling and production (Nabity et al. 2013). In spite of great theoretical importance of this problem, the effects of major and minor insect damage on net plant photosynthesis have never been compared directly.

Effects of Sap-Feeders

Detrimental impacts of sap-feeders on plant photosynthesis have been consistently demonstrated in numerous studies (reviewed by Welter (1989) and Zvereva et al. (2010a)), whereas compensatory increase in photosynthesis was found in a single study (Retuerto et al. 2004). Photosynthesis in plants damaged by sap-feeders can be affected in different ways. They include reduction in chlorophyll content (Kaakeh et al. 1992; Cabrera et al. 1994), decreased stomatal conductance

(Schaffer and Mason 1990; Nabity et al. 2009), alterations in water and nutrient transport (Nabity et al. 2009) and detrimental effects of salivary injection (Hodkinson and Hughes 1982). In addition, the honeydew produced by aphids and some species of scale insects serves as a substrate for sooty moulds, which may decrease photosynthesis by reducing light transmission (Tedders and Smith 1976).

Effects of Root-Feeders

Root herbivory reduces root biomass and disrupts water and nutrient uptake (Johnson et al. 2016). Water deficit can considerably reduce rates of photosynthesis through different mechanisms: decreased CO_2 availability caused by diffusion limitations through the stomata and the mesophyll, and alterations of photosynthetic metabolism or oxidative stress (Chaves et al. 2009). In general, the detrimental effects of root herbivory on plant photosynthesis are more severe than those of leaf-feeders (Johnson et al. 2016), and may appear as strong as the effects of sap-feeders.

4.2.2 Growth

Effects of Leaf-Feeders

The increased photosynthesis in plants damaged by defoliating insects may be accompanied by compensatory growth, by which plants replace some or all of the biomass consumed by insects, or even produce more biomass than was consumed (Trumble et al. 1993; Hawkes and Sullivan 2001). This regrowth is usually explained by increased sink demands in damaged areas (Trumble et al. 1993). Similar to compensatory photosynthesis, compensatory growth is observed mostly after severe defoliation. For example, mountain birch responds to high levels of defoliation by compensatory growth due to formation of additional long shoots from dormant buds on the stem and branches (Tenow and Bylund 2000; Kaitaniemi and Ruohomäki 2006), whereas low (2–16%) levels of leaf damage did not cause either regrowth or activation of dormant buds (Zvereva et al. 2012).

Detrimental effects of minor foliar damage on plant growth may result not only from the direct removal of photosynthetic area but also from the indirect decrease of photosynthesis in undamaged leaf tissues (Nabity et al. 2009) or premature abscission of damaged leaves (Sect. 4.2.4), leading to a reduced supply of assimilates to the developing buds, which therefore produce smaller leaves. Smaller leaves, in turn, provide less nutritive support to their primordial buds and thus change the sink strength; as the result, the shoots formed by these buds in the next season are shorter than in control trees (Zvereva et al. 2012).

Thus, minor damage typical for BIH, in contrast to severe damage, does not trigger compensatory growth in plants. In the long term, this results in stronger detrimental impact of BIH on plants relative to episodic bouts of severe damage (Table 1). In particular, the ability of the mountain birch to tolerate minor but chronic herbivory is lower than its ability to tolerate major defoliation or browsing occurring during one or two consecutive growth seasons (Zvereva et al. 2012). However, mountain birch has another mechanism of tolerance to BIH, which is expressed in the maintenance of the total photosynthetic area at the expense of the growth of woody plant parts. The removal of 2–8% of leaf area decreased the length of long shoots in birches, but these shoots formed the same number of buds as the longer long shoots in the control plants, and therefore the total number of leaves in damaged birches remained similar to control (Zvereva et al. 2012). This pattern may indicate that plants experiencing an increase in BIH reallocate resources from the growth of the woody parts to the maintenance of the photosynthetic area, primarily in terms of leaf number. This reallocation can be seen as a strategy of tolerance of minor herbivory, whereas compensatory responses typical of severe herbivory (increased photosynthesis rates and shoot regrowth) have not been detected in response to BIH.

Effects of Sap-Feeders

Plants only rarely compensate their losses to sap-feeders: only 8.1% of plant growth responses to damage imposed by sap-feeders were non-negative (Zvereva et al. 2010a). Consistent negative effects of sap-feeders on plant growth and reproduction result mostly from suppression of photosynthesis. Moreover, similar to major damage by defoliators, sap-feeders increase sink demands in plants by creating additional sinks (Crawley 1989). However, in the first case plant nutrients attracted to sinks are used for plant regrowth, while sap-feeders utilize attracted nutrients for their own growth, thus reducing plant abilities for compensative responses. Importantly, plant regrowth was not observed after either major or minor damage by sap-feeders (Zvereva et al. 2010a).

Effects of Root-Feeders

Compensatory growth in response to root herbivory can sometimes occur via lateral root proliferation (Brown and Gange 1990), similarly to increased branching following stem and leaf herbivory (Zvereva and Kozlov 2001; Stephens and Westoby 2015). However, the responses of plants to root-feeders are more similar to plant responses to sap-feeders than to leaf-feeders, because plants demonstrate exact compensation or overcompensation of root and sap herbivory much less frequently (8.1 and 17.2% of case studies, respectively: Zvereva et al. 2010a; Zvereva and Kozlov 2012) than of leaf herbivory (34.6%: Hawkes and Sullivan 2001). We suggest that this similarity is explained by a consistent reduction of photosynthesis in response to damage imposed by both root- and sap-feeders (Zvereva et al. 2010a; Zvereva and Kozlov 2012). The meta-analysis (Zvereva and Kozlov 2012) revealed the urgent need for the exploration of plant responses to the background levels of root herbivory. This research gap does not allow for comparison of the mechanisms behind the effects of major and minor root damage on plants.

4.2.3 Costs of Induction of Antiherbivore Defences

In addition to decrease in photosynthetic capacity, resource limitation in damaged plants may occur due to production of induced antiherbivore defences, which imposes a fitness cost on plants (Herms and Mattson 1992; Koricheva 2002). Although induced defences are mostly studied for major damage, corresponding to outbreak levels of herbivory (Karban and Baldwin 1997), costs of induced defences have been detected also following minor damage. For example, pronounced decreases in biomass production after minor foliar herbivory in wild parsnip (8.6% reduction after 2% damage) and grey willow (20.6% reduction after 3.3% damage) were attributed to the development of inducible defences, expressed in synthesis of defensive compounds and/or production of trichomes on leaves flushing after the damage (Zangerl et al. 1997; Björkman et al. 2008). Root herbivory may impose even greater costs, because root damage elicits equally strong responses in both leaves and roots of the damaged plant, whereas foliar damage generates strong responses in leaves, but much weaker responses in roots (Kaplan et al. 2008).

4.2.4 Premature Abscission of Damaged Leaves

BIH may reduce photosynthetic capacity of the whole plant and, consequently, plant performance by induction of premature leaf abscission (Zvereva and Kozlov 2014). For example, minor local damage imposed by several leaf-mining and leaf-rolling insect species reduced the leaf lifespan of downy birch, *B. pubescens*, by 12.6 days on average. This reduction was greater when the damage was imposed on growing leaves compared with mature leaves and increased with the leaf area consumed by a herbivore (Kozlov and Zvereva 2014).

Insect feeding had a greater effect on leaf lifespan than the same levels of mechanical wounding (Zvereva and Kozlov 2014), because many plant responses to herbivory are elicited by compounds present in insect oral secretions and thus do not develop following simulated herbivory (Kessler and Baldwin 2002). In particular, caterpillar feeding, but not wounding alone, induces a burst of ethylene (Kahl et al. 2000), which acts as an accelerator of leaf abscission (Roberts et al. 2002).

4.3 Productivity of Plant Communities

We are not aware of any whole-ecosystem experiment manipulating BIH in forest or tundra ecosystems. However, the effects of herbivore removal with insecticides are reasonably well documented for temperate herbaceous communities: the meta-analysis by Coupe and Cahill (2003) is based on 66 effect sizes extracted from 24 studies. Analysis of these data showed that application of insecticide increased community-wide plant biomass in grasslands experiencing BIH on average by $17.8 \pm 5.4\%$ ($N = 20$), whereas during insect outbreaks this increase was as large as 60.5 ± 12.2% ($N = 5$). The effect of BIH removal ranged from 4.2% in native grasslands ($N = 3$) to 32.3% in mature forb communities ($N = 5$); still, variation in the effect of BIH removal among community types was not significant ($F_{4,15} = 0.94$, $P = 0.47$). However, ranking of herbaceous communities by loss of productivity due to BIH is consistent with the conclusion by Turcotte et al. (2014) that forbs suffer threefold higher losses to defoliating insects than graminoids.

In herbaceous communities, the average loss of biomass due to BIH inflicted by all feeding guilds (17.8%: Coupe and Cahill 2003) exceeds the average losses of foliage of non-woody plants to defoliating insects (7.32%: Zhang et al. 2016) more than twofold. This difference can result from direct losses of plant biomass to sap-feeding insects combined with indirect losses expressed in growth reduction of damaged plants. Similarly, modelling of impacts of BIH on Northern European forests demonstrated that overall losses in birch biomass are around 1.8 greater than the removal of foliage by insect herbivores (Wolf et al. 2008) due to reduction in tree growth.

Stein et al. (2010) suppressed invertebrate herbivory at 14 grassland sites over a period of 5 years. The response of aboveground biomass to pesticide application was positively correlated with plant species richness, suggesting that the background invertebrate herbivory had a positive effect on plant biomass in low-diversity grasslands and a negative effect in more diverse sites (Stein et al. 2010). As long as effects of herbivory on growth of individual plants are generally negative, the increase in plant biomass under the impact of BIH may have been achieved due to changes in plant community structure.

4.4 *Structure of Plant Communities*

The studies exploring effects of insect herbivory on structure and composition of plant communities commonly address outbreak-level damage (e.g. Bach 2001) or do not report the level of herbivory (e.g. Brown and Gange 1989); the studies that can unequivocally be attributed to BIH remain scarce. The long-term herbivore suppression experiment (Stein et al. 2010) revealed that the background invertebrate (insect and mollusc) herbivory decreased plant species richness and evenness in grassland communities and altered plant community composition. In general, the effects of aboveground herbivores were relatively small, whereas belowground herbivores were important drivers of plant community composition: exclusion of belowground herbivores promoted grasses at the expense of herbs (Stein et al. 2010). Modelling the impacts of a realistic increase in background losses of downy birch foliage to defoliating insects also suggested that BIH, by modifying the strength of interspecific competition, may cause changes in structure of forest communities (Wolf et al. 2008).

4.5 *Ecosystem Carbon and Nutrient Cycling*

Herbivore-mediated processes can change both the quantity and quality of organic material transferred to the soil, with variable net consequences for soil processes and/or plant species composition (Hunter 2001; Metcalfe et al. 2014; and references therein). The loss of photosynthetic area to insects and the metabolic costs associated with repairing damage reduce total plant carbon uptake and leaf litterfall (Metcalfe et al. 2014). At the same time, the plant material consumed by herbivores is deposited as frass (arthropod excrement) and insect carcasses. Additionally, some defoliators produce greenfall, i.e. unconsumed leaf fragments that fall from trees, and damaged leaves may abscise prematurely (Sect. 4.2.4). All these materials, which generally are of higher nutritional quality than senesced leaf litter (Hunter 2001; Lovett et al. 2002), increase the movement of labile, nutrient-rich material through numerous pathways in terrestrial ecosystems, including uptake by plants and assimilation into foliage, utilization by microbial communities, and export from the system via surface and groundwater (Hunter 2001; Meehan et al. 2014; and references therein). BIH may also influence carbon and nutrient fluxes via changes in the decomposition rates of damaged leaves relative to undamaged ones (Kozlov et al. 2016a). All these effects may both increase and decrease nutrient cycling, depending on the level of herbivory and composition of plant communities (Belovsky and Slade 2000). Importantly, the estimated flux of foliar nitrogen and phosphorus from plants to soil mediated by defoliating insects at background levels of their densities was found to be similar to, or greater than, other major sources of these nutrients in tropical forests (Metcalfe et al. 2014). This finding stresses the need in further studies of BIH impacts on ecosystem-level processes.

5 Variation in Background Insect Herbivory

5.1 *Variation Among Plant Species and Functional Groups*

Despite a long history of research, we lack a comprehensive understanding of the reasons behind pronounced variation in foliar losses to insects among plant species. For a long time, this variation was attributed primarily to differences in composition and concentrations of plant secondary metabolites. However, plant biochemistry appeared to be poor predictor of the extent of plant damage caused by defoliating insects under natural conditions (Schuldt et al. 2012; Cárdenas et al. 2014; but see Agrawal and Weber 2015). This finding stimulated a search for other plant traits that could explain variation in the levels of BIH among plant species.

The growth rate hypothesis (frequently called resource availability hypothesis: Coley et al. 1985) is commonly used to explain variations in herbivory among plant species. This hypothesis suggests that the optimal level of constitutive defence for fast-growing species is low, even though this leads to high rates of herbivory, and explains this pattern by the ability of fast-growing species to replace plant tissues lost to insects at relatively low cost. A recent meta-analysis (Endara and Coley 2011) confirmed that fast-growing plants experience higher rates of herbivory than do slow-growing ones when different plant species were compared within the same locality. Furthermore, the research synthesis based on both published and original data on defoliating and leaf-mining insects generalized this conclusion for data combined across localities outside the tropics (Kozlov et al. 2015c).

Other researchers linked variation in herbivory among plant species with the size of a plant. Feeny (1976) argued that large plants (trees, in particular) are more readily located by herbivores and therefore suffer more persistent attacks when compared to small plants (i.e. shrubs, and especially dwarf shrubs), which are less apparent to herbivores (the apparency hypothesis). A recent review discovered that woody plants, on average, experienced 64% higher leaf herbivory when compared to non-woody plants, and attributed this difference to higher apparency of woody plants (Turcotte et al. 2014). Within woody plants, BIH varied among Raunkiaer plant life forms, increasing from dwarf shrubs to large trees outside the tropics and thus supporting the apparency hypothesis (Kozlov et al. 2015b). However, the conclusion that herbivory increases with plant size may appear applicable only to broadleaved woody plants. The coniferous trees, which are the most apparent woody plants of the boreal forest zone, showed background foliar losses to insects ranging from 0.1% to 0.7%, i.e. smaller than the average foliar losses of dwarf shrubs (Kozlov et al. 2015c). To resolve this contradiction, further representative data are required for BIH on conifers, as these are clearly under-represented among the published data.

Shade tolerance and responses to disturbance, which represent important dimensions of plant ecological strategies, were also suggested to affect the levels of BIH. In particular, shade-tolerant species were reported to experience both lower (Coley and Barone 1996; Villagra et al. 2013) and higher (Norghauer et al. 2014) foliar losses to insects when compared to shade-intolerant plants. Foliar losses to insects were predicted to be higher in early successional (pioneer) species than in late successional (climax) species (Lawton and McNeill 1979; Godfray 1985). However, exploration of diverse communities of woody plants revealed no relationships between plant successional status and herbivory (Vasconcelos 1999; Poorter et al. 2004). Furthermore, some studies reported the highest levels of herbivory at intermediate stages of succession (Neves et al. 2014) or in primary undisturbed forests (Newbery and de Foresta 1985). Consequently, at the global scale, BIH appeared independent of either shade tolerance or successional status of woody plants (Kozlov et al. 2015c).

Several studies that have explored the relationships of herbivory with a number of leaf functional traits have arrived at different conclusions regarding the relative importance of these traits for insect herbivores (Poorter et al. 2004; Cárdenas et al. 2014; Schuldt et al. 2014), and no general pattern has emerged. A higher herbivory was reported on woody plants with larger species-specific leaf size (Cárdenas et al. 2014; Zava and Cianciaruso 2014; Zhang et al. 2017); however, we found no support for this pattern in comparisons between woody plant species within a site (M.V.K. and E.L.Z., unpublished data). Similarly, the association between herbivory and leaf thickness (toughness, specific leaf weight or specific leaf area) can vary from negative (Coley 1983; Kozlov et al. 2015c) to non-significant (Cárdenas et al. 2014; Zhang et al. 2017) or even positive (Poorter et al. 2004).

Finally, leaf lifespan was recently identified as a good predictor of the levels of BIH across plant species and biomes, explaining 20% of interspecific variation in herbivory (Zhang et al. 2017). Importantly, BIH was found to show hump-shaped relationship with leaf lifespan, and this pattern explained the variation among plant life forms detected by Kozlov et al. (2015c). It also resolved the contradiction between low BIH on apparent conifers, because needle lifespan in conifers is exceptionally large.

The discrepancies between the outcomes of individual studies suggest that any attempt to predict plant losses to herbivores must account for several dimensions of plant ecological strategies simultaneously with leaf functional traits. Furthermore, the associations between foliar losses to insects and plant characteristics may vary among biomes (Kozlov et al. 2015c) and among localities due to variations in climate, soil fertility and the availability of other resources (Wise and Abrahamson 2005).

5.2 *Geographical Variation*

There exist several dozen partially overlapping or competing hypotheses to explain the pronounced spatial and temporal variation in insect herbivory through the action of various abiotic and/or biotic factors (Cronin et al. 2010). One of these hypotheses, arising from the seminal works by Coley and Aide (1991) and Coley and Barone (1996), predicts a decrease in herbivory from the equator to the poles. More specifically, this latitudinal herbivory hypothesis was erected not for herbivory in general but for losses of woody plant foliage to defoliating and, partially, mining insects, and the subsequent discussions, with rare exceptions (Hysell et al. 1996; Andrew and Hughes 2005b; Kozlov et al. 2015a), did not consider sap-feeding insects. The geographical variation in belowground herbivory remains virtually unknown.

The latitudinal herbivory hypothesis had not been questioned for a long time, until meta-analysis of 38 published studies (Moles et al. 2011) concluded that plant losses from herbivory were independent of latitude. However, subsequent publications reported significant latitudinal changes in herbivory in different study systems and in different geographical regions (Garibaldi et al. 2011; Kozlov et al. 2013, 2015a; Więski and Pennings 2014), and the existence of the latitudinal gradient in herbivory became the subject of heated debate. The recent research syntheses yielded variable conclusions: the latitudinal herbivory hypothesis was supported for evergreen species but not for species with short leaf lifespans (Lim et al. 2015) or exclusively for the Northern Hemisphere (Zhang et al. 2016).

To date, the most extensive data set, which includes both published and blindly collected original data (3,482 point estimates of the percentage of leaf area consumed or otherwise damaged by defoliating, leaf-mining and gall-forming insects, collected from 941 species of woody plants in 836 localities worldwide), was compiled and analysed by Kozlov et al. (2015b). This study for the first time suggested that latitudinal pattern in BIH may be non-linear and therefore fitted the primary data with both linear and quadratic regression models. The detected pattern was dome-shaped: BIH was found to peak in temperate zones, slightly decrease towards the equator and strongly decrease towards the poles. This pattern was consistent between published and original data (Fig. 4) and between the Northern and Southern Hemispheres, indicating the robustness of the detected relationship between herbivory and latitude. Within temperate and polar zones (but not in the tropics) BIH generally increased with ambient air temperatures (Kozlov et al. 2015b).

Zhang et al. (2016), by analysing another large data set, revealed that herbivory was on average higher in Southern than in Northern Hemisphere. Furthermore, these authors concluded that herbivory decreased with an increase in latitude and increased with an increase in mean annual temperature (MAT) only in the Northern Hemisphere, whereas in the Southern Hemisphere the herbivory did not change with latitude and decreased with MAT. Hemispheric asymmetries in latitudinal gradients were earlier reported for diversity of different taxa (Dunn et al. 2009), and most of this asymmetry was explained by differences in contemporary climate (Chown et al. 2004). However, the opposite effects of MAT on herbivory in different hemispheres look surprising.

Zhang et al. (2016) tested the hypothesis on the differences in the absolute levels of herbivory between the hemispheres separately from the hypothesis on the hemispheric differences in latitudinal or climatic patterns of herbivory. When both these hypotheses were tested with the same statistical model using the second estimate of plant losses from Moles and Westoby (2000), no differences in herbivory were found between the hemispheres (Kozlov and Klemola 2017).

The mechanisms behind the latitudinal patterns in BIH remain poorly understood. It was suggested that, along with direct effects of climate on insect performance (Bale et al. 2002) and on the number of generations in some species (Yamamura and Kiritani 1998), latitudinal gradient in BIH may be produced by changes in the mechanical properties of leaves (Onoda et al. 2011), in plant defensive chemistry (Moles et al. 2011) and in predator pressure (Björkman et al. 2011). However, all these hypotheses remain speculative for the moment due to shortage of data: we are not aware of the published studies that simultaneously explored host plant quality for herbivores, plant losses to herbivores and predation control on herbivores along a large-scale climatic gradient.

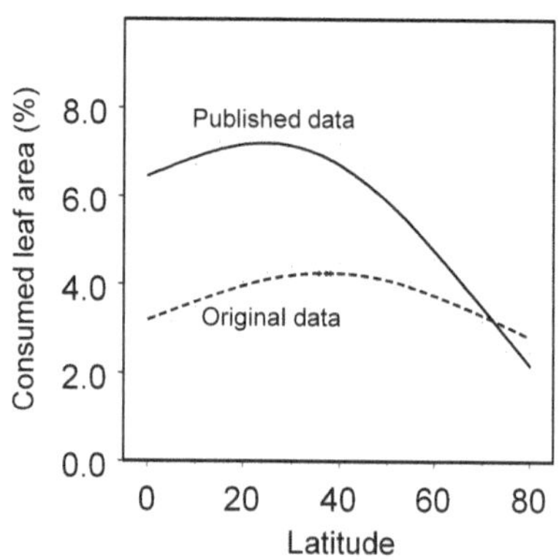

Fig. 4 Geographical variation of the background losses of woody plant foliage to insect herbivores: quadratic functions of latitude based on published data and on original data. Regression equations after Kozlov et al. (2015b)

In spite of a generally clear global latitudinal pattern in BIH, the regional latitudinal patterns considerably vary among individual studies. This variation may at least partly emerge from studying of different parts of the dome-shaped curve describing global latitudinal pattern in BIH (Fig. 4; Kozlov et al. 2015b). This may also explain an overall lack of latitudinal patterns in herbivory in meta-analysis by Moles et al. (2011). Moreover, the insect herbivores, even those closely related to each other and feeding on the same plant species, often show different latitudinal patterns. For example, only 14 of 23 birch-feeding leafminers demonstrated significant poleward decrease in abundance in Northern Europe (from 59 to 70°N); eight taxa showed no latitudinal changes, and abundance of one taxon increased with latitude (Kozlov et al. 2013). Furthermore, plant species may demonstrate different patterns in BIH in the same latitudinal gradients: the load of sap-feeding insects on four species of forest trees (*P. abies*, *P. sylvestris*, *B. pendula* and *B. pubescens*) showed different relationships with midsummer temperature (Kozlov et al. 2015a), and infestation of leaves by gall-forming eriophyoid mites increased towards the north in *B. pendula* but decreased in *B. pubescens* (Kozlov et al. 2016b).

Last but not least, at regional scales latitudinal patterns in BIH may differ between study years, and this variation may influence global patterns. In typical summers, abundance of leafminers on birches and load of sap-feeders on four species of forest trees in Northern Europe significantly decreased with latitude. In contrast, during the warm year, these variables were independent of latitude, while during an exceptionally hot summer, they increased towards the north (Kozlov et al. 2013, 2015a). This variation suggests that uncovering the mechanisms behind latitudinal or climatic patterns in BIH requires collection of long-term data on abundance and fitness of multiple herbivore species from a variety of environmental gradients across the globe.

5.3 Variation Related to Composition of Plant Community

In addition to plant functional traits and climate, structure of plant communities was currently recognized as an important source of variation in BIH experienced by individual plant species (Castagneyrol et al. 2014; Abdala-Roberts et al. 2015; Moreira et al. 2016). Two phenomena are generally identified: associational resistance, when a given plant suffers less damage from herbivores when growing with heterospecific neighbours than among conspecific plants, and associational susceptibility, when a given plant suffers more damage in tree mixtures as compared to monocultures.

A rich body of theory (reviewed by Moreira et al. (2016)) has been developed to explain the effects of plant diversity on communities at higher trophic levels and elucidate the mechanisms underpinning such effects. Although these theories predict both higher and lower BIH in diverse communities relative to monocultures, still mixed forests are frequently thought to be less prone to insect damage than monocultures (Castagneyrol et al. 2014). Main mechanisms underlying this effect include resource dilution, reduced host apparency and impact of natural enemies (Guyot et al. 2016). However, several studies discovered absence of the associational resistance (Koricheva et al. 2006; Vehviläinen et al. 2007; Haase et al. 2015) or even an increase in BIH with tree diversity (Schuldt et al. 2010).

The pronounced variation in outcomes of individual studies was so far explained by different factors, including specific composition of mixed stands (Castagneyrol et al. 2014), height of the focal plant species relative to its neighbours (Castagneyrol et al. 2013; Damien et al. 2016), size of patches of plant mixtures (Bommarco and Banks 2003), herbivore feeding guild (Castagneyrol et al. 2013) and herbivore specialization (Castagneyrol et al. 2014; Abdala-Roberts et al. 2015). Furthermore, the effects of plant diversity on herbivory were reported to change as focal plants grow (Damien et al. 2016), or after biotic disturbance, e.g. moose browsing (Muiruri et al. 2015). Thus, although effects of focal plant's neighbourhood on BIH have been frequently reported, the general pattern had not yet emerged, and it well may appear that these effects are idiosyncratic and depend primarily on composition of both plant and insect communities, as well as on multiple characteristics of abiotic environment.

6 Global Change Impacts on Background Insect Herbivory

6.1 Pollution

Industrial pollution has repeatedly been claimed to enhance plant damage by insects, in particular due to increase in probability of insect outbreaks in polluted areas (Smith 1974; Führer 1985). The supporting evidence is primarily based on data collected during insect outbreaks (reviewed by Zvereva and Kozlov (2010) and Kozlov (2015)), whereas only a handful of studies measured impacts of pollution on BIH. Several case studies (Kozlov 1985, 2015; Nevalainen et al. 1994) found that BIH varied independently of pollution. The meta-analysis of data collected from woody plants growing in impact zones of 16 industrial polluters of the Northern Hemisphere demonstrated that pollution did not influence average consumption per damaged leaf, but slightly decreased percentage of leaves bearing insect feeding marks (Kozlov et al. 2009). Similarly, background foliar damage to aspen, *Populus tremula*, from defoliating and leaf-mining insects near a large copper smelter in Revda, Middle Urals, was reduced to 56% of control (Belskaya and Vorobeichik 2013).

The discrepancy between the conclusions based on meta-analysis of published data, which demonstrated significant increase in insect herbivory in polluted areas (Zvereva and Kozlov 2010), and data on BIH (Kozlov 1985, 2015; Kozlov et al. 2009; Belskaya and Vorobeichik 2013), can be explained, at least partly, by research and publication biases. These biases may have affected the published data through preferential documentation of patterns that fit the expectations of the researcher (an increase in herbivory near the polluter), who tends to initiate the study when an expected pattern is observed. In turn, editors and reviewers select against studies reporting effects that contradict the currently favoured hypothesis (Zvereva and Kozlov 2010; Kozlov 2015). However, it is also possible that industrial pollution imposes different impacts on plant-feeding insects with non-eruptive and eruptive population dynamics: it may increase the probability and/or severity of outbreaks in eruptive species but not influence or even suppress BIH. In particular, substantial decrease in pollution terminated the outbreaks of two willow feeding leaf beetle species, *Chrysomela lapponica* and *Gonioctena pallida*, near the smelter at Monchegorsk (Zvereva et al. 2016), thus supporting the hypothesis that pollution favours outbreaks of herbivorous insects.

6.2 Fragmentation

Effects of fragmentation on BIH were generally studied in forests. These effects are expressed in differences in plant damage between edges of a forest fragment and its interior, and in changes in herbivory with the size of a forest fragment. The loss of habitat area generally results in a decrease in BIH (Valladares et al. 2006; Simonetti et al. 2007; Ruiz-Guerra et al. 2010), whereas the impacts of habitat edges on herbivory are diverse.

Meta-analysis revealed that edge plants exhibited 70% more insect damage than interior plants, and that this effect was primarily due to plant damage imposed by defoliators (Guimarães et al. 2014). On the other hand, Rossetti et al. (2014) found that impacts of chewing and sap-feeding insects on plants growing at the forest edge were lower compared with plants growing at the interior. Similarly, herbivory by leaf miners at the edges of woodlands was reported to be both higher (Rossetti et al. 2014) and lower (Valladares et al. 2006) than at the interior. Galling insects were found to occur more frequently in edge habitats than in forest interior (Altamirano et al. 2016).

The proximity to the forest edge also affects the levels of BIH in low-stature plants (grasses, forbs and shrubs) growing in open habitats surrounding the forest fragment, although the direction of the effect varied among plants: herbivory increased with distance to forest edge in six plant species and decreased in three plant species (Levey et al. 2016). This variation in the edge effects on BIH is not surprising, because insect species show different preferences for sun or shade, as well as for many other environmental factors that differ between the edge and interior of the habitat. As the result, overall plant losses to insects, which depend on the composition of the local insect community, change at the forest edge idiosyncratically. In our opinion, we can hardly expect to discover the general pattern in edge effects on BIH.

6.3 Urbanization

The existing case studies provide contradictory evidence, showing either increased or decreased plant damage by defoliating insects in urban habitats relative to rural habitats (Nuckols and Connor 1995; Rickman and Connor 2003; Matter et al. 2012; Cuevas-Reyes et al. 2013). The current review (Raupp et al. 2010) concluded that insect herbivores show idiosyncratic responses to urban environment; however, small arthropods with sucking mouthparts, limited mobility and multiple generations on the same host plant are generally favoured by urbanization.

The comparison between urban and rural habitats associated with 16 European cities demonstrated that the average background losses of tree foliage to defoliating insects were 16.5% lower in urban than in rural habitats. The overall adverse effect of urbanization on BIH increased with an increase in the size of the urban area: it was significant in large cities (city population 1–5 million) but not significant in medium-sized and small towns. These differences were tentatively attributed to the increased mortality of herbivorous insects from birds (Kozlov et al. 2017).

6.4 Climate Warming

The existing scenarios generally predict that increasing temperature will cause larger increases in herbivory than in plant productivity (Ayres and Lombardero 2000; O'Connor et al. 2009; DeLucia et al. 2012). A highly cited paper (Logan et al. 2003) warns that 'all aspects of insect outbreak behaviour will intensify as the climate warms', and this prediction has a solid support. However, along with numerous evidences for an increase in outbreak frequency, severity and duration during the past decades (reviewed by Kneeshaw et al. (2015)), climate warming was found to disrupt or dampen outbreaks of some insect species (Johnson et al. 2010; Haynes et al. 2014). Furthermore, many researchers (Zvereva and Kozlov 2006; Dukes et al. 2009; Haynes et al. 2014) appreciate the high variability in the relationships between the abiotic drivers of global changes and herbivory and, consequently, the uncertainty in predicting community-wide changes in plant foliar losses to insects.

An analysis of the published data on the background foliar losses of woody plants to insects in natural ecosystems across the globe from 1952 to 2013 provided no support for the hypothesis of an increase in BIH with climate warming (Kozlov and Zvereva 2015). Moreover, while BIH did not change with study year within the temperate climate zone, it decreased significantly in the tropics. From 1964 to 1990, BIH in the tropics was 39% higher than in the temperate region, but these differences disappeared by the beginning of the 2000s (Kozlov and Zvereva 2015). Similarly, no temporal changes in insect herbivory from 1960 to 2014 were discovered by analysing another, independently created data set (Zhang et al. 2017).

One possible reason for the discrepancy between these findings and the earlier predictions of increasing plant losses to insects with climate warming could be that these predictions were generally based on the results of short-term laboratory (climate chamber) experiments, which tend to overestimate the effects relative to field experiments (Zvereva et al. 2010a; Zvereva and Kozlov 2012) and to observations made in natural ecosystems (Kozlov et al. 2009). Furthermore, experiments conducted in controlled environments fail to account for interactions between abiotic drivers of global change, as well as for multiple biotic interactions within an ecosystem. For example, analysis of long-term monitoring data revealed that the positive relationships between performance of a leaf beetle, *C. lapponica*, and temperature discovered in laboratory experiments did not translate into an increase in beetle abundance following a 3°C elevation in mean spring temperature during the observation period (1993–2014). This discrepancy resulted from an increase in leaf beetle mortality from natural enemies (Zvereva et al. 2016). Therefore, the multiple interactions within an ecosystem should be accounted for when predicting climate change effects on BIH. And again, like in the case of industrial pollution, climate change may differently affect background and outbreak insect herbivory.

6.5 *Increase in Ambient CO_2 Concentration*

Carbon dioxide (CO_2) is one of the major abiotic drivers of global change. Currently, the global concentration of CO_2 has increased from 270 to 400 μL/L and is expected to nearly double from preindustrial levels by the end of the century (IPCC 2001). Changes in the concentration of atmospheric CO_2 induce changes in plant chemistry, potentially affecting herbivorous insects and, consequently, the levels of BIH. The increase in carbon-to-nitrogen ratio typically observed under elevated CO_2 decreases the nutritional quality of the foliage for insects due to both increase in concentrations of carbon-based defensive compounds and decrease in nitrogen (Zvereva and Kozlov 2006; Lindroth 2012; Robinson et al. 2012). This effect may in turn result in compensatory feeding by insects, which must consume more low-quality food in order to meet dietary requirements (Lincoln et al. 1993). However, large-scale free-air CO_2 enrichment experiments yielded diverse results: BIH was reported to increase (Couture et al. 2015), remain unaffected (Gherlenda et al. 2016) or decrease (Stiling et al. 2009; Knepp et al. 2005) under elevated CO_2. The observed effects of CO_2 elevation on BIH often differ between plant species within a community and between study years (Stiling et al. 2009; Knepp et al. 2005; Couture et al. 2015).

In line with the earlier meta-analysis (Zvereva and Kozlov 2006), several studies (Murray et al. 2013; Niziolek et al. 2013) demonstrated that the increase in plant consumption by insects, observed under elevated levels of CO_2, was mitigated by temperature increase. Therefore the outcomes of free-air fumigation experiments cannot be extrapolated to a more realistic climate change scenario, when ambient temperature increases simultaneously with CO_2 elevation. Similarly, the results of studies conducted in controlled environment should be interpreted with caution, especially when these studies did not account for herbivore mortality. The meta-analysis (Stiling and Cornelissen 2007) found that elevated CO_2 decreased herbivore abundance to a larger extent (−21.6%) than it increased total consumption (+9.2%); thus the resulting plant damage is likely to decrease. Still, BIH in temperate forest ecosystems did not change during the past 60 years (Kozlov and Zvereva 2015), in spite of a substantial (+26%) increase in ambient CO_2 level.

To conclude, the multiple lines of evidence suggest that BIH is unlikely to change substantially due to the expected increase in ambient CO_2 level.

7 Modelling Impacts of Background Insect Herbivory on Terrestrial Ecosystems

Ecosystem modelling is one of the key tools in modern environmental research. Models are increasingly used to investigate the effects of different kinds of environmental changes, primarily of climate change, on ecosystem structure and functions and to compare the properties of future ecosystems under different scenarios of environmental changes (Seidl et al. 2011; Kissling et al. 2012). In this way, modelling is crucial for the development of management support and policy advice (Revesz et al. 2014). Still the models involving components of biosphere remain oversimplified; they do not consider many functional groups of biota and do not incorporate effects of the abiotic drivers of global change on interactions among different groups of organisms (Sutherst et al. 2007; Lindner et al. 2010).

The majority of models that consider herbivory predict either the vulnerability of forests to insect attack or the distribution of eruptive species that can cause major disturbances in natural ecosystems, primarily in forests (Seidl et al. 2011). Much less attention has been paid to models that estimate losses of plant biomass and productivity to insects based on stand and environmental characteristics. The existing regional models generally consider changes in forest growth and/or timber yield caused by selected insect species of significant economic importance (Davidson et al. 2001; Komonen and Kouki 2008), i.e. they suffer from the same research biases as other studies of insect herbivory. Although these models may have good predictive power in particular situations, they cannot be extended to other regions and/or other insect species, and they do not predict the behaviour of the entire herbivore community.

Only a handful of models (Wolf et al. 2008; Jönsson et al. 2012; Couture et al. 2015) have considered the effects of insect herbivory on structure and productivity of plant communities, and only one of these models (Wolf et al. 2008) has explicitly addressed the consequences of climate-driven changes in BIH levels for ecosystem structure and functions. Notably, these impacts appeared stronger than expected, particularly due to their second order effect on the competition between tree species (Wolf et al. 2008).

The current level of knowledge on BIH allows empirical parameterizations of losses of plant foliage to defoliating insects at the global scale, both omitting variation among plant functional types (Kozlov et al. 2015b) and accounting for it (Kozlov et al. 2015c). This information allows modelling of BIH imposed by defoliating insects at the global scale. The empirical parameterization of losses of plant biomass to sap-feeding insects was made for forest trees in the Northern Europe (Kozlov et al. 2015a), whereas no data are yet available for modelling belowground BIH.

8 Methodology of Studies Addressing Background Insect Herbivory

8.1 General Considerations

Plants are usually damaged by insects belonging to several feeding guilds, and all this damage should be accounted for to quantify the overall background losses of plant biomass to insects. To our knowledge, studies of this kind have never been conducted in natural ecosystems, in particular because the measurements of direct losses of plant biomass to sap- and root-feeding insects represent significant methodological challenge. Moreover, the methodology of studies of foliar damage by defoliating insects also requires improvement and standardization.

To date, there exists no commonly accepted protocol for measurements of BIH, and the diversity of methods hampers generalization of the results. In particular, it was currently suggested that heterogeneity in methodology and problems with study design and interpretation represent the major sources of controversy surrounding latitudinal patterns in insect herbivory and plant defence (Anstett et al. 2016). Furthermore, research methodology was identified as the important source of variation in outcomes of experimental studies addressing impacts of insect herbivory on plant growth and reproduction (Zvereva et al. 2010a; Zvereva and Kozlov 2012). Therefore we pay special attention to the details of methodology that critically influence the quality of the data on BIH.

8.2 Selecting Plant Species

Randomization is commonly seen as a prerequisite for making valid generalizations from the sample to the population; still the selection of plant species for herbivory-related research is never random. Therefore the question arises on whether the existing species-specific data (summarized by Coley and Barone (1996), Schowalter and Lowman (1999), Turcotte et al. (2014), Kozlov et al. (2015b) and Zhang et al. (2016)) are representative of entire plant communities.

The broadleaved tree species are generally preferred to needle-leaved species in studies that involve measurements of plant losses to insects. Assessment of foliar damage in needle-leaved plants is more difficult, not only due to complications in measuring the missing area of a needle, but also because of premature abscission of damaged needles (Zvereva and Kozlov 2014). This explains why, despite the high ecological and economic importance of conifers, only 0.38% of the published data on the levels of BIH were collected from these plants (Kozlov et al. 2015c). In contrast, in studies of pollution impacts on plant growth and reproduction, the object selection bias acts in the opposite direction: nearly one-third of published data were obtained from two species of conifers, Scots pine and Norway spruce (Zvereva et al.

2010b), because they are economically important and, at the same time, more susceptible to environmental contamination than broadleaved trees.

Furthermore, the data on the levels of BIH, at least in the tropics, are prone to confirmation bias, i.e. to the tendency of humans to seek out evidence in a manner that confirms their existing ideas and hypotheses (Rosenthal 1976). It was found (Kozlov et al. 2014) that the published data on insect herbivory, which were collected by non-blind methods (i.e. the observer had some expectations concerning the results), yielded nearly tenfold greater average loss of leaf area of woody plants to defoliating insects in Brazil relative to the data collected by blind method, from nature and wildlife photographs found in the internet (when the photographer did not consider possible use of photographs for measurements of herbivory) (Fig. 5). This difference was attributed to the unconsciously preconceived selection of plant species with higher-than-average levels of herbivory, as a subconscious agreement with the dominant opinion (Coley and Barone 1996; Schowalter and Lowman 1999) that insect herbivory in tropical regions is high. The existence of this bias in tropics was further confirmed by comparing latitudinal patterns in BIH derived from published data and data collected in a blind way (Kozlov et al. 2015b): the difference between the approximating functions was highest in the tropics, sharply decreased within temperate zone and disappeared in subpolar regions (Fig. 4).

Based on these findings, we urge for caution in obtaining community-wide characteristics from the results of single-species studies. More generally, we argue that more attention should be paid by ecologists to the problem of biases occurring at the pre-publication phases of the scientific research and, consequently, to the development and the wide application of methods that avoid these biases.

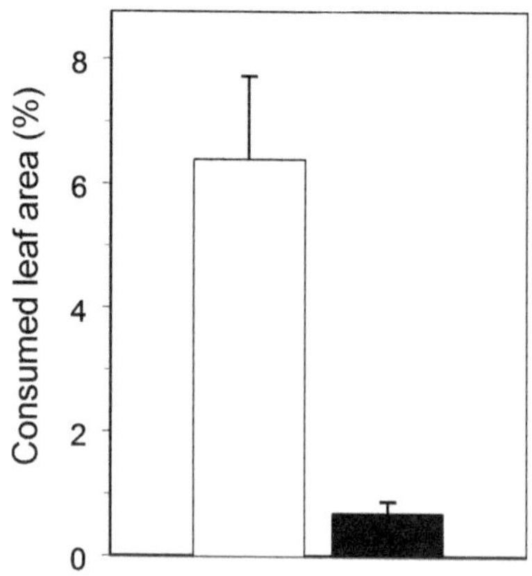

Fig. 5 Total foliar damage of woody plants in Brazil (mean + standard error) measured by non-blind (published data: *open bar*; $N = 10$) and blind (data collected from nature and wildlife photographs found in the WWW: *black filled bar*; $N = 30$) methods (modified from Kozlov et al. (2014))

8.3 *Selecting Target Groups of Herbivores*

Many published studies do not specify which groups of insects caused the reported leaf damage, thus complicating interpretation of the results. We suggest that the damage imposed by defoliating, leaf-mining and gall-forming insects is measured and reported separately – even when plant damage by leaf-mining and gall-forming insects is low. Earlier we found that the differences between the published and original data on plant damage were largest for miners and gallers (Kozlov et al. 2015b), possibly because plant losses to these herbivores were recorded and reported only when their occurrence was well manifested.

On the other hand, several researchers included feeding marks of sap-feeders in their measures of plant damage. Such a combined index of herbivory is impossible to interpret, because plant losses to sap-feeders cannot be measured from their feeding marks. Moreover, not all sap-feeders leave recognizable feeding marks. If counting of feeding marks of sap-feeders (or leaves bearing these marks) is important for achieving the goals of the study, these data should be reported separately from data on leaf damage by tissue feeders. Similarly, it is undesirable to include fungal damage into a measure of herbivory.

8.4 *Sampling Methods*

8.4.1 Aboveground Tissue Feeders

Two methods are commonly used to measure BIH by defoliating insects: (1) sampling, resulting in a snapshot of herbivory (sometimes termed standing herbivory), and (2) monitoring of individual leaves during a certain period of time, that provides the data on both the standing herbivory and the rates of leaf consumption. The first method is often criticized for not capturing complete consumption of leaves (Anstett et al. 2016); however, this critique can only be valid for measurements conducted during outbreaks of large-sized defoliators, who may consume leaves completely with their petioles. When populations of insect herbivores are at their background (low) levels, complete consumption of leaves is observed very rarely. For example, only 0.34% of 67,695 leaves analysed by Kozlov et al. (2015b) were lost to insects from 75 to 100% of their area, and these losses were measured reliably, because at the background levels of damage the petioles of totally consumed leaves usually remain on branches until the end of the growth season (M.V.K., pers. obs.). Similar distribution of insect damage among individual leaves was reported in many studies conducted both in tropics (Blundell and Peart 2000; Metcalfe et al. 2014) and in temperate regions (Mazía et al. 2012); thus, at least for woody plants, the likelihood that BIH levels are systematically underestimated as a result of complete consumption of some leaves is rather low.

Anstett et al. (2016) also questioned the applicability of single-time measurements of herbivory to evergreen species. However, if herbivory is measured from a random sample of leaves, which includes leaves of all age classes in the same proportion as they occur on a study plant, then the levels of standing herbivory in evergreen and deciduous trees can be compared directly.

In the database on BIH created by Kozlov et al. (2015b), only 11.1% of published data were obtained by monitoring of marked leaves. In contrast to the opinion by Anstett et al. (2016), this proportion did not differ between tropical and temperate regions ($\chi^2 = 0.80$, d.f. $= 1, P = 0.37$), and the levels of BIH measured by monitoring of marked leaves did not differ from ones obtained by single-time sampling either within tropical or temperate regions ($F_{1,585} = 1.83$, $P = 0.18$ and $F_{1,728} = 1.00$, $P = 0.32$, respectively). Thus, as long as the data on standing herbivory fit the study goals, there is no reason to prefer the labour-consuming monitoring of marked leaves to a single-time sampling. Furthermore, only the single-time sampling can be applied to evaluate plant damage by sap- and root-feeding insects.

8.4.2 Aboveground Sap-Feeders

The calculation of plant losses to sap-feeding insects requires the data on insect load, i.e. on the insect biomass per unit of dry weight of foliage. To obtain these data, insects should be quantitatively collected from a known amount of foliage. It can be done by cutting a randomly selected branch and weighing both foliage and all herbivorous insects found on it. The relative losses of plant biomass can then be calculated by using the published values of the coefficients of conversion of ingested food into insect biomass (Kozlov et al. 2015a).

8.4.3 Belowground Tissue Feeders

The data on root losses to insects generally refer to the entire plant community, because it is rather difficult to sort roots by plant species and to learn the diet of each collected specimen of root-feeding insects. In practice, root losses to insects can be calculated from insect load, i.e. the insect biomass per unit of dry weight of roots. To obtain these data, insects and roots should be quantitatively collected from a known amount of soil. We achieved this goal by extracting insects (macrofauna) by sieving and hand-sorting of soil samples 25 × 25 × 30 cm (depth) and extracting plant roots by hand-washing the soil samples 36 mm diameter × 30 cm depth over a mesh screen. The biomass of root-feeding insects was then converted to losses of root biomass by using the coefficients of conversion of ingested food into insect biomass (M.V.K. and E.L.Z., unpublished data).

8.5 Methods to Manipulate Background Insect Herbivory

Mechanical damage is commonly used to explore impacts of outbreak-level herbivory on plant performance (Baldwin 1990). However, in studies addressing BIH this method has several shortcomings. First, mechanical removal of a small percentage of plant tissues may not cause any effect on plants in the absence of specific elicitors present in insect oral secretions (Koo and Howe 2009; Poveda et al. 2010). For example, minor (<10%) mechanical damage did not cause premature abscission of damaged leaves, whereas the same levels of insect herbivory significantly decreased the leaf lifespan in woody plants (Zvereva and Kozlov 2014). Second, mechanical damage cannot simulate plant losses to sap-feeding insects. Third, although mechanical damage can be used to simulate root herbivory, it is rather difficult to control for the percent of roots removed in this way. Therefore, manipulations of BIH in natural environments have considerable advantages over simulated herbivory.

The plant- and community-wide levels of BIH can be reduced by application of insecticides, including those that selectively affect insects feeding on above- and belowground plant parts (Stein et al. 2010), and increased (at least for leaf-feeders) by securing insects from their major vertebrate predators, such as birds and bats, with a net (Böhm et al. 2011). The levels of BIH inflicted by all groups of insects should be measured in both unmanipulated and experimental plants, because insecticide never reduces the herbivory to the zero level.

The local effects of BIH on plants (e.g. on individual leaves or shoots) and mechanisms of BIH impact on plant growth and reproduction can be explored by either comparing naturally damaged and undamaged plant modules or infesting some of randomly selected leaves or shoots by insects. The introduced insects could be isolated to force them to feed on the selected plant part until the desired level of damage is achieved.

8.6 Selecting the Sampling Time

The losses of biomass accumulate during the entire lifetime of a leaf (Wint 1983; Torp et al. 2010), although in many plants the larger part of these losses occurs when leaves are young (Coley 1983; Aide 1993; and references therein). The researchers working with deciduous plants in temperate regions often measure standing herbivory near the end of the growing season, assuming that this sampling strategy yields lifetime foliar damage (Anstett et al. 2016). However, this sampling may underestimate the level of herbivory due to premature abscission of leaves that were severely damaged early in the season (Kozlov et al. 2013, 2016b; Zvereva and Kozlov 2014). On the other hand, densities of sap-feeding insects are likely to peak in the middle of the growth season. Therefore we recommend to measure BIH twice in the season. In temperate and polar climates, the sampling could be conducted in early summer, soon after termination of leaf growth, and in late summer, but well before the beginning of leaf fall in deciduous trees (Kozlov et al. 2013, 2015a, 2016b). In tropical climates, the sampling could be associated e.g. with dry and wet seasons to represent most different environmental conditions.

8.7 Selecting Plant Individuals and Sampling Units

The procedure of sampling may also introduce a bias, and therefore this procedure needs to be developed in advance and adequately described in the publications. Randomization, in the strict sense, is only rarely implemented in studies of BIH. Its reasonable substitute is the selection of plant individuals on a 'first found, first sampled' basis (Kozlov et al. 2013). To avoid unconscious selection of a branch within the tree crown, resulting from the collector's opinion on the 'typical' level of herbivory, the collector should choose a branch from a distance that does not allow visual evaluation of foliar damage, e.g. 5–10 m (Kozlov et al. 2015b). This bias is unlikely to occur in studies of root herbivores, because their density cannot be estimated before the sample is processed; still some kind of random choice, e.g. by throwing a stone at random over the study area, is desirable.

8.8 Choosing Sample Size

The majority of published studies do not provide any arguments for selection of a certain number of leaves to measure BIH. The published data, extracted from 309 papers (listed by Kozlov et al. 2015b), are each based on measurements from 5 to 60,000 leaves, with the median value of 200 leaves.

We used 67,695 values of the percentage of the leaf area consumed or damaged by insects, measured from at least 192 species of woody plants from 147 localities worldwide (published by Kozlov et al. (2015b)), to explore relationships between the number of evaluated leaves and the relative accuracy of measurements of plant losses to insects. Analysis of samples of different size, randomly selected from this database, demonstrated that the 95% confidence interval decreased from ±25% in a sample of 10 leaves to ±2.8% in a sample of 1,000 leaves (Fig. 6). Based on this analysis, we recommend to measure a site-specific BIH from at least 250 leaves of each plant species. However, when BIH is extremely low, like in Arctic tundra (Kozlov et al. 2015d), at least 500 leaves should be measured.

The leaves for measurements of BIH should be sampled from several plant individuals. In the published studies (summarized by Kozlov et al. (2015b)), the leaves were collected from one to 480 plants, with the median value of five plants. We think that sampling of leaves from five plant individuals is sufficient for most of the descriptive studies addressing BIH.

The shortage of information does not allow to quantitatively estimate the numbers of branches and soil samples for measurements of BIH inflicted by sap- and

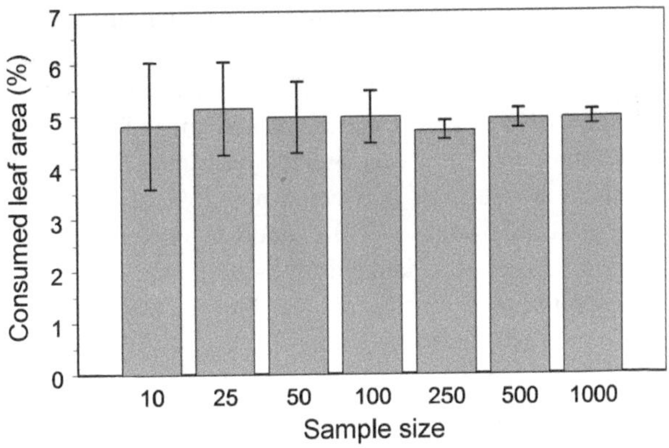

Fig. 6 Relationships between the accuracy of the mean estimate of the consumed leaf area (shown by its 95% confidence interval) and sample size, i.e. the number of leaves

root-feeding insects. However, our experience shows that numbers of sap- and root-feeding insects collected from five branches and from ten soil samples were sufficient to explore latitudinal patterns in BIH (Kozlov et al. 2015a; M.V.K. and E.L.Z., unpublished data).

Studies addressing effects of BIH on plant growth and reproduction should involve large numbers of experimental units: a sample of 20 plants per treatment appeared too low to confirm the statistical significance of the effects of 2 and 4% foliar damage in a 7-year long experiment, although growth characteristics in experimental plants were reduced by 15–35% relative to control values (Zvereva et al. 2012).

8.9 *Obtaining Accurate Measurements of Consumed Leaf Area*

The percentage of leaf area that was eaten away or otherwise damaged by insects is generally measured by three methods: visual estimation of missed leaf area or attribution of each leaf to one of the damage classes based on this estimation (42.7%), image processing (36.2%) and measurements with a grid or dot grid (20.9%; calculated from database created by Kozlov et al. (2015b)). The data collected by these methods yielded nearly equal values of BIH ($F_{2,\ 2140} = 1.67$, $P = 0.19$), suggesting a similar accuracy of these methods. Several intercalibration projects concluded that image processing, which is significantly more laborious than the visual assessment (in particular, due to the need to manually reconstruct the edge of each damaged leaf), has no advantages over visual estimation of the missed leaf area in terms of the accuracy of measurements (Wint 1983; Schaffer et al. 1997); the relative difference between the mean values obtained by these two methods ranged from 2 to 5% (Basset 1991; Blundell and Peart 2000). Moreover, the results of visual estimations made by different evaluators were quite similar to each other (Schaffer et al. 1997).

The accuracy of data obtained by visual classification of leaves depends on the number of damage classes used by a researcher. The vast majority (88.7%) of studies that employed visual method discriminated between damaged and undamaged leaves, whereas 6.3% of studies combined undamaged leaves with leaves which lost less than 1% of their area to insects, and the remaining 5.0% of studies set the upper border of the first damage class between 5 and 25% of the missed leaf area. We strongly recommend that in studies of BIH undamaged leaves are counted separately from leaves bearing the smallest damage by tissue-feeding insects, because even minor damage can affect photosynthesis and leaf lifespan (see Sect. 4.2).

The upper borderline of the first class of damaged leaves is also set differently by the researchers, and the data reported in studies using different upper limits for the first class of damaged leaves yielded different values of BIH ($F_{2,\,604} = 12.8$, $P < 0.0001$). Researchers who used 6–33% as the upper borderline reported, on average, 42% higher levels of herbivory than the researchers using 1% as the upper borderline. Our data set of 67,695 leaves (Kozlov et al. 2015b) yields a mean loss of 4.91% of leaf area to insects when the upper limit of the first damage class is set at 1%, but the mean loss increased to 5.28 and 9.04% if this limit is set at 5 and 25%, respectively. Thus, the use of low number of damage classes results in substantial (nearly twofold) overestimation of plant losses. We therefore recommend that in studies of BIH imposed by tissue-feeding insects the researchers use the following damage classes: intact leaves, 0.01–1, 1–5, 5–25, 25–50, 50–75 and 75–100%.

8.10 *Testing for Non-linearity of Patterns*

Many relationships between biological processes and environmental factors are non-linear. Non-linearity in herbivory-related patterns was in particular detected in relationships between BIH and latitude (Kozlov et al. 2015a, b). Currently, Zhang et al. (2017) concluded that herbivory has dome-shaped relationships with species-specific leaf longevity. Still, most ecological theories and models consider only linear regularities. We suggest that researchers check for possible dome-shaped patterns in the data prior to applying linear models for their analysis.

8.11 *Reporting Sufficient Information*

We suggest that the authors of future studies clearly express their opinion on whether the reported levels of herbivory can be classified as ‘background’ or ‘outbreak’. Furthermore, the studies addressing one feeding guild of insects should explicitly mention to which extent the other feeding guilds damaged their study plants. This especially concerns field studies on belowground herbivory, which generally provide no information on damage of aboveground plant parts by insects.

In addition to the 309 publications summarized by Kozlov et al. (2015b), we identified 137 papers, which either did not report the values of herbivory (i.e. provided only the results of statistical tests) or reported the data in a form that could not be used in research synthesis. First, some researchers summed BIH with other types of leaf damage, such as necroses and fungal infestation. Second, the use of so-called damage indices, calculated by averaging the sequential numbers of damage classes, made the back-transformation of the reported values to actual percentages of leaf damage impossible. Third, the data were often averaged across plant species and/or localities, which precluded their use for detecting patterns associated with characteristics of individual plant species (e.g. plant functional traits) or localities (e.g. climate). Fourth, many authors did not mention explicitly whether leaf miners and gallers were included in the published measures of herbivory.

Less frequent but still common problems were caused by absence of the exact information about location of study site(s) and the year(s) of data collection. Only 51.7% of the published data on BIH were accompanied by the information on the numbers of both plant individuals and leaves, on which these data were based. We recommend that the authors prefer direct measures of herbivory, primarily the percentage of leaf area or biomass consumed or otherwise damaged by insects, to any indices and pay more attention to complete reporting of primary data on BIH in forthcoming publications.

In studies of sap- and root-feeding insects, as well as in studies comparing impacts of different feeding guilds of herbivores on plants, the biomass of insects should be preferred to the number of individuals. Insects from different feeding guilds are often of a very different size; therefore the numbers of phloem-feeders and defoliators (as, e.g. in Balciunas and Burrows (1993) and Takahashi and Huntly (2010)) do not allow to compare impacts of these feeding guilds on plants.

9 Conclusions

Ubiquitous BIH is not yet widely recognized as a specific phenomenon, although its ecological significance has been unequivocally demonstrated. The differences between background herbivory (imposing minor but chronic damage on plants) and outbreak herbivory (imposing severe but short-term damage) are convincingly grounded, and distinguishing these two phenomena has important implications for the development of both basic and applied ecology.

BIH results in relatively minor losses of plant biomass: all groups of insect herbivores (above- and belowground tissue and sap-feeders) jointly remove 1–20% of NPP, depending on the plant community, region and study year. However, only the losses of leaf biomass to defoliating insects are relatively well documented; the omission of sap- and root-feeding insects from the majority of studies addressing BIH represents a considerable research gap that hinders the recognition of the importance of BIH in the functioning of terrestrial ecosystems. The indirect losses of NPP due to the adverse impacts of BIH on plant growth and reproduction may be of the same magnitude as the direct losses (i.e. biomass consumption by insects), but these indirect losses have only rarely been quantified.

Despite the ecological and evolutionary significance of BIH, only 8% of experimental studies addressing plant responses to herbivory have used removal of less than 15% of foliage, and only 6% of studies have examined the consequences of damage occurring over 3 years (Zvereva et al. 2012). Therefore, not surprisingly, the ecological theories explaining plant responses to herbivory have been primarily based on studies that simulated the rare and short events of outbreaks, but not the BIH continuously acting on plants. At the same time, accumulating data indicate that plants respond differently to background and outbreak herbivory. In particular, plants can tolerate short-term major damage better than chronic minor damage due to their opposite effects on photosynthesis and compensatory growth. Furthermore, plants may respond to minor damage by local induction of defences (Kozlov et al. 2012), whereas major damage usually triggers a systemic response. Direct comparisons of plant responses to background and outbreak herbivory are likely to yield exciting results, potentially leading to modifications of current theories on plant–herbivore interactions.

The limited knowledge of spatial and temporal BIH patterns and of its impacts on plant communities is certainly a scientific issue, but it is also a political and economic one because this knowledge is needed for the development, testing and implementation of strategies for mitigation of and adaptation to global climate change. For example, the prevailing scenarios consistently predict increases in outbreak herbivory with climate warming, and these predictions are supported by numerous examples of an increase in outbreak frequency, severity and duration over the past decades. Conversely, analysis of the published data reveals no increase in BIH during the past 60 years. Similarly, aerial pollution seems to favour insect outbreaks but causes no effects on BIH. Thus, the two types of insect herbivory are likely to respond to environmental changes in different ways; this hypothesis calls for comparative studies of the impacts of the abiotic drivers of global change (primarily temperature, precipitation, ambient CO_2 and UV-B) on the background and outbreak damage to plants by insects.

To conclude, we suggest moving the focus of studies of plant–insect interactions from rare bouts of severe plant damage to the ubiquitous phenomenon of BIH. A versatile investigation of this phenomenon is likely to provide great advancements in our understanding of the impacts of plant-feeding insects on the structure and functions of terrestrial ecosystems, while opening up new horizons for fundamental and applied ecological research.

Acknowledgements The study was supported by the Academy of Finland (projects 122133, 268124, 276671 and 311929). We thank V. Zverev for providing a photograph and R. Matyssek and an anonymous reviewer for inspiring comments to an earlier draft of the manuscript.

References

Abdala-Roberts L, Mooney KA, Quijano-Medina T et al (2015) Comparison of tree genotypic diversity and species diversity effects on different guilds of insect herbivores. Oikos 124:1527–1535
Agrawal A, Weber M (2015) On the study of plant defence and herbivory using comparative approaches: how important are secondary plant compounds. Ecol Lett 18:985–991
Aide TM (1993) Patterns of leaf development and herbivory in a tropical understory community. Ecology 74:455–466
Altamirano A, Valladares G, Kuzmanich N et al (2016) Galling insects in a fragmented forest: incidence of habitat loss, edge effects and plant availability. J Insect Conserv 20:119–127
Andrew NR, Hughes L (2005a) Herbivore damage along a latitudinal gradient: relative impacts of different feeding guilds. Oikos 108:176–182
Andrew NR, Hughes L (2005b) Diversity and assemblage structure of phytophagous Hemiptera along a latitudinal gradient: predicting the potential impacts of climate change. Glob Ecol Biogeogr 14:249–262
Anstett DN, Nunes KA, Baskett C et al (2016) Sources of controversy surrounding latitudinal patterns in herbivory and defense. Trends Ecol Evol 31:789–802
Anttonen S, Piispanen R, Ovaska J et al (2002) Effects of defoliation on growth, biomass allocation, and wood properties of *Betula pendula* clones grown at different nutrient levels. Can J For Res 32:498–508
Ayres MP, Lombardero MJ (2000) Assessing the consequences of global change for forest disturbance from herbivores and pathogens. Sci Total Environ 262:263–286
Bach CE (2001) Long-term effects of insect herbivory on responses by *Salix cordata* to sand accretion. Ecology 82:397–409
Bagchi R, Gallery RE, Gripenberg S et al (2014) Pathogens and insect herbivores drive rainforest plant diversity and composition. Nature 506:85–88
Balciunas JK, Burrows DW (1993) The rapid suppression of the growth of *Melaleuca quinquenervia* saplings in Australia by insects. J Aquat Plant Manag 31:265–270
Baldwin IT (1990) Herbivory simulations in ecological research. Trends Ecol Evol 5:91–93
Bale JS, Masters GJ, Hodkinson ID et al (2002) Herbivory in global climate change research: direct effects of rising temperature on insect herbivores. Glob Chang Biol 8:1–16
Basset Y (1991) The spatial distribution of herbivory, mines and galls within an Australian rainforest tree. Biotropica 23:271–281
Belovsky GE, Slade JB (2000) Insect herbivory accelerates nutrient cycling and increases plant production. Proc Natl Acad Sci U S A 97:14412–14417
Belskaya EA, Vorobeichik EL (2013) Responses of leaf-eating insects feeding on aspen to emissions from the Middle Ural copper smelter. Russ J Ecol 44:108–117
Berryman AA (1987) The theory and classification of outbreaks. In: Barbosa P, Schultz JC (eds) Insect outbreaks. Academic, San Diego, CA, pp 3–30
Bigger DS, Marvier MA (1998) How different would a world without herbivory be? A search for generality in ecology. Integr Biol 1:60–67
Björkman C, Dalin P, Ahrné K (2008) Leaf trichome responses to herbivory in willows: induction, relaxation and costs. New Phytol 179:176–184
Björkman C, Berggren A, Bylund H (2011) Causes behind insect folivory patterns in latitudinal gradients. J Ecol 99:367–369
Blundell AG, Peart DR (2000) High abscission rates of damaged expanding leaves: field evidence from seedlings of a Bornean rain forest tree. Am J Bot 87:1693–1698
Böhm SM, Wells K, Kalko EKV (2011) Top-down control of herbivory by birds and bats in the canopy of temperate broad-leaved oaks (*Quercus robur*). PLoS One 6:e17857
Bommarco R, Banks JE (2003) Scale as a modifier in vegetation diversity experiments: effects on herbivores and predators. Oikos 102:440–448
Brown VK, Gange AC (1989) Differential effects of above-ground and below-ground insect herbivory during early plant succession. Oikos 54:67–76

Brown VK, Gange AC (1990) Insect herbivory below ground. Adv Ecol Res 20:1–58
Cabrera HM, Argandona VH, Corcuera LJ (1994) Metabolic changes in barley seedlings at different aphid infestation levels. Phytochemistry 35:317–319
Cárdenas RE, Valencia R, Kraft NJB et al (2014) Plant traits predict inter- and intraspecific variation in susceptibility to herbivory in a hyperdiverse Neotropical rain forest tree community. J Ecol 102:939–952
Castagneyrol B, Giffard B, Péré C, Jactel H (2013) Plant apparency, an overlooked driver of associational resistance to insect herbivory. J Ecol 101:418–429
Castagneyrol B, Jactel H, Vacher C et al (2014) Effects of plant phylogenetic diversity on herbivory depend on herbivore specialization. J Appl Ecol 51:134–141
Chaves MM, Flexas J, Pinheiro C (2009) Photosynthesis under drought and salt stress: regulation mechanisms from whole plant to cell. Ann Bot 103:551–560
Chown SL, Sinclair BJ, Leinaas HP et al (2004) Hemispheric asymmetries in biodiversity – a serious matter for ecology. PLoS Biol 2:1701–1707 (e406)
Coley PD (1983) Herbivory and defensive characteristics of tree species in a lowland tropical forest. Ecol Monogr 53:209–233
Coley PD, Aide TM (1991) Comparison of herbivory and plant defenses in temperate and tropical broad-leaved forests. In: Price PW, Lewinsohn TM, Fernandes GW, Benson WW (eds) Plant-animal interactions: evolutionary ecology in tropical and temperate regions. Wiley, New York, NY, pp 25–49
Coley PD, Barone JA (1996) Herbivory and plant defenses in tropical forests. Annu Rev Ecol Syst 27:305–335
Coley PD, Bryant JP, Chapin FS (1985) Resource availability and plant antiherbivore defense. Science 230:895–899
Coupe MD, Cahill JF (2003) Effects of insects on primary production in temperate herbaceous communities: a meta-analysis. Ecol Entomol 28:511–521
Couture JJ, Meehan TD, Kruger EL, Lindroth RL (2015) Insect herbivory alters impact of atmospheric change on northern temperate forests. Nat Plants 1:15016
Crawley MJ (1983) Herbivory: the dynamics of animal–plant interactions. Blackwell, Oxford
Crawley MJ (1985) Reduction of oak fecundity by low-density herbivore populations. Nature 314:163–164
Crawley MJ (1989) Insect herbivores and plant population dynamics. Annu Rev Entomol 34:531–564
Cronin JP, Tonsor SJ, Carson WP (2010) A simultaneous test of trophic interaction models: which vegetation characteristic explains herbivore control over plant community mass? Ecol Lett 13:202–212
Cuevas-Reyes P, Gilberti L, González-Rodríguez A, Fernandes GW (2013) Patterns of herbivory and fluctuating asymmetry in *Solanum lycocarpum* St. Hill (Solanaceae) along an urban gradient in Brazil. Ecol Ind 24:557–561
Cyr H, Pace ML (1993) Magnitude and patterns of herbivory in aquatic and terrestrial ecosystems. Nature 361:148–150
Damien M, Jactel H, Meredieu C et al (2016) Pest damage in mixed forests: disentangling the effects of neighbor identity, host density and host apparency at different spatial scales. For Ecol Manag 378:103–110
Davidson CB, Johnson JE, Gottschalk KW, Amateis RL (2001) Prediction of stand susceptibility and gypsy moth defoliation in coastal plain mixed pine-hardwoods. Can J For Res 31:1914–1921
DeLucia EH, Nabity PD, Zavala JA, Berenbaum MR (2012) Climate change: resetting plant–insect interactions. Plant Physiol 160:1677–1685
Dukes JS, Pontius J, Orwig D et al (2009) Responses of insect pests, pathogens, and invasive plant species to climate change in the forests of northeastern North America: what can we predict? Can J For Res 39:231–248
Dunn RR, Agosti D, Andersen AN et al (2009) Climatic drivers of hemispheric asymmetry in global patterns of ant species richness. Ecol Lett 12:324–333

Endara MJ, Coley PD (2011) The resource availability hypothesis revisited: a meta-analysis. Funct Ecol 25:389–398
Faeth SH (1987) Community structure and folivorous insect outbreaks: the roles of vertical and horizontal interactions. In: Barbosa P, Schultz JC (eds) Insect outbreaks. Academic, San Diego, CA, pp 135–171
Feeny P (1976) Plant apparency and chemical defense. Recent Adv Phytochem 10:1–40
Führer E (1985) Air pollution and the incidence of forest insect problems. Z Angew Entomol 99:371–377
Garibaldi LA, Kitzberger T, Ruggiero A (2011) Latitudinal decrease in folivory within *Nothofagus pumilio* forests: dual effect of climate on insect density and leaf traits? Glob Ecol Biogeogr 20:609–619
Gherlenda AN, Moore BD, Haigh AM et al (2016) Insect herbivory in a mature *Eucalyptus* woodland canopy depends on leaf phenology but not CO_2 enrichment. BMC Ecol 16:art 47
Godfray HCJ (1985) The absolute abundance of leaf miners on plants of different successional stages. Oikos 45:17–25
Guimarães CDDC, Viana JPR, Cornelissen T (2014) A meta-analysis of the effects of fragmentation on herbivorous insects. Environ Entomol 43:537–545
Guyot V, Castagneyrol B, Vialatte A et al (2016) Tree diversity reduces pest damage in mature forests across Europe. Biol Lett 12:20151037
Haase J, Castagneyrol B, Cornelissen JHC et al (2015) Contrasting effects of tree diversity on young tree growth and resistance to insect herbivores across three biodiversity experiments. Oikos 124:1674–1685
Hawkes CV, Sullivan JJ (2001) The impact of herbivory on plants in different resource conditions: a meta-analysis. Ecology 82:2045–2058
Haynes KJ, Allstadt AJ, Klimetzek D (2014) Forest defoliator outbreaks under climate change: effects on the frequency and severity of outbreaks of five pine insect pests. Glob Chang Biol 20:2004–2018
Herms DA, Mattson WJ (1992) The dilemma of plants – to grow or defend. Q Rev Biol 67:283–335
Hodkinson ID, Hughes MK (1982) Insect herbivory. Chapman and Hall, London/New York, NY
Hoogesteger J, Karlsson PS (1992) Effects of defoliation on radial stem growth and photosynthesis in the mountain birch (*Betula pubescens* ssp. *tortuosa*). Funct Ecol 6:317–323
Hunter MD (2001) Insect population dynamics meets ecosystem ecology: effects of herbivory on soil nutrient dynamics. Agric For Entomol 3:77–84
Hysell MT, Wagner MR, Grier CC (1996) Pattern of herbivore abundance along a climatic gradient: guild shift of leaf-feeding insects and changes in leaf characteristics. Environ Entomol 25:977–982
IPCC (2001) Climate change 2001: the scientific basis. Contribution from working group I to the third assessment report of the intergovernmental panel for climate change. Cambridge University Press, Cambridge
Johnson DM, Büntgen U, Frank DC et al (2010) Climatic warming disrupts recurrent alpine insect outbreaks. Proc Natl Acad Sci U S A 107:20576–22058
Johnson SN, Erb M, Hartley SE (2016) Roots under attack: contrasting plant responses to below- and aboveground insect herbivory. New Phytol 210:413–418
Jönsson AM, Schroeder LM, Lagergren F et al (2012) Guess the impact of *Ips typographus* – an ecosystem modelling approach for simulating spruce bark beetle outbreaks. Agric For Meteorol 166:188–200
Kaakeh W, Pfeiffer DG, Marini RP (1992) Combined effects of spirea aphid (Homoptera, Aphididae) and nitrogen fertilization on net photosynthesis, total chlorophyll content, and greenness of apple leaves. J Econ Entomol 85:939–946
Kahl J, Siemens DH, Aerts RJ et al (2000) Herbivore-induced ethylene suppresses a direct defense but not a putative indirect defense against an adapted herbivore. Planta 210:336–342
Kaitaniemi P, Ruohomäki K (2006) Uncaged larvae elicit a combination of local and integrated growth responses within mountain birch crown. Oikos 115:537–548

Kaplan I, Halitschke R, Kessler A et al (2008) Constitutive and induced defenses to herbivory in above- and belowground plant tissues. Ecology 89:392–406
Karban R, Baldwin IT (1997) Induced responses to herbivory. University of Chicago Press, Chicago, IL/London
Kessler A, Baldwin IT (2002) Plant responses to insect herbivory: the emerging molecular analysis. Annu Rev Plant Biol 53:299–328
Kissling WD, Dormann CF, Groeneveld J et al (2012) Towards novel approaches to modelling biotic interactions in multispecies assemblages at large spatial extents. J Biogeogr 39:2163–2178
Kneeshaw D, Sturtevant BR, Cooke B et al (2015) Insect disturbances in forest ecosystems. In: Peh KS-H, Corlett RT, Bergeron Y (eds) Routledge handbook of forest ecology. Routledge, London/New York, NY, pp 93–113
Knepp RG, Hamilton JG, Mohan JE et al (2005) Elevated CO_2 reduces leaf damage by insect herbivores in a forest community. New Phytol 167:207–218
Komonen A, Kouki J (2008) Do restoration fellings in protected forests increase the risk of bark beetle damages in adjacent forests? A case study from Fennoscandian boreal forest. For Ecol Manag 255:3736–3743
Koo AJK, Howe GA (2009) The wound hormone jasmonate. Phytochemistry 70:1571–1580
Koricheva J (2002) Meta-analysis of sources of variation in fitness costs of plant antiherbivore defenses. Ecology 83:176–190
Koricheva J, Vehvilainen H, Riihimaki J et al (2006) Diversification of tree stands as a means to manage pests and diseases in boreal forests: myth or reality? Can J For Res 36:324–336
Kozlov MV (1985) Changes in phytophagous insect communities affected by airborne pollutants near the non-ferrous smelter. In: Kastrel TN (ed) Problems of biosphere (information bulletin), vol 11. Scientific Board on Biosphere Problems, Academy of Sciences, Moscow, pp 76–90 (in Russian)
Kozlov MV (2008) Losses of birch foliage along geographical gradients in Northern and Central Europe: a climate-driven pattern? Clim Change 87:107–117
Kozlov MV (2015) Insect herbivory on two willow species in northern Europe is independent of pollution load. Boreal Environ Res 20:423–430
Kozlov MV, Klemola T (2017) Hemispheric asymmetries in herbivory: do they exist? J Ecol (in press)
Kozlov MV, Zvereva EL (2014) Variations in the effects of local foliar damage on life span of individual leaves of downy birch (*Betula pubescens*). Botany 92:477–484
Kozlov MV, Zvereva EL (2015) Changes in the background losses of woody plant foliage to insects during the past 60 years: are the predictions fulfilled? Biol Lett 11:20150480
Kozlov MV, Zvereva EL, Zverev V (2009) Impacts of point polluters on terrestrial biota: comparative analysis of 18 contaminated areas. Springer, Dordrecht
Kozlov MV, Lanta V, Zverev VE, Zvereva EL (2012) Delayed local responses of downy birch to damage by leafminers and leafrollers. Oikos 121:428–434
Kozlov MV, van Nieukerken EJ, Zverev V, Zvereva EL (2013) Abundance and diversity of birch-feeding leafminers along latitudinal gradients in northern Europe. Ecography 36:1138–1149
Kozlov MV, Zverev V, Zvereva EL (2014) Confirmation bias leads to overestimation of losses of woody plant foliage to insect herbivores in tropical regions. PeerJ 2:e798
Kozlov MV, Stekolshchikov AV, Soderman G et al (2015a) Sap-feeding insects on forest trees along latitudinal gradients in northern Europe: a climate-driven patterns. Glob Chang Biol 21:106–116
Kozlov MV, Lanta V, Zverev V, Zvereva EL (2015b) Global patterns in background losses of woody plant foliage to insects. Glob Ecol Biogeogr 24:1126–1135
Kozlov MV, Lanta V, Zverev V, Zvereva EL (2015c) Background losses of woody plant foliage to insects show variable relationships with plant functional traits across the globe. J Ecol 103:1519–1528
Kozlov MV, Filippov BY, Zubrij NA, Zverev V (2015d) Abrupt changes in invertebrate herbivory on woody plants at the forest-tundra ecotone. Polar Biol 38:967–974

Kozlov MV, Zverev V, Zvereva EL (2016a) Shelters of leaf-tying herbivores decompose faster than untied leaves damaged by free-living insects. Sci Total Environ 568:946–951
Kozlov MV, Skoracka A, Zverev V, Lewandowski M, Zvereva EL (2016b) Two birch species demonstrate opposite latitudinal patterns in infestation by gall-making mites in Northern Europe. PLoS One 11(11):e0166641
Kozlov MV, Lanta V, Zverev V et al (2017) Decreased losses of woody plant foliage to insects in large urban areas are explained by bird predation. Glob Chang Biol 23 (in press)
Labandeira CC (2012) Evidence for outbreaks from the fossil record of insect herbivory. In: Barbosa P, Letourneau DK, Agrawal A (eds) Insect outbreaks revisited. Wiley-Blackwell, Chichester, pp 269–290
Lamarre GPA, Baraloto C, Fortunel C et al (2012) Herbivory, growth rates, and habitat specialization in tropical tree lineages: implications for Amazonian beta-diversity. Ecology 93: S195–S210
Lawton JH, McNeill S (1979) Between the devil and the deep blue sea: on the problem of being a herbivore. In: Anderson RM, Turner BD, Taylor LR (eds) Population dynamics. Blackwell, Oxford, pp 223–244
Lehndal L, Agren J (2015) Herbivory differentially affects plant fitness in three populations of the perennial herb *Lythrum salicaria* along a latitudinal gradient. PLoS One 10:e0135939
Lempa K, Agrawal AA, Salminen JP et al (2004) Rapid herbivore-induced changes in mountain birch phenolics and nutritive compounds and their effects on performance of the major defoliator, *Epirrita autumnata*. J Chem Ecol 30:303–321
Levey DJ, Caughlin TT, Brudvig LA et al (2016) Disentangling fragmentation effects on herbivory in understory plants of longleaf pine savannah. Ecology 97:2248–2258
Lim JY, Fine PVA, Mittelbach GG (2015) Assessing the latitudinal gradient in herbivory. Glob Ecol Biogeogr 24:1106–1112
Lincoln DE, Fajer ED, Johnson RH (1993) Plant-insect herbivore interactions in elevated CO_2 environments. Trends Ecol Evol 8:64–68
Lindner M, Maroschek M, Netherer S et al (2010) Climate change impacts, adaptive capacity, and vulnerability of European forest ecosystems. For Ecol Manag 259:698–709
Lindroth RL (2012) Atmospheric change, plant secondary metabolites and ecological interactions. In: Iason GR, Dicke M, Hartley SE (eds) The ecology of plant secondary metabolites: from genes to global processes. Cambridge University Press, Cambridge, pp 120–153
Linhart YB (1991) Disease, parasitism and herbivory: multidimensional challenges in plant evolution. Trends Ecol Evol 6:392–396
Logan JA, Regniere J, Powell JA (2003) Assessing the impacts of global warming on forest pest dynamics. Front Ecol Environ 1:130–137
Lovett GM, Christenson LM, Groffman PM et al (2002) Insect defoliation and nitrogen cycling in forests. Bioscience 52:335–342
Markkola A, Kuikka K, Rautio P et al (2004) Defoliation increases carbon limitation in ectomycorrhizal symbiosis of *Betula pubescens*. Oecologia 140:234–240
Marquis RJ (1992) A bite is a bite is a bite – constraints on response to folivory in *Piper arieianum* (Piperaceae). Ecology 73:143–152
Marquis RJ, Whelan CJ (1994) Insectivorous birds increase growth of white oak through consumption of leaf-chewing insects. Ecology 75:2007–2014
Maschinski J, Whitham TG (1989) The continuum of plant responses to herbivory: the influence of plant association, nutrient availability, and timing. Am Nat 134:1–19
Matter SF, Brzyski JR, Harrison CJ et al (2012) Invading from the garden? A comparison of leaf herbivory for exotic and native plants in natural and ornamental settings. Insect Sci 19:677–682
Mazía N, Chaneton EJ, Dellacanonica C et al (2012) Seasonal patterns of herbivory, leaf traits and productivity consumption in dry and wet Patagonian forests. Ecol Entomol 37:193–203
Meehan TD, Couture JC, Bennett AE, Lindroth RL (2014) Herbivore-mediated material fluxes in a northern deciduous forest under elevated carbon dioxide and ozone concentrations. New Phytol 204:397–407
Metcalfe DB, Asner GP, Martin RE et al (2014) Herbivory makes major contributions to ecosystem carbon and nutrient cycling in tropical forests. Ecol Lett 17:324–332

Moles AT, Westoby M (2000) Do small leaves expand faster than large leaves, and do shorter expansion times reduce herbivore damage? Oikos 90:517–524
Moles AT, Bonser SP, Poore AG et al (2011) Assessing the evidence for latitudinal gradients in plant defence and herbivory. Funct Ecol 25:380–388
Morales-Castilla I, Matias MG, Gravel D, Araújo MB (2015) Inferring biotic interactions from proxies. Trends Ecol Evol 30:347–356
Moreira X, Abdala-Roberts L, Rasmann S et al (2016) Plant diversity effects on insect herbivores and their natural enemies: current thinking, recent findings, and future directions. Curr Opin Insect Sci 14:1–7
Muiruri EW, Milligan HT, Morath S et al (2015) Moose browsing alters tree diversity effects on birch growth and insect herbivory. Funct Ecol 29:724–735
Murray TJ, Ellsworth DS, Tissue DT et al (2013) Interactive direct and plant-mediated effects of elevated atmospheric [CO_2] and temperature on a eucalypt-feeding insect herbivore. Glob Chang Biol 19:1407–1416
Nabity PD, Zavala JA, DeLucia EH (2009) Indirect suppression of photosynthesis on individual leaves by arthropod herbivory. Ann Bot 103:655–663
Nabity PD, Zavala JA, DeLucia EH (2013) Herbivore induction of jasmonic acid and chemical defences reduce photosynthesis in *Nicotiana attenuata*. J Exp Bot 64:685–694
Nevalainen S, Niemelä P, Arkhipov P, Fedorets N (1994) Männiköiden terveydentila Karjalan aluella. Metsäntutkimuslaitoksen Tiedonantoja 489:25–30
Neves FS, Silva JO, Espírito-Santo MM, Fernandes GW (2014) Insect herbivores and leaf damage along successional and vertical gradients in a tropical dry forest. Biotropica 46:14–24
Newbery DM, de Foresta H (1985) Herbivory and defense in pioneer, gap and understory trees of tropical rain forest in French Guiana. Biotropica 17:238–244
Niziolek OK, Berenbaum MR, DeLucia EH (2013) Impact of elevated CO_2 and increased temperature on Japanese beetle herbivory. Insect Sci 20:513–523
Norghauer JM, Glauser G, Newbery DM (2014) Seedling resistance, tolerance and escape from herbivores: insights from co-dominant canopy tree species in a resource poor African rain forest. Funct Ecol 28:1426–1439
Nowak RS, Caldwell MM (1984) A test of compensatory photosynthesis in the field: implications for herbivory tolerance. Oecologia 61:311–318
Nuckols MS, Connor EF (1995) Do trees in urban or ornamental plantings receive more damage by insects than trees in natural forests? Ecol Entomol 20:253–260
Nykänen H, Koricheva J (2004) Damage-induced changes in woody plants and their effects on insect herbivore performance: a meta-analysis. Oikos 104:247–268
O'Connor MI, Piehler MF, Leech DM et al (2009) Warming and resource availability shift food web structure and metabolism. PLoS Biol 7:e1000178
Onoda Y, Westoby M, Adler PB et al (2011) Global patterns of leaf mechanical properties. Ecol Lett 14:301–312
Pachzelt A, Forrest M, Rammig A et al (2015) Potential impact of large ungulate grazers on African vegetation, carbon storage and fire regimes. Glob Ecol Biogeogr 24:991–1002
Poorter L, van de Plassche M, Willems S, Boot RGA (2004) Leaf traits and herbivory rates of tropical tree species differing in successional status. Plant Biol 6:746–754
Poveda K, Jimenez MIG, Kessler A (2010) The enemy as ally: herbivore-induced increase in crop yield. Ecol Appl 20:1787–1793
Price PW (1975) Insect ecology. Wiley, New York, NY
Raupp MJ, Shrewsbury PM, Herms DA (2010) Ecology of herbivorous arthropods in urban landscapes. Annu Rev Entomol 55:19–38
Reichman OJ, Smith SC (1991) Responses to simulated leaf and root herbivory by a biennial, *Tragopogon dubius*. Ecology 72:116–124
Reingold AL (1998) Outbreak investigations – a perspective. Emerg Infect Dis 4:21–27
Retuerto R, Fernandez-Lema B, Rodriguez R, Obeso JR (2004) Increased photosynthetic performance in holly trees infested by scale insects. Funct Ecol 18:664–669

Revesz RL, Howard PH, Arrow K et al (2014) Global warming: improve economic models of climate change. Nature 508:173–175
Rickman JK, Connor EF (2003) The effect of urbanization on the quality of remnant habitats for leaf-mining lepidoptera on *Quercus agrifolia*. Ecography 26:777–787
Roberts JA, Elliott KA, Gonzalez-Carranza ZH (2002) Abscission, dehiscence, and other cell separation processes. Annu Rev Plant Biol 53:131–158
Robinson EA, Ryan GD, Newman JA (2012) A meta-analytical review of the effects of elevated CO_2 on plant–arthropod interactions highlights the importance of interacting environmental and biological variables. New Phytol 194:321–336
Root RB (1996) Herbivore pressure on goldenrods (*Solidago altissima*): its variation and cumulative effects. Ecology 77:1074–1087
Rosenthal R (1976) Experimenter effects in behavioral research. Wiley, New York, NY
Rossetti MR, Gonzalez E, Salvo A et al (2014) Not all in the same boat: trends and mechanisms in herbivory responses to forest fragmentation differ among insect guilds. Arthropod Plant Interact 8:593–603
Ruiz-Guerra B, Guevara R, Mariano NA et al (2010) Insect herbivory declines with forest fragmentation and covaries with plant regeneration mode: evidence from a Mexican tropical rain forest. Oikos 119:317–325
Schaffer B, Mason LJ (1990) Effects of scale insect herbivory and shading on net gas exchange and growth of a subtropical tree species (*Guaiacum sanctum* L.). Oecologia 84:468–473
Schaffer B, Pena JE, Colls AM et al (1997) Citrus leafminer (Lepidoptera: Gracillariidae) in lime: assessment of leaf damage and effects on photosynthesis. Crop Prot 16:337–343
Schowalter TD, Lowman MD (1999) Forest herbivory: insects. In: Walker LR (ed) Ecosystem of the World 16: ecosystems of disturbed ground. Elsevier, Amsterdam, pp 253–269
Schuldt A, Baruffol M, Böhnke M et al (2010) Tree diversity promotes insect herbivory in subtropical forests of south-east China. J Ecol 98:917–926
Schuldt A, Bruelheide H, Durka W et al (2012) Plant traits affecting herbivory on tree recruits in highly diverse subtropical forests. Ecol Lett 15:732–739
Schuldt A, Assmann T, Bruelheide H et al (2014) Functional and phylogenetic diversity of woody plants drive herbivory in a highly diverse forest. New Phytol 202:864–873
Seidl R, Fernandes PM, Fonseca TF et al (2011) Modelling natural disturbances in forest ecosystems: a review. Ecol Model 222:903–924
Selikhovkin AV (2009) Can outbreaks of dendrophagous insects make a considerable impact on the biosphere? Biosphere 1(1):72–81 (in Russian, English summary)
Simonetti JA, Grez AA, Celis-Diez JL et al (2007) Herbivory and seedling performance in a fragmented temperate forest of Chile. Acta Oecol 32:312–318
Smith WH (1974) Air pollution – effects on structure and function of temperate forest ecosystem. Environ Pollut 6:111–129
Stein C, Unsicker SB, Kahmen A et al (2010) Impact of invertebrate herbivory in grasslands depends on plant species diversity. Ecology 91:1639–1650
Stephens AEA, Westoby M (2015) Effects of insect attack to stems on plant survival, growth, reproduction and photosynthesis. Oikos 124:266–273
Stiling P, Cornelissen T (2007) How does elevated carbon dioxide (CO_2) affect plant-herbivore interactions? A field experiment and meta-analysis of CO_2-mediated changes on plant chemistry and herbivore performance. Glob Chang Biol 15:1823–1842
Stiling P, Moon D, Rossi A et al (2009) Seeing the forest for the trees: long-term exposure to elevated CO_2 increases some herbivore densities. Glob Chang Biol 15:1895–1902
Sutherst RW, Maywald GF, Bourne AS (2007) Including species interactions in risk assessments for global change. Glob Chang Biol 13:1843–1859
Takahashi M, Huntly N (2010) Herbivorous insects reduce growth and reproduction of big sagebrush (*Artemisia tridentata*). Arthropod Plant Interact 4:257–266
Tedders WL, Smith JS (1976) Shading effect on pecan by sooty mold growth. J Econ Entomol 69:551–553

Tenow O, Bylund H (2000) Recovery of a *Betula pubescens* forest in northern Sweden after severe defoliation by *Epirrita autumnata*. J Veg Sci 11:855–862
Torp M, Olofsson J, Witzell J et al (2010) Snow-induced changes in dwarf birch chemistry increase moth larval growth rate and level of herbivory. Polar Biol 33:693–702
Trumble JT, Kolodnyhirsch DM, Ting IP (1993) Plant compensation for arthropod herbivory. Annu Rev Entomol 38:93–119
Turcotte MM, Davies TJ, Thomsen CJM, Johnson MTJ (2014) Macroecological and macroevolutionary patterns of leaf herbivory across vascular plants. Proc R Soc Lond B Biol Sci 281:20140555
Tylianakis JM, Didham RK, Bascompte J, Wardle DA (2008) Global change and species interactions in terrestrial ecosystems. Ecol Lett 11:1351–1363
Valladares G, Salvo A, Cagnolo L (2006) Habitat fragmentation effects on trophic processes of insect-plant food webs. Conserv Biol 20:212–217
Vasconcelos HL (1999) Levels of leaf herbivory in Amazonian trees from different stages in forest regeneration. Acta Amaz 29:615–623
Vehviläinen H, Koricheva J, Ruohomäki K (2007) Tree species diversity influences herbivore abundance and damage: meta-analysis of long-term forest experiments. Oecologia 152:287–298
Villagra M, Campanello PI, Montti L, Goldstein G (2013) Removal of nutrient limitations in forest gaps enhances growth rate and resistance to cavitation in subtropical canopy tree species differing in shade tolerance. Tree Physiol 33:285–296
Vogt KA, Vogt DJ, Palmiotto PA et al (1996) Review of root dynamics in forest ecosystems grouped by climate, climatic forest type, and species. Plant Soil 187:159–219
Welter SC (1989) Arthropod impact on plant gas exchange. In: Bernays EA (ed) Plant-insect interactions. CRC, Boca Raton, FL, pp 135–150
Więski K, Pennings S (2014) Latitudinal variation in resistance and tolerance to herbivory of a salt marsh shrub. Ecography 37:763–769
Wilson EO (1987) The little things that run the world (the importance and conservation of invertebrates). Conserv Biol 1:344–346
Wint W (1983) The role of alternative host-plant species in the life of a polyphagous moth, *Operophtera brumata* (Lepidoptera, Geometridae). J Anim Ecol 52:439–450
Wise MJ, Abrahamson WG (2005) Beyond the compensatory continuum: environmental resource levels and plant tolerance of herbivory. Oikos 109:417–428
Wolf A, Kozlov MV, Callaghan T (2008) Impact of non-outbreak insect damage on vegetation in northern Europe will be greater than expected during a changing climate. Clim Change 87:91–106
Yamamura K, Kiritani K (1998) A simple method to estimate the potential increase in the number of generations under global warming in temperate zones. Appl Entomol Zool 33:289–298
Zangerl AR, Arntz AM, Berenbaum MR (1997) Physiological price of an induced chemical defense: photosynthesis, respiration, biosynthesis, and growth. Oecologia 109:433–441
Zangerl AR, Hamilton JG, Miller TJ et al (2002) Impact of folivory on photosynthesis is greater than the sum of its holes. Proc Natl Acad Sci U S A 99:1088–1091
Zava PC, Cianciaruso MV (2014) Can we use plant traits and soil characteristics to predict leaf damage in savanna woody species? Plant Ecol 215:625–637
Zhang S, Zhang Y, Ma K (2016) Latitudinal variation in herbivory: hemispheric asymmetries and the role of climatic drivers. J Ecol 104:1089–1095
Zhang S, Zhang Y, Ma K (2017) The association of leaf lifespan and background insect herbivory at the interspecific level. Ecology 98:425–432
Zvereva EL, Kozlov MV (2001) Effects of pollution-induced habitat disturbance on the response of willows to simulated herbivory. J Ecol 89:21–30
Zvereva EL, Kozlov MV (2006) Consequences of simultaneous elevation of carbon dioxide and temperature for plant–herbivore interactions: a meta-analysis. Glob Chang Biol 12:27–41
Zvereva EL, Kozlov MV (2010) Responses of terrestrial arthropods to air pollution: a meta-analysis. Environ Sci Pollut Res 17:297–311
Zvereva EL, Kozlov MV (2012) Sources of variation in plant responses to belowground insect herbivory: a meta-analysis. Oecologia 169:441–452

Zvereva EL, Kozlov MV (2014) Effects of herbivory on leaf life span in woody plants: a meta-analysis. J Ecol 102:873–881

Zvereva EL, Lanta V, Kozlov MV (2010a) Effects of sap-feeding insect herbivores on fitness of woody plants: a meta-analysis of experimental studies. Oecologia 163:949–960

Zvereva EL, Roitto M, Kozlov MV (2010b) Growth and reproduction of vascular plants in polluted environments: a synthesis of existing knowledge. Environ Rev 18:355–367

Zvereva EL, Zverev VE, Kozlov MV (2012) Little strokes fell great oaks: minor but chronic herbivory substantially reduces birch growth. Oikos 121:2036–2043

Zvereva EL, Hunter MD, Zverev V, Kozlov MV (2016) Factors affecting population dynamics of leaf beetles in a subarctic region: the interplay between climate warming and pollution decline. Sci Total Environ 566–567:1277–1288

Erratum to: Interplay of Circadian Rhythms and Light in the Regulation of Photosynthesis-Derived Metabolism

Matthew A. Jones

Erratum to:

Progress in Botany, DOI 10.1007/124_2017_2

The Figure 1 in this chapter was inadvertently published with a wrong version. The correct version of the Figure 1 is shown below.

The updated online version for this chapter can be found under
http://dx.doi.org/10.1007/124_2017_2

Communicated by Ulrich Lüttge

M.A. Jones (✉)
School of Biological Sciences, University of Essex, Colchester, UK
e-mail: matthew.jones@essex.ac.uk

Progress in Botany (2018) 79: 357–358, DOI 10.1007/124_2017_12,
Published online: 3 May 2017

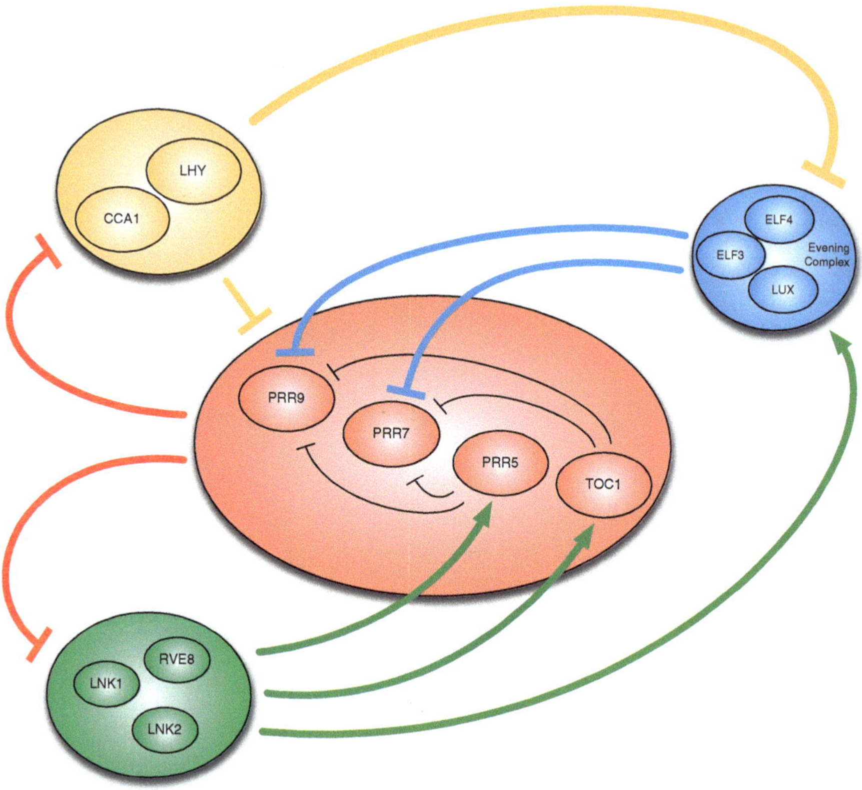

The original article was corrected.

FSC
www.fsc.org
MIX
Papier | Fördert
gute Waldnutzung
FSC® C083411

Kolibri 360 GmbH
Ferdinand-Jühlke-Straße 7
99095 Erfurt, Deutschland
produktsicherheit@kolibri360.de